A SURVEY OF THE SCIENTIFIC MANUSCRIPTS IN THE EGYPTIAN NATIONAL LIBRARY

ARCE

AMERICAN RESEARCH
CENTER IN EGYPT

CATALOGS / Volume 5

Published under the auspices of
THE AMERICAN RESEARCH CENTER IN EGYPT, INC.

هذا كتاب لو يباع بوزنه
ذهبا لكان البايع المغبونا

أوما من الخسران أنك آخذ
ذهبا وتارك جوهرا مكنونا

(ṬM 243, fol. 1r)

A SURVEY OF THE SCIENTIFIC MANUSCRIPTS IN THE EGYPTIAN NATIONAL LIBRARY

by

David A. King

Published for
THE AMERICAN RESEARCH CENTER IN EGYPT, INC.
by
EISENBRAUNS
Winona Lake, Indiana 1986

Abstract

This work surveys over 2,500 Arabic, Persian and Turkish manuscripts relating to mathematics, astronomy and astrology preserved in the Egyptian National Library. It is arranged as a supplement to the standard bio-bibliographical reference works for the history of Islamic science (Suter, Brockelmann, Krause, Renaud, Storey, Sezgin and Matvievskaya & Rosenfeld).

The manuscripts cover all aspects of the exact sciences in medieval Islamic civilization. Many of them are copies of works which were compiled during the period from the eighth to the fifteenth century when Muslim scientists were the leading scholars in their field. Not a few of these were not previously known to modern scholarship, and call for a reappraisal of various aspects of Islamic science, notably astronomical timekeeping and instrumentation.

This English survey contains cross-references to the author's new Arabic *Catalogue of the Scientific Manuscripts in the Egyptian National Library*, 2 vols. (Cairo, 1981 and 1986?), in which details are given on each individual manuscript (Part I) and extracts are given from each work represented in the collection (Part II). Indications are given to manuscripts of particular importance which have or will cast new light on the Islamic achievement in the exact sciences. Over 100 plates illustrate extracts, tables, diagrams, instruments, signatures, *etc.*, of particular historical interest.

The author is Professor of the History of Science at the Johann Wolfgang Goethe University in Frankfurt-am-Main and retains his former association with the Department of Near Eastern Languages and Literatures at New York University. During the period 1972–79, he directed the Project in Medieval Islamic Astronomy based at the American Research Center in Egypt and funded mainly by the Smithsonian Institution and the National Science Foundation, Washington, D.C. This survey was prepared during the summer of 1976, when the cataloging of the Cairo manuscripts (1974–76) was completed. Prof. King has worked extensively in manuscript libraries all over Europe, the Near East, India and Soviet Central Asia. His publications include *Mathematical Astronomy in Medieval Yemen* (Undena, 1983); *Spherical Astronomy in Medieval Islam* (to appear, Frankfurt ?); *Studies in Astronomical Timekeeping in Medieval Islam* (to appear, Springer-Verlag); and numerous papers on different aspects of Islamic science currently being reprinted as *Islamic Mathematical Astronomy* and *Islamic Astronomical Instruments* by Variorum Reprints, London.

Library of Congress Cataloging-in-Publication Data

King, David A.
A survey of the scientific manuscripts in the Egyptian National Library.

(Catalogs / American Research Center in Egypt; v. 5)
"Published under the auspices of the American Research Center in Egypt, Inc.—Ser. t.p.
1. Science—Manuscripts—Catalogs. 2. Science—History—Sources—Bibliography—Catalogs. 3. Egypt. Dār al-Kutub wa-al-Wathā'iq al-Qawmīyah—Catalogs. I. Title. II. Series: Catalogs (American Research Center in Egypt); v. 5.
Z6611.S4K56 1986 [Q125] 016.5 85-29369
Paper ISBN 0-936770-12-0
Cloth ISBN 0-936770-14-7

Printed in the United States of America

To Prof. Fuat Sezgin
in appreciation

TABLE OF CONTENTS

ACKNOWLEDGEMENTS

It is a pleasure to record my gratitude to the Egyptian National Library for the privileges afforded to me in the Library.

The research on medieval Islamic science conducted at the American Research Center in Egypt during the period 1972–1980 was supported by the Smithsonian Institution and National Science Foundation (1972–1980), the American Philosophical Society (1972–1974), and the Ford Foundation (1976–1978). This support is gratefully acknowledged. To Prof. Owen Gingerich of the Harvard-Smithsonian Center for Astrophysics goes much of the credit for assuring the continued support of the Smithsonian Institution and the National Science Foundation.

I could not have included the Persian and Turkish manuscripts had it not been for the assistance of Dr. Cornell Fleischer. It was he who prepared the Persian and Turkish extracts for Vol. II of the catalog published in Cairo and who identified numerous titles and authors.

The typescript of this survey was prepared by Mrs. Nabila Hajj, whose faithful devotion and exceptional ability have maintained the Smithsonian Institution Project since its inception.

Publication of this volume in its present form was made possible by a generous grant from the Egyptian American Bank and American Express International Banking Corporation.

D.A.K.

FOREWORD

According to the thirteenth-century historian of science Ibn al-Qifṭī, the books in the library of Cairo in the mid-eleventh century were being cataloged and the Egyptian astronomer Ibn al-Nabdī went along to have a look at the works relating to his field. He found six thousand, five hundred volumes on astronomy, mathematics, and philosophy. Not one of these books is contained in the Egyptian National Library today. However, the present Library collection of two thousand, five hundred medieval Islamic astronomical and mathematical manuscripts is the largest single collection in the world, and from the point of view of the history of Islamic science was, until recently, untapped.

As a result of a collaborative venture between the Egyptian National Library and the Smithsonian Institution, an Arabic catalog has been prepared for the entire collection. This English survey of the Cairo scientific manuscripts, compiled during the summer of 1976 using a set of index cards to the Cairo collection, complements the Arabic catalog and is arranged so as to supplement the standard bibliographical works on the history of Islamic science. It contains cross-references to the Arabic catalog in which the reader will find further information on the structure and contents of a given work and further details about a given manuscript.

D.A.K.
Agami, Egypt
September, 1976

In preparing the final version of this survey for publication, I have mainly added references to Sezgin's volumes on astronomy and astrology and inserted some remarks about research on individual manuscripts recently completed or currently in progress.

D.A.K
Crystal Beach, Florida
August, 1984

BIBLIOGRAPHICAL ABBREVIATIONS

For manuscript catalogs, see *GAS*, VI

AHES	*Archive for History of Exact Sciences.*
AIHS	*Archives Internationales d'Histoire des Sciences.*
Aleppo Conference Proceedings	A. Y. al-Hassan *et al.*, eds., *Proceedings of the First International Symposium for the History of Arabic Science, April 5–12*, 1976, vol. I (Arabic) and vol. II (European languages), Aleppo: Institute for the History or Arabic Science, 1978.
Azzawi	A. El-Azzawi, *History of Astronomy in Iraq and Its Relations with Islamic and Arab Countries in the Post Abbasid Periods* (in Arabic), Baghdad: Iraq Academy Press, 1959.
Berggren	J. L. Berggren, "History of Mathematics in the Islamic World: The Present State of the Art," to appear in *Bulletin of the Middle East Studies Association of North America.*
Brockelmann	See *GAL*.
Centaurus	*Centaurus: International Magazine of the History of Mathematics, Science and Technology.*
Djebbar	A. Djebbar, *Enseignement et recherche mathématiques dans le Maghreb . . .*, Orsay, 1981.
DSB	*Dictionary of Scientific Biography*. New York: Charles Scribner's sons. 16 vols., 1970–80.
EI_1	*Encyclopedia of Islam*, 1st ed., 4 vols., Leiden: E. J. Brill, 1913–1934.
EI_2	———, 2nd ed., 4 vols. to date, Leiden: E. J. Brill, 1960 to present.
GAL	C. Brockelmann. *Geschichte der arabischen Litteratur*, 2 vols. (2nd ed.), Leiden: E. J. Brill, 1943–49, and *Supplementbände*. 3 vols., Leiden: E. J. Brill, 1937–42.

GAS F. Sezgin, *Geschichte des arabischen Schrifttums*, Band V: Mathematik, Band VI: Astronomie, Band VII: Astrologie, Meteorologie und Verwandtes; Leiden: E. J. Brill, 1975, 1978, and 1979.

GHA *A General History of Astronomy*, to be published by the International Astronomical Union.

Graf G. Graf, *Geschichte der christlichen arabischen Literatur*, 5 vols., Rome: Biblioteca Apostolica Vaticana, 1944–1953.

Ḥājjī Khalīfa Ḥājjī Khalīfa, *Kashf al-ẓunūn ᶜan asāmi l-kutub wa-l-funūn*, 2 vols., Istanbul: Bahiya Press, 1941.

Hartner Festschrift W. G. Saltzer and Y. Maeyama, eds., *Prismata: Festschrift für Willy Hartner*, Wiesbaden: Franz Steiner Verlag, 1977.

HES S. K. Sirazhdinov, ed., *Iz istorii tochnykh nauk na srednevekovom blizhnem i srednem vostoke* (*From the History of the Exact Sciences in the Medieval Near and Middle East*), Tashkent: Academy of Sciences of the Uzbek S.S.R., 1972 (reviewed by E. S. Kennedy in *HM*, 2 [1975], pp. 86–90).

HM *Historia Mathematica.*

HS *History of Science.*

Istanbul Congress Proceedings *Proceedings of the 1st International Congress of the History of Turkish-Islamic Science and Technology,* Istanbul: ITU Press, 1981.

Istanbul Symposium Proceedings *Proceedings of the International Symposium on the Observatories in Islam, (Istanbul, 1977),* Istanbul: Kandilli Observatory, 1979.

JA *Journal Asiatique.*

JAOS *Journal of the American Oriental Society.*

JHA *Journal for the History of Astronomy.*

JHAS *Journal for the History of Arabic Science.*

Kennedy E. S. Kennedy, "A Survey of Islamic Astronomical Tables," *Transactions of the American Philosophical Society*, N.S., 46:2 (1956), pp. 123–177.

Kennedy et al. E. S. Kennedy, Colleagues and Former Students, *Studies in the Islamic Exact Sciences*, Beirut: American University of Beirut Press, 1983.

Kennedy Festschrift D. A. King and G. Saliba, eds., *From Deferent to Equant: A Volume of Studies in the History of Science in the Near East in Honor of E. S. Kennedy*, to be published by the New York Academy of Sciences in 1986.

King 1 D. A. King, "The Exact Sciences in Medieval Islam: Some Remarks on the Present State of Research," *Bulletin of the Middle East Studies Association of North America*, 4 (1980), pp. 10–26.

2 ———, "The Astronomy of the Mamluks," *ISIS*, 74 (1983), pp. 531–555, with a summary in *Muqarnas*, 2 (1984), pp. 73–84.

See also *MAY* and *SATMI.*

Krause	M. Krause, "Stambuler Handschriften islamischer Mathematiker," *Quellen und Studien zur Geschichte der Mathematik Astronomie und Physik*, B3:4 (1936), pp. 437–532.
MAY	D. A. King, *Mathematical Astronomy in Medieval Yemen: A Bio-Bibliographical Survey*, Publications of the American Research Center in Egypt, Malibu, California: Undena Publications, 1983.
Mayer	L. A. Mayer, *Islamic Astrolabists and Their Works*, Geneva: Ernst Kundig, 1956.
OPNE	*Occasional Papers on the Near East* (New York University, Hagop Kevorkian Center for Near Eastern Studies).
Pingree 1	D. Pingree, "The Greek Influence on Early Islamic Mathematical Astronomy," *Journal of the American Oriental Society*, 93 (1973), pp. 32–43.
2	________, "Indian Influence on Sasanian and Early Islamic Astronomy and Astrology," *Journal of the Oriental Research, Madras*, 34–35 (1964–66/1973), pp. 118–126.
Renaud	H. J. P. Renaud, "Additions et corrections à Suter 'Die Mathematiker und Astronomen der Araber'," *ISIS*, 18 (1932), pp. 166–183.
Sarton	G. Sarton, *Introduction to the History of Science*, 3 vols., Baltimore, 1927–48.
SATMI	D. A. King, *Studies in Astronomical Timekeeping in Medieval Islam*, to appear (Springer-Verlag).
Sayili	A. Sayili, *The Observatory in Islam*, Ankara: Turkish Historical Society (Series VII, No. 38), 1960.
Sezgin	See *GAS*.
SM	*Studia Mathematica*.
Storey	C. A. Storey, *Persian Literature: A Bio-Bibliographical Survey*, Vol. II, London: Luzac and Co., 1958.
Suter	H. Suter, "Die Mathematiker und Astronomen der Araber und ihre Werke," *Abhandlungen zur Geschichte der mathematischen Wissenschaften*, 10 (1900), and "Nachträge und Berichtigungen," *ibid.*, 14 (1902), pp. 157–185, reprinted Amsterdam: Oriental Press, 1982.
Ullmann	M. Ullmann, *Die Natur- und Geheimwissenschaften im Islam*, Leiden: E. J. Brill, 1972.
Wiedemann	E. Wiedemann, *Aufsätze zur arabischen Wissenschaftsgeschichte*, 2 vols., Hildesheim: G. Olms, 1970 ed. (originally published in the *Sitzungsberichte der physikalisch-medizinischen Sozietät zu Erlangen*, 1902–1928).
ZDMG	*Zeitschrift der Deutschen Morgenländischen Gesellschaft*.
ZGAIW	*Zeitschrift für Geschichte der Arabisch-Islamischen Wissenschaften*.

MISCELLANEOUS ABBREVIATIONS

The abbreviation "b." is used in personal names for *ibn*, "son of." A comma in a name means "known as," and precedes the name by which the individual is most commonly known.

In titles of treatises, the letters "*K.*," "*R.*" and "*M.*" refer to "*Kitāb* (= Book)," "*Risāla*" and "*Maqāla* (= Treatise)."

The abbreviation "MS" means "manuscript." For each Cairo manuscript cited, the collection (see pp. 2–4) and call number are given, followed by—in parentheses—the folios and date of copying.

The abbreviation "cop." means "copied in the year. . . ." Dates of copying are given in the Hijra calendar, abbreviated by the letter "H." To find the Gregorian year X corresponding to a Hijra year H, use the formula

$$X = 622 + H - H/33$$

(because 33 lunar years = 32 solar years, and the Hijra calendar begins in A.D. 622). Occasionally dates in the Seleucid and Sasanian calendars are given; these are indicated by the letters A (= Alexander) and Y (= Yazdigird), respectively.

INTRODUCTION

1. The Documentation of the Sources for the History of Islamic Science

In 1900 the Swiss historian of science Heinrich Suter compiled his monumental work *Die Mathematiker und Astronomen der Araber* which, until recently, was the main reference work for the student of Islamic science. On the basis of the catalogs of manuscript collections which were then available and medieval Arabic biographical and bibliographic sources, Suter compiled a list of over 500 medieval Muslim mathematicians and astronomers and their works, including all known manuscripts thereof. In 1914 he published a short supplement to this list. Since Suter had translated the sections on mathematics and astronomy in the 1890 catalog of the manuscripts in the then Viceregal Library in Cairo, he was able to include some of the manuscripts mentioned in the present survey.

In 1932 the French historian of science H. P. J. Renaud published a list of additions and corrections to Suter's work, based mainly on the Arabic scientific manuscripts preserved in Rabat. In 1936 the German historian of science Max Krause published a list of further additions based on the Arabic scientific manuscripts preserved in Istanbul. During the years 1898 to 1942, the German orientalist Carl Brockelmann published his monumental survey of medieval Arabic literature entitled *Geschichte der arabischen Litteratur* (hereafter *GAL*), which contained further information on the Muslim scientists. For the period preceding the year 1050 A.D., this in now replaced by the much more comprehensive survey by the Turkish Islamicist Fuat Sezgin, entitled *Geschichte des arabischen Schrifttums* (*GAS*). Sezgin's volumes on mathematics (V), astronomy (VI), astrology and folk astronomy (VII) incorporate material from manuscript libraries all over Europe, the Near East, India and Central Asia, and they raise the level (and standard) of documentation of the manuscript sources to heights which Suter and Brockelmann could never have dreamed of. For Persian scientific literature, we have the valuable bibliographical work of the British orientalist C. A. Storey, published in 1958. There is no corresponding work for Turkish scientific literature. For the Islamic astronomical handbooks known as *zīj*es, we have the valuable survey which the American historian of science E. S. Kennedy published in 1956. A revised version of this, containing information on some two hundred *zīj*es, is planned by Kennedy in collaboration with the present author. In 1983, G. P. Matvievskaya and B. A. Rosenfeld published their *Matematiki* *i astronomi musulmanskogo srednevekovya i ikh trudi (VIII–XVIIvv.)*, incorporating information from published sources and adding considerable new material.

The purpose of the present work is to supplement all of these reference works with information on the scientific manuscripts in the Egyptian National Library in Cairo. The authors are arranged chronologically as far as this is possible. For the period covered by Sezgin, the authors are not divided according to their geographical location. For the later centuries, I consider that the most useful arrangement is chronological according to location; such an arrangement was adopted by Brockelmann. Much of the material which is new to the modern literature relates to astronomical timekeeping; this is analysed in my monograph on astronomical timekeeping in medieval Islam (*SATMI*). There is a great deal of other material in the Cairo collection which merits the attention of historians of science and Islamicists alike. In view of the difficulty of obtaining microfilms, Cairo copies of manuscripts that are available in Europe are of limited use to researchers outside Cairo unless the manuscripts are particularly good copies and the researchers are unusually fortunate. The main advantage of the Cairo collection is its sheer size and the availability in one place of so many works of which the other known copies are scattered all over Europe and the Near East.

2. The Scientific Manuscript Collections in the Egyptian National Library

This survey embraces all the astronomical, astrological, and mathematical manuscripts in the Egyptian National Library. These manuscripts are preserved in different collections within the stacks, as follows.

(1) The main collection (*Dār al-Kutub, al-rasīd al-ᶜāmm*) in which the manuscripts are arranged by subject (*mīqāt* = "astronomy," *hayʾa* = "theoretical astronomy," *riyāḍa* = "mathematics," *ḥurūf* = "magic," *etc.*). The private library of Muṣṭafā Fāḍil was incorporated in this collection. Manuscripts of mixed contents are stacked separately (*majāmīᶜ* = mixed contents). In this collection, the manuscripts acquired by the Egyptian National Library between 1936 and 1955 are classified differently for a reason which is not at all clear (K = "astronomy and mathematics," S = "oriental languages (!)," B = "religion," *etc.*). The only private collection which was incorporated in this addition to *al-rasīd al-ᶜāmm* is the library of Muḥammad ᶜAbdū.

(2) The former private libraries of Ṭalᶜat, Taymūr, Qawala, Ḥalīm, Khalīl Āghā, and Aḥmad Zakī. The first five of these are subdivided according to subject matter in the same way as the original Dār al-Kutub collection (*mīqāt*, *riyāḍa*, *hayʾa*, *majāmīᶜ*, *etc.*). The Zakīya collection is not arranged according to subject matter.

(3) Persian and Turkish manuscripts are contained in each of the above collections and are stacked separately.

For each collection uncritical handlists (*sijillāt*) are available inside the Library, though not for the general reader.

In view of the large number of individual collections into which the scientific manuscripts in the Egyptian National Library are divided, I have adopted an alphabetic notation to abbreviate the names of the collections. All of the Arabic collections are indicated by one or two Arabic letters and the Persian and Turkish collections by three.

Single letters are used for collections acquired by the Egyptian National Library between 1930 and 1950 (and cataloged by F. Sayyid). This convention is that of the Library. The most frequently used letter is

K ("astronomy and mathematics")

The first letter of a two- or three-letter symbol refers to the main theme of the collection, thus:

D = Dār al-Kutub
M = Muṣṭafā Fāḍil
Ṭ = Ṭalᶜat
T = Taymūr
Q = Qawala
Kh = Khalīl Āghā
Ḥ = Ḥalīm

In two exceptional cases two letters are used:

ZK = Zakīya
KA = K (Muḥammad ᶜAbdū)

The second letter of a two- or three-letter symbol refers to the internal subject classification of the collections. The most frequently used symbols are:

M *mīqāt* (astronomy)
H *hayʾa* (theoretical astronomy)
F *falak* (astronomy, Persian and Turkish)
R *riyāḍa* (mathematics)
J *majāmīᶜ* (collections of works on different subjects)

The third letter of a three-letter symbol refers to the Persian and Turkish collections, thus:

F = *fārisī* (Persian)
T = *turkī* (Turkish)

A complete list of symbols used follows, arranged in the order of the entries in the Arabic handlist:

DM	Dār al-Kutub	*mīqāt*
DMF	"	*mīqāt fārisī*
DMT	"	*mīqāt turkī*
DH	"	*hayʾa*
DHF	"	*hayʾa fārisī*
DHT	"	*hayʾa turkī*
DR	"	*riyāḍa*
DRF	"	*riyāḍa fārisī*
DRT	"	*riyāḍa turkī*
DNT	"	*handasa turkī*
DḤ	"	*ḥurūf*
DṢ	"	*ṣināᶜa*
DṬ	"	*ṭabīᶜīyāt*
DSh	"	*ᶜulūm maᶜāshīya*
DO	"	*funūn mutanawwaᶜa*
DJF	"	*majāmīᶜ fārisī*

B	*ʿulūm dīnīya*
J	*ʿaqāʾid*
S	*lughāt sharqīya*
Sh	*ḥurūf wa-awfāq*
K	*falak wa-riyāḍa*
KA	*falak wa-riyāḍa* (Muḥammad ʿAbdū)
L	*ṭibb*
W	*falsafa*
Y	*funūn mutanawwaʿa*

MM	Muṣṭafā Fāḍil	*mīqāt*
MMF	″	*mīqāt fārisī*
MMT	″	*mīqāt turkī*
MH	″	*hayʾa*
MHF	″	*hayʾa fārisī*
MR	″	*riyāḍa*
MRT	″	*riyāḍa turkī*
MNT	″	*handasa turkī*
MḤ	″	*ḥurūf*
MJ	″	*majāmīʿ*
MJF	″	*majāmīʿ fārisī*

ṬM	Ṭalʿat	*mīqāt*
ṬMF	″	*mīqāt fārisī*
ṬMT	″	*mīqāt turkī*
ṬFF	″	*falak fārisī*
ṬFT	″	*falak turkī*
ṬH	″	*hayʾa*
ṬR	″	*riyāḍa*
ṬRF	″	*riyāḍa fārisī*
ṬRT	″	*riyāḍa turkī*
ṬḤ	″	*ḥurūf*
ṬḤF	″	*maʿārif-i-asrār fārisī*
ṬK	″	*ḥikma*
ṬOT	″	*maʿārif mutanawwaʿa turkī*
ṬJ	″	*majāmīʿ*
ṬJF	″	*majāmīʿ fārisī*

TR	Taymūr	*riyāḍa*
TGh	″	*ghaybīyāt*
TKh	″	*akhlāq*
TK	″	*ḥikma*
TṢ	″	*ṣināʿa*
TJ	″	*majāmīʿ*

QM	Qawala	*mīqāt*
QH	″	*hayʾa*
QR	″	*riyāḍa*
QJ	″	*majāmīʿ*

KhM	Khalīl Āghā	*mīqāt*
KhMT	″	*mīqāt turkī*
KhR	″	*riyāḍa*
KhJ	″	*majāmīʿ*

ḤM	Ḥalīm	*mīqāt*
ḤH	″	*hayʾa*
ḤR	″	*riyāḍa*
ḤF	″	*falsafa*
ḤJ	″	*majāmīʿ*

ZK	Zakīya

Of the present collection of about 2,500 scientific manuscripts, only about 250 manuscripts from the Dār al-Kutub and about 300 from the Muṣṭafā Fāḍil *mīqāt*, *hay*ʾ*a*, and *riyāḍa* collections were listed in the first catalog of the former Khedival Library (1890), and about 275 manuscripts from the Dār al-Kutub "K" collection, acquired between the years 1936 and 1955, were listed in the catalog of Fuad Sayyid (1961–63). Some of the Turkish manuscripts were listed in the 1890 catalog and some of the Persian manuscripts were listed in a catalog published in the 1950's, although this was not used by Storey. The remainder was cataloged for the first time only recently. The most important uncataloged batches of scientific manuscripts were those added to the main Dār al-Kutub collections between 1890 and 1936, but the fact that these were never cataloged is not generally known either outside the Library or inside. Likewise important are the Ṭalʿat and Taymūr scientific manuscripts, also uncataloged until now.

The cataloging situation is far better for religious, legal, and historical manuscripts, but not for the rich collections of medical or magical manuscripts. However, some of those catalogs that have been prepared are not easily available outside the Library or are not available even inside the Library.

3. The New Arabic Catalog of the Scientific Collections

In this English survey, each work listed is accompanied by a cross-reference to Part II of the Arabic catalog where details about its structure may be found. Likewise, Part I of the Arabic catalog must be consulted for information on the size of the manuscripts, the style of handwriting, the date of copying, the name of the copyist, the provenance, *etc.*

Part I of the Arabic catalog consists of a critical handlist of all the scientific manuscripts in the Egyptian National Library, arranged according to the Library classification system (see Section 2 above). The handlist gives for each manuscript or part of a manuscript such basic information as title and author as given in the manuscript, folio or page numbers, size, type of script, date of copying and name of copyist, and provenance. The reader should bear in mind that in many cases there is no title, no author, no date of copying or name of copyist, and no indication of provenance. As many as possible of the titleless and anonymous works have been identified by comparison with other manuscripts in other libraries. Most of the bound photographs bear no indication of the location of the original manuscript: this has also been established where possible. All of the undated manuscripts have been dated to the nearest fifty years. Although the provenance of the manuscripts is generally not stated, only in the case of a very few manuscripts has it proved impossible to ascertain the major cultural area where they originated.

The handlist is intended to be critical in the sense that it was prepared in three main stages. In the first, the basic information about each manuscript was recorded on index cards. In the second, all of the copies of a given work were inspected simultaneously, and the index cards corrected where necessary. In the third, the manuscripts and cards were inspected once again in the order of the library classification system and checked for details. For each work listed in Part I a three number cross-reference is given to Part II of the Arabic catalog.

The Egyptian National Library possesses a number of astronomical instruments, all of them quadrants, and a brief description of each is included in Part I. Indexes are also given for copyists, owners, readers, and dated manuscripts. A list of manuscripts available on microfilm from the Arab League Institute of Arabic manuscripts is included.

Part II of the Arabic catalog contains extracts from the various works represented in the Cairo scientific manuscripts, arranged chronologically according to subject matter. Each of the works has been classified according to its contents using a two-digit numerical scheme. The classification system adopted is as follows:

1 GENERAL ASTRONOMICAL WORKS

- 1.1 Translations and recensions of Greek astronomical works
- 1.2 General astronomy (*kutub al-hayʾa*)
- 1.3 Constellation figures
- 1.4 Folk astronomy (*kutub al-anwāʾ, etc.*)
- 1.5 Religious aspects of astronomy
- 1.6 Miscellaneous works of consequence
- 1.7 Miscellaneous late works on astronomy

2 *ZĪJ*ES AND EPHEMERIDES

- 2.1 Miscellaneous early *zīj*es and related treatises
- 2.2 Miscellaneous late *zīj*es
- 2.3 Ephemerides and related treatises
- 2.4 Miscellaneous tables
- 2.5 Miscellaneous *zīj* topics

3 ASTRONOMICAL TIMEKEEPING

- 3.1 Tables for timekeeping
- 3.2 Treatises on timekeeping
- 3.3 Treatises on the determination of the qibla
- 3.4 Treatises on lunar crescent visibility
- 3.5 Miscellaneous

4 ASTRONOMICAL INSTRUMENTS

- 4.1 General treatises
- 4.2 Armillary sphere
- 4.3 Astrolabe
- 4.4 Almucantar quadrant
- 4.5 Sine quadrant
- 4.6 Miscellaneous astrolabes and quadrants
- 4.7 Sundials
- 4.8 Miscellaneous instruments

5 ASTROLOGY

- 5.1 Hermetic material and translations of Greek, Sasanian, and Indian works
- 5.2 Early Islamic works
- 5.3 Late Islamic works attributed to their authors
- 5.4 Late anonymous Islamic works
- 5.5 Miscellaneous

6 MATHEMATICS

- 6.1 Translations and recensions of Greek works and commentaries
- 6.2 Early Islamic mathematics
- 6.3 Later algebra and geometry
- 6.4 Later arithmetic
- 6.5 Sexagesimal arithmetic
- 6.6 Sexagesimal multiplication tables
- 6.7 Surveying
- 6.8 Weights and measures
- 6.9 Inheritance
- 6.10 Miscellaneous late works on mathematics

7 PHYSICS

- 7.1 Mechanics
- 7.2 Optics
- 7.3 Miscellaneous

8 NON-MATHEMATICAL TOPICS

Within each of these sections, the individual works have been arranged chronologically as far as this is possible and assigned a third number. Different works on the same subject by a given author have been grouped together, and commentaries and abridgements have been grouped with the original work. This system has been adopted for convenience, and no conclusions should be drawn about the relative importance of a given work or its author from the fact that it has been designated with an individual three-"digit" number.

The extracts given in Part II of the Arabic catalog are adequate to convey a limited idea of the nature of each work and to identify related works in other libraries. They are not intended as a substitute for an analysis of the various works; such a task is clearly beyond the scope of a catalog. I have thought it worthwhile to include *incipit*s, chapter titles, and *explicit*s, even in the case of treatises of very little historical interest; thus the lengths of the extracts should not be taken as a measure of the importance of a particular work for the history of Islamic science.

The format of each entry in Part II of the Arabic catalog was dictated by the need to separate the Arabic and English parts of the catalog for printing purposes. Each extract begins with a title and author where these are secure. The extract is taken from the best available manuscript, which is the first of those indicated in the lists following the extract. Where the title and author are not secure or there are substantial variants between the available copies, some indication is given. The criteria used for judging the relative merits of one copy of a particular work over another include the date of copying relative to the date of compilation, and an estimate of the mentality of the copyist as reflected in his work. In cases where the title or name of the author is in question, the reader must have recourse to Part I of the catalog where these are given as they appear in the manuscript, if they appear at all.

The reader will find many of the extracts deficient in references to folio or page numbers. The reason for this is simply that many of the Cairo manuscripts are either improperly foliated or not foliated at all. In the case of important works, folio or page references are given.

Most of the extracts were copied from the manuscripts by Miss Khulud Mustafa and Miss Manar Rateb under my direction. We did not attempt to "edit" the extracts in any way. The reader will observe many instances of total disregard for the grammar of the Arabic language; the extracts have been copied from the best available copies of each work. The extracts from the Persian and Turkish manuscripts were copied by Dr. Cornell Fleischer, who also identified for me various obscure titles and mysterious authors. Part II of the Arabic catalog also contains indexes of authors and titles.

I have not thought it worthwhile to present lists of Arabic titles of tables, not least because some of the most interesting tables have no titles at all. Rather, I have set aside all those works containing tables for study in separate publications.

In the case of the astronomical handbooks known as *zīj*es, which often consist of several hundred pages of text and tables, I have not felt obliged to describe them in the catalog. There are so many *zīj*es of consequence which merit detailed investigation and on which there is no published material, that in preparing the catalog I restricted my purpose to simply identifying new material so as to return to more important tasks. There is, for example, no point in describing fragments of Ibn Yūnus' works in the catalog when his major work has not yet been adequately studied; I preferred to continue my research on the *Ḥākimī Zīj*. Again, rather than work through the *Īlkhānī Zīj* of al-Ṭūsī and the *Sulṭānī Zīj* of Ulugh Beg for the catalog, I have postponed this task for the new survey of Islamic astronomical handbooks being undertaken by Dr. Kennedy and myself. Likewise, I felt it was hardly worth describing in the catalog half a dozen late Egyptian copies of astrolabe tables when the astrolabe tables of al-Khwārizmī and al-Farghānī had never been described in the modern literature. The proper place to discuss the late Egyptian material would be at the end of a monograph describing the earlier tables.

Thus the Arabic catalog, judged as a catalog of Arabic manuscripts, in several ways falls short of the standards set by orientalists of previous generations who have devoted some time to cataloging. However, the conditions under which it was prepared should be taken into consideration in any evaluation. At no time between 1972 and 1978 was it at all clear that I was permitted to conduct research on the manuscripts. My purpose in investigating the Cairo manuscripts was, in fact, not primarily to catalog them but to sift them for information which would add to our knowledge of the history of the exact sciences in medieval Islam.

The survey was conducted under the constant threat by the Library administration of immediate termination, at least until mid-1975. It took one year (1972–73) to negotiate an agreement, another year (1973–74) to put that agreement into effect so that work could begin. No effective assistance was made available to the cataloger until the cataloging project was officially over: requests were refused for assistance even for someone to number folios. The cataloger was never officially allowed access to the stacks: even during the fifth year of his presence in Cairo questions were still being asked in the manuscript reading room about whether he had "permission" to copy material for his own research, and in the Library security office about whether he had "permission" to be in the Library at all.

The copying of extracts for Part II of the Arabic calalog was begun *after* the cataloging project had officially finished, when for the first time some effective help was afforded to the project. No reference works whatsoever were available in the manuscript reading room, and all the material gathered for the catalog was organized outside the Library. Often it was impossible to refer again to manuscripts that had been consulted and then returned to the stacks. One of the reasons for this is that the stack numbers are indicated inside the manuscripts, not on the covers. Another is a particular problem of the Egyptian National Library: any reader who has ordered manuscripts there in the past few years has heard the phrase *mush mawgūd* (= it's not there) used in relation to some invaluable manuscript which he/she wishes to consult.

The first three scholars to consult the catalog in 1976, respectively specialists in Islamic arithmetic, algebra, and technology, and all Arabs, were refused microfilms of the material which they subsequently requested. Fortunately, the policies of the administration of the Egyptian National Library have since improved radically, but similar situations persist in other libraries. I plead with the administrations of libraries with collections of Arabic manuscripts for greater understanding of the needs of individual scholars trying to contribute to the documentation of the history of Islamic civilization. More generally, I plead with all library administrations to deal kindly with those who wish to read any of their books or manuscripts, and particularly with anyone who, for any reason, is willing to catalog them.

LIST OF AUTHORS

A. Greek, Sasanian, and Indian Authors in Arabic

A1	Pythagoras
A2	Socrates
A3	Aristotle
A4	Autolycus
A5	Euclid
A6	Aristarchus
A7	Archimedes
A8	Apollonius
A9	Hipparchus
A10	Hypsicles
A11	Heron
A12	Theodosius
A13	Antiochus
A14	Diodorus
A15	Dorotheus
A16	Menelaus
A17	Nichomachus
A18	Ptolemy
A19	Diocles
A20	Aristus / Eutochius
A21	Vettius Valens
A22	Hermes / Daniel / Alexander
A23	Teukrus (Tankalusha)
A24	Kīrkis b. Fīlūs al-Baṭrak (?)
A25	Rymbr (?) al-Yūnānī
A26	Sarjis (al-Raʾsʿaynī)
A27	Zarathustra (Zarādasht)
A28	Jāmasp al-Ḥakīm
A29	Anonymous (= Jbhr?)
A30	Jnh al-Hindī
A31	Ṭumṭum al-Hindī
A32	Kanaka (?)
A33	Yuḥannā (?) al-Hindī
A34	Jaʿfar al-Hindī

B. Early Muslim Authors (to *ca.* 1100)

B1	ʿAlī b. Abī Ṭālib
B2	al-Imām al-Shāfiʿī
B3	Ibrāhīm b. Muḥammad al-Fazārī
B4	Yaʿqūb b. Ṭāriq
B5	Māshāʾallāh
B6	ʿUmar b. al-Farrukhān al-Ṭabarī
B7	al-Ṭabarī al-Munajjim (= ʿUmar b. al-Farrukhān?)
B8	Yaḥyā b. Abī Manṣūr
B9	al-Maʾmūn
B10	Abū Zakarīyā Yaḥyā b. Māsawayh
B11	Sanad b. ʿAlī
B12	ʿAlī b. ʿĪsā
B13	Abū ʿAlī al-Khayyāṭ
B14	Muḥammad b. Mūsā al-Khwārizmī
B15	Sahl b. Bishr al-Isrāʾīlī
B16	Banū Mūsā
B17	Jābir b. Ibrāhīm al-Ṣābiʾ
B18	Abu l-ʿAbbās Aḥmad b. Muḥammad b. Kathīr al-Farghānī
B19	Yaʿqūb b. Isḥāq al-Kindī
B20	Ḥunayn b. Isḥāq
B21	Abū Yūsuf Yaʿqūb b. ʿAlī al-Qaṣrānī
B22	Abū Maʿshar
B23	Abu l-ʿAnbas al-Ṣaymarī
B24	Abū Saʿīd al-Ḍarīr
B25	Abū Muḥammad ʿAbd Allāh b. Muslim, Ibn Qutayba al-Dīnawarī
B26	ʿAbd al-Raḥmān b. Isḥāq al-Zajjājī
B27	Ibn al-Qāṣṣ

B28 Daʾūd b. Sulaymān
B29 Abu l-Qāsim al-Munajjim
B30 Thābit b. Qurra
B31 Muḥammad b. ʿAbd Allāh b. ʿUmar b. al-Bāzyār
B32 Aḥmad b. ʿUmar al-Karābīsī
B33 Abū Daʾūd Sulaymān b. ʿIṣma
B34 Isḥāq b. Ḥunayn
B35 Ḥabash
B36 Abū l-Qāsim b. Āmājūr
B37 ʿAbd Allāh b. Masrūr al-Ḥāsib
B38 Qusṭā b. Lūqā
B39 Abu l-ʿAbbās al-Faḍl b. Ḥātim al-Nayrīzī
B40 Aḥmad b. Yūsuf Kātib Āl-Ṭūlūn (Ibn al-Dāya)
B41 al-Ḥasan b. Aḥmad b. Yaʿqūb al-Hamdānī
B42 Ibrāhīm b. Sinān
B43 Abū Jaʿfar al-Khāzin
B44 Sinān b. al-Fatḥ
B45 ʿAbd al-ʿAzīz b. ʿUthmān al-Qabīṣī
B46 Abu l-Ḥusayn Aḥmad b. al-Ḥusayn al-Ahwāzī
B47 Yaʿqūb b. Muḥammad al-Sijistānī
B48 Abū Isḥāq al-Ṣābiʾ
B49 al-Ḥasan b. Jaʿfar al-Anṣārī
B50 Abū Maḥmūd Ḥāmid b. al-Khiḍr al-Khujandī
B51 Ghulām Zuḥal
B52 Abu l-Ḥusayn ʿAbd al-Raḥmān b. ʿUmar al-Ṣūfī
B53 Abu Sahl Wayjan b. Rustam al-Qūhī / al-Kūhī
B54 Abu l-Wafāʾ Muḥammad b. Muḥammad al-Būzajānī
B55 Nūr al-Din Abu l-Qāsim ʿAlī b. Aḥmad al-Balkhī
B56 Aḥmad b. Muḥammad b. ʿAbd al-Jalīl al-Sijzī
B57 Aḥmad b. Ibrāhīm al-Sanjarī (?)
B58 Abu l-Fatḥ al-Samarqandī
B59 Abu l-Ḥasan ʿAlī b. ʿAbd al-Raḥmān b. Aḥmad, Ibn Yūnus
B60 Abū Naṣr al-Ḥasan b. ʿAlī al-Qummī
B61 Yūsuf b. ʿAbd Allāh b. Muḥammad, Ibn ʿAbd al-Barr
B62 Abū Naṣr Yaḥyā b. Jarīr al-Takrītī
B63 Abū Manṣūr Sulaymān b. al-Ḥusayn b. al-Yazdawīya (?) al-Abralī (?)
B64 Abū ʿAlī b. Abi l-Ḥusayn al-Ṣūfī
B65 Ikhwān al-Ṣafāʾ
B66 Abū Naṣr b. ʿIrāq
B67 Abu l-Ḥasan ʿAlī b. Aḥmad al-Nasawī
B68 Muḥammad b. Aḥmad al-Qummī
B69 Abū Bakr Muḥammad b. al-Ḥasan al-Karajī
B70 Kūshyār b. Labbān
B71 Abū ʿAbd Allāh Muḥammad b. Aḥmad al-Shannī
B72 Abu l-Jūd Muḥammad b. al-Layth
B73 Abū Saʿd al-ʿAlāʾ b. Sahl
B74 Abu l-Qāsim Aḥmad b. ʿAbd Allāh al-Ṣaffār
B75 ʿAbd al-Qāhir b. Ṭāhir al-Baghdādī
B76 ʿAlī b. Abi l-Rijāl al-Shaybānī
B77 Abū ʿAlī al-Ḥasan b. al-Ḥasan, Ibn al-Haytham
B78 Abu l-Rayḥān Muḥammad b. Aḥmad al-Bīrūnī
B79 Anonymous
B80 al-Ḥasan b. Aḥmad b. ʿAbd Allāh al-Ṣūfī al-Kirmānī
B81 Abū ʿAlī al-Ḥusayn b. ʿAbd Allāh, Ibn Sīnā
B82 Abū ʿUbayd ʿAbd al-Wāḥid b. Muḥammad al-Jūzajānī
B83 Anonymous
B84 Anonymous
B85 ʿAlī b. Riḍwān
B86 Abū Bakr Aḥmad b. ʿAlī b. Thābit, al-Khaṭīb al-Baghdādī
B87 Abū Isḥāq Ibrāhīm b. Yaḥyā al-Zarqāllu al-Ṭulayṭulī
B88 Abu l-Yusr ʿAlī b. Muḥammad al-Bazdawī
B89 Abu l-Ṣalt Umayya b. ʿAbd al-ʿAzīz b. Abi l-Ṣalt
B90 Mnʿby (?) b. Farīghūn
B91 Shāh-Mardān b. Abi l-Khayr Rāzī
B92 Abū Jaʿfar Muḥammad b. Ayyūb al-Ṭabarī
B93 Abu l-Fatḥ Muḥammad b. ʿAbd al-Malik al-Dawwānī

B94 Muḥammad b. Ḥasan al-Juyūbī
B95 Abū ʿAlī al-Shāhid
B96 Abū ʿAbd Allāh Muḥammad b. al-Ḥusayn al-Ḥāsib
B97 Abū Yūsuf Aḥmad b. al-Ḥasan or al-Ḥasan b. Yūsuf
B98–114 Anonymous
B115 Ibn al-Samḥ

C. Egyptian and Syrian Authors (*ca.* 1100 to *ca.* 1600)

C1 Riḍwān b. Muḥammad al-Khurāsānī
C2 Abu l-ʿIzz Ismāʿīl b. al-Razzāz al-Jazarī
C3 Najm al-Din Ayyūb b. ʿAyn al-Dawla b. Naṣr Allāh al-Ikhlāṭī
C4 Sirāj al-Dunyā wa-l-Dīn
C5 Abu l-Ghanāʾim Muslim b. Maḥmūd b. Niʿma b. Arslān al-Shayzarī
C6 Shams al-Dīn Ismāʿīl b. Ibrāhīm b. Ghāzī al-Māridīnī, Ibn Fallūs
C7 Muḥammad b. ʿUmar b. ʿAlī al-ʿUthmānī al-Fāriqī
C8 Ibn Khallikān
C9 Baylak b. ʿAbd Allāh al-Qipjāqī
C10 al-Asʿad ibn al-ʿAssāl
C11 Jamāl al-Dīn Aḥmad b. Thibāt
C12–13 Anonymous
C14 Ḍiyāʾ al-Dīn (?) ʿAbd al-ʿAzīz b. Aḥmad b. Saʿīd al-Dīrīnī
C15 Abu l-ʿAbbās Shihāb al-Dīn Aḥmad b. ʿUmar b. Ismāʿīl al-Maqsī
C16 Najm al-Din Muḥammad b. Muḥammad al-Miṣrī
C17 Sharaf al-Dīn Abū ʿAlī al-Ḥasan b. ʿUmar al-Marrākushī
C18 Anonymous (al-Marrākushī?)
C19 Najm al-Dīn Abu l-ʿAbbās Aḥmad b. Muḥammad b. ʿAlī b. al-Rifʿa al-Anṣārī
C20 Abū ʿAlī al-Fārisī
C21 ʿUmar b. Daʾūd b. Sulaymān al-Fārisī
C22 Tāj al-Dīn Abu l-Ḥasan ʿAlī b. Abī Muḥammad ʿAbd Allāh b. al-Ḥasan b. Abī Bakr al-Tabrīzī
C23 Badr al-Dīn Muḥammad b. Ibrāhīm b. Saʿd Allāh b. Jamāʿa al-Kinānī al-Ḥamawī
C24 Nāṣir al-Dīn Muḥammad b. Simʿūn
C25 Anonymous (treatise on the *musātara*)
C26 Shihāb al-Dīn Aḥmad b. Abī Bakr b. ʿAlī b. al-Sarrāj
C27 Taqi l-Dīn ʿAlī b. ʿAbd al-Kāfī al-Subkī
C28 Shams al-Dīn Aḥmad b. Muḥammad b. Aḥmad al-Azharī al-Bakhāniqī
C29 Khalīl b. Isḥāq al-Jundī
C30 ʿAlāʾ al-Dīn ʿAlī b. Ibrāhīm, Ibn al-Shāṭir
C31 Anonymous (treatise on the "complete" quadrant)
C32 Muḥammad b. Muḥammad b. Muḥammad b. ʿAbd al-Qawī al-Qurashī, Ibn al-Kattānī
C33 Abū ʿAbd Allāh Muḥammad b. Muḥammad, Ibn al-Ghazūlī
C34 Shams al-Dīn Muḥammad b. Aḥmad b. ʿAbd al-Raḥīm al-Mizzī
C35 Zayn al-Dīn Abū Bakr Muḥammad b. Ayyūb al-Ṣūfī al-Karakī
C36 Abu l-Maʿālī Muḥyi l-Dīn Murtafaʿ b. Ḥasan b. Murtafaʿ al-Sāʿātī
C37 Shams al-Dīn Abū ʿAbd Allāh Muḥammad b. Muḥammad Al-Khalīlī
C38 Sharaf al-Dīn Abū ʿImrān Mūsā b. Muḥammad al-Khalīlī
C39 Shams al-Dīn Abū ʿAbd Allāh Muḥammad b. Burhān al-Dīn Ibrāhīm al-Rashīdī

Timurbāy

C95 Abū ^{c}Abd Allāh Muḥammad b. Muḥammad b. Abī Bakr al-Ḥalabī al-Tīzīnī

C96 Abū Bakr b. al-Imām

C97 Badr al-Dīn Abū ^{c}Abd Allāh Muḥammad b. Muḥammad b. Aḥmad, Sibṭ al-Māridīnī

C98 Shams al-Dīn Muḥammad b. Abi l-Fatḥ al-Ṣūfī al-Miṣrī

C99 Yaḥyā b. ^{c}Alī al-Rifā^{c}ī

C100 Anonymous

C101 Muhammad al-Wafā$^{\supset}$ī

C102 ^{c}Alī al-Mālaqī al-Andalusī

C103 Jalāl al-Dīn ^{c}Abd al-Raḥmān b. Abī Bakr al-Suyūṭī

C104 Zayn al-^{c}Ābidīn ^{c}Umar b. Ibrāhīm b. Nujaym al-Miṣrī al-Ḥanafī

C105 Ṣārim al-Dīn Ibrāhīm b. ^{c}Umar al-Bajalī / al-Najalī (?)

C106 Shihāb al-Dīn Aḥmad b. Mūsā b. ^{c}Abd al-Ghaffār

C107 Aḥmad b. ^{c}Alī Zunbul al-Maḥallī al-Munajjim

C108 Shams al-Dīn Muḥammad b. ^{c}Alī b. Muḥammad b. Ṭūlūn al-Dimashqī

C109 Anonymous

C110 ^{c}Abd al-Raḥmān b. Muḥammad (sometimes ^{c}Abd Allāh) b. Aḥmad al-Tājūrī

C111 ^{c}Abd Allāh b. Bahā$^{\supset}$ al-Dīn Muḥammad al-Shinshawrī

C112 Muḥammad b. Ibrāhīm al-Ḥanbalī al-Ḥalabī al-Rub^{c}ī

C113 Ibrāhīm al-Ḥalabī

C114 Shihāb al-Dīn Aḥmad b. Muḥammad b. Muḥammad . . . b. Jibrīl al-Ghazzī (?)

C115 Muḥammad b. Aḥmad b. Ḥasan al-Ghazzī

C116 Muḥammad b. ^{c}Alī b. Ibrāhīm, Ibn Zurayq al-Khayrī

C117 Sulaymān b. Ḥamza b. Bakhshīsh (?) al-Rūmī al-^{c}Uthmānī

C118 Muḥammad b. Muḥammad al-Ḥaṭṭāb (or al-Khaṭṭāb?)

C119 Yaḥyā b. Muḥammad al-Ḥaṭṭāb / al-Khaṭṭāb (?)

C120 Shams al-Dīn Muḥammad al-Minūfī

C121 ^{c}Abd al-Qādir b. Muḥammad al-Minūfī

C122 Ibrāhīm b. Qaytbāy al-Ashrafī al-Ḥanafī

C123 Miscellaneous anonymi

C124 Muḥammad b. Muḥammad (?) b. Abi l-Khayr al-Ḥusnī al-Urmayūnī

C124X Ibrāhīm b. Muḥammad al-Andalusī

C125 Dā$^{\supset}$ūd b. ^{c}Umar al-Anṭākī

C126 Shihāb al-Dīn Aḥmad b. Aḥmad b. ^{c}Abd al-Ḥaqq al-Sunbāṭī

C127 ^{c}Alī b. Muḥammad b. ^{c}Alī, Ibn Ghānim

C128 Anonymous

C129 Shams al-Dīn Abū ^{c}Abd Allāh Muḥammad al-Shāfi^{c}ī

C130 Khiḍr b. ^{c}Abd al-Qādir al-Birlisī al-Qabbānī

C131 ^{c}Abd al-Qādir b. ^{c}Alī al-Sakhkhāwī

C132 Shams al-Dīn Muḥammad b. Muḥammad al-Lādhiqī

C133 Muḥammad al-Hunaydī

C134 Shams al-Dīn Muḥammad al-Hunayd

C135 ^{c}Abd al-Raḥīm b. al-Bannā$^{\supset}$

C136 Muḥammad b. al-Qala^{c}ī

C137–164 Miscellaneous anonymi

C165 Abū Na^{c}īm Muslim b. Ismā^{c}īl al-Fāriqī

C166 Yūsuf b. al-Damīrī

C167 Aḥmad ^{c}Ubayd Allāh

C168 Shams al-Dīn Muḥammad b. al-^{c}Aṭṭār al-Badrī al-Shāfi^{c}ī

C169 Ibn Abi l-Qāsim al-Andalusī

C170 Mu$^{\supset}$ayyad al-Dīn al-^{c}Urḍī

C171 Abū ^{c}Abd Allāh Muḥammad b. ^{c}Alī al-Raḥbī, Ibn al-Mutaqqina

C172 Ya^{c}īsh b. Ibrāhīm b. Yūsuf al-Umawī

C173 al-Saḥtītī

D. Egyptian and Syrian Authors (after *ca.* 1600)

D1 ᶜUthmān b. ᶜAlāᵓ l-Dīn b. Yūnus b. Muḥammad, Ibn al-Malik al-Dimashqī
D2 ᶜUmar b. Muḥammad b. Abī Bakr al-Fāriskūrī
D3 Muḥammad b. Haykal al-Ikhṣāṣī
D4 Muḥammad b. Aḥmad b. Muḥammad b. Bīrī (?) al-Ḥanafī al-Makkī al-ᶜAlwānī
D5 ᶜAbd al-Raḥmān b. ᶜĪsā b. Murshid al-ᶜUmrī al-Ḥanafī
D6 Sālim b. Aḥmad b. Shaykhān
D7 Muḥammad b. ᶜAlī al-Shabrāmallisī
D8 Muḥammad b. Maḥmud al-Manāshīrī
D9 Anonymous
D10 Muḥammad b. ᶜUmar b. Ṣiddīq b. ᶜUmar al-Bakrī al-Fawānīsī
D11 ᶜAbd al-Qādir b. Aḥmad b. Ḥasan al-ᶜAjmāwī al-Azharī
D12 ᶜAbd al-Qādir b. Muḥammad al-Fayyūmī
D13 Zayn al-Dīn b. Sura l-Dīn . . . al-Durrī al-Mālikī
D14 Yaḥyā b. Taqi l-Dīn b. Ismāᶜīl al-Ḥalabī
D15 Anonymous (treatises on time-keeping)
D16 ᶜAbd al-Raḥmān b. ᶜAbd Allāh al-Iskirī al-Ṭūlūnī
D17 ᶜAbd Allāh b. ᶜAbd al-Raḥmān al-Ṭūlūnī
D18 ᶜAbd al-Raḥmān b. ᶜAbd Allāh al-Kātib al-Azharī
D19 Abu l-Khayr ᶜAbd al-Raḥmān al-Wafāᶜī al-Kātib
D20 Abū Bakr Muḥammad al-Kutāmī
D21 Muḥammad al-Ṣaḥrī / al-Ṣakhrī (?) al-Ḥarrī (?)
D22 Muḥammad b. Ibrāhīm al-Ḥulwānī
D23 Muṣṭafā b. Shams al-Dīn b. Aḥmad b. Khiḍrbāy al-Jarkashī al-Ṭāhirī
D24 Muṣṭafā al-Muqriᵓ al-Khalwatī al-Falakī al-Dimyāṭī
D25 Saᶜīd b. ᶜAbd Allāh al-Tanbaktī al-Janawī
D26 ᶜAlī b. ᶜAbd al-Qādir al-Nabtītī
D27 Muḥammad al-Kutubī
D28 ᶜAbd al-Qādir b. Maḥmūd al-Nabtītī al-Qādirī
D29 Ḥusayn b. Muḥammad al-Kilābī
D30 Shihāb al-Dīn Aḥmad b. Aḥmad al-Qalyūbī
D31 Shams al-Dīn Abu l-Ṣalāḥ Muḥammad al-Qaṭarī *thumma* al-Jawlānī
D32 ᶜAlī b. Rajab Mushāq al-Ḥanafī al-Rashīdī
D33 Mūsā b. Shāhīr (?) al-Abshādī (?) al-Muslimī (?) al-Ḥusaynī
D34 ᶜAlī (b. Muḥammad Zayn al-ᶜĀbidīn) al-Ujhūrī
D35 ᶜAbd Allāh b. Aḥmad al-Munāwī
D36 Muṣṭafā b. Sahrāb
D37 Quṭb al-Dīn Maḥmūd al-Maḥallī
D38 ᶜAbd al-Raḥīm al-ᶜAjamī
D39 Muḥammad b. Muḥammad b. Sulaymān al-Rūdānī al-Maghribī
D40 Aḥmad b. Tāj al-Dīn
D41 ᶜAbd al-Majīd al-Sāmūlī al-Hindī
D42 Badr al-Dīn Ḥusayn b. Shāmī al-Hattārī al-Faraḍī (?)
D43 Qāsim b. Ṣalāḥ al-Dīn al-Khānī
D44 ᶜAlī b. ᶜAbd al-Qādir al-Nabtītī al-Ḥanafī
D45 ᶜAbd Allāh b. Aḥmad al-Maqdisī al-Ḥanbalī
D46 ᶜAlī al-Zaᶜtarī al-Miṣrī
D47 Muḥammad b. Abī Bakr b. Aḥmad Shillī Bāᶜ Alawī
D48 Aḥmad b. ᶜAbd al-ᶜAzīz al-Sharafī al-Ṣafāqusī al-Azharī
D49 Muḥammad b. ᶜAlī b. Muḥammad al-Adfīnī al-Shāfiᶜī al-Faraḍī
D50 ᶜAbd al-Wahhāb al-Muqriᵓ al-Sarājī (?)
D51 Muḥammad b. Muḥyi l-Dīn al-Dimyāṭī, al-Nimra

D52 Yūsuf b. Muḥammad al-Maḥallī al-Masdī
D53 ᶜAbd al-Qādir b. Ibrāhīm al-Maḥallī
D54 ᶜUmar b. Aḥmad al-Māʾī wa-l-Jalī
D55 Shihāb al-Dīn Aḥmad al-Kutubī al-Khurfānī (?)
D56 Ḥasan al-Ṭablāwī
D57 Muḥammad b. ᶜAlī al-Ḥamīdī
D58 Riḍwān Efendī b. ᶜAbd Allāh al-Razzāz
D58X ᶜUthmān
D59 ᶜAbd al-Raḥmān al-Qaddūsī al-Manzilāwī
D60 ᶜAbd al-Laṭīf b. Ibrāhīm, Ibn al-Kayyāl
D61 al-Jamālī Yūsuf b. Yūsuf, Kalarjī of Ḥasan Āghā
D62 ᶜAbd al-Fattāḥ b. Ibrāhīm al-Daystī al-Mālikī
D63 Muḥammad al-Najāmī
D64 Muḥammad b. ᶜAbd al-Raḥmān b. Aḥmad, Ibn al-Jundī (al-Janadī?)
D65 ᶜAbd al-Muᶜṭī b. Sālim b. ᶜUmar al-Samalāwī al-Shāfiᶜī al-Azharī
D66 Muḥammad b. ᶜAbd al-Maḥmūd al-Ḥakīm al-Lādhiqī
D67 Aḥmad b. Muḥammad ᶜAyyād al-Maḥallī
D68 Miscellaneous (almanacs)
D69 Shams al-Dīn Maḥmūd b. Aḥmad al-Awfī al-Ḥijāzī
D70 Yaḥyā al-Birshamsī
D71 Mūsā b. Muḥammad b. Mūsā al-Qulaybī al-Mālikī al-Ghamrī
D72 Anonymous
D73 ᶜAlī b. Muḥammad al-Dāghistānī
D74 Muḥammad b. Muḥammad al-Fallātī al-Kishnawī al-Ashᶜarī al-Mālikī al-Sūdānī
D75 Muḥammad al-Ghamrī al-Falakī al-Shāfiᶜī
D76 Abu l-Riḍā ᶜAbd al-Laṭīf b. Aḥmad al-Kutubī al-Dimashqī
D77 Aḥmad b. ᶜUmar al-Dayrabī al-Shāfiᶜī al-Ghunaymī
D78 Ramaḍān b. Ṣāliḥ al-Khawānikī
D79 Anonymous
D80 Ḥusayn al-Qabbānī
D81 Muṣṭafā Ṣidqī b. Ṣāliḥ
D82 Yūsuf b. Aḥmad al-Jūrdī (?) al-Wanā (?) al-Shāfiᶜī al-Azharī
D83 Muṣṭafā b. Muḥammad al-Ṭāʾī al-Ḥanafī
D84 Muḥammad Abū Shahya al-Manfalūṭī al-Shāfiᶜī al-Shādhilī al-Azharī
D85 Ḥusayn b. Muḥammad al-Maḥallī
D86 Aḥmad b. Aḥmad al-Sajjāᶜī
D87 ᶜAlī b. Saᶜd al-Baysūsī (?)
D88–89 Anonymous
D90 Muḥammad b. ᶜAbd al-Raḥīm . . . al-Mukhallalātī
D91 Ḥasan b. Ibrāhīm al-Jabartī
D92 ᶜAbd al-Raḥmān b. Ḥasan al-Jabartī
D93 Muḥammad Chelebī al-Shuraybī
D94 Muḥammad al-Nafrawī
D95 ᶜAbd al-Ṣamad
D96 Maḥmūd Efendī b. Sulaymān Jumliyān (?)
D97 Muḥammad b. ᶜAlī al-Ṣabbān
D98 Aḥmad b. Ismāᶜīl Āghā al-Ḥanafī, Balghiyār (?) al-Falakī
D99 Aḥmad b. ᶜAbd al-Munᶜim al-Damanhūrī
D100 Ibrāhīm b. Muḥammad al-Zamzamī
D101 Abu l-Itqān Muṣṭafā al-Wafāʾī al-Khayyāṭ
D102 Muḥammad ᶜUthmān
D103 Ḥasan Chelebī
D104 Muḥammad al-Ḥifnī
D105 Muḥammad al-Munīr al-Samānūdī al-Azharī al-Shāfiᶜī
D106 ᶜAbd al-Raḥmān al-Ṭanānī
D107 ᶜUthmān b. Sālim al-Wardānī
D108 Muḥammad Sannār
D109 Muḥammad b. Ḥusayn al-ᶜAṭṭār
D110 Muḥammad Amīn b. ᶜUmar . . . b. ᶜĀbidīn
D111 Muḥammad al-Khuḍrī
D112 Shihāb al-Dīn Ibrāhīm b. Ḥasan
D113 Shihāb al-Dīn Aḥmad al-Marzūqī

D114 Muḥammad b. Ibrāhīm al-Ibrāshī
D115 Muḥammad ʿAbd al-Ḥayy al-Shībīnī
D116 ʿAbd al-Ghanī
D117 Muṣṭafā al-Dhahabī al-Shāfiʿī
D118 Aḥmad b. Qāsim
D119 Muḥammad b. ʿAbd al-Raḥmān al-Nābalī (= al-Nābulsī (?))
D120 ʿAlī al-Khashshāb
D121 Abū ʿAyyāsha Muḥammad al-Shāfiʿī al-Damanhūrī
D122 ʿAbd al-Razzāq al-Munbāwī
D123 Muḥammad b. Muṣṭafā al-Ṭanṭāwī
D124 ʿAlī Pāshā Mubārak
D125 Muḥammad ʿAbd al-ʿAzīz b. ʿAlī al-Sāʿātī al-Shubrāwī
D126 ʿAbd al-Raḥīm Manṣūr al-Ḥalabī
D127 Īlyā al-Maṭrān
D128 ʿAbd Allāh b. ʿAbd al-Raḥmān b. Aḥmad al-Ḥanbalī al-Ḥalabī
D129 Awīs al-Qwjwy (?)
D130 Muḥammad b. Shuʿayb al-Ḥalabī Zāʾir al-Qarāfatayn al-Shāfiʿī al-Ibshīhī
D131 Aḥmad b. Muḥammad b. ʿAbdū al-Jadalī
D132 Abu l-Fatḥ b. ʿAbd al-Raḥmān al-Danūshīrī
D133 Shams al-Dīn Muḥammad b. Muḥammad al-Shamsī al-Khawānikī
D134 Khalīl b. Ibrāhīm b. Ḥasan al-ʿAzzāzī
D135 Muḥammad b. Muḥammad al-Ḥanafī
D136 al-Najdī
D137 Yūnus b. ʿAbd al-Qādir al-Rashīdī
D138 al-Fawzī
D139 ʿAbd al-Raḥmān ʿUthmān
D140 Muḥammad ʿAbd Allāh Bey
D141 Abu l-Barakāt Muḥammad b. Muḥammad b. ʿĀmir al-Daljamūnī al-Wafāʿī
D142 Abu l-Khayr al-Shukr al-Sāmī (?)
D143 Yūsuf b. ʿAlī Abi l-ʿĪd
D144 Ḥasan b. Muḥammad al-ʿAṭṭār
D145 Muḥammad Surūr
D146 Ḥusayn
D147 Shihāb al-Dīn Aḥmad b. Muḥammad al-Ujhūrī
D148 Badr al-Dīn Ḥasan b. ʿAlī b. Aḥmad al-Ujhūrī
D149 Dāʿūd b. Muḥammad al-Qāriḥ (?)
D150 ʿAbd al-Ṣamad al-Azharī
D151 Shams al-Dīn Muḥammad al-Ḥanafī
D152 Ḥusayn b. ʿAlī al-Suyūṭī, al-Dīk
D153 ʿUmar b. Maḥmūd b. al-Ṣārim
D154 Muḥammad b. ʿAbd al-Raḥmān b. Muḥammad al-Dādī (?)
D155 Ḥasan al-Ghawrī
D156 Muḥammad b. Aḥmad b. Muḥammad al-Qabbānī
D157 Muṣṭafā al-Manzilī
D158 ʿAbd Allāh al-Muwaqqit
D159 Muṣṭafā
D160 Yūsuf Efendī al-ʿAqqād al-Makkī
D161 Aḥmad al-Drdyr (?)
D162 Shams al-Dīn b. ʿAbd Allāh Fatḥ al-Farghalī al-Sabrabāwī
D163 Ṣāliḥ Manṣūr al-Yāfī al-Ḥanafī
D164 Yūsuf (??)
D165 Aḥmad b. ʿAbd al-Raḥmān
D166 Muṣṭafā b. Muḥammad b. Muṣṭafā al-Muʿtamadāwī
D167 Mūsā b. Shāhīn al-Abshādī
D168 Muḥammad al-Shamsī al-Muqriʾ
D169 ʿAbd Allāh b. Aḥmad b. al-Munāwī
D170 Muḥammad b. Muḥammad al-Badīrī al-Dimyāṭī
D171 Muḥammad Efendī
D172 Aḥmad b. Muḥammad al-Janājī
D173 Aḥmad al-Aṣbaḥī
D174 Muḥammad b. Ḥasan al-Bayṭār
D175 Sirāj al-Dīn ʿUmar al-Fāriskūrī
D176 Muḥammad Efendī Darandar
D177 Ṣadr al-Dīn
D178 al-Manbāwī (?)
D179 Imām al-Qalʿa
D180 Muḥammad b. ʿĀbidīn
D181 Muḥammad Ḥattāta (?) al-Fāriskūrī
D182 Muḥammad b. Muḥammad b. Muḥammad b. Bahādur al-Mawlawī al-Shāfiʿī
D183 Baṭrus Ḥannā ʿAnjūrī al-Dimashqī
D184 Nūr al-Dīn b. ʿAbd Allāh al-Wāʿiẓ
D185 Yūsuf al-Talīnī (?)
D186 Muḥammad Ṣāliḥ

D187 Ismāʿīl b. Sharaf, Ibn al-Muqriʾ
D188 Yaḥyā al-Khazrajī al-Anṣārī
D189 Abu l-Ḥasan ʿAlī b. Abī Maḥmūd al-Maghribī al-Manzilī
D190 Muḥammad b. Muḥammad b. Muḥammad al-Mubashshir
D191 Ibrāhīm al-Ḍābiṭ
D192 Muḥammad b. Muḥammad al-Marāghī al-Jirjāwī
D193 Abū Saʿīd Muḥammad al-Darwīsh
D194 Muḥammad Efendī Ḥifẓī
D195 ʿUmar Niʿmat Allāh
D196 Aḥmad al-Jālī
D197 ʿAbd al-Ghanī b. Muḥammad Hilāl
D198 Maḥmūd al-Anṭākī
D199 Ibrāhīm Jawda al-Damanhūrī
D200 Aḥmad al-Maktabī (?)
D201 Zayn al-Dīn ʿAbd al-Raḥmān al-Shabrīsī (?)
D202 ʿAbd al-Fattāḥ al-Bannāʾ
D203 Aḥmad b. Muḥammad al-Shāfiʿī al-Janājī
D204 Abū ʿAbd Allāh al-Kawmī
D205–238 Miscellaneous
D239 Tāj al-Dīn

E. Yemeni Authors

E1 Aḥmad (or Isḥāq?) b. Yūsuf al-Ṣardafī
E2 Nashwān b. Saʿīd al-Ḥimyarī
E3 Abu l-Ḥasan Aḥmad b. Muḥammad al-Ashʿarī al-Yamanī
E4 al-Faḍl b. Abī Saʿd al-ʿUṣayfirī
E5 Ibrāhīm b. ʿAlī b. Muḥammad b. Manṣūr al-Aṣbaḥī
E6 Abū ʿAbd Allāh Muḥammad b. Aḥmad b. ʿUmar al-Khuzāʿī
E7 Muḥammad b. Abī Bakr al-Fārisī
E8 ʿUmar b. Yūsuf b. ʿUmar, the Rasulid Sultan al-Ashraf
E9 Ibrāhīm b. Mamdūd
E10 Ḥasan b. ʿAlī al-Fihrī
E11 Anonymous (annual ephemerides)
E12 ʿAlī b. Rāshid b. Aḥmad b. Mūsā b. Yaḥyā b. Nasr al-Nasrī
E13 Abu l-Qāsim ʿIzz al-Dīn Ibrāhīm b. Muḥammad b. Sulaymān al-Bawsī
E14 Abū Muḥammad ʿAbd Allāh b. Asʿad b. ʿAlī al-Yāfiʿī
E15 Abū Bakr b. Abi l-Maʿālī
E16 Aḥmad b. Muḥammad al-Khālidī
E17 Ibrāhīm b. Khālid b. Aḥmad al-ʿAlfī
E18 ʿImād al-Dīn Yaḥyā b. Muḥammad b. Ḥasan
E19 ʿAfīf al-Dīn ʿAbd Allāh b. ʿUmar b. ʿAbd Allāh b. Aḥmad Bā Makhrama
E20 Abū Ḥamad Maḥfūẓ b. ʿAbd al-Raḥmān al-Ḥaḍramī
E21 Ṣārim al-Dīn Ibrāhīm b. Yaḥyā Jaḥḥāf
E22 Aḥmad b. Yaḥyā al-Ṣaʿdī al-Dawwārī
E23 Nūr al-Dīn ʿAlī b. ʿAbd Allāh al-Ṭawāshī
E24 ʿAbd Allāh b. Ḥamza al-Qāḍī al-Dawwārī
E25 Anonymous

F. Andalusian and the Maghribi Authors

F1 Jābir b. Aflaḥ
F2 Abu l-Walīd Ismāʿīl b. Aḥmad al-Qurṭubī
F3 Ibn Muʿādh?
F4–8 Anonymous
F9 Muḥammad b. ʿAlī b. Yaḥyā al-Naṭṭāḥ
F10 Abū ʿAbd Allāh Muḥammad b. Aḥmad b. Hishām al-Lakhmī
F11 Aḥmad b. Yūsuf, Ibn al-Kammād

F12 Abū Muḥammad ^{c}Abd Allāh b. al-Ḥajjāj, Ibn al-Yasamīn

F13 Aḥmad b. ^{c}Alī b. Yūsuf al-Būnī

F14 Abū ^{c}Abd Allāh Muḥammad b. ^{c}Alī, Ibn Tūmart al-Andalusī

F15 Abū Muḥammad ^{c}Abd al-Ḥaqq b. Ibrāhīm, Ibn Sab^{c}īn

F16 Anonymous

F17 Abū Miqra^{c}

F18 Abū ^{c}Abd Allāh Muḥammad b. ^{c}Umar, Ibn Badr

F19 Abū ^{c}Alī al-Ḥusayn b. Aḥmad b. Yūsuf b. Bāṣo al-Aslamī (?)

F20 ^{c}Abd Allāh b. Muḥammad al-Tujībī

F21 Abu l-Qāsim al-Zubayr b. Aḥmad b. Ibrāhīm b. al-Zubayr al-Thaqafī al-Qāḍī

F22 Abū ^{c}Abd Allāh Muḥammad b. Ibrāhīm al-Awsī, Ibn al-Raqqām

F23 Abu l-^{c}Abbās Aḥmad b. Muḥammad, Ibn al-Bannā$^{\supset}$

F24 Anonymous

F24X ^{c}Abd al-^{c}Azīz b. ^{c}Alī b. Dā$^{\supset}$ūd al-Hawarī al-Miṣrī (?)

F25 Ibn al-Qunfudh

F26 Abū Zayd ^{c}Abd al-Raḥmān b. Abī Ghālib Muḥammad al-Jādarī

F27 Aḥmad b. Abī Ḥumayda al-Muṭarrifī

F28 Abū ^{c}Abd Allāh Muḥammad b. Aḥmad al-Ḥabbāk

F29 Abū ^{c}Abd Allāh Muḥammad b. Yūsuf al-Muqri$^{\supset}$ al-Sanūsī

F30 Anonymous

F31 Abraham Zacuto

F32 Abu l-Ṣalāḥ Jābir b. ^{c}Abd Allāh b. al-Ḥājj al-Ghiyāthī

F33 Aḥmad b. Qāsim al-Jaḥdarī al-Andalusī

F34 Abu l-Ḥasan ^{c}Alī b. Muḥammad al-Qurashī al-Basaṭī al-Qalaṣādī

F35 Abū ^{c}Abd Allāh Muḥammad b. Muḥammad b. ^{c}Abd al-Wahhāb Abī Muḥammad b. ^{c}Abd Allāh al-Jannād al-Anṣārī

F36 Abu l-Ḥasan ^{c}Alī, Ibn al-Maghribī

F37 ^{c}Alī b. ^{c}Abd Allāh b. Muḥammad b. Haydūr

F38 Abū Arawī (?) Muḥammad b. ^{c}Imrān b. Abī Muḥammad al-Quṭb Sayyidī ^{c}Abd al-Salām al-Fītūrī (?)

F39 Abū Ja^{c}far ^{c}Umar b. ^{c}Abd al-Raḥmān b. Abi l-Qāsim al-Tūnisī al-Tūzarī

F40 Muḥammad b. Aḥmad . . . b. Ghāzī al-^{c}Uthmānī

F41 ^{c}Abd al-Raḥmān b. Abī Muḥammad al-Ṣaghīr al-Akhḍarī

F42 ^{c}Abd al-^{c}Azīz b. Aḥmad b. Muslim al-Akhḍarī (??)

F43 Abū Zayd ^{c}Abd al-Raḥmān b. ^{c}Umar b. Aḥmad al-Sūsī al-Jazūlī al-Bā^{c}Uqaylī, Ibn al-Muftī

F44 Ibn al-Najjār

F45 Abū ^{c}Abd Allāh Muḥammad Kwynkh (?)

F46 Muḥammad b. Muḥammad b. Muḥammad al-Idrīsī = Abū Muḥammad Ḥamūda al-Idrīsī (?)

F47 ^{c}Abd Allāh Ḥusayn Quṣ^{c}a al-Tūnisī, Ibn Quṣ^{c}a

F48 ^{c}Alī b. Muḥammad al-Dādisī

F49 Abū Zayd Muḥammad b. Sa^{c}īd b. Muḥammad al-Sūsī al-Marghīthī

F50 ^{c}Abd al-Raḥmān b. ^{c}Abd al-Qādir al-Fāsī

F51 Abū Zayd ^{c}Abd al-Raḥmān al-Akhḍarī = (??) Abū Zayd ^{c}Abd al-Raḥmān b. ^{c}Abd al-Qādir al-Fāsī

F52 Muḥammad Abu l-Amdād (?) al-Maghribī al-Tūnisī

F53 Abū ^{c}Abd Allāh Muḥammad al-Sharīf al-Tūnisī, Sanjaq Dār

F54 ^{c}Alī b. Māmī (??) al-Ḥanafī

F55 Aḥmad b. Muḥammad Bū Daydaḥ (?) al-Qādirī (?) al-Qayrawānī

F56 ^{c}Umar al-Maghribī

F57 Muḥammad b. ^{c}Abd al-Salām b. Ḥamdūn al-Bannānī

F58 Abū Muḥammad ^{c}Abd al-^{c}Azīz b. ^{c}Abd al-Salām b. Aḥmad al-Wazkānī al-Wazzānī

F59 Muḥammad b. ^{c}Abd al-^{c}Azīz al-Jazūlī al-Rasmūkī

F60 al-Jazūlī
F61 Saḥnūn b. ᶜUthmān b. Sulaymān . . . al-Wansharīshī
F62 ᶜAbd al-Salām b. Muḥammad b. Aḥmad al-Ḥusnī al-ᶜlmy (?)
F63 ᶜAlī al-Bāṣo (?) al-Tūnisī
F64–73 Anonymous
F74 Aḥmad b. ᶜAlī, Ibn Isḥāq
F75 Abū ᶜAbd Allāh Muḥammad al-Bārī al-Tūnisī

G. Authors from Iraq and Iran

G1 Abū Jaᶜfar Muḥammad b. ᶜAbd Allāh al-Sharīfī
G2 Abu l-Fatḥ ᶜUmar b. Ibrāhīm al-Khayyāmī
G3 Abū Ḥātim al-Muẓaffar b. Ismāᶜīl al-Isfazārī
G4 Hibat Allāh b. al-Ḥusayn al-Asṭurlābī
G5 Maḥmūd b. Qāsim b. al-Faḍl al-Iṣfahānī
G6 Abū Muḥammad ᶜAbd al-Jabbār b. ᶜAbd al-Jabbār b. Muḥammad al-Thābitī al-Kharaqī
G7 Abu l-Futūḥ Aḥmad b. Muḥammad b. al-Sārī, Ibn al-Ṣalāḥ
G8 ᶜAlī b. Muḥammad Sharīf al-Bakrī al-Mawṣilī al-Munajjim
G9 Samawᵓal b. Yahūdā al-Isrāᵓīlī = Samawᵓal b. Yaḥyā al-Maghribī
G10 Muḥyi l-Dīn ᶜAbd al-Qādir b. ᶜAbd Allāh al-Kīlānī
G11 Sirāj al-Dīn Abū Ṭāhir Muḥammad b. Muḥammad al-Sajāwandī
G12 Fakhr al-Dīn Muḥammad b. ᶜUmar b. Ḥusayn al-Rāzī
G13 Kamāl al-Dīn Mūsā b. Yūnus
G14 Muḥammad (b?) ᶜAbd al-Karīm al-Ghaznawī
G15 Sharaf al-Dīn al-Muẓaffar b. Muḥammad b. al-Muẓaffar al-Ṭūsī
G16 Athīr al-Dīn al-Mufaḍḍal b. ᶜUmar al-Abharī
G17 Maḥmūd b. Muḥammad b. ᶜUmar al-Jaghmīnī
G18 ᶜIzz al-Dīn al-Zanjānī
G19 al-Bayhaqī (?)
G20 Zakarīyā b. Muḥammad b. Maḥmūd al-Qazwīnī
G21 Muḥyi l-Dīn Yaḥyā b. Muḥammad b. Abi l-Shukr al-Maghribī
G22 Ḥusām al-Dīn ᶜAlī b. Faḍl Allāh al-Sālār
G23 Naṣīr al-Dīn Muḥammad b. Muḥammad al-Ṭūsī
G24 ᶜAbd Allāh b. Muḥammad, Ibn al-Khawwām
G25 Jamāl al-Dīn Abu l-Qāsim b. Maḥfūẓ al-Baghdādī
G26 Shams al-Dīn Muḥammad b. Ashraf al-Samarqandī
G27 Quṭb al-Dīn Maḥmūd b. Masᶜūd al-Shīrāzī
G28 Anonymous
G29 Jamāl al-Dīn Muḥammad b. ᶜUmar b. Aḥmad b. Hibat Allāh, Ibn Abī Jarāda
G30 Kamāl al-Dīn Muḥammad b. al-Ḥasan al-Fārisī
G31 Nāṣir al-Dīn Aḥmad b. Ḥaydar b. Muḥammad al-Shīrāzī
G32 ᶜAlī Shāh b. Muḥammad b. Qāsim al-Khwārizmī, ᶜAlāᵓ al-Dīn al-Bukhārī
G33 Abu l-Maḥāmid Muḥammad b. Masᶜūd b. Muḥammad Zakī al-Ghaznawī
G34 Muwaffaq al-Qayṣarī
G35 Anonymous
G36 Shams al-Aᵓimma Muḥammad b. ᶜAlī al-Wābkanawī
G37 Shams al-Dīn Muḥammad b. Sharaf

al-Kallāʾī

G38 Niẓām al-Dīn al-Ḥasan b. Muḥammad al-Nīsāpūrī

G39 Ḥamza b. ʿAlī al-Bayhaqī

G40 Ḥasan b. Ḥusayn b. Ḥasan Shāhinshāh al-Simnānī

G41 Rukn al-Dīn b. Sharaf al-Dīn al-Ḥusaynī al-ʿĀmulī

G42 ʿAlī b. Muḥammad, al-Sayyid al-Sharīf al-Jurjānī

G43 Muḥammad al-Ḥusaynī, Sayyid al-Munajjim

G44 Maḥmūd b. Muḥammad b. Qiwām al-Wālishtānī

G45 ʿAbd al-Wājid b. Muḥammad

G46 Muḥammad b. Muḥammad, Humām al-Ṭabīb

G47 Sayyid Ḥasan b. Sayyid ʿAlī al-Qumnātī

G48 Ghiyāth al-Dīn Jamshīd b. Masʿūd b. Maḥmūd al-Kāshī

G49 Ulugh Beg

G50 Mūsā b. Muḥammad b. Maḥmūd, Qāḍī Zāde al-Rūmī

G51 ʿAlāʿ al-Dīn ʿAlī b. Muḥammad al-Qūshjī

G52 Abu l-Fatḥ Muḥammad b. al-Hādī Abī Naṣr b. Abī Saʿīd al-Ḥusaynī al-ʿIrāqī, Tāj al-Saʿīdī

G53 Ḥaydar b. Aḥmad al-Kurdī al-Ḥusaynābādī

G54 Ḥusām al-Dīn al-Tūqātī

G55 Anonymous

G56 Sharaf al-Dīn ʿAlī al-Yazdī

G57 Ḥusayn al-Wāʿiẓ b. ʿAlī al-Bayhaqī, al-Kāshifī

G58 ʿAlī/ʿAbd al-ʿAlī b. Muḥammad b. al-Ḥusayn al-Birjandī

G59 ʿImād (al-Dīn) b. Jamāl (al-Dīn) Bukhārī

G60 Ḥusayn b. Muʿīn al-Dīn al-Maybūdī

G61 Anonymous

G62 Abu l-Khayr Muḥammad al-Taqī b. Muḥammad al-Fārisī

G63 Muṣliḥ al-Dīn Muḥammad b. Ṣalāḥ al-Lārī al-Anṣārī

G64 Malik-Muḥammad b. Sulṭān-Ḥusayn Iṣfahānī

G65 Īrānshāh b. ʿAlī al-Nīsābūrī

G66 Qiwām al-Dīn Ḥusayn b. Shams al-Dīn Muḥammad al-Khafrī

G67 Muẓaffar Munajjim b. Muḥammad Qāsim Junābadī

G68 Bahāʾ al-Dīn Muḥammad b. al-Ḥusayn al-ʿAmulī Burhān al-Dīn

G69 Ḥusayn al-Ḥusaynī al-Khalkhālī

G70 Jalāl al-Dīn Muḥammad b. ʿAbd Allāh Yazdī

G71 Muḥammad Bāqir b. Zayn al-ʿĀbidīn Yazdī

G72 Ibrāhīm b. Zayn al-Dīn al-Shīrāzī

G73 Jawād b. Saʿd b. Jawād

G74 Malik Maḥmūd Karrāmī b. Malik Aḥmad Khwānsārī

G75 Ramaḍān b. Abī Hurayra al-Jazarī al-Qādirī

G76 Muḥammad Zamān b. Sharaf al-Dīn Ḥusayn al-Mashhadī

G77 Ḥaydara b. ʿAbd al-Raḥmān al-Ḥusaynī al-Jazarī

G78 Muḥammad Bāqir b. Muḥammad Muʾmin Khurāsānī Sabzawārī

G79 Aḥmad b. Muḥammad al-Mahdī al-Sharīf al-Iṣfahānī al-Khātūnabādī

G80 Abū Muḥammad ʿAbd Allāh b. Fakhr al-Dīn al-Ḥusaynī al-Mawṣilī

G81 Muḥammad b. Ādam al-Rustāʾī

G82 Ḥasan b. Shujāʿ b. Muḥammad al-Ḥasan al-Ḥāfiẓ al-Tūnī (?)

G83 Muḥammad b. al-Ḥurr al-ʿĀmūlī

G84 Abu l-Yumn ʿAbd al-Bāqī b. Abi l-Thanāʿ Maḥmūd al-Ālūsī

G85 Maḥmūd Munajjim Kīlānī

G86 Ḥusām b. Shams al-Dīn al-Khaṭīb al-Khiṭābī al-Ahjānī

G87 Karīm b. Ibrāhīm

G88 ʿAlī Tūrī Zāde

G89 Quṭb al-Dīn ʿIzz al-Dīn (b.?) ʿAbd al-Ḥayy al-Zāhidī al-Bakrī al-Ḥusaynī al-Lārī

G90 Saʿd al-Dīn Asʿad b. Saʿīd al-Hamadānī

G91 ʿAbd Allāh b. Naṣr Allāh Qaragozlu

G92 Kamāl al-Dīn al-Ḥusayn al-Fārisī

G93 Muḥammad Bāqir b. Muḥammad Taqī
G94 La^c^lī Zāde ^c^Abd al-Bāqī al-Ḥusaynī
G95 Rustam Bey al-Ḥārithī al-Badakhshī
G96 Ma^c^būdī b. Ibrāhīm al-Shāhrūdī al-Bisṭāmī
G97 Khōja ^c^Aṭā^ɔ^ Allāh
G98 Menlā Chelebī al-Kurdī
G99 Ḥasan b. Muḥammad al-Faṣīḥī al-Niẓāmī, Qāḍī Ḥasan
G100 Muḥammad Ḥakīm b. ^c^Alā^ɔ^ al-Dawla al-Ḥusaynī
G101 ^c^Abd al-Raḥīm b. Abī Bakr al-Mar^c^īshī
G102 ^c^Abd al-Qādir b. ^c^Uthmān al-Ādanī
G103 Muḥammad al-Bāqir al-Bānī al-Kurdī
G104 Ibn Abi l-Qāsim al-Ḥusnī Muḥammad Bāqir al-Tabrīzī
G105 ^c^Umar ^c^Abd al-^c^Azīz
G106 Jalāl al-Dīn Faḍl Allāh al-^c^Ubaydī
G107–125 Anonymous
G126 Quṭb al-Dīn ^c^Alī al-Tabrīzī
G127 Muḥammad b. ^c^Umar Asdafānī (?)/Asadabādī
G128 ^c^Abd al-Mun^c^im al-^c^Āmulī

H. Turkish Authors

H1 Darendelī
H2 Muṣṭafā b. Aḥmad al-Ṣayrawī al-^c^Īsāwī, Shaykh Vefā
H3 Mūsā Gālīnus
H4 Maḥmūd b. Muḥammad b. Qāḍī Zāde, Mīram Chelebī
H5 Yūsuf b. Khiḍr Beg, Sinān Pāshā
H6 Ḥamza b. Arslān
H7 Muḥyi l-Dīn Muḥammad b. al-Ḥājjī Ātemjeh al-Kātib
H8 Muḥammad b. Kātib Sinān
H9 Aḥmad b. Sulaymān, Kamāl Pāshā Zāde
H10 Muṣṭafā b. ^c^Alī
H11 Sayyid ^c^Alī b. Ḥusayn
H12 Taqi l-Dīn Muḥammad b. Ma^c^rūf
H13 Nuqṭa (?) b. Ma^c^rūf
H14 Anonymous
H15 Ḥusayn b. Khalīl, Qrlfj (?) Zāde
H16 ^c^Alī b. Walī b. Ḥamza al-Jazā^ɔ^irī al-Maghribī
H17 al-Knjdy (?)
H18 Ṣalāḥ al-Dīn Yāzījī
H19 Ibrāhīm Jūrī (?)
H20 Muḥammad b. ^c^Alī
H21 Muḥammad Efendī
H22 Ibrāhīm al-Qaramānī *thumma* al-Āmudī
H23 Abū ^c^Abd Allāh Muḥammad b. Aḥmad b. Muḥammad b. Ḥasan b. Ḥusayn, Munajjimak = Muḥammad Fazā^ɔ^ī(?), Ra^ɔ^īs al-Munajjimīn Munajjimak
H24 Ibrāhīm al-Ḥalabī
H25 ^c^Abd al-Ḥalīm Suwaylam Zāde al-Qayṣarī
H26 Qōja Sā^c^atjī
H27 Muṣṭafā Zekī
H28 Aḥmad Efendī
H29 Badr al-Dīn Muḥammad b. As^c^ad al-Islāmbūlī
H30 As^c^ad Efendī b. ^c^Alī b. ^c^Uthmān al-Yāniyawī
H31 Ṣubḥī Zāde ^c^Abd al-^c^Azīz
H32 Darwīsh ^c^Abbās Wasīm
H33 Aḥmad
H34 Muḥammad b. ^c^Alī al-Ḥamīdī
H35 ^c^Alī al-Rābiṭ al-Mūytābī (?)
H36 Ṣāliḥ Efendī Mi^c^mārī
H37 Muḥammad Ṣādiq Jihāngīrī
H38 Ibrāhīm Ḥaqqī
H39 Ismā^c^īl Fahīm b. Ibrāhīm Ḥaqqī
H40 Ismā^c^īl b. Muṣṭafā Kalambawī
H41 Ismā^c^īl Chinārī, Khalīfa Zāde
H42 Sulaymān Maqāmī b. Aḥmad
H43 ^c^Uthmān b. ^c^Abd al-Manān

H44	ᶜUthmān b. Muḥammad
H45	Mūsā Zāde Muḥammad ᶜUbayd Allāh
H46	Ibrāhīm Khōja al-Muhandisī
H47	Ḥusayn Ḥusnī
H48	Aḥmad Muḥsin
H49	Ḥusayn Zāᵓid
H50	Muḥammad ᶜĀrif
H51	Muḥammad Raḥmat al-Makkī
H52	Ḥusayn b. Jamāl (?)
H53	Muᵓmin Zāde
H54	Ibrāhīm Qaṣṣāb Bāshī Zāde
H55	ᶜUmar b. ᶜAbd Allāh, Nuzhat
H56	Ākovalī Zāde Khātim (?)
H57	al-Bursawī / al-Brusawī
H58	Aḥmad
H59	Muḥammad b. Ibrāhīm al-Istānbūlī al-Ḥanafī
H60	ᶜAlī Chelebī al-Rūmī
H61	Pīr Ḥasan
H62	Muṣṭafā b. Muṣṭafā Ḥisān
H63	Maḥmūd Efendī Qarshahrī
H64	Najm al-Dīn b. Sayyid Muḥammad al-Munajjim
H65	Sulaymān Riyāḍī
H66	Aḥmad al-Dāᶜī
H67	Yūnus Efendī
H68	Muḥammad Amīn b. ᶜAbd al-Raḥīm
H69	Mawlūd b. al-Ṣiddīq al-Ḥāfiẓī al-Yaᶜlāwī
H70–102	Miscellaneous anonymi
H103	Shāh Efendī al-Fanārī

J. European Authors

J1	Clavius
J2	Cassini
J3	Lalande
J4	Edward Abbott
J5	Miscellaneous
J6	Rudolf Walp

X. Miscellaneous Authors

X1	Abū Muḥammad ᶜUmar b. Hārūn al-Saqalī
X2	Aḥmad b. ᶜAbd al-Ghanī al-Sarūjī
X3	Ḥātim
X4	Muṣṭafā b. Nūr al-Dīn
X5	ᶜAbd al-Wahhāb
X6	ᶜAbd al-ᶜAzīz Aḥmad
X7	Shams al-Dīn Muḥammad b. Muḥammad al-Baghdādī
X8	Faṣīḥ al-Dīn Muḥammad b. (?) ᶜAbd al-Karīm al-Niẓāmī
X9	Aḥmad b. Mājid
X10	Mullā ᶜAlī Efendī al-Dāghistānī
X11	Ibrāhīm b. Muḥammad b. Shukr
X12	Imām Badr al-Dīn
X13	Abū Isḥāq
X14	Faṣīḥ al-Dīn al-Hindī (?)
X15	Shams al-Dīn Muḥammad
X16	al-Imām al-Amīr al-Mālikī
X17	ᶜAbd Allāh b. Ḥasan al-ᶜAfīf al-Kāzūnī
X18	Muḥammad Shaykh Zāde
X19	ᶜAlī b. ᶜAbd Allāh al-ᶜAshshāqī
X20	Ibrāhīm b. ᶜAbd al-Qādir b. Ibrāhīm al-ᶜAlāᵓī
X21	Aḥmad al-Ḥusaynī
X22	Aḥmad b. Ḥasan b. ᶜAbd al-Karīm Ḥamza
X23	Muḥammad b. Arnūṭ al-Qāḍī
X24	Isḥaq al-Zarkānī
X25	Kamāl al-Dīn Muḥammad b. al-Nāsikh al-Mālikī
X26	Sharaf al-Dīn Abu l-Jūd Daᵓūd b. Sulaymān
X27	Sulaymān b. Aḥmad al-Mahrī

Z. Anonymous Authors

Z1–41	Anonymous

"Nous avons la ferme persuasion de n'avoir rien omis qui puisse avoir la moindre importance."

J.-B. Delambre, *Histoire de l'astronomie ancienne*, 1817

LIST OF SCIENTIFIC WORKS IN THE EGYPTIAN NATIONAL LIBRARY

A. Greek, Sasanian, and Indian Authors in Arabic

For Arabic works on the "astronomy of the ancients" Sezgin's *GAS*, vols. V, VI, and VII constitute the basic sources. On the transmission of these early works, see *Pingree* 1 and 2.

A1 PYTHAGORAS

GAS, V, pp. 75–76.

See Ibn Fallūs (no. C6).

A2 SOCRATES

GAS, V, p. 77.

See Thābit b. Qurra (no. B30).

A3 ARISTOTLE

GAS, V, pp. 79–81; VI, pp. 70–72; VII, pp. 60–63, 212–216, and 309–310.

5.1.2 ***K. al-Thamra fī ᶜilm al-nujūm li-Arisṭāṭālis = K. Asrār al-nujūm***, on astrology in 365 unnumbered *qaḍiyas*: ṬM 243,4 (fols. 111r–129r, 644H in Dayr Bashīr), unique (?), not listed in *GAS*, VII, pp. 61–62, nos. 1 and 2. The copies listed by Sezgin apparently contain 100 or 122 sections. See Pl. LXII.

5.1.2 ***(Qiṭᶜa) fī ḥisāb al-ghālib wa-l-maghlūb istikhrāj al-ḥakīm Arisṭāṭālīs li-l-Iskandar***: ṬJ 811,10 (fols. 80r–82v, 1135H). This title is not listed in *GAS*, VII.

5.1.2 ***K. fī Tadbīr fī manāzil al-qamar***: ḤM 11,2 (fols. 6r–14r, 1134H), not listed in *GAS*, VII, p. 62, no. 3.

5.1.2. A letter from Aristotle to Jesus (!) and the latter's reply (!!): MM 59,2 (fols 8r–8v, *ca.* 1200H).

See also no. H30.

A4 AUTOLYCUS

GAS, V, pp. 81–82, and VI, pp. 73–74.

1.1.1 ***K. al-Ṭulūᶜ wa-l-ghurūb***
Taḥrīr by NAṢĪR AL-DĪN AL-ṬŪSĪ (no. G23): MR 41,1 (fols. 1v–10r, 1146H), not listed in *GAS*, VI, pp. 73–74, no. 1.

6.1.1 ***K. al-Kura al-mutaḥarrika***
Taḥrīr by NAṢĪR AL-DĪN AL-ṬŪSĪ (no. G23): TR 347,2 (pp. 58–66, 895H in Shiraz); MR 40,4 (fols. 72v–75r, 1146H); DR 704,2 (fols. 16v–22r, 1011H); *etc.* (3 copies). None of these is listed in *GAS*, V, p. 82, no. 1.

A5 EUCLID

GAS, V, pp. 83–120, and VI, p. 74.

1.1.2 ***K. Ẓāhirāt al-falak.***
Taḥrīr by NAṢĪR AL-DĪN AL-ṬŪSĪ (no. G23): MR 41,30 (fols. 177v–179r, 1146H, incomplete). *Cf. GAS*, V, p. 119.

6.1.2 ***K. al-Muᶜṭayāt***
Taḥrīr by NAṢĪR AL-DĪN AL-ṬŪSĪ (no. G23): MR 40,7 (fols. 91v–103v, 1146H); DR 704,3 (fols. 22v–48v, 1011H in Shiraz), not listed by Sezgin. *Cf. GAS*, V. p. 116.

6.1.3. *K. al-Uṣūl.*

THĀBIT B. QURRA (no. B30): *M. fī Burhān al-muṣādara al-mashhūra* and *R. fi l-ᶜIlla allatī lahā rattaba Uqlīdis ashkāl kitābihi dhālika l-tartīb.*

ABŪ JAᶜFAR AL-KHĀZIN (no. B30): *Tafsīr ṣadr al-maqāla al-ᶜāshira min K. Uqlīdis.*

SULAYMĀN B. ᶜIṢMA (no. B33): *Fī dhawāt al-ismayn wa-l-munfaṣilāt . . .*

AḤMAD B. AL-ḤUSAYN AL-AHWĀZĪ (no. B46): *Sharḥ al-maqāla al-ᶜāshira min K. Uqlīdis* and an abridgement thereof.

ABU SAHL AL-QŪHĪ (no. B53): *al-Maqālatān al-ūlā wa-l-thāniya min K. Uqlīdis fi l-Uṣūl.*

IBN AL-HAYTHAM (no. B77): *Ḥall shukūk K. Uqlīdis* . . . and *Sharḥ al-muṣādarāt.*

ḤUSĀM AL-DĪN ᶜALĪ B. FAḌL ALLĀH AL-SĀLĀR (no. G22): *Ikhtiṣār daᶜāwa l-maqālatayn al-ūlā wa-l-thāniya min K. al-Uṣūl li-Uqlīdis* and *Muqaddimāt li-tabyīn al-maṣādara fi l-maqāla al-ūlā li-Uqlīdis.*

NAṢĪR AL-DĪN AL-ṬŪSĪ (no. G23): *Taḥrīr K. Uṣūl al-handasa*: ṬR 107 (137 fols., 789H, with a *dhayl*); DR 703,2 (fols. 2r–129v, 1006H); MR 8 (135 fols., *ca.* 1100H); ṬR 152 (192 fols., diff. hands, the last of which dated 1014H in Damascus); *etc.* (6 copies). None of these manuscripts is listed by Sezgin. See Pl. LXXV.

KAMĀL AL-DĪN AL-ḤUSAYN AL-FĀRISĪ (no. G92): *Taᶜlīq* on the thirteenth *maqāla* of al-Ṭūsī's *Taḥrīr*, not mentioned by Sezgin.

AL-SAYYID AL-SHARĪF AL-JURJĀNĪ (no. G42): *Ḥāshiya* on al-Ṭūsī's *Taḥrīr*, not mentioned by Sezgin.

ABŪ ISḤĀQ (no. X13): A commentary on the first four *maqāla*s of al-Ṭūsī's *Taḥrīr*, not mentioned by Sezgin.

6.1.4 AL-SAMARQANDĪ (no. G26): *Ashkāl al-taᵓsīs* and various commentaries thereon.

6.1.3 ᶜABD ALLĀH B. MUḤAMMAD, IBN AL-KHAWWĀM (no. G24): *Sharḥ al-maqāla al-ᶜāshira min K. al-Uṣūl.*

6.1.3 ANONYMOUS (no. B108): *al-ᶜAmalīyāt min K. Uqlīdis fi l-Uṣūl.*

6.1.5 *K. al-Zawāya l-ḥāditha fi l-dāᵓira*: MR 41,21 (fols. 138v–143v, 1153H), unique. This treatise, which comprises 10 constructions, contains several references to Euclid's *Elements*. *Cf. GAS*, V, p. 120.

6.2.27 *K. al-Qisma*: see no. B98.

7.2.1 *K. al-Manāẓir*
An anonymous translation, perhaps by THĀBIT B. QURRA (no. B30): DR 720 (photos of a manuscript in Damietta Institute, 77 pls., 751H); DR 260,2 (fols. 59r–92r, *ca.* 600H, lacks first page or two); neither listed in *GAS*, V, p. 117. Neither of these copies bears the name of the translator. See Pl. XCVb.

7.2.1 *Iṣlāḥ al-Manāẓir* by ABŪ YŪSUF AL-KINDĪ (no. B19).

7.2.1 *Taḥrīr* by NAṢĪR AL-DĪN AL-ṬŪSĪ (no. G23): MR 40,6 (fols. 80v–88r, 1146H); DR 704,1 (fols. 1v–15v, 1011H, with additional notes by the editor not in the published text); ṬR 102,5 (fols. 115r–126r, *ca.* 1200H). None of these is listed in *GAS*, V, p. 117. See also DR 897,3 (photos of MS Leiden 897,3) (fols. 93v–99r, *ca.* 900H).

7.2.1 *Tajrīd Uqlīdis fi l-manāẓir* by MUḤAMMAD B. ᶜUMAR B. AḤMAD IBN ABĪ JARĀDA (no. G29): DR 638 (12 fols., *ca.* 1100H), unique (?), not mentioned by Sezgin.
See also no. Z28.

A6 ARISTARCHUS

GAS, VI, pp. 74–75.

1.1.3 *K. Jarmayi l-nayyirayn wa-buᶜdayhimā*: *Taḥrīr* by NAṢĪR AL-DĪN AL-ṬŪSĪ (no. G23): MR 41,31 (fols. 181v–187v, 1146H). *Cf. GAS*, VI, p. 75.

A7 ARCHIMEDES

GAS, V, pp. 121–136.

6.1.6 *K. al-Kura wa-l-usṭuwāna*
ABU SAHL AL-QŪHĪ (no. B53): *R. li-Taṣḥīḥ al-maqāla al-thāniya min K. al-Kura wa-l-usṭuwāna li-Arshimīdis.*

6.1.6 IBN AL-HAYTHAM (no. B77): *R. li-Tawḍīḥ qismat al-khaṭṭ fi l-maqāla al-thāniya min K. al-Kura wa-l-usṭuwāna li-Arshimīdis.*

6.1.7 *K. al-Maᵓkhūdhāt*
Taḥrīr by NAṢĪR AL-DĪN AL-ṬŪSĪ (no. G23) (?) based on the *tafsīr* by ᶜALĪ B. AḤMAD AL-NASAWĪ (no. B67): MR 41,3 (fols. 15v–20v, 1146H), not mentioned in *GAS*, V, p. 133.

6.1.8 *R. fi ᶜAmal al-dāᵓira al-maqsūma bi-sabᶜat aqsām mutasāwiya,* translated by THĀBIT B. QURRA (no. B30) and apparently edited by the copyist MUṢṬAFĀ ṢIDQĪ (no. D81): MR 41,14 (fols. 105v–110r, 1153H), unique. *Cf. GAS*, V, pp. 133–134, and *AHES* 30 (1984), pp. 285–290.

6.1.9 *K. al-Mafrūḍāt*
Taḥrīr by NAṢĪR AL-DĪN AL-ṬŪSĪ (no. G23); MR 41,4 (fols. 21v–25v, 1146H); DR 898,24 (photos of MS Leiden 14, pp. 470–481, *ca.* 1050H). *Cf. GAS*, V, p. 135, no. 7.

6.2.28 See also the anonymous ***M. fī Tarkīb li-taḥlīl muqaddimat al-musabba***^{c} ... in MR 41,15 (fols 111v–112v, 1153H) listed below (no. B111).

See also no. H30!

A8 APOLLONIUS

GAS, V, pp. 136–143.

6.1.10 *K. al-Makhrūṭāṭ*
BANŪ MŪSĀ (no. B16): *Muqaddimāt K. al-Makhrūṭāṭ.*

MAḤMŪD B. QĀSIM B. AL-FAḌL AL-IṢFAHĀNĪ (no. G5): *Talkhīṣ al-Makhrūṭāṭ.*

MUḤYI L-DĪN YAḤYĀ B. ABĪ SHUKR AL-MAGHRIBĪ (no. G21): *Sharḥ K. al-Makhrūṭāṭ*

Taḥrīr by NAṢĪR AL-DĪN AL-ṬŪSĪ (no. G23): DR 783 (photos of a manuscript of unknown provenance, 195 pls., *ca.* 1000H); DR 897,1 (photos of MS Leiden 513, fols. 1v–87r, *ca.* 900H). Neither copy listed in *GAS*, V, p. 141. See also DR 898,1 (photos of MS Leiden 14, pp. 1–163, 1063H).

A9 HIPPARCHUS

GAS, VI, pp. 77–80, especially p. 78.

On the attribution to him of the invention of the astrolabe, see no. B79.

A10 HYPSICLES

GAS, V, pp. 143–145.

1.1.4 *K. al-Maṭāli*^{c}
Taḥrīr by NAṢĪR AL-DĪN AL-ṬŪSĪ (no. G23), based on the translation of Qusṭā b. Lūqā (no. B38) improved by al-Kindī (no. B19): MR 41,2 (fols. 11v–12v, 1146H). *Cf. GAS*, V, p. 145, no. 2.

A11 HERON

GAS, V, pp. 151–154.

7.1.1 *K. fī Raf*^{c} *al-ashyā*$^{\supset}$ *al-thaqīla*, in 3 *maqālas*, translated by Qusṭā b. Lūqā (no. B38): DR 668 (47 fols., *ca.* 1000H, fine diagrams); ṬR 123 (31 fols., 973H by Ibn Abi l-Khayr al-Ḥusnī (no. C124), fine diagrams); MR 39,5 (fols. 42v–95r, 1151H, lacks diagrams). Only the last of these is listed in *GAS*, V, pp. 153–154. See Pl. XCIV.

A12 THEODOSIUS

GAS, V, pp. 154–156, and VI, pp. 80–81.

1.1.5 *K. al-Masākin*
Taḥrīr by NAṢĪR AL-DĪN AL-ṬŪSĪ (no. G23): MR 40,5 (fols. 76v–79v, 1146H); DR 897,2 (photos of MS Leiden 513, fols. 91r–93v, *ca.* 900H). *Cf. GAS*, V, p. 155–156, no. 2.

6.1.11 *K. al-Ukar*
Taḥrīr by NAṢĪR AL-DĪN AL-ṬŪSĪ (no. G23): TR 347,1 (pp. 2–54, 895H in Shiraz, with notes taken from the handwriting of Quṭb al-Dīn al-Shīrāzī (no. G27)); DR 704,4 (fols. 49v–88r, 1011H); *etc.* (3 copies). None of these is listed in *GAS*, V, pp. 154–155, no. 1.

A13 ANTIOCHUS

GAS, VI, p. 31.

See Pīr Ḥasan (no. H61).

A14 DIODORUS

GAS, V, pp. 156–157.

6.2.5 *Kitāb al-Analīmā*: see Abū Sa^{c}īd al-Ḍarīr (no. B24).

A15 DOROTHEUS

GAS, VII, pp. 32–38.

See Pīr Ḥasan (no. H61).

A16 MENELAUS

GAS, V, pp. 158–164.

6.1.12 *K. al-Ashkāl al-kurīya*
Taḥrīr by NAṢĪR AL-DĪN AL-ṬŪSĪ (no. G23): DR 704,5 (fols. 88v–185v, 1011H), not listed in *GAS*, V, p. 163.

A17 NICHOMACHUS

GAS, V, pp. 164–166.

See Ibn Fallūs (no. C6).

A18 PTOLEMY

GAS, V, pp. 166–174; VI, pp. 83–96; and VII, pp. 41–48.

1.1.6 *K. al-Majisṭī*
Translation by ḤUNAYN B. ISḤĀQ (no. B20) improved by THĀBIT B. QURRA (no. B30): DH 71 (photos of MS Paris B.N. ar. 2482, 131 pl., 618H, first 6 *maqāla*s only). *Cf. GAS*, VI, p. 89.

1.1.6 *Qawl fī īḍāḥ al-wajh alladhī dhakara Baṭlamiyūs anna bihi stakhraja man taqaddamahu masīrāt al-qamar al-dawrīya wa-hiya l-mustawiya* by THĀBIT B. QURRA (no. B30).

Talkhīṣ al-Majisṭī by IBN SĪNĀ (no. B81).

1.1.6 *Taḥrīr al-Majisṭī* by NAṢĪR AL-DĪN AL-ṬŪSĪ (no. G23): ṬH 43 (91 fols., *ca.* 650H by Muᵓayyad al-Dīn al-ᶜUrḍī (no. C170)); DH 17 (128 fols., *ca.* 1150H); K 3822 (191 fols., 1116H); K 8530 (224 fols., 1252H). *Cf. GAS*, VI, p. 93. See Pl. I.

Tafsīr al-Taḥrīr by AL-ḤASAN B. MUḤAMMAD AL-NĪSĀBŪRĪ (no. G38).

1.1.7 A fragment from the beginning of the first *maqāla* of the ***K. al-Iqtiṣāṣ (Planetary Hypotheses)***: TR 238 (5 pp., *ca.* 1300H). *Cf. GAS*, VI, pp. 94–95. The fragment corresponds to the first 1⅓ pages of the text of MS London B.L. Add. 7473, published in B. R. Goldstein, "The Arabic Version of Ptolemy's *Planetary Hypotheses*," *Transactions of the American Philosophical Society*, N.S., 57:4 (1967), pp. 1–55. MS Hyderabad Osmania University 306/520-RH (94 fols., *ca.* 650H) is an extensive fragment of a treatise on theoretical astronomy bearing the spurious title *K. al-Iqtiṣāṣ li-Baṭlamyūs* on the cover. Both beginning and end are missing but the work is arranged in at least 4 *fann*s of 14 + 14 + 40 + at least 7 *faṣl*s, and the author mentions both Ptolemy's *Almagest* and *Hypotheses* in the text. Sezgin (*ibid.*) lists this as another copy of the *Hypotheses*.

4.3.37 An anecdote about the discovery of the astrolabe by Ptolemy is contained in a late Maghribi treatise on the astrolabe (no. F57): see TR 113, pp. 9–10.

5.1.3 *K. al-Thamra (Centiloquium)*
Commentary by AḤMAD B. YŪSUF KĀTIB ĀL ṬŪLŪN (no. B40). Arabic and Persian commmentaries by NAṢĪR AL-DĪN AL-ṬŪSĪ (no. G23).

5.1.3 *al-Kutub al-arbaᶜa (Tetrabiblos)*
Recension by ḤUNAYN B. ISḤĀQ and THĀBIT B. QURRA (nos. B20 and B30).
Quotations from the commentary by ᶜALĪ B. RIḌWĀN (no.B85).
Turkish translation by SULAYMĀN RIYĀḌĪ (no. H65).
See also ARISTUS/EUTOCHIUS (no. A20).

5.1.3 ***K. al-Majālis***: DM 122,1 (fols. 1r–2r, *ca.* 1150H). *Cf. GAS*, VII, p. 46, no. 9. See Pl. LXXIII and also no. A30.

5.1.3 A note on Ptolemy tracing his academic genealogy through Galen, Ylmyt's, Aristotle, Plato, Socrates, Bqrᵓṭys, Bqrᵓṭ, Pythagoras, to Luqmān: DJ 103,4B (fol. 57r, *ca.* 1200H).

5.2.8 On his *K. Dhawāt al dhawāᵓib* dealing with comets, see Ḥunayn b. Isḥāq (no. B20).

A19 DIOCLES

GAS, forthcoming volume on physics.

6.2.25 ***K. al-Marāya l-muḥriqa***: see no. B109.

A20 EUTOCHIUS (ARISTUS)

GAS, V, p. 188; and VII, p. 48.

5.1.4 ***al-Aḥkām al-nujūmīya***, an astrological work in 200 sections attributed to Eutochius (*ᵓwṭwqyws*) in DM 1058 (10 fols., *ca.* 800H), and to Aristus (*ᵓrsṭws*) in MM 105 (9 fols., 1182H). *Cf. GAS*, VII, p. 48. The texts differ slightly but in both cases the author is referred to as the commentator (*mufassir*) on Ptolemy's ***Tetrabiblos***—see Pls. LXXIVb–c. Another copy of this work attributed to Arṭwsws is in MS Princeton Mach 5049 = Yehuda 373 (fols. 249v–256v, *ca.*, 1050H): *Cf. GAS*, VII, p. 371.

A21 VETTIUS VALENS

GAS, VII, pp. 38–41.

5.2.6 A table for computing the length of a person's life stated to have been originally written in gold in a ***K. al-Kāmil fi l-nujūm*** in the treasury of ᶜAbd Allāh b. Ṭāhir b. al-Ḥusayn from (*ᶜan*) al-Maᵓmūn (no. B9) is contained in several manuscripts. See Pl. LXXIX. In ṬM 102,9 (fols. 83v–85v, *ca.* 1150H) the table is accompanied by a treatise and is refered to as *hashtaq* and attributed to Vettius Valens. The same table occasionally called *jadwal al-ḥayā* is contained in MM 58,3 (fol. 6r, *ca.* 850H). See also DM 157,1 (fol. 1r, *ca.* 900H); DM 637,2 (fols. 47v–48r, *ca.* 1000H); DM 1108,4B (fol. 20v, 1053H); DM 1205,2 (fol. 215v, *ca.* 1250H); ṬM 241,2 (34r–35v, *ca.* 1250H). See further O. Neugebauer and H. B. van Hosen, *Greek Horoscopes*, (Philadelphia, 1959), pp. 174–175 for references to Vettius Valens' table and the underlying theory.

A22 HERMES/DANIEL/ALEXANDER

Articles "Hirmis," "Daniyāl," and "Iskandar" in *EI₂*; *GAS*, IV, pp. 31–44; VII, 50–58 (Hermes); 64 (Alexander), and 312–317 (Daniel).

Note: The Egyptian National Library contains a remarkably rich collection of Hermetic material. The following classification of the available manuscripts will surely require modification at the hands of a specialist. Most of the material listed below is not recorded by Sezgin.

5.1.1 ***K. al-Sirr al-badīᶜ al-anwar***, in 3 *babs* and 8 unnumbered *bābs*: DM 1156,1 (fols. 1r–33v, *ca.* 800H), unique. Not mentioned by Sezgin. See Pl. CVIb.

5.1.1 A treatise on the astrological significance of eclipses, halos, *etc.*, in 22 *bābs*: DM 1156,4 (fols. 45v–87v, *ca.* 800H) — see Pl. LXXI; and two late copies, the first entitled *K. al-Sdᵓᶜ* and the second entitled *K. al-Shrᵓyᶜ*: DM 1092 (141 pp., 1296H, incomplete), and ṬM 159 (52 fols., 1221H). See *GAS*, IV, p. 42, no. 9 and VII, pp. 313–314, nos. 1 and 3.

5.1.1 A treatise on the astrological significance of conjunctions: DM 896,1 (photos. of MS Paris B.N. ar. 2487, fols. 32v–36r, 699H!). See *GAS*, IV, pp. 42–43, no. 10.

5.1.1 ***K. Hirmis***, a compendium of astrological lore usually bearing a title relating to the first treatise, which deals with the heliacal risings of Sirius and which is attributed to Hermes, but consisting mainly of sections attributed to Daniel: K3852 (68 fols., *ca.* 1100H); DM 132 (92 fols., *ca.* 1100H, Christian): DM 575 (48 fols., *ca.* 1150H): DḤ 16,2 (fols. 173v–117v, 1133H); K 3852 (68 fols., *ca.* 1100H); *etc.* (about a dozen late copies and fragments).

5.1.1 A treatise on the astrological significance of the heliacal risings of Sirius: DM 59,2 + 3 (fols. 1v–6v, *ca.* 1000H); DJ 289,2 (fols. 56r–95v, 1222H); DM 994 (16 fols., 1271H); DM 1056,1 (fols. 1r–33r, 1308H); DM 788 (40 fols., *ca.* 1050H, defective); ḤM 2,2 (fols. 203v–209r, 1290H); DM 79,2 (fols. 4r–13r, *ca*, 1150H); Sh 72 (13 fols., 1301H). In some copies the title *Bulūgh al-umniya* or *al-amāniya* occurs. In DM 994 the treatise is attributed to al-Shaykh ᶜAlī al-Ḥākim al-Baṣīr; in DM 1056,1 to ᶜAlī al-Ḥātimī al-Baṣrī; in DM 788 to ᶜAlī al-Ḥākimī al-Basrī. This confusion with Ibn Yūnus (no. B59) derives from a corruption of *al-ḥakīm al-baṣīr*. See also a treatise on the same subject compiled in 1130H: Sh 73 (7 fols., 1283H), and Sh 71 (30 fols., 1283H). *Cf. GAS*, IV, pp. 41–42, no. 3, and VII, p. 173.

5.1.1 A *Malḥama* attributed to Alexander and Abū Maᶜshar (no. B22): S 4467,5 (fols. 59r–74v, *ca.* 1200H). See also K 3862,1 (fols. 1v–10r, *ca.* 1300H). A *malḥama* in Persian arranged in 24 *bābs*: ṬJF 12,3 (fols. 29r–56v, *ca.* 1250H).

5.1.1 A *faṣl* on celestial cycles, *tasyīr*, and *intihāᵓāt*: K 3979,1 (fols. 1v–4r, *ca.* 1150H).

5.1.1 A *namūdār* of Hermes: K 3979,2 (fols. 4v–12v, *ca.* 1150H). This manuscript contains a calculation for Baghdad, 684H.

5.1.1 A treatise on the increase of the level of the Nile and astrological prediction: MM 89 (4 fols., *ca.* 1050H) and a copy thereof in ZK 326.

5.1.1 *K. al-Khaṣāᵓiṣ al-malakīya fi l-qawāᶜid al-falakīya*: DM 180,1 (fols. 1v–14v, 1238H, defective). *Cf. GAS*, IV, p. 43, no. 23, and VII, p. 56, no. 14.

5.1.1 *K. al-Qirānāt wa-l-ittiṣālāt wa-l-mumāzajāt*, in 16 *bābs*: ṬM 96 (55 fols., 1184H);

ṬM 93,1 (fols. 1r-67r, *ca.* 1250H). *Cf. GAS*, IV, p. 42, no. 6, and VII, p. 55, no. 5.

5.1.1 *K. al-Asās*, in 45 + 15 *bāb*s (although 85 are promised in the introduction): ṬJF 21,4 (fols. 139v–147v, 1206H). *Cf. GAS*, IV, p. 43, no. 19, and VII, p. 55, no. 7.

5.1.1 Other Hermetic material in Arabic is contained in ṬM 241,4 (fol. 39v, 1245H); Sh 70,3 (fols. 17v–20v, 1282H); TR 141,8 (pp. 166–167, *ca.* 1100H); ṬJ 298,6 (fols. 52r–54r, *ca.* 1085H); ṬM 97 (149 fols., 1324H); ZK 453,2 (pp. 295–341, 1263H); MḤ 72,3 (fols. 75v–89v, 1220H); DM 122,3 (fols. 36v–70v, *ca.* 1150H); Sh 26 (57 fols., 1315H); DMT 4,2 (fols. 78v–87r, *ca.* 1200H). See the Arabic catalog for further details. See also al-Samalāwī (no. D65).

A23 TEUKRUS (TANKALŪSHĀ)

GAS, IV, pp. 112–116; VII, pp. 71–73.

5.1.9 ***K. Ṣuwar daraj al-falak***: MM 84 (107 fols., *ca.* 700H, slightly damaged) and photos thereof in DM 1181. This fine copy is not listed in *GAS*, IV, p. 116, no. 1 or VII, p. 73, no. 2. See Pl. LXXVIIIa.

A24 KĪRKIS B. FĪLŪS AL-BAṬRAK (?) AL-YŪNĀNĪ

Not identified.

5.1.11 A treatise on astrology in about 20 unnumbered *bāb*s and sections: DM 111 (84 fols., *ca.* 1100H); DM 502 (20 fols., *ca.* 1250H, lacks introduction), no other known copies. The first copy bears the spurious title ***K. Qiyās al-shams wa-l-qamar fī ᶜilm al-falak***. This work is currently being investigated by J. Samsó Moya of the University of Barcelona.

A25 RYMBR (?) AL-YŪNĀNĪ

GAS, VII, pp. 73–74.

5.1.10 An astrological work beginning *hādhā kitāb allafahu Rymbr al-Yunānī fī ᶜilm al-nujūm wa-tarjamahu talmīdhuhu Sirjis* . . . in Sh 76,1 (fols. 1v–5r, *ca.* 1300H), unique. See Pl. LXXIVa. *Cf. GAS*, VII, p. 74, where Sezgin suggests the name Rūsūs for the author.

A26 SARJIS (AL-RAᵓSᶜAYNĪ)

GAS, IV, pp. 106–107.

See Rymbr (no. A25).

A27 ZARATHUSTRA (ZARĀDASHT)

GAS, VII, pp. 81–86.

5.1.8 ***K. al-Qirānāt***: DM 898 (photos of MS Paris B.N. ar. 2487,4, fols. 56–106, 679H). This Arabic version was compiled in Amud in 544H. *Cf. GAS*, VII, p. 86, no. 2. See also al-Sijzī (no. B56).

A28 JĀMASP AL-ḤAKĪM

GAS, IV, pp. 59–60, and VII, pp. 86–88.

5.1.7 ***K. fī Aḥkām al-qirānāt***: DM 897 (photos of MS Paris B.N. ar. 2487,3, fols. 39–56, 679H). *Cf. GAS*, VII, p. 88, no. 1.

A29 ANONYMOUS (= JBHR?)

Not listed in *GAS*, VII.

5.1.5 An astrological work beginning . . . *hādhā kitāb waḍaᶜahu l-Hind fī amr al-amṭār wa-l-asᶜār wa-yusamma l-jahr (?)*: DM 943,1 (fols. 1v–43r, *ca.* 1050H); ṬM 227,7 (fols. 134v–150v, 1145H). See Pl. LXXXIIa. No other copies known to me. The author of this work may be identical to the Indian Jbhr mentioned by Ibn al-Nadīm (ed. Flügel, p. 271).

A30 JNH AL-HINDĪ

GAS, VII, p. 94.

2.4.7 Two diagrams for the councils of kings according to the opinion of Jnh al-Hindī: MM 54,2 (fols. 9v–10r, *ca.* 1200H). *Cf. GAS*, VII, no. 94, no. 2. See also Pl. LXXIII and no. A18 (*sub* 5.1.3), as well as DM 122,2 (fols. 2v–3r, *ca.* 1100H) and the Yemeni almanacs listed under no. E11, in which these diagrams also occur.

A31 ṬUMṬUM AL-HINDĪ

GAS, IV, pp. 118–119, and VII, pp. 96–97.

5.1.6 *K. Ṭumṭum*: DḤ 71 (60 fols., *ca.* 1000H, lacks end). This should be compared with MS Paris ar. 2595. *Cf. GAS*, VII, pp. 96–97 on both manuscripts.

5.1.6 An anonymous *Sharh Urjūzat Ṭumṭum al-Hindī fī ᶜilm al-zāʾiraja*: ṬJ 604,2 (fols. 65r–72r, *ca.* 1150H). Not listed by Sezgin.

A32 KANAKA (?)

See the article "Kanaka" in *DSB* by D. Pingree.

5.2.10 See the *Maqāla fi l-Aᶜdād al-mutaḥābba* of Abū Maᶜshar (no. B22).

A33 YUḤANNĀ (?) AL-HINDĪ

5.2.10 See Abū Maᶜshar (no. B22).

A34 JAᶜFAR AL-HINDĪ

5.2.10 See Abū MaᶜShar (no. B22).

B. Early Muslim Authors (to *ca.* 1100)

For this period Sezgin's *GAS*, vols. V, VI, and VII supercede all other bio-bibliographical sources. No satisfactory account of early Islamic astronomy exists. For a brief survey, see my chapter in the forthcoming *GHA*. See also the more specialized studies of D. Pingree listed in the introduction to Section A.

B1 ᶜALĪ B. ABĪ ṬĀLIB, the fourth Caliph

Article "ᶜAlī b. Abī Ṭālib" in EI_2 by L. Vecchia Vaglieri. For other scientific material attributed to him, see D. A. King, "A Medieval Arabic Report on Algebra before al-Khwārizmī," to appear.

2.5.14 A short treatise on the climates attributed to the *Imām* ᶜAlī: ṬM 72,2 (fols. 14r–15v, *ca.* 1300H, incomplete).

2.5.15 A rule for finding the *notae* of the beginnings of the Muslim months, attributed to the *Imām* ᶜAlī b. Abī Ṭālib: TR 321,1 (p. 1, 1300H). See Pl. LXIVa.

B2 AL-IMĀM AL-SHĀFIᶜĪ

GAS, I, pp. 484–490, *etc.*

3.5.1 A poem on the prayer-times attributed to the Imām al-Shāfiᶜī: DM 119,3 (fols. 8v–9r, 1109H); DM 702,4 (fols. 27r.27v, 1160H, Maghribi script). In DM 181,4 (fols. 46v–47v, *ca.* 1200H) the same poem is attributed to Ibn Yūnus (no. B59). This poem is published in *Aleppo Conference Proceedings*, I, pp. 393–394. See Pl. XLIII.

B3 IBRĀHĪM B. MUḤAMMAD AL-FAZĀRĪ (the historian)

GAS, I, p. 292. This individual is not to be confused with the astronomer al-Fazārī, on whom see the article "al-Fazārī" by D. Pingree in *DSB*; *GAS*, II, p. 438; V, pp. 216–217; VI, pp. 122–124; and VII, p. 101.

1.4.16 A poem on the Syrian months attributed to Ibrāhīm b. Muḥammad al-Fazārī is contained in a treatise on folk astronomy by the late Egyptian astronomer al-Nabtītī (no. D28) in MM 108,1 (fols. 1v–23r, *ca.* 1150H). Similar extracts are given in the almanac of al-Shayzarī (no. C5) and in an anonymous Maghribī almanac (no. F64).

B4 YAᶜQŪB B. ṬĀRIQ

Article "Yaᶜqūb b. Ṭāriq" by D. Pingree in *DSB*; *GAS*, V, pp. 217–218; VI, pp. 124–127; and VII, pp. 101–102.

5.2.1 Quotations of his recorded by Abū Maᶜshar (no. B22) on the astrological significance of the time of conception and gestation: DM 623,1 (fols. 1v–9v, *ca.* 1150H).

B5 MĀSHĀʾALLĀH

Article "Māshāʾallāh" by D. Pingree in *DSB*; *GAS*, VI, pp. 127–129; VII, pp. 102–108 and 324.

5.2.2 *K. al-Asᶜār*: ṬM 157,3 (fols. 27v–31v, *ca.* 1150H), not listed in *GAS*, VII, p. 104, no. 3.

5.2.2 A note attributed to Māshāʾallāh dealing with the auspicious times for blood-letting is in K 4314,2 (fols. 19r–19v, *ca.* 1200H). Material on astrology attributed to him occurs amidst various notes in K 4003,2 (fols. 27v–29v, *ca.* 1050H).

B6 ᶜUMAR B. AL-FARRUKHĀN AL-ṬABARĪ

GAS, VII, pp. 79 and 111–113.

5.2.3 *K. al-Masāʾil*: DM 165 (77 fols., *ca.* 1150H); ṬM 119,2 (fols. 1r–68r, 1185H); DM 1214 (69 fols., *ca.* 1050H, arranged in 140 *masāʾil* rather than the usual 138); *etc.* (2 late copies and 1 fragment). *Cf. GAS*, VII, p. 112, no. 2.

5.2.3 *K. fi l-Mawālīd*: DM 943,2 (fols. 43v–74r, *ca.* 1050H); ṬM 227,8 (fols. 151r–163r, 1145H), not listed in *GAS*, VII, p. 112, no. 3.

5.2.3 A book on astrology with no specific title, stated to be translated from Syriac; ṬM 139,5 (fols. 66r–87r, *ca.* 1100H). *Cf. GAS*, VII, p. 79.

B7 AL-ṬABARĪ AL-MUNAJJIM (= ᶜUMAR B. AL-FARRUKHĀN?)

5.2.34 A *faṣl* on the astrology of the Sabians (of Harran) copied from the published text of R. Dozy (1885) (!): W 2314 (15 fols., 1355H).

B8 YAḤYĀ B. ABĪ MANṢŪR

Kennedy, no. 51; *GAS*, V, p. 227; VI, pp. 136–137; and VII, pp. 116–117.

1.6.6 A fragment on the determination of the distance corresponding to one degree of terrestrial latitude taken from Ibn Khallikān (no. C8): ZK 782,12 (fol. 70r, 1121H).

2.1.2 The solar eclipse tables from the *Mumtaḥan Zīj* (on which E. S. Kennedy and N. Faris, "The Solar Eclipse Technique of Yaḥyā b. Abī Manṣūr," *JHA*, 1 (1970), pp. 20–38) are found in various later Egyptian *zīj*es and collections of tables; see, for example: MM 19 (fol. 9v, 1092H) (see no. D50); DM 618,2 (fols. 24r–33v, *ca.* 1200H); DM 639,14 (fols. 102r–109r, *ca.* 1150H); and also DM 678 and 1198. There is no indication in any of these sources of the origin of the tables.

2.1.2 *(Qiṭᶜa) fī maᶜrifat taqwīm al-kawākib al-khamsa*, on computing planetary positions using simple Ptolemaic theory: DM 895,1 (photos of MS Paris B.N. ar. 2487, fol. 1v, 679H). *Cf.* MS Paris B.N. supp. pers. 1488, fol. 229v, of the *Ashrafī Zīj*.

B9 AL-MAʾMŪN, the Abbasid Caliph

GAS, VII, p. 118.

5.2.6 + *K. al-Qurᶜa al-Maʾmūnīya*, a handbook of astrological prognostications: TGh 140 (96 pp., *ca.* 1200H). See Pl. LXXVI. Sezgin (*GAS*, VII, p. 118) lists a much earlier copy in Oxford, which merits investigation.

See also no. A21 on the *K. al-Kāmil fi l-nujūm*.

B10 ABŪ ZAKARĪYĀ YAḤYĀ B. MĀSAWAYH

GAS, III, pp. 231–236; and VII, p. 326.

1.4.1 *K. al-Azmān*: MM 4 (34 fols., *ca.* 900H). *Cf. GAS*, III, p. 234, no. 10, where this copy is not listed.

B11 SANAD B. ᶜALĪ

GAS, V, pp. 242–243; VI, p. 138; VII, pp. 119–120.

5.2.3+ A fragment from the beginning of his *R. fi l-Quṭūᶜ* on the astrological doctrine of *qaṭᶜ*: MMF 8,4 (fol. 77v, *ca.* 700H). a complete copy of this short treatise is MS Istanbul Hamidiye 1446 (fols. 220r–220v, *ca.* 900H); *cf. GAS*, VII, p. 119, no. 1.

B12 ᶜALĪ B. ᶜĪSĀ

GAS, VI, pp. 143–144.

4.8.1 An anonymous ***Bāb fi l-ᶜAmal bi-l-ṣafīḥa al-qamarīya*** on the use of an eclipse plate in ZK 706,6 (fols. 20r–22r, *ca.* 1100H) is based on the treatise attributed to ᶜAlī b. ᶜĪsā in MS Istanbul Topkapi A 3509, fols. 304r–309v, listed in *GAS*, VI, p. 144, no. 2. Merits publication and analysis.

B13 ABŪ ᶜALĪ AL-KHAYYĀṬ

GAS, VII, pp. 120–121.

5.2.4 *Aḥkām wa-khtiyārāt*: ṬM 243,1 (fols. 1v–22v, 644H), and an anonymous fragment containing additional *bābs*: MJ 39,4 (fols. 59v–80v, *ca.* 1200H).

5.2.4 (*K. fī Jumlat al-adilla ʿala l-mawālīd min aḥkām al-nujūm*): MM 204,3 (fols. 22v–49v, 1052H); *cf. GAS*, VII, pp. 120–121, no. 1.

5.2.4 *Fawāʾid min kalām Abī ʿAlī al-Khayyāṭ*: DM 180,3 (fols. 55v–73r, 1238H); *cf. GAS*, VII, p. 121, no. 2.

5.2.4 An anonymous Persian treatise based on the writings of Abū ʿAlī al-Khayyāṭ: MMF 7,2 (fols. 5v–8v, 1060H). Not mentioned by Storey.

B14 MUḤAMMAD B. MŪSĀ AL-KHWĀRIZMĪ

Article "al-Khwārizmī" in *DSB* by G. Toomer; *GAS*, V, pp. 228–241, VI, pp. 140–143; VII, pp. 128–129. See also D. A. King, "al-Khwārizmī and New Trends in Mathematical Astronomy in the Ninth Century," *OPNE*, 2 (1983).

2.1.3 For Ibn Masrūr's commentary on the *Zīj* of al-Khwārizmī, see no. B37.

3.3.6 The qibla table in ṬR 103,2 (pp. 100–104, *ca.* 1300H) is attributed to Abū Jaʿfar Muḥammad b. Mūsā al-Khwārizmī (see no. Z25). This attribution is not to be trusted. It is more probable that al-Khwārizmī computed a qibla table extant in various other medieval manuscripts (none of them in Cairo).

6.2.1 *al-Kitāb al-mukhtaṣar fī Ḥisāb al-jabr wa-l-muqābala*: *cf. GAS*, V, p. 240, no. 1. Commentary by Sinān b. al-Fatḥ (no. B44): DR 260,4 (fols. 95r–105v, *ca.* 600H), unique, listed in *GAS*, VII, p. 301.

Chapter from the commentary of al-Khuzāʿī (no. E6): ṬJ 207,4 (fols. 52r–61v + 65r–69v, 734H). al-Khuzāʿī calls al-Khwārizmī by the *kunya* Abū Bakr, confusing him with a tenth-century secretary called Abū Bakr al-Khwārizmī (*cf.* the article "al-Khuwārizmī" by J. Vernet in *DSB*). See Pl. LXXXVI.

See Plate XLIVb for al-Khwārizmī's table for predicting lunar crescent visibility.

B15 SAHL B. BISHR AL-ISRĀʾĪLĪ

GAS, V, p. 245, and VII, pp. 125–128.

5.2.5 *al-Masāʾil fi l-aḥkām*: ṬM 139,1 (fols. 1v–46v, *ca.* 1100H); K 4314,1 (fols. 1r–18v, *ca.* 1200H); DM 963 (65 fols., *ca.* 1300H); *etc.* (3 copies and 3 fragments). *Cf. GAS*, VII, p. 125, no. 1.

5.2.5 *R. fi l-Awqāt*: MM 190,1 (fols. 1r–8r, *ca.* 1100H); MM 200,3 (fols. 40v–49r, 1182H); ṬM 139,3 (fols. 59v–62v, *ca.* 1100H). *Cf. GAS*, VII, p. 127, no. 4.

5.2.5 *K. al-Ikhtiyārāt ʿala l-buyūt al-ithnā ʿashar*: ṬM 139,2 (fols. 47r–58v, *ca.* 1100H), unique. This work was previously known only in Latin translation. *Cf. GAS*, VII, p. 128, where this treatise is not identified.

5.2.5 *(R. fi) l-Fuṣūl*: ṬM 139,4 (fols. 62v–65r, *ca.* 1100H). *Cf. GAS*, VII, p. 127, no. 5.

B16 BANŪ MŪSĀ

GAS, V, pp. 246–252; VI, pp. 147–148; and VII, pp. 129–130.

6.1.10 *Muqaddimāt K. al-Makhrūṭāt*, on the premises of Apollonius' *Conica* (see no. A8): J 3626,33 (photos. of MS Istanbul Aya Sofia 4832, fols. 71v–74v, *ca.* 650H), unique. *Cf. GAS*, V, p. 139. A facsimile edition was published by N. Terzioğlu, *Das Vorwort des Astronomen* [!!] *Bani Musa* . . . , Istanbul, 1974. See also Ibn al-Haytham (no. B77).

6.2.2 The *taḥrīr* of their *R. fī Maʿrifat misāḥat al-ashkāl al-basīṭa wa-l-kurīya* by al-Ṭūsī (no. G23): MR 41,5 (fols. 26v–33v, 1153H, incomplete). An anonymous fragment is in DR 898,25 (photos. of MS Leiden Or. 14, pp. 481–487, *ca.* 1050H). Printed Hyderabad, 1940. *Cf. GAS*, V, p. 251, no. 1.
See also al-Dawwānī (no. B93).

7.1.2 *K. al-Ḥiyal*, on mechanical devices: TṢ 69 (photos of MS Vatican ar. 317, 148 pls., *ca.* 600H). Another incomplete anonymous copy is MS Istanbul Topkapi A 3474. A new translation of this work and a commentary based on all the available manuscripts has been published as D. R. Hill, *The Book of Ingenious Devices . . . by the Banū . . . Mūsā . . .* , Dordrecht (Holland), Boston (U.S.A.), and London (U.K.): D. Reidel, 1979.

B17 JĀBIR B. IBRĀHĪM AL-ṢĀBIᵓ

GAS, V, p. 254.

6.2.4 *Īḍāḥ al-burhān ᶜalā ḥisāb al-khaṭaᵓayn*, with additions by Aḥmad al-Sārī (no. G7): DR 898,4 (photos of MS Leiden 14, pp. 218-223, *ca.* 1050H). *Cf. GAS*, V, p. 254.

B18 ABU L-ᶜABBĀS AḤMAD B. MUḤAMMAD B. KATHĪR AL-FARGHĀNĪ

GAS, V, pp. 259–260, and VI, pp. 149–151.

1.2.1 *K. fī Jawāmiᶜ ᶜilm al-nujūm*: MM 194,1 (fols. 1v–31r, 876H); MJ 47,1 (fols. 1r–18v, *ca.* 900H, incomplete); DM 944 (155 pp., 1329H). *Cf. GAS*, VI, p. 150, no. 1.

1.2.1 A fragment from the end of ***al-Kalām ᶜalā ḥisāb al-aqālīm al-sabᶜa***, a treatise by al-Farghānī on the seven climates: MM 194,3 (fol. 32r, 876H). This is not part of the ***K. fī Jawāmiᶜ ᶜilm al-nujūm*** as stated in *GAS*, V, p. 260.

1.2.1 The first page only of an anonymous ***Bāb fī maᶜrifat al-awqāt allatī yakūn fīha l-qamar fawq al-arḍ***, on finding the times when the moon will be above the horizon: MM 194,2 (fol. 31v, 876H). This treatise follows a copy of al-Farghānī's main treatise (see above) and may be by him.

4.3.1 Anonymous fragments of ***al-Kāmil fi l-asṭurlāb*** on the construction of the astrolabe: DM 103,2 (fols. 29r–56v, *ca.* 750H in the hand of the copyist of MS Paris B.N. ar. 2513 of the ***Muṣṭalaḥ Zīj***, lacks beginning, end, and some tables); DM 106,3 (fols. 13r–33v, tables only, *ca.* 1200H); K 3815,6 (fols. 183r–198r, *ca.* 1100H). See also DM 644,3 for an isolated table of al-Farghānī's *aṣl* function. *Cf. GAS*, VI, pp. 150–151, no. 2 for complete copies of this work. See Pl. LV for an extract.

B19 YAᶜQŪB B. ISḤĀQ AL-KINDĪ

GAS, V, pp. 255–259; VI, pp. 151–155; VII, pp. 130–134, 241–261, and 326–327. On his improvement of the translation by Qusṭā b. Lūqā of Hypsicles' *K. al-Maṭāliᶜ*, see no. A10.

2.5.1 ***R. fī Maᶜrifat ṣuᶜūd al-kawākib fī ruᵓūs jawzaharātihā wa-hubūṭihā minhā***, a short treatise on planetary nodes: MM 1,2 (fols. 143r–144v, 1184H), unique, not listed by Sezgin. Merits investigation.

5.2.9 *K. al-Dwᵓr dhmzj* (?), an astrological treatise: DḤ 15 (72 pp., *ca.* 1100H). *Cf. GAS*, VII, p. 132, no. 7. See Pl. LXXVIIIb.

5.2.9 ***R. fi l-Qaḍāᵓ ᶜala l-kusūf***, on prognostications from eclipses: J 3626,28 (photos of MS Istanbul Aya Sofia 4832, fol. 57r, *ca.* 450H, unique). *Cf. GAS*, VII, p. 131, no. 1.

7.2.1 A fragment from his *Iṣlāḥ al-Manāẓir*: MR 40,2 (fols. 88r, 1146H). The fragment, which is not mentioned in *GAS*, V, p. 117, relates to the 60th section of Euclid's *Optics*. Other such fragments are in MSS Paris B.N. ar. 2467,2 (*GAS*, V, p. 117) and Manchester Rylands 350,4 (fols. 60v–80r, 1196H).

B20 ḤUNAYN B. ISḤĀQ

GAS, VI, p. 89; VII, pp. 43, 134, 261–267, and 327–328.

1.1.6 On his translation of the *Almagest* of Ptolemy, see no. A18.

5.1.3 A recension of Ptolemy's *Tetrabiblos*: DM 1054 (103 fols., *ca.* 1100H, lacks part of 4th *maqāla*). A different version is in DM 123 (75 fols., *ca.* 1250H); KhM 5 (52 fols., *ca.* 1250H). *Cf. GAS*, VII, p. 43.

5.2.8 ***(R.) fī Dhawāt al-dhawāᵓib (?) wa-mā dhukira fīhā min al-ᶜajāᵓib***, a treatise on the astrological qualities of comets, quoting Ptolemy (no. A18): MM 204,6 (fols. 75v–76r, 1052H); ṬM 157,4 (fols. 32r–34v, *ca.* 1150H). *Cf. GAS*, VII, p. 328, no. 2. See Pl. LXXX and also no. D45.

B21 ABŪ YŪSUF YAᶜQŪB B. ᶜALĪ AL-QAṢRĀNĪ

GAS, VII, pp. 138–139.

5.2.12 *K. al-Masāᵓil*: ṬM 247 (126 fols., 1069H); DM 410 (2/vols., 340 + 178 fols., *ca.* 1150H); *etc.* (9/copies); plus selected chapters in ṬM 148 (*bis*),1 (fols. 1v–7v, *ca.* 1150H). *Cf. GAS*, VII, pp. 138–139, no. 1.

B22 ABŪ MAᶜSHAR

Article "Abū Maᶜshar" in *DSB* by D. Pingree; *GAS*, V, pp. 274–275; VI, pp. 156–157; VII, pp. 139–151 and 328–329.

3.4.1 A note on the determination of crescent visibility using an astrolabe, attributed to Abū Maᶜshar: DM 926,2 (photos of MS London B.L. Add. 9599, fol. 13r, late), unique.

5.2.1 A treatise on the astrology of the time of conception (*masqaṭ al-nuṭfa*) and the period of gestation (*makth al-mawlūd fī baṭn ummihi*) quoting Yaᶜqūb b. Ṭāriq (no. B4) and including remarks by one Ramaḍān: DM 623,1 (fols. 1v–9v, *ca.* 1100H). Even if the Ramaḍān mentioned in the text is the late Egyptian astronomer al-Khawānikī (no. D78), the treatise is of considerable interest.

5.2.10 Parts of a treatise which may be from the *K. al-Ṭabāʾiᶜ al-kabīr* (*Pingree*, no. 13): ṬM 157,1 (fols. 1v–25r, *ca.* 1150H).

5.2.10 Part of the 14th bāb of *al-Madkhal al-ṣaghīr* (*Pingree*, no. 2): K 8527,2 (fols. 88r–89v, *ca.* 1100H). According to Pingree this work was arranged in 7 *faṣl*s.

5.2.10 A fragment of the *K. al-Haylaj wa-l-kadkhudāh*: ṬM 243,2 (fols. 23v–24r, 644H), not listed in *GAS*, VII, pp. 148–149, no. 22.

5.2.10 A fragment of the *K. al-Nukat* (this title is not listed by Pingree, but see *GAS*, VII, pp. 142–143, no. 7) dealing with price fluctuation (*al-ghilāʾ wa-l-rakhṣ*): ṬM 157,2 (fols. 25r–27v, *ca.* 1150H). The fragment concludes with a quotation "from the book of Jaᶜfar al-Hindī" (no. A34).

5.2.10 The beginning of a work entitled *K. Asrār al-nujūm*: MM 248 (fols. 1v–8v, *ca.* 750H in Maghribī script). Another copy of the same fragment apparently copied from this very manuscript: ṬFF 4,2 (fols. 105r–108r, *ca.* 1200H). Another copy of the beginning of the same work: MM 204,4 (fols. 50v–57v, 1052H). *Cf. GAS*, VII, p. 146, no. 13, where only this last copy is listed.

5.2.10 ***K. fī Aḥkām al-nujūm wa-l-ittiṣālāt***, in 8 *bāb*s subdivided in 12 + 7 + 7 + 5 + 2 + 8 + 12 + 2 *faṣl*s: ṬM 161,1 (fols. 1r–16v, 1231H). There are two *bāb*s labelled as the sixth. The first, in 8 *faṣl*s, is attributed to Ghulām Zuḥal (see no. B51); the second is not subdivided.

5.2.10 A ***K. fī Aḥkām al-nujūm*** in 10 *bāb*s subdivided into *faṣl*s: ṬM 240,2 (fols. 24v–39v, 1303H).

5.2.10 Treatises on the astrological significance of conjunctions, without subdivisions: DM 884, (23 fols., *ca.* 1100H), ṬM 161,2 (fols. 17v–32r, 1231H), TR 141,1 (pp. 2–22, *ca.* 900H).

5.2.10 A late copy of a treatise on the same subject arranged in 6 + 5 + 4 + 7 + 2 + 1 + 6 + 4 *bāb*s: ṬM 111 (89 fols., 1342H).

5.2.10 A Persian version entitled *R. fi l-Qirānāt*: ṬJF 21,12 (fols. 188v–195r, 1206H); ṬJF 12,1 (fols. 1v–22r, *ca.* 1250H); S 4467,4 (fols. 35r–58r, *ca.* 1200H). A fragment on the hours of night and day: DJF 22,2 (fols. 35r–40v, *ca.* 1200H).

5.2.10 A short work on *majālis*: ḤM 11,1 (fols. 1v–4r, 1134H).

5.2.10 A treatise entitled ***K. al-Sirr***: MM 204,5 (fols. 58r–74v, 1052H). A *faṣl* on prices stated to be taken from ***K. al-Sirr*** in DM 63,2 (fols. 12v–18r, *ca.* 1250H) and ṬJ 370,3 (fols. 55r–62v, 1243H) does not occur in MM 204,5. *Cf. GAS*, VII, p. 148, no. 19.

5.2.10 A fragment stated to be taken from ***K. al-Asrār***: K 3774,3 (fol. 21r, 1269H).

5.2.10 A treatise on the indications of the lunar nodes: DM 59,4 (fols. 6v–10r, *ca.* 1000H).

5.2.10 A treatise on the zodiacal signs: DḤ 66 (58 fols., 1192H). The author mentions Yuḥannā al-Hindī (no. A33).

5.2.10 Treatises on nativities of men and women: DḤ 65,1 (fols. 1v–43r, 1191H); DḤ 65,2 (fols. 43v–86r, 1191H); DḤ 66,2 (fols. 30v–58v, 1192H). *Cf. GAS*, VII, p. 145, no. 8.

5.2.10 ***Maqāla fi l-Aᶜdād al-mutaḥābba***, a treatise on the properties of the amicable numbers 220 and 284; DṬ 124,5 (pp. 151–156, 1257H), unique. The Indian king Knkh (no. A32) is mentioned in the introduction. *Cf. GAS*, V, p. 275.

5.2.10 ***K. al-Qawāriᶜ***, a book of prognostications in tabular form: Sh 25 (24 fols., *ca.* 1200H). See Pl. LXXVII. *Cf. GAS*, VII, p. 149, no. 25.

5.2.10 ***Bughyat al-ṭālib fī maᶜrifat al-ḍamīr*** . . . : TGh 127,2 (pp. 20–365, 1281H).

Notes: The only copy of the *K. Aṣl al-uṣūl* . . . (*Pingree*, no. 26) attributed to Abū Maᶜshar is K 3790,2 (fols. 100v–225v, *ca.* 1200H). The remainder are attributed to al-Ṣaymarī (no. B23).

Two copies of ***K. al-Madkhal fī ᶜilm al-nujūm*** by Abu l-Qāsim al-Balkhī (no. B55) are falsely attributed to Abū Maᶜshar, namely DM 1204 (30 fols., 1121H) and ṬJ 425,1 (fols. 1r-22v, *ca.* 1250H).

5.2.22 Some quotations from the *K. al-Amṭār* (*Pingree*, no. 34) are found together with some notes attributed to ᶜAlī b. Riḍwān (no. B85) in DM 632,4 (fols. 8v–12r, *ca.* 1050H); MJ 213,1 (fols. 1r–3v, *ca.* 1150H); *etc.* (2 copies). *Cf. GAS*, VII, p. 328, no. 4.

5.3.35 See also ᶜAbd al-Raḥīm Manṣūr (no. D126).

B23 ABU L-ᶜANBAS AL-ṢAYMARĪ[1]

GAS, V, p. 262; VII, pp. 152–153.

5.2.11 *K. Aṣl al-uṣūl*: DM 1008 (photos of a manuscript of unknown provenance, 117 pls., 1105H); DM 11 (108 fols., *ca.* 1200H); ṬM 110 (167 fols., *ca.* 1150H in two hands); *etc.* (3 copies, one attributed to Abū Maᶜshar (no. B22). *Cf. GAS*, VII, pp. 152–153, no. 1. In DM 11, fol. 1r it is stated that al-Ṣaymarī was born in Ramaḍān, 213H: *cf. GAS*, VII, p. 152, note 1.

B24 ABŪ SAᶜĪD AL-ḌARĪR

GAS, V, pp. 263–264.

6.2.5 ***Masāᵓil handasīya***: MR 41,9 (fols. 69v–71r, 1153H), unique. *Cf. GAS*, V, p. 263, no. 1.

6.2.5 ***Istikhrāj khaṭṭ niṣf al-nahār min K. Anālīmā wa-l-burhān ᶜalayhi***: MR 41,25 (fols. 153v–154v, 1153H), unique. *Cf. GAS*, V, p. 264, no. 2.

Both of these treatises merit publication.

1. In DM 1008, DM 11, and ṬM 236 he is called Abu l-ᶜAnbas al-Ṣaymarī (or incorrectly al-Ṣamyarī). In ṬM 110 and ṬM 217 he is called Abu l-ᶜAnbas Aḥmad b. Muḥammad al-Ṣaymarī.

B25 ABŪ MUḤAMMAD ᶜABD ALLĀH B. MUSLIM, IBN QUTAYBA AL-DĪNAWARĪ

GAS, VII, pp. 305–321, and the article "Ibn Qutayba" in *DSB* by P. Kunitzsch.

1.4.2 *K. al-Anwāᵓ*: ZK 239 (171 pp., 1338H), and photos thereof in DM 1080. This manuscript was used by C. Pellat in his critical edition of the *K. al-Anwāᵓ* (published in Hyderabad, 1956). On the various manuscripts and the problems of the colophon of this Cairo manuscript, see *ibid.*, pp. xxxi–xxxix. *Cf. GAS*, VII, p. 351, no. 1.

See also no. B26.

B26 ᶜABD AL-RAḤMĀN B. ISḤĀQ AL-ZAJJĀJĪ

GAL, I, p. 112; *GAS*, VII, p. 354.

14

1.4.3 Two fragments on *anwāᵓ* stated to be taken from ***K. Adab al-kātib*** by Abū (*sic*) Isḥāq al-Zajjājī: MM 122 (8 fols., 850H); MM 198,2 (fols. 19v–27r, *ca.* 1150H). *Cf. GAS*, VII, p. 354, where only the second copy is listed. al-Zajjājī is known to have written a commentary on the ***K. Adab al-kātib*** of Ibn Qutayba (no. B25) (see *GAL*, p. 126, no. 7).

B27 IBN AL-QĀṢṢ

GAS, I, pp. 496–497.

1.4.4 ***K. Dalāᵓil al-qibla***, a treatise on the determination of the qibla by non-technical means: DM 1201 (27 fols., *ca.* 1100H, lacks end). Sezgin (*GAS*, I. p. 497, no. 2) lists two copies in Beirut and Istanbul. The present whereabouts of the Beirut manuscript is not known to me (or to the Librarian of the Université de St. Joseph, Beirut) but from its description in the catalogue its contents resembled those of the Istanbul copy, in that it dealt mainly with geography. The Cairo copy, on the other hand, deals mainly with stars and lunar mansions used for finding the qibla. These manuscripts deserve detailed investigation.

B28 DĀᵓŪD B. SULAYMĀN

Not mentioned in the modern sources.

4.2.1 *K. Dhāt al-ḥalaq* and *(K.) Ṣifat al-ᶜamal bi-l-ḥalaq*, two treatises on the construction and use of armillary sphere in 4 + 6 unnumbered *bābs*: DM 969,1 (fols. 1v–3v, *ca.* 1200H), unique. Merits study.

B29 ABU L-QĀSIM AL-MUNAJJIM

Not mentioned in the modern sources.

4.2.2 *(K. fi) l-ᶜAmal bi-l-bayḍa wa-l-kura*, a treatise on the celestial globe, commpiled in 287H in Jurjān: DM 969,2 (fols. 3v–5r, *ca.* 1200H), unique. Merits study.

B30 THĀBIT B. QURRA

Article "Thābit ibn Qurra" in *DSB* by B. A. Rosenfeld and A. T. Grigorian; *GAS*, V, pp. 264–272; VI, pp. 163–170; and VII, pp. 151–152.

1.1.1 On the recension of his *iṣlāḥ* of Autolycus' *K. al-Ṭulūᶜ wa-l-ghurūb* by al-Ṭūsī, see no. A4.

1.1.6 On his improvement of the translation of Ptolemy's *Almagest* by Ḥunayn b. Isḥāq, see no. A18.

1.1.6 *Qawl fī īḍāḥ al-wajh alladhī dhakara Baṭlaymūs annā bihi stakhraja man taqaddamahu masīrat al-qamar al-dawrīya wa-hiya l-mustawiya*: DM 1047,2 (photos. of MS Istanbul Köprülü 948, pp. 91–107, 370H!). *Cf. GAS*, VI, pp. 167–168, no. 7.

1.2.2 *R. fī Dhikr al-aflāk wa-ḥalaqihā wa-ᶜadad ḥarakātihā wa-miqdār masīrihā*, on theoretical astronomy: ṬJ 337,2 (fols. 67r–68r, *ca.* 1300H, purportedly 1160H). Sezgin (*GAS*, VI, p. 166, no. 3 and p. 293) lists other copies in Istanbul and Leningrad.

4.3.6 For his attribution of the astrolabe to Hipparchus (no. A9), see no. B79 and Pl. LVII.

4.7.1 *K. fi Ālāt al-sāᶜāt allatī tusammā rukhāmāt*, a treatise on sundials: DM 1047,1 (photos of MS Istanbul Köprülü 948,1, pp. 1–89, 370H!, unique). Other photos of this precious manuscript in DM 1048, K 8532, K 8533, and TR 356. *Cf. GAS*, V, p. 270, no. 14, and VI, p. 168, no. 9.

5.1.3 On his recension of Ḥunayn b. Isḥāq's version of Ptolemy's *Tetrabiblos*, see no. A18.

6.1.2 On his improved translation of Euclid's *Data* later edited by al-Ṭūsī, see no. A5.

6.1.3 *M. fī Burhān al-muṣādara al-mashhūra min K. Uqlīdis*: MR 40,17 (fols. 200v–202r, 1159H), unique (MS Damascus Ẓāhirīya 5648 is based on this copy). *Cf. GAS*, V, p. 271, no. 21, and p. 402.

6.1.3 *(R. fi) l-ᶜIlla allatī lahā rattaba Uqlīdis ashkāl kitābihi dhālika l-tartīb*: DR 898,22 (photos of MS Leiden 14, pp. 380–389, *ca.* 1050H). *Cf. GAS*, V, p. 271, no. 22.

6.1.8 On his translation of the *K. ᶜAmal al-dāʾira* . . . attributed to Archimedes, see no. A7.

6.1.9 On al-Ṭūsī's recension of his translation of the *K. al-Mafrūḍāt* attributed to Archimedes, see no. A7.

6.1.11 On al-Ṭūsī's recension of his improved translation of the *K. al-Ukar* of Theodosius, see no. A12.

6.2.6 *K. fi l-shakl al-mulaqqab bi-l-qaṭṭāᶜ*: MR 40,16 (fols. 191v–199r, 1159H). *Cf. GAS*, V, p. 268, no. 1.

6.2.6 *R. fī kayfa yanbaghī an yuslaka ilā nayl al-maṭlūb min al-maᶜānī al-handasīya*: MR 40,11 (fols. 155v–159r, 1159H from a copy by Ibn Sīnā (no. B81)). *Cf. GAS*, V, p. 268, no. 4.

6.2.6 *R. fi l-Ḥujja al-manṣūba ilā Suqrāṭ fi l-murabbaᶜ wa-quṭrihi*: MR 40,12 (fols. 161v–164v, 1159H). *Cf. GAS*, V, p. 269, no. 7.

6.2.6 *Maqāla fī ᶜAmal shakl mujassam dhī arbaᶜ ᶜashara qāᶜida*: DM 1047,3 *etc.* (photos. of MS Istanbul Köprülü 948, pp. 108–115, 370H!, unique). *Cf. GAS*, V, p. 269, no. 8.

6.2.6 A recension of his *K. Quṭūᶜ al-usṭuwāna wa-basīṭuhā* by Jamāl al-Dīn Muḥammad b. ᶜUmar . . . Ibn Abī Jarāda (no. G29), compiled in 691H: MR 41,6 (fols. 36v–64v, 1153H). I have not compared this with the other copy listed by Sezgin, MS Istanbul Aya Sofia 4832,2 (fols. 4r–26r, 5th century H). *Cf. GAS*, V, p. 269, no. 9. A brief description of the contents of the Cairo manuscript was published in L. Karpova and B. A. Rosenfeld, "A treatise of Thābit ibn Qurra on sections of a cylinder, and on its surface," *AIHS*, 24 (1974), pp. 66–77.

6.2.6 *K. fī Misāḥat qaṭ*ᶜ *al-makhrūṭ alladhī yusamma l-mukāfī*: MR 40,13 (fols. 165v–181r, 1159H). *Cf. GAS*, V, p. 269, no. 10.

B31 MUḤAMMAD B. ᶜABD ALLĀH B. ᶜUMAR B. AL-BĀZYĀR

GAS, VII, p. 154.

5.2.7 *K. fī Jumal min dalālāt al-ashkhāṣ al-ᶜālīya*, in 8 *maqālas* of 4 + 8 + 6 + 12 + 7 + 12 + 12 + 2 *faṣls*: K 3790,1 (fols. 1v–225v, *ca.* 1200H), not listed in *GAS*, VII, p. 146, no. 15.

B32 AḤMAD B. ᶜUMAR AL-KARĀBĪSĪ

GAS, V, p. 277.

6.2.7 *K. Misāḥat al-ḥalaq*: MR 41,26 (fols. 155v–158r, 1148H). *Cf. GAS*, V, p. 277, no. 1.

B33 ABŪ DAᵓŪD SULAYMĀN B. ᶜIṢMA

GAS, V, p. 106, no. 18, and pp. 337–338.

6.1.3 *K. fī Dhawāt al-ismayn wa-l-munfaṣilāt alladhī min al-maqāla al-ᶜāshira min K. Uqlīdis*: DR 898,21 (photos. of MS Leiden 14,20, pp. 358–380, *ca.* 1050H). *Cf. GAS*, V, p. 338, no. 1.

B34 ISḤĀQ B. ḤUNAYN

GAS, V, pp. 272–273; VI, p. 171.

6.1.2 For his translation of Euclid's *Data* later edited by al-Ṭūsī, see no. A5.

B35 ḤABASH

GAS, V, pp. 275–276; and VI, pp. 173–175.

See Ibn Amājūr (no. B36).

B36 ABU L-QĀSIM B. ĀMĀJŪR

GAS, V, p. 282; and VI, p. 177–178.

On the title folio of TR 99 of the *ᶜIlal al-zījāt* of Ibn Masrūr (see no. B37), the title indicates that the manuscript once contained *rasāᵓil zīj Ḥabash taṣḥīḥ Abu (sic) l-Qāsim b. Mājūr* . . . , that is, Ibn Āmājūr's corrections to the text of one of the *zīj*es of Ḥabash (no. B35). See Pl. XI.

B37 ᶜABD ALLĀH B. MASRŪR AL-ḤĀSIB

GAS, VI, pp. 205–206.

2.1.3 *K. ᶜIlal al-zījāt*, a commentary on the *Zīj* of al-Khwārizmī (no. B14): TR 99 (114 pp., *ca.* 600H), unique. *Cf. GAS*, VI, pp. 205–206, no. 1. Merits detailed analysis: a preliminary translation has been made by E. S. Kennedy. See also nos. B35 and B36. For extracts, see Pl. XI.

B38 QUSṬĀ B. LŪQĀ

GAS, V, pp. 285–286; and VI, pp. 180–182.

1.1.4 On his translation of ***K. al-Maṭāli***ᶜ of Hypsicles, see no. A10.

1.1.5 On his translation of Theodosius' ***K. al-Masākin***, see no. A12.

4.2.3 *K. al-Kura al-falakīya* = *K. al-ᶜAmal bi-l-kura al-falakīya* in three versions, one in 67 *bābs* with a confused dedication to Abu l-Ṣaqr Ismāᶜīl b. Bulbul or (*wa-qīla*) Abu l-Ḥasan ᶜAbd Allāh b. Yaḥyā: DM 528 + 529 (8 + 12 fols., 941H, lacks end of introduction and 1st ch.); DM 1223 (37 fols., *ca.* 1250H); ḤM 7 (11 fols., *ca.* 1250H, lacks introduction); and the second in 65 *bābs* with a dedication to Abu l-Ḥasan ᶜAbd Allāh b. Yaḥyā: DM 702,3 (fols. 9r–26r, 1160H, Maghribi script); TR 131,7 (pp. 130–164, 1186H); and the third with an abridged introduction: MJ 180,21 (fols. 176r–195r, 1112H); K 3824,13 (fols. 107v–122r, 1150H). *Cf. GAS*, VI, p. 181, no. 1.

6.1.11 On al-Ṭūsī's recension of a translation by him of part of the ***K. al-Ukar*** of Theodosius, see no. A12.

6.2.8 Remarks on his ***Burhān ᶜalā ᶜamal ḥisāb al-khaṭaᵓayn*** by Samawᵓal b. Yahūdā (no. G9): DR 702,3 (fols. 38v–39v, *ca.* 1100H), not listed in *GAS*, V, p. 286, no. 1.

7.1.1 On his translation of ***K. Īrūn (Heron) fī rafᶜ al-ashyāᵓ al-thaqīla***, see no. A11.

B39 ABU L-ᶜABBĀS AL-FAḌL B. ḤĀTIM AL-NAYRĪZĪ

GAS, V, pp. 283–285; VI, pp. 191–192; and VII, p. 156.

5.2.13 (*R. fi Uṣūl aḥkām al-qirānāt*), a treatise on astrology prepared for the Caliph al-Muktafī bi-llāhi: Sh 69,3 (fols. 28r–45v, *ca.* 1300H). In this treatise the author is called . . . al-Brwry (= al-Nayrīzī!) al-Daḥdānī; in a commentary on an anonymous treatise which follows in Sh 69,1 (fols. 1r–25v, *ca.* 1300H) the author is named simply as al-Dajdānī. Both parts of this manuscript merit investigation. *Cf. GAS*, VII, p. 156.

B40 AḤMAD B. YŪSUF KĀTIB ĀL ṬŪLŪN, IBN AL-DĀYA

GAS, V, pp. 288–290; VI, p. 193; and VII, p. 157.

5.1.3 ***Sharḥ K. al-Thamra li-Baṭlamiyūs***, a commentary on the ***Centiloquium*** attributed to Ptolemy (see no. A18): DM 962 (48 fols., *ca.* 1200H, anonymous); TR 141,4 (pp. 129–141, *ca.* 1100H, falsely attributed to Aḥmad b. Yūsuf al-Kammād); plus a fragment in TKh 290,13 (pp. 302–303, *ca.* 600H) amidst philosophical treatises. *Cf. GAS*, VII, p. 157.

6.2.9 ***Risāla fi l-Nisba wa-l-tanāsub***: MR 39,1 (fols. 1r–25r, 1151H) [*n.b.* this manuscript was originally numbered 6]; ṬJ 635,3 (fols. 11r–20v, 1170H, incomplete), not listed by Sezgin. *Cf. GAS*, V, p. 289, no. 1.

B41 AL-ḤASAN B. AḤMAD B. YAᶜQŪB AL-HAMDĀNĪ

GAS, VII, pp. 164–165, and 272–273; *MAY*, II, no. 1.

5.2.14 The tenth *maqāla* of al-Hamdānī's ***Sarāʾir al-ḥikma fī ᶜilm al-nujūm*** is extant in K 7012 (33 plates, photos of a manuscript in a private library in the Yemen, copied 1091H). This fragment, in 33 *faṣl*s, deals with mathematical astrology and contains several references to early Abbasid astronomers. Merits detailed investigation. *Cf. GAS*, VII, pp. 164–165, no. 1. The text of this manuscript is now published in Muḥammad b. ᶜAli b. al-Ḥusayn al-Akwaᶜ al-Ḥiwāli, ed., *al-Maqāla al-ᶜāshira min Sarāʾir al-ḥikma li-l- . . . Hamdānī*, Sanaa: al-Maktaba al-Yamanīya, n.d.

B42 IBRĀHĪM B. SINĀN

GAS, V, pp. 292–295; and VI, pp. 193–195. Seven treatises by him are now published in A. S. Saidan, *Rasāʾil Ibn Sinān*, (Kuwait, 1983).

6.2.10 ***K. fī Misāḥat al-qaṭᶜ al-mukāfi***: MR 40,14 (fols. 182v–186v, 1159H), *cf. GAS*, V, p. 293, no. 1; and ***K. Misāḥat al-qaṭᶜ al-mukāfi***: MR 41,28 (fols. 162v–164v, 1153H), not mentioned by Sezgin. These treatises are slightly different from each other and also from the text printed in Hyderabad, 1947.

6.2.10 ***K. fī Ṭarīq al-taḥlīl wa-l-tarkīb wa-l-aᶜmāl al-handasīya***: MR 40,10 (fols. 130v–153v, 1159H); TR 323,2 (pp. 82–132, *ca.* 1250H, incomplete). *Cf. GAS*, V, p. 294, no. 2.

6.2.10 ***Maqāla fī Rasm al-quṭūᶜ al-thalātha***: DR 690,2 (photos. of MS London B.L. Add. 14332,3, pp. 62–72, 1832 A.D., lacks diagrams). *Cf. GAS*, p. 294, no. 3.

B43 ABŪ JAᶜFAR AL-KHĀZIN

GAS, V, p. 106, no. 16, and pp. 298–299; and VI, pp. 189–190.

6.1.3 ***Tafsīr ṣadr al-maqāla al-ᶜāshira min K. Uqlīdis***: DR 898,19 (photos of MS Leiden 14, pp. 327–340, *ca.* 1050H). *Cf. GAS*, V, p. 299, no. 1 and VII, p. 406.

6.2.25 ***Zīj al-ṣafāʾiḥ***: see no. B109. See also *Centaurus*, 23 (1980), pp. 105–117.

B44 SINĀN B. AL-FATḤ

GAS, V, p. 301; VI, p. 207; and VII, p. 406.

6.2.11 ***K. fīhi l-Kaᶜb wa-l-māl wa-l-aᶜdād al-mutanāsiba***, a commentary on the *K. al-Jabr wa-l-muqābala* by al-Khwārizmī (no. B14): DR 260,4 (fols. 95r–105v, *ca.* 600H), unique. *Cf. GAS*, VII, p. 406.

7.2.2 ***Qawl fi l-misāḥāt al-manāẓirīya***, a short treatise on optics: DR 260,3 (fols. 92v–94r, *ca.* 600H), unique. *Cf. GAS*, VII, p. 406.

B45 ᶜABD AL-ᶜAZĪZ B. ᶜUTHMĀN AL-QABĪṢĪ[1]

GAS, V, pp. 311–312; VI, pp. 208–210; and VII, pp. 170–171.

1.2.3 *Masāʾil yamtaḥanu biha l-munajjimūn*, 28 questions and answers on astronomical topics: DM 447,1 (fols. 1v–11r, *ca.* 900H). This copy differs slightly from the Damascus manuscript listed in *GAS*, VI, p. 210, no. 4.

5.2.15 *K. al-Madkhal ilā/fī ṣināᶜat aḥkām al-nujūm*, a treatise on astrology in 5 *faṣl*s: DM 144 (35 fols., *ca.* 950H, lacks part of 5th *faṣl*); Sh 89,2 (fols. 10v–29r, 1025H, anonymous); DM 139 (35 fols. *ca.* 1150H); *etc.* (4 copies). *Cf. GAS*, VII, pp. 170–171, no. 1.

B46 ABU L-ḤUSAYN AḤMAD B. AL-ḤUSAYN AL-AHWĀZĪ

GAS, V, p. 106, no. 19 and pp. 312–313.

6.1.3 *Sharḥ al-maqāla al-ᶜāshira min K. Uqlīdis*: K 4528,1 (fols. 1r–13r, *ca.* 1150H); DR 300,2 (fols. 4r–17r, 1150H), not listed by Sezgin; and a shorter version thereof, also by al-Ahwāzī: DR 898,20 (photos of MS Leiden 14, pp. 341–349, *ca.* 1050H). *Cf. GAS*, V, pp. 312–313, no. 1; and VII, p. 407.

B47 YAᶜQŪB B. MUḤAMMAD AL-SIJISTĀNĪ

GAS, V, p. 313.

6.7.1 *K. Maᶜrifat al-misāḥa*, a treatise on surveying in 14 unnumbered *bāb*s: TR 278,1 (pp. 1–32, 1057H), unique. The title folio on this manuscript indicates that al-Sijistānī was a scholar of the 4th Hijra century. He is otherwise unknown to the literature.[2]

1. DM 447,1 has ᶜAbd al-ᶜAzīz b. ᶜUthmān al-Qabīṣī; DM 144 has ᶜAbd al-ᶜAzīz b. ᶜAlī b. ᶜUthṃān al-Qibṭī; Sh 89,2 is anonymous; DM 139 has ᶜAbd al-Raḥmān b. ᶜUthmān al-Qabīṣī; ṬJ 370,1 has Abu l-Ṣaqr ᶜAbd al-ᶜAzīz b. ᶜUthmān al-Qabīṣī.

2. I have wondered whether the copyist might have confused the name of the author with the fourth-century Ismāᶜīlī author Abū Yaᶜqūb al-Sijistānī.

B48 ABŪ ISḤĀQ AL-ṢĀBIʾ

GAS, V, p. 314 and also p. 320, no. 23.

6.2.12 Correspondence between Abū Isḥāq and al-Qūhī (no. B53) (four treatises): MR 40,20 (fols. 209v–221v, 1159H). *Cf. GAS*, V, p. 314, no. 1. Investigated by J. L. Berggren in *JHAS* 7 (1986).

B49 AL-ḤASAN B. JAᶜFAR AL-ANṢĀRĪ

GAS, VII, p. 169.

5.2.17 *al-Dawr al-ᶜuṭāridī*, on the astrological significance of Mercury, compiled in 377H, no extant copies? An abridgement by Muḥammad b. ᶜUmar al-Fāriqī (no. C7) compiled in Aleppo in 641H survives in ṬM 150 (58 fols., 1181H); DM 54 (45 fols., *ca.* 1150H); *etc.* (1 copy). Merits detailed investigation. *Cf. GAS*, VII, p. 169 for other copies.

B50 ABŪ MAḤMŪD ḤĀMID B. AL-KHIḌR AL-KHUJANDĪ

GAS, V, pp. 307–308; and VI, pp. 220–222.

4.6.15 (*K. fi. l-Āla al-shāmila*), a treatise in 5 *maqāla*s on a "comprehensive" instrument invented by al-Khujandī: DM 970 (42 fols., *ca.* 1150H, anonymous, lacks part of last *maqāla*). See Pl. LVIa. This and other copies are listed in *GAS*, VI, p. 221, no. 1. Merits publication and analysis. Note that DM 969, a collection of early treatises on instruments (see nos. B28, B29, B102 and B105), and DM 970 are in the same hand.

6.2.27 An extract from a book by al-Khujandī is in the anonymous collection of geometrical problems in MR 41,29 (fols. 165v–170v, 1153H). *Cf. GAS*, V, p. 308, and read MR 41 for Sezgin's "40 m."
See also al-Juyūbī (no. B94).

B51 GHULĀM ZUḤAL

GAS, VII, p. 168.

5.2.10 See the *K. fī Aḥkām al-nujūm* of Abū Maᶜshar (no. B22).

B52 ABU L-ḤUSAYN ᶜABD AL-RAḤMĀN B. ᶜUMAR AL-ṢŪFĪ

GAS, V, pp. 309–310; VI, pp. 212–215; and VII, pp. 168–169; and *Storey*, pp. 41–42.

1.3.1 *K. Ṣuwar al-kawākib*, on the constellation figures: DM 904 (photos of MS Paris B.N. ar. 2488, 191 pls., *ca.* 650H, Maghribi script); DM 907 (photos of MS Paris B.N. ar. 2490, 155 pls., 922H); DM 390 (180 fols., *ca.* 700H, lacks end); DM 831,2 (fols. 16v–135v, *ca.* 1200H) and photos thereof in ṬM 258 and DM 1094,2; *etc.* (1 late copy, quite wretched, 1 set of miserable illustrations of the constellation figures, and 1 fragment containing some star coordinates). See Pl. II. *Cf. GAS*, VI, p. 214, no. 1 for other copies. See also no. Z28.

An anonymous Persian edition (no. G108): MMF 9 (136 fols., 1043H), not listed in Storey, pp. 41–42. This magnificent copy is vastly superior to the Istanbul copy recently published by photo-offset in Iran (Tehran, n.d.). See Pl. III.

See also nos. B91 and H71 for other manuscripts containing illustrations of constellation figures.

Other such works that have come to my attention are an Arabic treatise entitled *al-Taqrīr* (?), written *ca.* 985/1577 in Iran by Abū Muḥammad Muḥsin b. ᶜAlī b. Manṣūr al-Husaynī al-Dashtakī and extant in MS Dublin Chester Beatty 4220 (127 fols., *ca.* 1000H), and an astrological treatise with illustrations entitled *K. al-Burhān*, on which see *GAS*, VII, pp. 24–25.

4.3.2 *R. fi l-ᶜAmal bi-l-asṭurlāb*, on the use of the astrolabe in 386 *bābs*: DM 908 (photos of MS Paris B.N. 2493, 183 pls., 1283H); DM 913 (photos of MS Paris B.N. ar. 2498, 178 pls., incomplete, late). *Cf. GAS*, VI, p. 215, no. 2.

4.6.1 The title page of DM 647 of al-Zarqāllu's treatise on the *ṣafīḥa* (see no. B87) indicates that the manuscript once contained al-Ṣūfī's *R. fī Maᶜrifat mā maḍā li-layl min sāᶜa bi-qiyās al-kawākib al-thābita wa-l-ṭāliᶜ*, on timekeeping by night.

B53 ABU SAHL WAYJAN B. RUSTAM AL-QŪHĪ / AL-KŪHĪ

GAS, V, pp. 314–321, and VI, pp. 218–219.

4.3.3 *K. Ṣināᶜat al-asṭurlāb bi-l-burhān*, on the construction of the astrolabe: DR 898,12 (photos. of MS Leiden 14, pp. 254–282, *ca.* 1050H). Commentary by Abū Saᶜd al-ᶜAlāʾ b. Sahl (no. B73): DR 898,13 (photos. of MS Leiden 14, pp. 282–293, *ca.* 1050H). *Cf. GAS*, VI, p. 219, no. 1.

4.8.2 *K. fi l-Āla allatī summiyat al-birkār al-tāmm*, a treatise on the "complete compass" in 2 *maqālas*: MR 41,13 (fols. 95v–104v, mainly 1146H); ṬJ 239,2 (fols. 33v–67v, *ca.* 1000H). *Cf. GAS*, V, pp. 317–318, no. 1.

6.1.3 *al-Maqālatān al-ūlā wa-l-thāniya min K. Uqlīdis fi l-Uṣūl*: MR 41,12 (fols. 83v–94v, 1153H), unique. *Cf. GAS*, V, p. 319, no. 14.

6.1.6 *R. li-Taṣḥīḥ al-maqāla al-thāniya min K. al-Kura wa-l-usṭuwāna li-Arshimīdis*: DR 898,26 (photos of MS Leiden 14, pp. 487–497, *ca.* 1050H). *Cf. GAS*, V, p. 320, no. 25.

6.2.12 Correspondence with Abū Isḥāq al-Ṣābiʾ: see no. B48.

6.2.13 *R. fi stikhrāj ḍilᶜ al-musabbaᶜ*: MR 40,21 (fols. 222v–225r, 1159H). *Cf. GAS*, V, p. 318, no. 2.

6.2.13 *R. fi ᶜAmal mukhammmas mutasāwi l-aḍlāᶜ fī murabbaᶜ maᶜlūm*: MR 40,18 (fols. 203v–205v, 1159H). *Cf. GAS*, V, p. 318, no. 3. Edited by J. P. Hogendijk in *ZGAIW* 1 (1985), pp. 100–144.

6.2.13 *M. fi stikhrāj al-khaṭṭayn bayn al-khaṭṭayn*: MR 40,22 (fols. 226v–227r, 1159H). *Cf. GAS*, V, p. 318, no. 4.

6.2.13 *R. fi stikhrāj misāḥat al-mujassam al-mukāfi*: MR 40,15 (fols. 187v–190v, 1159H) and *K. Misāḥat al-mujassam al-mukāfi*: MR 41,20 (fols. 135v–137r, 1153H). *Cf. GAS*, V, p. 318, no. 5. These two treatises are not identical.

6.2.13 *Masʾalatān handasīyatān = Masāʾil handasīya*: MR 40,19 (fols. 206v–208r, 1159H). *Cf. GAS*, V, p. 319, nos. 11 and 12.

B54 ABU L-WAFĀʾ MUḤAMMAD B. MUḤAMMAD AL-BŪZAJĀNĪ

GAS, V, pp. 321–325; VI, pp. 222–224.

2.1.4 *al-Majisṭī*, a *zīj* (?) with an introduction in 7 *maqālas* of 6 + 10 + 11 + 16 + 14 + 6 + 6 *nawᶜ*s subdivided into *faṣl*s: DH 73 (photos of MS Paris B.N. 2494, 108 pl., *ca.* 800H, unique). This manuscript lacks all the tables which originally accompanied the explanatory text. *Cf. GAS*, VI, p. 223, no. 1. On the anonymous *Shāmil Zīj*, which is based on al-Buzajānī's *Zīj*, see no. B100.

6.2.14 *K. al-Manāzil al-sab*ᶜ: MR 42,1 (230 fols. except fol. 69, 487H, lacks beginning) [*n.b.* this manuscript was previously numbered 9]. *Cf. GAS*, V, p. 323, no. 1. See Pl. LXXXVIIa.

6.2.14 *K. al-Najāra fī ᶜamal al-masṭara wa-l-birkār wa-l-kūnīyā = K. fīmā yaḥtāj ilayhi l-ṣāni*ᶜ *min a*ᶜ*māl al-handasa*: DR 260,1 (fols. 1r–58v, *ca.* 600H!), and photos thereof in K 6965 and K 8890; DR 366 (46 fols., 831H, defective, anonymous). *Cf. GAS*, V, p. 324, no. 2 and VII, p. 408, where only the first copy is listed.

The attribution to al-Būzajānī of a treatise on the rainbow (see *GAS*, VI, p. 224, no. 4) appears to result from a confusion with al-Tūqātī (no. G54).

B55 NŪR AL-DĪN ABU L-QĀSIM ᶜALĪ B. AḤMAD AL-BALKHĪ

Krause, p. 514 (no. 5); *Storey*, no. 107; *GAS*, VII, pp. 176–177.

5.2.24 *K. al-Madkhal fī ᶜilm al-nujūm*, a treatise dealing mainly with astrology arranged in 60 *bāb*s: DM 143 (43 fols., *ca.* 1100H); DM 1204 (30 fols., 1121H); MM 207 (32 fols., 1180H); TR 184 (106 pp., *ca.* 1200H); and selected chapters in DM 876,2 (fols. 18v–46r, 1051H); and ṬJ 425,1 (fols. 1r–22v, 1200H). Of these various copies DM 1204 and ṬJ 425,1 are falsely attributed to Abū Maᶜshar al-Balkhī (no. B22). See *GAS*, VII, p. 177, no. 2, and p. 372 for other copies. On a Persian translation of this work, see *Krause*, p. 514 and *Storey*, no. 107. This work merits investigation.

B56 AḤMAD B. MUḤAMMAD B. ᶜABD AL-JALĪL AL-SIJZĪ

GAS, V, pp. 329–334; VI, pp. 224–226; and VII, pp. 177–182. See also B57.

2.1.5 A fragment of an anonymous *zīj* containing tables for oblique ascensions and some unusual astrological tables, as well as a table for determining the position of the moon at nativities specifically attributed to al-Sijzī: ṬFF 18,3 (fols. 230r–246r, *ca.* 800H), unique. Merits detailed study. al-Sijzī is not known to have compiled a *zīj* himself, and the tables are not related to those in his *al-Jāmi*ᶜ *al-Shāhī* (see below). See also *Storey*, p. 106, no. 19 and p. 107, no. 31, for two other manuscripts containing tables attributed to al-Sijzī.

5.2.19 A complete copy of his *al-Jāmi*ᶜ *al-Shāhī*, an astrological compendium made up of 13 separate treatises: DM 887 (149 fols., 1186H). *Cf. GAS*, VII, pp. 178–181, nos. 1–13, where this copy is not listed. A late copy of treatises 6 to 13 is Sh 79 (79 fols., 1335H). Also "a *faṣl* from *K. al-Ulūf*": K 3979,3 (fols. 12v–13v, *ca.* 1150H). al-Sijzī's astrological works merit detailed investigation. See already D. Pingree, *The Thousands of Abū Ma*ᶜ*shar*, (London, 1968), pp. 21–22, *etc.*

6.2.16 *K.* ᶜ*Amal al-musabba*ᶜ *fi l-dā*ᵓ*ira* . . . : MR 41,16 (fols. 113v–116r, 1153H). *Cf. GAS*, V, p. 331, no. 8. Edited in *AHES* 30 (1984), pp. 290–316.

6.2.16 *Muqaddima fi l-handasa*: TR 140, 1 + 2 (pp. 19–20 + 3–18 + 21–27, *ca.* 1150H, lacks end). *Cf. GAS*, V, p. 333, no. 20.

6.2.16 *Ta*ᶜ*līqāt handasīya*: DR 699 (35 fols., 1342H from a copy dated 699H). Sezgin (*GAS*, V, p. 333, no. 27) mentions only MS Chester Beatty 3045,14 of this treatise, copied in 699H, and perhaps the original of the Cairo copy.

6.2.16 A treatise on asymptotes: DR 898,6 (photos of MS Leiden 14, pp. 226–231, *ca.* 1050H), not listed in *GAS*, V. In this work al-Sijzī mentions a treatise of his entitled *Tashīl al-subul*, also not listed by Sezgin.

See also no. B57.

B57 AḤMAD B. IBRĀHĪM AL-SANJARĪ

Not mentioned by Suter or Brockelmann. See *GAS*, V, p. 333, no. 26.

4.8.3 *Muqaddima li-Ṣinā*ᶜ*at āla tu*ᶜ*raf biha l-ab*ᶜ*ād*, a treatise on the construction of an instrument for measuring celestial distances: DR 898,5 (photos of MS Leiden Or. 14, pp. 223–226, *ca.* 1050H not 589H as in Sezgin). Merits detailed investigation. Sezgin (*GAS*, V, p. 333, no. 26), falsely attributes this, and another copy in MS New York Columbia Or. 45,11, to al-Sijzī (no. B56).

B58 ABU L-FATḤ AL-SAMARQANDĪ

GAS, VI, pp. 216–217.

2.4.1 Cotangent tables to base 12, computed for each minute of argument to 3 digits: DM 136,4 (fols. 67r–82r, 1147H). *Cf. GAS*, VI, p. 217, no. 3. Another anonymous set of the same tables is in DM 715 (16 fols., *ca.* 1050H). The question of the authenticity of the attribution of these tables to an astronomer of the tenth century merits investigation.

B59 ABU L-ḤASAN ᶜALĪ B. ᶜABD AL-RAḤMĀN B. AḤMAD, IBN YŪNUS

GAS, V, pp. 342–343; VI, pp. 228–231; and VII, p. 173 (all somewhat confused). See also my article "Ibn Yūnus" in *DSB* for lists of manuscripts.

2.1.6 Inconsequential anonymous fragments from ***al-Zīj al-Ḥākimī al-kabīr***, originally in 81 chapters: DM 639,23 (fols. 196v–205r, part of Ch. 10, *ca.* 1150H) and DM 736,2 (fols. 2r–5v, Chs. 16 and 17, *ca.* 1150H), both in the same hand; also MM 44,3 (fols. 28r–28v, part of Ch. 28, *ca.* 850H by al-Dalāmī).

2.1.6 Planetary tables taken originally from the *Ḥākimī Zīj*: DM 593 (6 fols., *ca.* 1000H); DM 468,1 (fols. 1r–2v, *ca.* 1000H); MM 31,2 (fol. 28, *ca.* 1150H); MM 116,1 (fols. 1r–7r, *ca.* 1150H); K4032 (6 fols., *ca.* 1150H). See also no. D207. An extract is shown in Pl. XIVa.

2.1.6 Oblique ascension tables: K4003,1 (fols. 1v–27r, *ca.* 1100H).

2.1.6 Fragment on solar and lunar distances: DM 718,1 (fols. 1r–3r and 4v–6v, 917H). Merits investigation.

2.1.6 Text and tables for solar eclipse computation, together with a treatise on the same subject by Baylak al-Qipjāqī (no. C9): DM 639,14 (fols. 102r–109r, *ca.* 1150H), unique. Merits investigation.

2.1.6 Miscellaneous disordered chapters from the *Ḥākimī Zīj*: DM 909,4 (photos of MS Paris B.N. ar. 2496, fols. 10r–107v, *ca.* 1000H (?)).

2.1.6 Assorted tables from the *Zīj*es of Ibn Yūnus and Kūshyār (no. B70) collected by Aḥmad b. Timurbāy (no. C94): MM 188,2 (fols. 77v–83r, *ca.* 875H by Aḥmad b. Timurbāy). This collection includes some tables by Ibn Yūnus not attested in other sources, such as his star catalog—see Pls. XIVb and also XV.

2.1.6 A note on lunar crescent visibility attributed to Ibn Yūnus: MM 93,2 (fols. 2v–5r, *ca.* 1100H). See also no. D79.

2.1.6 *K. Ḥabṭaq ḥall al-shams wa-l-qamar*, extensive double-argument solar and planetary equation tables attributed to Ibn Yūnus: DM 29 (49 fols., *ca.* 850H). no solar tables, unique—see Pl. XVI. An anonymous fragment of the same table is in MS Gotha A1410. Analyzed in *Centaurus*, 18 (1974), pp. 129–146. On the meaning of *ḥabṭaq*, see no. C10.

2.1.6 The ***Khātima fī ᶜamal dustūr muqawwam kull kawkab li-sana kāmila***, on the compilation of ephemerides stated to be taken from the *Zīj* of Ibn Yūnus in MM 116,2 (fols. 8r–9r, *ca.* 1150H) is in fact taken from the small *zīj* entitled ***Tashīl al-ᶜibāra*** . . . which is falsely attributed to Ibn Yūnus on the title folio in MS Aleppo Awqāf 947—see no. D66. For other material due to Ibn Yūnus, see also the *Muṣṭalaḥ Zīj* (nos. C12 and C13).

2.1.10 On the attribution to Ibn Yūnus of a copy of the *Shāmil Zīj*, see no. B100 and Pl. XVIII.

3.1.1 ***K. al-Samt***, solar azimuth tables for latitude 30°, Cairo: MM 64 (34 fols., *ca.* 700H, anonymous, the main tables are preceded by part of a solar longitude table for a four Coptic year cycle, not contained in any other known copies); MM 137 (34 fols., *ca.* 900H); *etc.* (2 late copies). These tables are also contained in the edition of the Cairo corpus of tables for timekeeping by al-Bakhāniqī (no. C28). Another anonymous copy of the same tables arranged in a different format is DM 1108,9 (fols. 35v-59r, 1053H). See *SATMI* for further details.

3.1.21 The tables for timekeeping for latitude 31° in TR 354 (142 pp., *ca.* 1100H) are falsely attributed to Ibn Yūnus al-Mawṣilī: see no. C143 and Pl. XXXV.

3.5.1 A poem on the prayer-times atributed to the Imām al-Shāfiᶜī (no. B2) in DM 119,3 (fols. 8v–9r, 1109H) and DM 702,4 (fols. 27r–27v, 1160H, Maghribī script) is attributed to Ibn Yūnus in DM 181,4 (fols. 46v–47v, *ca.* 1200H).

The attribution of this poem to Ibn Yūnus occurs also in the longer commentary called *al-Mumti*ᶜ on a poem called *al-Muqni*ᶜ, both compiled by the seventeenth-century Moroccan astronomer Muḥammad b. Saᶜīd al-Sūsī (F49) and based on the teachings of a thirteenth-century Moroccan astronomer named Abū Miqraᶜ (no. F17). I have examined MS Cairo DM 415 (55 fols., *ca.* 1100H) of the *Mumti*ᶜ, where the poem is attributed to Ibn Yūnus on fol. 41r. This poem is now published—see no. B2 above.

5.1.1 The attribution of the astrological work *K. Bulūgh al-umniya* to Ibn Yūnus in several sources (*cf. GAS*, VII, p. 173) results from a corruption of *al-ḥakīm al-baṣīr* into *al-Ḥākimī al-Baṣrī* and hence *ᶜAlī al-Ḥākimī al-Miṣrī*. See further no. A22.

B60 ABŪ NAṢR AL-ḤASAN B. ᶜALĪ AL-QUMMĪ

GAS, VII, pp. 174–175.

5.2.16 *al-Madkhal ilā ᶜilm aḥkām al-nujūm*, a treatise on theoretical astronomy and on astrology in 5 *maqāla*s of 12 + 21 + 12 + 12 + 7 *faṣl*s, compiled in 365H: ṬM 222,2 (fols. 60r–177r, mainly 619H); MM 208 (91 fols., *ca.* 1150H, lacks end); *etc.* (2 copies). Merits study. *Cf. GAS*, VII, pp. 174–175.

B61 YŪSUF B. ᶜABD ALLĀH B. MUḤAMMAD, IBN ᶜABD AL-BARR

GAL, I, pp. 453–454, and SI, pp. 628–629; *GAS*, I, p. 461; article "Ibn ᶜAbd al-Barr" by C. Pellat in EI_2.

3.5.2 A fragment from his legal work *al-Tamhīd (li-mā fi l-Muwaṭṭaᵓ . . .)* dealing with the Caliph ᶜUmar's use of an instrument (*āla*), doubtless a gnomon, for finding the time of the *ᶜaṣr*: DM 119,2 (fol. 8v,1109H). See Pl. XLIII and *SATMI*, IV. A work entitled *K. al-Ṣawm* by Ibn Yūnus is mentioned in the fragment: this is not listed by Sezgin and the author may be the father of the celebrated astronomer Ibn Yūnus (see further *GAS*, I, 357–358, and article "Ibn Yūnus" in EI_2). See further *SATMI*, IV.

B62 ABŪ NAṢR YAḤYĀ B. JARĪR AL-TAKRĪTĪ

Suter, no. 230; *Krause*, no. 230; *GAL*, SI, pp. 862–863; *GAS*, VI, pp. 19–20.

2.1.7 + A fragment from "the mathematical collection which he gathered from *zīj*es in the year 400" dealing with the ratios of the diameters of the planets: TR 140,18 (pp. 86–87, *ca.* 1150H).

6.2.30 A fragment on the complete quadrilateral: TR 140,15 (p. 83, *ca.* 1150H).

On the astrological treatise *K. Ikhtiyārāt falakīya* attributed to him in some copies, see no. B63 below.

B63 ABŪ MANṢŪR SULAYMĀN B. AL-ḤUSAYN B. AL-YAZDAWĪYA (?) AL-ABRALĪ (?) AL-MAWṢILĪ

Suter, no. 230; *Krause*, no. 230; *GAL*, SI, p. 863; *GAS*, VII, pp. 19–20.

5.2.28 *al-K. al-Mukhtār min kutub al-ikhtiyārāt*, a treatise on astrology in 31 *bāb*s subdivided into *faṣl*s: ṬM 251 (163 fols., 1080H in two different hands), *etc.* (1 very late copy). Merits detailed investigation. See *GAS*, VI, pp. 19–20, where the Cairo copy is not listed.

B64 ABŪ ᶜALĪ B. ABI L-ḤUSAYN AL-ṢŪFĪ

GAS, VI, p. 232.

1.3.2 *Urjūza fī Ṣuwar al-kawākib*, a poem on the constellation figures: DM 831,1 (fols. 1v–15v, *ca.* 1200H) and photos thereof in ṬM 257 and DM 1094,1; DM 163 (23 fols., *ca.* 1100H); DM 417,2 (fols. 1v–13v, *ca.* 1000H). *Cf. GAS*, VI, p. 232.

B65 IKHWĀN AL-ṢAFĀᵓ

GAS, V, pp. 348–352; VI, pp. 234–239; and VII, pp. 284–287.

1.6.1 Anonymous fragments of the *Rasāᵓil Ikhwān al-Ṣafāᵓ*: DR 301,1 (fols. 1v–4v, *ca.* 700H) and DM 122,3 (fols. 23r–29v, *ca.* 1150H) on arithmetic, MM 166,1 (fol. 1v, *ca.* 1100H) on astronomy, and DR 301,2 (fols. 5r–8v, *ca.* 700H) on geography. There are several complete copies of the *Rasāʾil* in the Egyptian National Library, the finest of which may be ṬK 383 (224 fols., *ca.* 600H). This particular copy contains a passage on lunar crescent visibility that is not in the published text of the *Rasāʾil*.

B66 ABŪ NAṢR B. ᶜIRĀQ

GAS, V, pp. 338–341; and VI, pp. 242–245.

See al-Juyūbī (no. B94).

B67 ABU L-ḤASAN ᶜALĪ B. AḤMAD AL-NASAWĪ

GAS, V, pp. 345–348; VI, pp. 245–246; and VII, p. 182.

5.2.30 The Persian work *Rawḍat al-munajjimīn* by al-Nasawī's student Shāh-Mardān Rāzī (no. B91) contains material from al-Nasawī's *al-Zīj al-Fākhir* (*Kennedy*, no. 44) no longer extant in the original Arabic—see Pl. XIIIa. See also *Storey*, p. 113, no. 91 for another Persian treatise whose author was familiar with al-Nasawī's work.

6.1.7 On al-Ṭūsī's *taḥrīr* of his *tafsīr* of the *K. al-Maʾkhūdhāt* attributed to Archimedes, see no. A7.

B68 MUḤAMMAD B. AḤMAD AL-QUMMĪ[1]

GAS, V, pp. 336 and 403; VII, p. 410.

6.2.17 ***R. fī Imkān wujūd al-khaṭṭayn*** . . . : K 4528,2 (fols. 13v–15v, *ca.* 1150H); DR 898,7 (pp. 232–235, photos. of MS Leiden 14,7, pp. 232–235, *ca.* 1050H). *Cf. GAS*, V, p. 336, no. 1, where only the second copy is mentioned. The text of the treatise in these two copies differs slightly.

1. K 4528,2 gives the name as Rashīd al-Dīn M. b. A. b. Shadd al-Bannāʾ al-Qummī. The Leiden MS has M. b. A. b. M. b. K-sh-n-h al-Qummī.

B69 ABŪ BAKR MUḤAMMAD B. AL-ḤASAN AL-KARAJĪ

GAS, V, pp. 325–329; VII, pp. 408–409.

6.2.15 *al-Fakhrī fī ṣināᶜat al-jabr wa-l-muqābala*: MR 23 (74 fols., 1111H). *Cf. GAS*, V, p. 328, no. 2.

6.2.15 *K. al-Muqniᶜ fi l-misāḥa*, on surveying: DR 1098 (19 fols., 1174H). Neither title nor manuscript is mentioned in *GAS*.

See also no. B96.

B70 KŪSHYĀR B. LABBĀN

GAS, V, pp. 343–345; VI, pp. 246–249; and VII, pp. 182–183.

2.1.7 *al-Zīj al-Jāmiᶜ wa-l-balīgh*, arranged in 4 *maqāla*s, the first containing an introduction in 84 *bāb*s, the second containing 44 tables, the third containing 30 *bāb*s on theoretical astronomy and one additional (*mufrad*) *bāb* on definitions of terms in spherical astronomy (see below), and the fourth containing the proofs of the propositions of the first *maqāla* in 66 *bāb*s: DM 400 (85 fols., *ca.* 650H, 1st and 2nd *maqāla*s only, with annotations by a later Yemeni astronomer—see Pl. XII); DM 691 (45 fols., *ca.* 700H, anonymous, parts of 1st and 2nd *maqāla*s only); DM 188,2 (fols., 11r–49v, *ca.* 1200H, anonymous, part of 2nd *maqāla* only); MM 213,1 (fols. 1r–26r, 1169H, first *maqāla* only with 85 *bāb*s divided into 8 *faṣl*s); ṬR 102,3 (fols. 49v–88v, 1128H, fourth *maqāla* only). *Cf. GAS*, VI, pp. 247–248 and 294, no. 1 for other copies.

See also no. C13.

2.1.7 ***al-Bāb al-mufrad fī jawāmiᶜ ᶜilm al-hayʾa***, 135 definitions of terms in spherical astronomy: MM 158,2, fols. 21v–28r, 1182H); ṬJ 811,18 (fols. 198v–202v, 1063H, incomplete, anonymous). Another anonymous copy is in MS Princeton Mach 4898 = Yehuda 373, fols. 320b–325r, *ca.* 1050H. This is the last chapter of the third *maqāla* of the *Zīj*: *cf.* MS Istanbul Yeni Cami 784, fols. 336r–339v.

2.1.7 Assorted tables from the *zīj*es of Ibn Yūnus (no. B59) and Kūshyār, collected by Aḥmad b. Timurbāy (no. C94): MM 188,2 (fols. 77v–83r, *ca.* 875H by Aḥmad b. Timurbāy).

4.3.4 *K. al-Asṭurlāb*, on the construction and use of the astrolabe in 4 *faṣl*s of 24 + 12 + 12 + 12 *bāb*s: DM 895,2 (photos of MS Paris B.N. ar. 2487, 31 pls., 679H); MM 158,1 (fols. 1r–21v, 1182H). *Cf. GAS*, VI, pp. 248–249, no. 3.

5.2.20 *K. fī Uṣūl ṣinā^c^at al-aḥkām*, a treatise on astrology in 4 *maqāla*s of 22 + 12 + 21 + 3 *bāb*s: ṬM 222,1 (fols. 1r–59v, 619H); MM 8 (77 fols., *ca.* 750H); K 8531,1 (fols. 1r–33v, *ca.* 900H); *etc.* (17 copies), plus an anonymous abridgement in DM 859,2 (fols. 20v–38r, *ca.* 1300H). *Cf. GAS*, VII, pp. 182–183, no. 1.

6.2.19 *ᶜUyūn al-uṣūl fi l-ḥisāb al-Hindī*, a treatise on Indian arithmetic in 12 *bāb*s: MMF 8,3 (fols. 73r–77v, *ca.* 700H), not listed in *GAS*, V, p. 345, no. 2.

B71 ABŪ ᶜABD ALLĀH MUḤAMMAD B. AḤMAD AL-SHANNĪ

GAS, V, p. 352.

6.2.20 *Kashf tamwīh Abi l-Jūd* . . . : MR 41,19 (fols. 129v–134v, 1153H). *Cf. GAS*, V, p. 352, no. 1. See also no. B72.

6.2.20 *K. Misāḥat kull muthallath mukhtalif al-aḍlāᶜ min jihat aḍlāᶜihi*: MR 41,23 (fols. 148v–150r, 1153H). *Cf. GAS*, V, p. 352, no. 2. This treatise was analyzed in E. S. Kennedy and Y. Id, "A Medieval Proof of Heron's Formula," *The Mathematics Teacher*, 62 (1969), 585–587, using the Beirut MS listed by Sezgin.

6.2.20 *K. Misāḥat kull muthallath min jihat aḍlāᶜihi*: MR 41,24 (fols. 151v–152v, 1153H), unique. This treatise, which is not listed by Sezgin, differs from the one listed immediately above.

B72 ABU L-JŪD MUḤAMMAD B. AL-LAYTH

GAS, V, pp. 353–355.

6.2.21 *K. ᶜAmal al-musabbaᶜ fi l-dāʾira*: MR 41,17 (fols. 117v–120r, 1153H), unique. *Cf. GAS*, V, p. 354, no. 2, and *AHES* 30 (1984), p. 278.

6.2.21 *Maqāla (fī Thalāth masāʾil handasīya)*: MR 41,10 (fols. 72v–73v, 1153H), unique. *Cf. GAS*, V, p. 354, no. 3. Analyzed by C. Schoy in *Isis*, 7 (1925), pp. 5–8.

See also nos. B73 and G2.

B73 ABŪ SAᶜD AL-ᶜALĀʾ B. SAHL

GAS, V, pp. 341–342; and VI, pp. 232–233.

4.3.3 Commentary on *K. Ṣināᶜat al-asṭurlāb bi-l-burhān* by al-Qūhī: see no. B53.

6.2.18 *K. Tarkīb al-masāʾil allatī ḥallalahā Abū Saᶜd* . . . collected by Abu l-Jūd: MR 41,18 (fols. 121v–128v, 1153H), unique. *Cf. GAS*, V, p. 342, no. 2, and *AHES* 30 (1984), pp. 257–261.

B74 ABU L-QĀSIM AḤMAD B. ᶜABD ALLĀH AL-ṢAFFĀR

GAS, V, pp. 356–357; and VI, pp. 250–251.

4.3.7 *K. fi l-ᶜAmal bi-l-asṭurlāb*, in 40 unnumbered *bāb*s: TR 163,1 (fols. 1v–25r, *ca.* 1100H); DM 928 (photos. of MS London B.L. 22672, 25 pls., late); DM 639,8 (fols. 63v–79v, *ca.* 1150H in two hands), *etc.* (2 copies). *Cf. GAS*, VI, pp. 250–251, no. 2.

B75 ᶜABD AL-QĀHIR B. ṬĀHIR AL-BAGHDĀDĪ

GAS, V, p. 357.

6.2.22 *K. al-Takmila fī ᶜilm al-ḥisāb*: DR 793,1 (fols. 1r–85r, 1112H), unique complete copy, not mentioned in *GAS*, V, p. 357, no. 1. See Pl. LXXXVIIb and also A. S. Saidan, "The *Takmila fi 'l-ḥisāb* of ᶜAbd al-Qāhir al-Baghdādī," in *Kennedy Festschrift*.

B76 ᶜALĪ B. ABI L-RIJĀL AL-SHAYBĀNĪ

GAS, VII, pp. 186–188.

5.2.18 *K. al-Bāriᶜ*, a treatise on astrology in 8 *juzʾ*: ṬM 246 (183 fols., 744H, pts. 5–8 only); ṬM 100 (247 fols., 1127H); ṬM 231 (415 fols., *ca.* 1150H); ṬM 149 (119 + 147 fols. in 2 vols., *ca.* 1100H); *etc.* (4 copies); plus commentary by Aḥmad b. Aḥmad Timurbāy (no. C94) entitled *al-Barq al-sāṭiᶜ* . . . : MM 13,2 (fols. 1r–99v, 894H in author's hand); ṬM 112 (159 fols. 1089H); *etc.* (2 copies). *Cf. GAS*, VII, pp. 187, no. 1.

5.2.18 A poem on astrology: DM 939 (photos of MS Paris B.N. ar. 2541, 8 pls., *ca.* 1150H, lacks beginning); plus a commentary by Ibn Qunfudh (no. F25): ṬM 252 (58 fols., 1067H); KhM 9 (58 fols., *ca.* 1150H); ṬM 183 (165 pp. 1091H); *etc.* (6 copies). *Cf. GAS*, VII, pp. 187-188, no. 2.

B77 ABU ᶜALĪ AL-ḤASAN B. AL-ḤASAN, IBN AL-HAYTHAM

Article "Ibn al-Haytham" by A. I. Sabra in *DSB*; *GAS*, V, pp. 358–374; VI, pp. 251–261; VII, p. 288.

1.4.7 On his *qaṣīda* on the sun's entry into the lunar mansions, see Abu ᶜAbd Allāh al-Lakhmī (no. F10).

1.6.2 *M. fi stikhrāj irtifāᶜ al-quṭb ᶜalā ghāyat al-taḥqīq*: DR 898,11 (photos of MS Leiden 14, pp. 246–254, *ca.* 1050H). *Cf. GAS*, V, p. 366, no. 5.

1.6.2 *R. fī Māhīyat al-athar alladhī yaẓhar fī wajh al-qamar*: DṬ 425 (15 fols., 1335H from the unique MS Alexandria Municipal Library 2096D). Published by A. I. Sabra in *JHAS*, 1 (1977), pp. 166–180. *Cf. GAS*, VI, pp. 256–257, no. 5 and p. 295.

3.3.1 *Qawl fi samt al-qibla bi-l-ḥisāb*: K 3823,3 (fols. 14v–18v, *ca.* 900H from a copy by Qāḍī Zāde al-Rūmī (no. G50)), not listed by Sezgin. This treatise, which has never been studied, merits publication and analysis, and comparison with Ibn al-Haytham's other treatise on the qibla which has been translated and analyzed by C. Schoy. *Cf. GAS*, V, p. 368, nos. 15 and 16; and VI, p. 259, nos. 18 and 19 for the available copies of both treatises.

6.1.3 *K. fī Ḥall shukūk K. Uqlīdis fi l-Uṣūl wa-sharḥ maᶜānīhi*: KhR 1 (159 fols., 712H), not mentioned by Sezgin; DR 891 (photos. of MS Leiden 516, 114 plates, *ca.* 800H, incomplete). *Cf. GAS*, V, p. 370, no. 27.

6.1.3 A fragment from his *Sharḥ al-muṣādarāt*: DR 703,1 (fol. 1v, 1006H). *Cf. GAS*, V, pp. 370–371, no. 28, to which add also MS Princeton Mach 4849 = Yehuda 1039 (90 fols., 1028H).

6.1.6 *R. li-Tawḍīḥ qismat al-khaṭṭ fi l-maqāla al-thāniya min K. al-Kura wa-l-usṭuwāna li-Arshimīdis*: DR 898,27 (photos of MS Leiden 14, pp. 498–501, *ca.* 1050H). *Cf. GAS*, V, p. 371, no. 31.

6.2.23 *M. fī Tarbīᶜ al-dāʾira*: J 3626,21 (photos of MS Istanbul Aya Sofya 4832, pls. 85–88, *ca.* 500H); TR 140,5 (pp. 36–37, *ca.* 1150H). *Cf. GAS*, V, p. 365, no. 1.

6.2.23 *M. fi l-Taḥlīl wa-l-tarkīb*: TR 323,1 (pp. 1–68, *ca.* 1100H). *Cf. GAS*, V, p. 368, no. 17.

6.2.23 *R. fī Maᶜrifat irtifāᶜ al-ashkhāṣ al-qāʾima wa-aᶜmidat al-jibāl wa-rtifāᶜ al-ghuyūm*: DR 898,8 (photos of MS Leiden 14, pp. 236–237, *ca.* 1050H). *Cf. GAS*, V, p. 370, no. 25. This copy ends with some notes by Saᶜd al-Dīn Asᶜad b. Saᶜīd al-Hamadānī (no. G90).

6.2.23 *R. fī Shakl Banī Mūsā*: DR 690,1 (photos of MS London B.L. Add. 14332, pp. 42–62, 1832 A.D.). *Cf. GAS*, V, p. 372, no. 32.

7.2.3 A note on the visible and true horizons by Ibn al-Haytham stated to be taken from *al-Tuḥfa al-Shāhīya* by Quṭb al-Dīn al-Shīrāzī (no. G27): TR 140,20 (p. 115, *ca.* 1150H). See Pl. Cb.
al-Shīrāzī indeed mentions in the *Tuḥfa* (*cf.* DH 55 (= MS Paris B.N. ar. 2516), fol. 12r) that Ibn al-Haytham had established that the visible part of the sky was more than the invisible part by 4 minutes and 26 seconds (0;4,26) if the height of the individual was 3½ cubits. al-Shīrāzī adds that Ibn al-Haytham had stated this in a treatise entitled *R. fī anna l-ẓāhir min al-samāʾ akthar min niṣfihā*. For copies of this, see *GAS*, VI, p. 260, no. 24.

7.3.1 *R. fi l-Makān*: K 3823,1 (fols. 1r–5v, *ca.* 900H). (This treatise was published in Hyderabad, 1357H.)

See also no. B112.

B78 ABU L-RAYḤĀN MUḤAMMAD B. AḤMAD AL-BĪRŪNĪ

Article "al-Bīrūnī" by E. S. Kennedy in *DSB*; *GAS*, V, pp. 375–383; VI, pp. 261–276; VII, pp. 188–192 and 288–292.

1.6.3 *K. fi Tasṭīḥ al-ṣuwar wa-tabṭīḥ al-kuwar*: DR 898,16 (photos of MS Leiden 14, pp. 300–314, *ca.* 1050H, in bad shape). Sezgin (*GAS*, V, pp. 381–382, no. 10 and VI, p. 268, no. 5) lists another copy of this work in MS

Tehran Danishgah 5469,3 (fols. 7v–13v, *ca.* 550H), and notes that Suter published a translation of the defective Leiden copy in 1922. The Arabic text of this work (edited from the Leiden manuscript only) has been published by A. Saidan in *Dirāsāt* (Jordanian University), 4 (1977), pp. 7–22, and the treatise has been studied by J. L. Berggren and L. Richter-Bernburg in *JHAS*, 5 (1981).

1.6.3 *K. Taḥdīd nihāyāt al-amākin wa-taṣḥīḥ masāfāt al-amākin*: photos of the unique MS Istanbul Fatih 3386 (171 fols., 416H in Ghazna) in DH 84 and 85 (4 vols. each, 340 pls.). This work is now edited, translated, and analyzed: see *GAS*, VI, pp. 267–268, no. 4.

2.1.9 *al-Qānūn al-Masᶜūdī*, a *zīj* in 11 *maqāla*s: DM 866 (268 fols., 673H, magnificent copy), used for the published edition by Max Krause (Hyderabad, 1956); plus a fragment in DM 874 (121 fols., *ca.* 700H). *Cf. GAS*, VI, pp. 265-266, no. 1. See Pl. XVII.

4.3.5 *K. al-Istīᶜāb fi l-wujūh al-mumkina fī ṣināᶜat al-asṭurlāb*, a treatise on the theory and construction of different kinds of astrolabes: K 8528 (75 fols., *ca.* 1300H, wretched), not listed in *GAS*, VI, p. 268, no. 6.

4.3.5 *M. fi l-Taṭrīq ila stiᶜmāl funūn al-asṭurlābāt*, a treatise on the use of different kinds of astrolabes, arranged in 20 subdivided *bāb*s: DM 914 (photos. of MS Paris B.N. ar. 2498, 63 pls., *ca.* 1000H, defective). This treatise is not the same as the one in MS Berlin Ahlwardt 5794, as stated by Suter. *Cf. GAS*, VI, p. 269, nos. 8 and 11.

4.3.5 *Riyāḍat al-fikr wa-l-ᶜaql fi stikhrāj* (read: *ikhrāj*?) *mā fī quwwat al-asṭurlāb ila l-fiᶜl*, a treatise on the use of the astrolabe in unnumbered sections: DM 262 (19 fols., lacks last section, *ca.* 1300H), and photos thereof in K 3929; K 3774,2 (fols. 14r–21r, 1249H). Another copy of this treatise, without the title but attributed to al-Bīrūnī, is MS Berlin Ahlwardt 5794 (fols. 1–43, *ca.* 1100H). On the title, see *Boilot*, nos. 49 and 170. See also *GAS*, V, p. 381, no. 8, and VI, pp. 268–269, nos. 7 and 8. An unidentified treatise on the astrolabe by al-Bīrūnī in MS Istanbul Aşir Reisülküttap 577 is listed in *Krause*, p. 480, no. 7, but not in *GAS*, VI. See also no. B79 below.

5.2.21 *K. al-Tafhīm li-awāᵓil ṣināᶜat al-tanjīm*: DM 450 (photos of a very fine manuscript now in the Chester Beatty Library in Dublin, no. 3910, 378 pls., 573H in Nishapur, lacks beginning); DM 848 (photos. of MS Istanbul Topkapi 3477, 307 pls., 871H); DM 901 (photos. of MS Paris B.N. ar. 2497, 86 pls., *ca.* 1000H, defective). *Cf. GAS*, VII, pp. 189–190, no. 1.

6.2.24 *K. Istikhrāj al-awtār fi l-dāᵓira bi-khawāṣṣ al-khaṭṭ al-munḥani al-wāqiᶜ fīhā*, "on chords": DR 897,5 (photos of MS Leiden 513, fols. 107v–129r, *ca.* 900H), analysed by Suter; and an edition by Muṣṭafā Ṣidqī (no. D81): MR 41,11 (fols. 74v–82r, 1153H by editor). Both of these differ from the text of the Bankipore manuscript published in Hyderabad. For other copies see *GAS*, V, p. 381, no. 3. The Cairo edition of A. Damirdash is without critical apparatus. The more recent edition by A. Ghorbani does not take account of all the available manuscripts.

B79 ANONYMOUS

GAS, VI, pp. 78 and 169.

4.3.6 *al-Miqyās al-murajjaḥ fi l-ᶜamal bi-l-asṭurlāb al-musaṭṭaḥ*: ṬM 155,1 (fols. 1r–15v, *ca.* 1050H, incomplete). This treatise is incorrectly attributed to al-Bīrūnī on the title folio; he is also mentioned in the text. See Pls. LVIb and LVII. The author refers to a work of Thābit b. Qurra (no. B30) on the astrolabe in which he stated that the astrolabe was invented by Hipparchus (no. A9). See already *GAS*, VI, pp. 78 and 169, and for further details, D. A. King, "The Origin of the Astrolabe According to the Medieval Islamic Sources," *JHAS*, 5 (1981), pp. 43–83, esp. pp. 51–52.

B80 AL-ḤASAN B. AḤMAD B. ᶜABD-ALLĀH AL-ṢŪFĪ AL-KIRMĀNĪ

GAS, VI, p. 282, and VII, pp. 193–194.

5.2.29 *K. al-Uṣūl*, a treatise on astrology in 8 *maqāla*s each subdivided into numbered *kalima*s: ṬM 188 (34 fols., 1301H). Another copy of this treatise is MS Princeton Mach 5067 = Yahuda 2501 (30 fols., 1302H). See Mach's summary of the preface of this work, in which al-Kirmānī mentions various early sources. A treatise on astrology by Abu l-Qāsim al-Kirmānī which may be identical to the above treatise is contained in MS Oxford Bod. Hunt 663 (fols. 127r–132r, copied 640H), listed in *GAS*, VII, p. 194. An individual named Ibn al-Ṣūfī al-Kirmānī is mentioned on fol. 214v of MS Paris B.N. pers. supp. 1488 of the early-fourteenth-century ***Ashrafī Zīj***.

B81 ABŪ ᶜALĪ AL-ḤUSAYN B. ᶜABD ALLĀH, IBN SĪNĀ

GAS, V, p. 108; VI, pp. 276–280; and VII, pp. 292–302.

1.1.6 An abridgement of Ptolemy's *Almagest* (see no. A18): DH 72 (photos. of MS Paris B.N. ar. 2484, 144 pls., 673H). *Cf. GAS*, VI, p. 278, no. 1.

1.2.5 ***R. fi l-Hay***ᵓ***a***, on theoretical astronomy, arranged in unnumbered *faṣl*s: DH 10,2 (fols. 25r–40r, 1163H); DH 49 (photos of MS London B.L. Add. 9599,11 pls., 1221H); DH 43 (38 pp., 1337H). Not listed by Sezgin.

1.2.5 The introduction of a *R. fi l-Hay*ᵓ*a*, in 9 *faṣl*s: DJ 863,15 (fols. 297v–298r, *ca.* 850H). Not listed by Sezgin.

1.2.5 ***R. fī Qiyām al-arḍ wasaṭ al-samā***ᵓ, a treatise on the central position of the earth in the universe, arranged in 10 *faṣl*s: DH 47 (photos of MS London B.L. Add. 16839, 3 pls., *ca.* 1150H). Several other copies are listed in *GAS*, VI, p. 279, no. 4.

1.4.5 A poem on the four seasons: TJ 25,2 (pp. 26–31, *ca.* 1300H). Not listed by Sezgin.

See next three entries.

B82 ABŪ ᶜUBAYD ᶜABD AL-WĀḤID B. MUḤAMMAD AL-JŪZAJĀNĪ

Suter, no. 425 (confused); *GAL*, I, p. 592; *GAS*, V, p. 108, and VI, pp. 280–281.

6.3.1 An abridgement of the *K. al-Arithmāṭīqī* of Ibn Sīnā (no. B81): DJ 863,13 (fols. 278v–290r, *ca.* 850H); ṬR 118,3 (fols. 69v–86v, 1075H). *Cf. GAL*, I, p. 592.

6.3.1 ***R. fi l-Arithmāṭīqī***, taken from the *K. al-Najāt* of Ibn Sīnā and attributed to Abū ᶜAbd Allāh: DJ 863,14 (fols. 290v–297r, *ca.* 850H). *Cf. GAL*, I, p. 592.

B83 ANONYMOUS (late)

6.3.1 ***Ghāyat al-ᶜudad fī ᶜilm al-ᶜadad***, an extensive treatise on number theory in 4 *maqāla*s of 6 + 2 + 3 + 10 *bāb*s: ṬR 130 (60 fols., 1209H from a copy dated 1112H, anonymous), unique. Not listed in the modern sources. The thirteenth-century scholar (Quṭb al-Dīn) al-Shīrāzī (no. G27) is mentioned in the introduction, as well as Ibn Sīnā (no. B81).

B84 ANONYMOUS

1.2.4 A fragment of an anonymous commentary on an anonymous treatise on theoretical astronomy in DH 48 (17 fols., *ca.* 1050H), containing the reports of Venus transits observed in Farīn (?) and discussed in B. R. Goldstein, "Some Medieval Reports of Venus and Mercury Transits," *Centaurus*, 14 (1969), esp. pp. 52–53. Merits investigation.

B85 ᶜALĪ B. RIḌWĀN

Suter, no. 232; *GAL*, I, pp. 638–639, and SI, p. 886; article "Ibn Riḍwān" in *DSB* by R. Arnaldez; *GAS*, VII, p. 44.

5.2.22 An extract from his commentary on the ***Tetrabiblos*** of Ptolemy (no. A18) relating to the Nile, together with quotations from the ***K. fi l-Amṭār*** of Abū Maᶜshar (no. B22) and a horoscope for the beginning of 855H: DM 632,4 (fols. 8v–12r, *ca.* 1050H); MJ 213,1 (fols. 1r–3v, *ca.* 1150H); *etc.* A Turkish translation by Sulaymān Riyāḍī (no. H65) of ᶜAlī b. Riḍwān's commentary on the *Tetrabiblos*: ṬFT 34 (298 fols., 1195H); ṬFT 16 (175 fols., 1278H). See *GAS*, VII, p. 44, no. (g) for manuscripts of the commentary.

B86 ABŪ BAKR AḤMAD B. ᶜALĪ B. THĀBIT, AL-KHAṬĪB AL-BAGHDĀDĪ

GAL, I, pp. 400–401, and SI, pp. 562–564 (esp. no. 8); *GAS*, VII, p. 29, no. 13.

5.2.25 A fragment from a treatise on astrology entitled *K. al-Quwā fi l-nujūm wa-bayān al-maḥmūd wa-l-madhmūm*: ṬFF 13,4 (fol. 122v, *ca.* 1000H). MS Istanbul Ašir 190 (18 fols., *ca.* 950H) is entitled *R. fī ᶜIlm al-nujūm hal al-shūrā fīhi maḥmūd aw madhmūm* appears to be a complete copy. This work merits investigation. *Cf. GAS*, VII, p. 29.

B87 ABŪ ISḤĀQ IBRĀHĪM B. YAḤYĀ AL-ZARQĀLLU AL-ṬULAYṬULĪ

Suter, no. 255; *GAL*, I, p. 623, and SI, p. 862; article "al-Zarqālī" by J. Vernet in *DSB*.

4.6.1 *(R. fi) l-ᶜAmal bi-l-ṣafīḥa al-mushtarika li-jamīᶜ al-ᶜurūḍ*, in 100 *bābs*: DM 647 (61 fols., *ca.* 600H). *Cf.* MS Escorial ar. 962, *i.a.* The anonymous MS Istanbul Esat 3804,3 (cop. 665H) listed in *Krause*, p. 526, is a copy of the same treatise. See Pl. LVIb.

4.6.1 *al-Shakkāzīya*, a treatise in 60 *bābs*: TR 131,4 (pp. 14–40, 1186H in Maghribi script, anonymous). The same work is attributed to al-Zarqāllu in MS Istanbul University Library A4800. No other known copies. See also next entry.

4.6.1 *(R. fi l-ᶜAmal bi-l-ṣafīḥa al-mushtarika bi-jamīᶜ al-ᶜurūḍ)*, in 61 *bābs*: DH 40 (54 fols., *ca.* 950H, anonymous), unique? The Hebrew translation of this treatise was published by M. Millás Vallicrosa. Commentary by Abu l-Walīd Ismāᶜīl b. Aḥmad al-Qurṭubī (no. F2) in 60 *bābs*: TR 131,6 (pp. 82–129, 1186H in Maghribī script). Another copy exists in MS Damascus Ẓāhirīya 9241 (cop. 1301H, *cf. Cat.*, pp. 227–228).

On these treatises, see now D. A. King, "On the Early History of the Universal Astrolabe in Islamic Astronomy . . . . . . ," *JHAS*, 3 (1979), pp. 244–257.

5.2.23 *R. fī Ḥarakāt al-kawākib al-sayyāra wa-tadbīrihā*: ṬJ 424,3 (fols. 51v–60v, *ca.* 1200H); DM 920 (photos of MS London B.L. add. 9599, 13 plates, incomp., late); Sh 124 (10 fols., *ca.* 1200H). The text of the three copies differs substantially. Another copy is MS Vienna Nationalbibliothek 1421 (*GAL*).

B88 ABŪ L-YUSR ᶜALĪ B. MUḤAMMAD AL-BAZDAWĪ (d. 482H in Samarqand)[1]

GAL, I, p. 460, and SI, pp. 637–638; *GAS*, I, pp. 412 and 428.

3.3.2 *R. fī Samt al-qibla*, a treatise on the qibla, with especial reference to the early practice in Transoxania: B19385 (12 pp., 1355H from a copy preserved in Sohag), no other known copies, not listed by Brockelmann. Edited and translated in D. A. King, "al-Bazdawī on the Qibla in Early Islamic Transoxania," *JHAS*, 7 (1986). See Pl. XLVIII.

B89 ABŪ L-ṢALT UMAYYA B. ᶜABD AL-ᶜAZĪZ B. ABI L-ṢALT

Suter, no. 272 and 272N; *Krause*, no. 272; *GAL*, I, p. 641, and SI, p. 889.

4.3.8 *K. fi l-ᶜAmal bi-l-asṭurlāb*, a treatise on the use of the astrolabe in 90 *bābs*: DM 773 (51 fols., *ca.* 700H); K4000 (24 fols., 1318H); plus a fragment of a treatise arranged in 88 *bābs*: DM 916 (photos of MS London B.L. 9599, fols. 167–172, late). Another fragment in TGh 197,12 (fols. 49r–54v, *ca.* 700H) bears a marginal note stating that Abu l-Ṣalt wrote this treatise whilst he was in prison in Egypt.

B90 MNᶜBY (?) B. FARĪGHŪN (?), student of Abū Zayd Aḥmad b. Zayd al-Balkhī[2]

GAL, SI, p. 435; *Suter*, no. 170, corrected in *Renaud*, no. 170; V. Minorsky in *A Locust's Leg: Studies in Honor of S. H. Taqizadeh*, (London, 1962), pp. 189–196, and the references there cited; *GAS*, I, p. 388. See now the *Habilitationsschrift* of Dr. H. H. Biesterfeldt (Bochum, 1985).

1.6.4 *K. Jawāmiᶜ al-ᶜulūm*, an encyclopaedia in two *maqālas*: DO 528 (photos of [MS Istanbul

---

1. In the Cairo manuscript the author is named as Ṣadr al-Islām Abu l-Yusr Muḥammad b. Muḥammad b. al-Ḥusayn b. ᶜAbd al-Karīm b. Mūsā b. Mujāhid al-Yazdawī and it stated that he died in 493H.

2. al-Balkhī's name is given thus in both MSS Cairo DO 528 and MS Escorial ar. 950, *contra GAL*, SI, p. 435, and *GAS*, I, p. 388.

Topkapi A 2768], 168 pls., 396H, beautiful *naskhī*); DO 527 (photos of [MS Istanbul Topkapi A 2675], 158 pls., *ca.* 600H, beautiful *naskhī*, anonymous). In DO 528 the author is named as Mnᶜby b. Fryᶜwn. Another copy of this work is MS Escorial ar. 950 (84 fols., cop. before 655H from a copy dated 393H): the author is named Shiᶜyā b. Farīghūn. The sections on science in this work are of very limited interest.

B91 SHĀH-MARDĀN B. ABI L-KHAYR RĀZĪ

Storey, no. 81; *GAS*, VI, p. 246.

5.2.30 ***Rawḍat al-munajjimīn***, a Persian treatise on astrology arranged in 15 *maqāla*s, compiled *ca.* 465H: ṬFF 11 (164 fols., anonymous, no title, mainly *ca.* 700H in fine *naskhī*, lacks part of last *maqāla*); K 3774,4 (fols. 22v–83v, 1269H, anonymous, no title, defective). Storey mentions several other copies. The work is mainly astrological, but also contains sections on planetary astronomy, the astrolabe, and the constellations (beautifully illustrated in ṬFF 11). Shāh-Mardān included material from his teacher al-Nasawī (no. B67), Kūshyār (no. B70), al-Bīrūnī (no. B78), as well as constellation figures from al-Ṣūfī (no. B52). A detailed table of contents has been prepared by Dr. E. S. Kennedy. The Ṭalᶜat manuscript contains some spurious material lifted from the *Īlkhānī Zīj* of al-Ṭūsī (no. G23). See Pls. IV, XIIIa and XLIVb.

B92 ABŪ JAᶜFAR MUḤAMMAD B. AYYŪB AL-ṬABARĪ

Suter, no. 360; *Krause*, no. 360; *GAL*, SI, p. 859; *Kennedy*, no. 65; *Storey*, nos. 5 and 79; *GAS*, V, pp. 385–386.

2.2.14 DM 678 (71 fols., *ca.* 1150H) is an anonymous Egyptian or Syrian collection of tables which includes some attributed to al-Ṭabarī (see no. D207). I have not checked that these are indeed from al-Ṭabarī's *Mufrad Zīj*, extant in Cambridge. See Pl. XV.

B93 ABU L-FAṬH MUḤAMMAD B. ᶜABD AL-MALIK AL-DAWWĀNĪ

GAS, V, p. 252.

6.2.3 ***(R. fī) Sahw waqaᶜa li-Banī Mūsā fi l-burhān ᶜala l-shakl al-ākhir***, on a problem in the last part of the treatise of the Banū Mūsā (no. B16): J 3626,33 (photos. of MS Istanbul Aya Sofia 4832, pls. 155–157, *ca.* 500H, unique). *Cf.* *GAS*, V, p. 252, no. 1(c).

B94 MUḤAMMAD B. ḤASAN AL-JUYŪBĪ (?)

GAS, VII, pp. 414–415.

6.3.4 ***Tashrīḥ al-kura***, a treatise on spherical trigonometry quoting Thābit b. Qurra (no. B30), Abū Naṣr b. ᶜIrāq (no. B66), Abu l-Wafāᵓ al-Būzajānī (no. B54), al-Khujandī (no. B50), Kūshyār (no. B70), al-Bīrūnī (no. B78), and Ibn Sīnā (no. B81): DM 1202 (58 fols., *ca.* 1350H, attributed to al-Juyūbī in pencil in still later hand), unique. *Cf.* *GAS*, VII, pp. 414–415. I place little reliance on the attribution to al-Juyūbī. Since the author mentions another treatise by him on *al-nisba al-muᵓallafa* the author may be Muḥyi l-Dīn al-Maghribī (no. G21), who wrote such a work, extant in MS Berlin Ahlwardt 5957. In any case the *Tashrīḥ al-kura* appears to predate al-Ṭūsī's treatise on spherical trigonometry, as do the *Maqālīd ᶜilm al-hayᵓa* of al-Bīrūnī (no. B78) and the treatise by Ḥusām al-Dīn Sālār (no. G22). This manuscript has been investigated by M.-Th. Debarnot: see her article "Introduction du Triangle Polaire par Abū Naṣr b. ᶜIrāq," *JHAS*, 2 (1978), pp. 126–136, esp. p. 130.

B95 ABŪ ᶜALĪ AL-SHĀHID

Not mentioned in the modern sources.

5.2.26 ***K. fīhi Asrār al-ᶜilm bi-aḥkām al-nujūm***, a book of questions and answers relating to astrology: MMF 8,2 (fols. 68r–72v, *ca.* 700H).

B96 ABŪ ᶜABD ALLĀH MUḤAMMAD B. AL-ḤUSAYN AL-ḤĀSIB[1]

Not mentioned in the modern sources.

---

1. If not for the *kunya* this might be Abū Bakr Muḥammad b. al-Ḥusayn al-Karajī (no. B69). See also al-Ṣaydanānī (*GAS*, V, p. 301).

6.2.31 A fragment of one folio from a treatise on arithmetic dealing with measures of length: MR 42,2 (fols. 69r–69v, 487H).

B97 ABŪ YŪSUF AḤMAD B. AL-ḤASAN or AL-ḤASAN B. YŪSUF[1]

GAL, SII, p. 1019, no. 21.

6.2.32 *K. fi l-Jabr wa-l-muqābala*, a treatise on algebra in 12 *bābs*: DR 100 (65 fols., *ca.* 850H) unique. Merits investigation.

B98 ANONYMOUS (perhaps named Abū Bakr)

GAS, V, p. 396.

6.2.27 Twelve geometrical problems from various sources, including the *K. al-Qisma* of Euclid (no. A5) and the writings of Abū Sahl al-Kūhī (no. B53), Ibn al-Haytham (no. B77), Abū Maḥmūd al-Khujandī (no. B50), and Thābit b. Qurra (no. B30): MR 41,29 (fols. 165v–170v, 1153H). The last problem begins *qāla Abū Bakr hādhā min al-basāʾiṭ fa-ammā . . .* , so that the name of the author may have been Abū Bakr. As noted in *GAS*, V, p. 396, some of these problems have been discussed by C. Schoy.

B99 ANONYMOUS

2.1.1 ***Jadwal ruʾyat al-ahilla fi-l-aqālīm al-sabʿa***, a table displaying the minimum elongation between the sun and moon for visibility in each of the zodiacal signs and each of the climates: ṬM 119,1 (fol. 1r, 1185H); MM 23,2 (fol. 9r, 1153H). Several copies in European libraries. The nature of the table and the profusion of copyists' errors seems to indicate its early origin. Analyzed in D. A. King, "Some Early Islamic Tables for Determining Lunar Crescent Visibility," to appear in *Kennedy Festschrift*. See already *JHAS*, 6 (1982), pp. 19–20.

1. The first name occurs at the beginning of the treatise in DR 100, the second on the title folio (in the same hand).

B100 ANONYMOUS

Kennedy, no. 29; *cf. GAS*, V, pp. 324–325, no. 9, where the *Shāmil Zīj* is erroneously attributed to Abu l-Wafāʾ al-Būzajānī (no. B54).

2.1.10 *al-Zīj al-Shāmil*, a *zīj* arranged in an introduction in ten *bābs* followed by tables: DM 902 (photos. of MS Paris B.N. ar. 2528, 73 pls., *ca.* 1000H); ṬM 138 (115 pp., *ca.* 900H, lacks introduction) and photos thereof in K 22519; TR 296,1 (pp. 1–160, *ca.* 1200H). See Pl. XVIII. A commentary entitled ***al-Kāmil fī sharḥ al-Zīj al-Shamil*** by Sayyid Ḥasan b. Sayyid ʿAlī al-Qumnātī (no. G47), compiled in 822H: DM 951 (photos of MS Paris B.N. ar. 2530, 118 pls., 822H). A detailed analysis of the ***Shāmil Zīj*** would be worthwhile.

B101 ANONYMOUS

3.2.1 An Egyptian almanac arranged according to the Coptic months and containing simple calendrical tables and prayer-tables: DM 187 (74 fols., *ca.* 1250H), unique. Parts of this almanac may be very old, and it certainly merits detailed investigation.

B102 ANONYMOUS

4.3.9 Fragment containing *bābs* 73 to 89 of an anonymous treatise on the astrolabe: DM 969,3 (fols. 5v–7v + 10v–14r, *ca.* 1200H). Since the author mentions (fol. 12v) the sixth and seventh *maqālas* of his treatise this is clearly part of a much larger work. The extant chapters include one "on the construction of a sundial from an astrolabe."

B103 ANONYMOUS

4.3.9 Fragment containing *bābs* 95 to 100 of an anonymous treatise on the astrolabe: DM 925,1 (photos of MS London B.L. Add. 9599, fols. 14v–33v, late).

B104 ANONYMOUS

4.6.1 An anonymous collection of selected chapters on the *shakkāzīya*, some taken directly from the treatise in TR 131,4 listed above (see no. B87): TR 159,3 (pp. 81–97, 1320H).

B105 ANONYMOUS

4.6.12 ***R. fī Rub*ᶜ *al-dā*ᵓ*ira***, a treatise on the quadrant with fixed or movable cursor, in 10 unnumbered sections: DM 969,4 (fols. 8v–9v, *ca.* 1200H). See Pl. LIVa. I date this work to the third or fourth Hijra century by virtue of the Arabic style and the nature of the treatise. If this is correct, the treatise predates the earliest known works on quadrants of this kind by several centuries. Analyzed in D. A. King, "On the Early History of the Quadrant in Islamic Astronomy," forthcoming.

B106 MISCELLANEOUS ANONYMI

5.2.31 Various early treatises on astrology: MḤ 78,1 (fols. 1v–24v, *ca.* 1100H, defective, contains references to *ᶜnṭyᶜws* (= Antiochus?) (no. A13) and Dorotheos (no. A15)); DM 122,2 (fols. 2v–22v, *ca.* 1150H); DM 1156,2 (fols. 34r–43v, *ca.* 800H); ṬM 148 *bis*, 2 (fols. 8v–64r, *ca.* 1150H, defective); ṬM 243,5 (fols. 129v–135v, 644H); ṬM 243,3 (fols. 24v–116v, 644H); Sh 70,4 (fols. 21v–45r, *ca.* 1250H). All of this material merits investigation at the hands of a specialist.

B107 ANONYMOUS

5.2.35 ***al-Raᶜdīya***, an astrological almanac arranged according to the (Western) Christian months (and probably of Andalusian origin): ṬM 214,1 (fols. 1r–31r, 1321H). Merits investigation.

B108 ANONYMOUS

Suter, p. 150, no. 11; *GAS*, V, pp. 114 & 400.

6.1.3 ***al-ᶜAmalīyāt min k. Uqlīdis fi l-Uṣūl***, a collection of 105 constructions (see no. A5): MR 40,8 (fols. 110v–122v, 1147H, anonymous), listed by Suter and Sezgin as a work of Naṣīr al-Dīn al-Ṭūsī (no. G23), unique but for the late Damascus copy made from this one (*GAS*, V, p. 400). A marginal note in the Cairo copy mentions Sayyid Sharīf (al-Jurjānī) (no. G42).

B109 ANONYMOUS

6.2.25 A fragment of a treatise on some geometrical problems occuring in the *Zīj al-Safāᵓiḥ* of Abū Jaᶜfar al-Khāzin (no. B43) and *al-Marāyā l-muḥriqa* of Diocles (no. A19): DR 898,14 (photos of MS Leiden 14, pp. 294–298, *ca.* 1050H). *Cf. GAS*, V, p. 299, no. 2.

B110 ANONYMOUS

Note added in proof: Dr. J. P. Hogendijk has identified this as a fragment of the *Kitab al-Istikmāl* of al-Muᵓtaman ibn Hūd: See *HM* 12 (1985).

6.2.26 A treatise on number theory apparently part of a more extensive work which I am unable to identify: MR 40,1 (fols. 1v–37v, 1146H by Muṣṭafā Ṣidqī (no. D81)). This part of MR 40 is not listed by Sezgin. The treatise begins: *al-nawᶜ al-awwal min al-jins al-awwal min jinsayi l-taᶜālīm al-riyāḍīya fī ᶜilm khawāṣṣ al-aᶜdād mufrada wa-muḍāfa wa-huwa yanqasim ilā arbaᶜa fuṣūl* . . . , and each of the four *faṣl*s is contained in the manuscript.

B111 ANONYMOUS

GAS, VII, p. 415.

6.2.28 ***M. fī Tarkīb li-taḥlīl muqaddimāt al-musabba*ᶜ *al-mutasāwi l-aḍlā*ᶜ *fi l-dā*ᵓ*ira***: MR 41,15 (fols. 111v–112v, 1153H). *Cf. GAS*, VII, p. 415, and J. P. Hogendijk in *AHES* 30 (1984), pp. 223–224.

B112 ANONYMOUS

6.2.29 ***R. fī Wujūd khaṭṭayn yaqrubān wa-lā yaltaqiyān***, on the existence of two lines which get closer together without meeting: K 4528,3 (fols. 15v–20r, *ca.* 1150H). The copyist wrote at the end of the treatise that "it can be understood from the expressions in the treatise that it was compiled by Ibn al-Haytham" (see no. B77). This manuscript is not listed by Sezgin. The preceding treatises in K 4528 are a treatise on the same subject by al-Qummī (no. B68) and a commentary on the 10th *maqāla* of Euclid's ***Elements*** by al-Ahwāzī (no. B46).

B113 ANONYMOUS

6.2.33 A treatise on sexagesimal arithmetic in 6 *bābs*: MMF 8,5 (fols. 78r–82v, *ca.* 700H).

B114 ANONYMOUS

6.2.34 Miscellaneous notes on algebra, ratios, music: see TR 140.

ADDENDUM

B115 IBN AL-SAMḤ

GAS, VI, p. 249.

See nos. F9, F21, and G18.

C. Egyptian and Syrian Authors (*ca.* 1100 to *ca.* 1600)

References to scientists from this period are to be found in *Suter*, Brockelmann's *GAL*, and *Azzawi*. The only survey of astronomy in this period is *King* 2. The Cairo corpus of tables for timekeeping attributed in many sources to Ibn Yūnus is analyzed in D. A. King, "Ibn Yūnus' *Very Useful Tables for Reckoning Time by the Sun*," *AHES*, 10 (1973), pp. 342–394, and the Damascus and Jerusalem corpuses of tables are analyzed in "Astronomical Timekeeping in Fourteenth-Century Syria," *Aleppo Conference Proceedings*, II, pp. 75–84.

C1 RIḌWĀN B. MUḤAMMAD AL-KHURĀSĀNĪ

Suter, no. 343; *GAL*, I, p. 625, and SI, p. 866; article "Ibn al-Sāᶜātī" in EI_2 by H. Suter and J. Vernet.

7.1.3 *K. ᶜIlm al-sāᶜāt wa-l-ᶜamal bi-hā*, a treatise on mechanical clocks in 5 *faṣl*s: DM 890 and DR 488 (two sets of photos of MS Istanbul Köprülü 949, 116 pls., 658H by Baylak al-Qipjāqī (no. C9)); plus a wretched copy thereof in TṢ 24 (199 pp., 1332H). Another copy is MS Gotha 1348 on which the translation by Wiedemann and Hauser was based. Now published by M. A. Dahman (Damascus, n.d.).

C2 ABU L-ᶜIZZ ISMĀᶜĪL B. AL-RAZZĀZ AL-JAZARĪ

Suter, no. 344; *Krause*, no. 344; *GAL*, I, pp. 902–903.

7.1.4 *al-Jāmiᶜ bayn al-ᶜilm wa-l-ᶜamal fī ṣināᶜat al-ḥiyal*, a treatise on mechanical devices: DR 487 (photos of [MS Topkapi A 3472], 356 pls., purportedly 602H); DR 486 (photos of [MS Topkapi A 3350], 500 pls., 863H, lacks beginning), plus a wretched recent copy thereof in TṢ 37 (529 pp., 1333H).
A translation of this work and a commentary based mainly on the Bodleian manuscript was published recently by D. R. Hill, *The Book of Knowledge of Ingenious Mechanical Devices* . . . , Dordrecht & Boston: Reidel, 1974. See my review in *HS*, 13 (1975), pp. 284–289. The Arabic text was published by A. Y. al-Hassan in Aleppo in 1979.

C3 NAJM AL-DĪN AYYŪB B. ᶜAYN AL-DAWLA B. NAṢR ALLĀH AL-IKHLĀṬĪ

GAS, VII, pp. 21–22.

5.3.9 *Iẓhār mā kāna mustakhfiyᵃⁿ fī aḥkām al-nujūm*, a treatise on astrology consisting of an introduction and non-numerical tables: DM 4 (74 fols., *ca.* 900H). A late copy of this work is in MS Istanbul Beyazit 4642, and a fragment in MS Berlin Ahlwardt 5880. Sezgin (*GAS*, VII, p. 22) lists another copy in Tehran. This work merits detailed investigation. See Pl. LXXXI.

C4 SIRĀJ AL-DUNYA WA-L-DĪN

Unidentified.

3.2.4 *Maṣābīḥ al-anwār wa-mafātīḥ al-asrār fī aᶜmāl al-layl wa-l-nahār*, a treatise on timekeeping in 19 *bāb*s, the last two on the almuqantar and sine quadrants in 11 + 20 *faṣl*s respectively: ṬM 127 (59 fols., 1255H), unique. The author mentions (fol. 50r) a work by another author who is not named entitled *K. Ḥall al-mushkilāt fī maᶜrifat al-daqāᵓiq wa-l-daraj wa-l-sāᶜāt*, no copies of which are known to me. MS Princeton Yahuda 4657 (= Mach 4983), fols. 40v–80r, *ca.* 750H, attributed simply to Sirāj, purports to be an abridgement of the treatise *Maṣābīḥ al-anwār*, but the two works are not so closely related. The treatise in the Princeton manuscript was composed in 607H (fol. 40v). If the *Maṣābīḥ al-anwār* were related to this, it would be the earliest dateable treatise in which the almucantar quadrant is described. However, the two chapters on quadrants were incorporated by a copyist: the first is identical with the treatise of Ibn al-Majdī (no. C62) (in a *muqaddima* and 10 *faṣls*) and the second is identical with the treatise of Jamāl al-Dīn al-Māridīnī (no. C47) (in 20 *bābs*).

6.4.33 *Hidāyat al-ṭullāb ilā ᶜilm al-ḥisāb*, a treatise on arithmetic in 3 *bābs*: K4004,1 (fols. 1r–36v, *ca.* 1200H). Another copy is in Alexandria (*cf. GAL*, SII, p. 1018, no. 9).

C5 ABU L-GHANĀᵓIM MUSLIM B. MAḤMŪD B. NIᶜMA B. ARSLĀN AL-SHAYZARĪ

GAL, I, p. 302, and SI, p. 460 (does not mention the work below); *MAY*, p. 22.

1.4.8 *ᶜĀdāt al-nujūm*, an almanac for an unspecified locality: K 4678 and K 16000 (two sets of photos of a manuscript in the Grand Mosque, Sanaa numbered *majāmīᶜ* 32, 58 pls., 1076H), other copies in the Biblioteca Ambrosiana in Milan. A brief description of this work is in *MAY*. See also no. B3, B107, and D28.

C6 SHAMS AL-DĪN ISMĀᶜĪL B. IBRĀHĪM B. GHĀZĪ AL-MĀRIDĪNĪ, IBN FALLŪS

Suter, no. 359; *Krause*, no. 359; *GAL*, I, p. 622, and SI, p. 860; *GAS*, V, pp. 76 and 166.

6.3.13 *K. Niṣāb al-jabr fī ḥisāb al-jabr*, a treatise on algebra in 5 *faṣls* compiled in Cairo in 636H: B 23317,2 (fols. 55r–61v, 789H); J 3964,3 (fols. 10r–21v, 862H); DR 122,3 (fols. 21r–24v, 707H, anon. fragment); *etc.* (1 late copy).

6.3.13 *Iᶜdād al-asrār fī asrār al-aᶜdād* on number theory, in 3 *bābs*: B 23317,3 (fols. 62r–72r, 789H).

6.3.13 *Ḥall ᶜaqd al-ishkāl fī misāḥat al-ashkāl* on geometry, in 2 *bābs*, compiled in Cairo in 629H: B23317,4 (fols. 72r–76v, 789H), *etc.* (3 late copies). Printed Cairo, 1310H (Brockelmann).

6.3.13 *Irshād al-ḥussāb ila l-maftūḥ min ᶜilm al-ḥisāb*, in 5 *bābs*: B 23317,5 (fols. 77r–81r, 789H). See Pl. LXXXVIIIb.

6.3.13 *Tarjamat Ibn Fallūs*, recorded by al-Ḥāfiz Jamāl al-Dīn Abu l-Muḥāsin Yūsuf b. Aḥmad al-Taghrī Birdī: K 8522,2 (fol. 44v, *ca.* 900H).

Each of these four works by Ibn Fallūs merits investigation. Brockelmann lists a fifth treatise *Mīzān al-ᶜulūm fī taḥqīq al-maᶜlūm*, extant in MS Ambrosiana C 217, iii.

C7 MUḤAMMAD B. ᶜUMAR B. ᶜALĪ AL-ᶜUTHMĀNĪ AL-FĀRIQĪ

GAS, VII, p. 169.

5.2.17 *Ikhtiṣār al-Dawr al-ᶜuṭāridī*, an abridgement of an astrological work (now lost) by al-Ḥasan b. Jaᶜfar al-Anṣārī (no. B49), compiled in Aleppo in 641H: ṬM 150 (58 fols., 1181H); DM 54 (45 fols., *ca.* 1150H); *etc.* (1 copy). Sezgin lists three other copies in Istanbul.

C8 IBN KHALLIKĀN (should be listed in Section G)

Article by J. W. Fück in EI_2.

1.6.6 A fragment from the *Wāfiyāt al-ayᶜān* on al-Maᵓmūn's determination of the terrestrial distance corresponding to one degree of latitude is in ZK 782,12 (fol. 70r, 1126H). See no. B8.

C9 BAYLAK B. ᶜABD ALLĀH AL-QIPJĀQĪ

GAL, SI, p. 904; *Ullmann*, p. 128; and the article "Baylak al-Qipjāqī" by M. Plessner in *DSB*.

2.1.6 On his treatise on Ibn Yūnus' eclipse tables, see no. B59.

3.1.24 *Jadāwil maṭāliᶜ al-falak al-mustaqīm min awwal al-Ḥamal maḥlūla daqīqa daqīqa*, tables of the right ascension for each minute of ecliptic longitude, prepared in Cairo in 674H: MM 41 (41 fols., 1191H, preceded by an introduction by Sayf (!) al-Dīn Ibn al-Mushrif (no. C43)), unique. See also no. C1 on a manuscript copied by him in 648H.

C10 AL-ASᶜAD IBN AL-ᶜASSĀL

Article "Ibn al-ᶜAssāl" by A. Atiya in EI_2; *Graf*, II, pp. 403–407.

2.5.3 A handbook of text and tables for finding the ᵓ*bṭqy* (elsewhere ᵓ*bqṭy*) of the sun and moon, that is, their longitudes, as well as for determining fasts in the Coptic calendar and for converting calendar: DM 910,1 (fol. 3r + 6r–82v, *ca.* 1200H). Merits detailed investigation. See Pls. XXa–b.

Graf (II, pp. 405–407) lists another copy in MS Vatican ar. 152 (copied in 1354). The term ᵓ*bṭqy* / ᵓ*bqṭy* derives from the Greek *epaktai*, epact (the number of days by which the solar year exceeds 12 lunar months). The curious Arabic term *ḥabṭaq*, used for double argument tables of the lunar equation (see, for example, no. B59, *sub* 2.1.6), appears to be related to this.

2.5.3 A poem on the ᵓ*bqṭy* (see above) and the Coptic fasts: DM 910,2 (fols. 3r–5v, *ca.* 1200H); plus an anonymous commentary: DM 853 (94 pp., 1203H).

C11 JAMĀL AL-DĪN AḤMAD B. THIBĀT

Suter, no. 146; *Krause* no. 146; *GAL*, SI, p. 860.

6.7.7 ***Ghunyat al-ḥussāb fī ᶜilm al-ḥisāb*** on arithmetic, extant in MSS Istanbul Aya Sofia 2728 (see *Krause*) and Damascus Ẓāhirīya 3075 (cop. 693H, *cf. Cat.*, pp. 112–113), extracted by Muḥammad b. Ibrāhīm al-Ḥalabī: see no. C112.

C12 ANONYMOUS[1]

2.1.15 *al-Zīj al-Muṣṭalaḥ* (*Kennedy*, no. 43): Anonymous fragments without mention of the name of the *zīj*, identified using two complete recensions of the ***Muṣṭalaḥ Zīj*** surviving in MSS Paris B.N. ar. 2513 and 2520:

(a) Lunar mean motion and equation tables in MM 39 (5 fols., *ca.* 900H).

(b) Solar mean motion and equation and declination tables in MM 168,1 (fols. 1r–5r + 17r–17v, *ca.* 800H) and DM 620,4B (fols. 35v–36r, *ca.* 950H).

(c) Ch. 2 of the introduction in DM 1106 (32 fols., *ca.* 1150H, falsely entitled ***Sharḥ Zīj Ibn Yūnus***).

(d) A list of planetary apogees from the ***Muṣṭalaḥ Zīj*** for the beginning of the year 740H: MM 241,1 (fol. 1r, *ca.* 850H).

See also the next entry.

1. Ḥājjī Khalīfa (*cf. Kennedy*, no. 43) attributes the ***Muṣṭalaḥ Zīj*** to Muḥammad b. Muḥammad al-Fāriqī al-Muḥāsib, on whom I have no further information.

C13 ANONYMOUS

2.1.16 Introduction to a *zīj* which I label *Pseudo-Muṣṭalaḥ*, without any tables: MM 38 (22 fols., *ca.* 1100H); ṬM 191 (30 fols., 1330H from MM 38); unique. The title reads: "This is ***al-Zīj al-Muṣṭalaḥ*** and its planetary equations, called ***Taysīr al-ṭālib*** (read: ***al-maṭālib***) ***fī tasyīr al-kawākib***, and its planetary equations are according to the method of the distinguished scholar Kūshyār and are always additive, written by Abū Bakr Ibn al-Mushrif, compiled by the scholar, the *muwaqqit* Aḥmad ᶜUbayd Allāh . . . " The text is introduced in the name of Aḥmad ᶜUbayd Allāh and consists of unnumbered *faṣl*s dealing with topics of planetary and spherical astronomy. Some of the material is attributed to Ibn Yūnus (no. B59) and Kūshyār (no. B70).

In passing we note firstly that a small *zīj* entitled *Taysīr al-maṭālib fī tasyīr al-kawākib* was compiled in Cairo *ca.* 1100H by Aḥmad b. Muḥammad ᶜAyyād (no. D67) which is unrelated to the work in MM 38. Secondly, the planetary equation tables of Kūshyār in his *Zīj* are not additive. Thirdly, Abū Bakr Ibn al-Mushrif (no. C43) is not otherwise known to have written on planetary astronomy. Fourthly, the *muwaqqit* Aḥmad ᶜUbayd Allāh (no. C167) is not otherwise known to the literature, but there may be some confusion with the Aḥmad ᶜAyyād mentioned above. See also no. D204.

C14 ḌIYĀᵓ AL-DĪN (?) ᶜABD AL-ᶜAZĪZ B. AḤMAD B. SAᶜĪD AL-DĪRĪNĪ

GAL, I, pp. 588–589, and SI, p. 810.

3.2.3 ***K. al-Yawāqit fī ᶜilm al-mawāqīt***, a poem on timekeeping with occasional passages in prose: DM 651,5 (fols. 41r–58r, 828H, anonymous, lacks beginning); DM 584 (19 fols., *ca.* 1150H, lacks end). These two copies seem to be based on two different versions. Brockelmann mentions a copy in Mosul; another fine copy is in MS Istanbul Hamidiye 1453. See *SATMI*, II for a summary.

C15 ABU L-ᶜABBĀS SHIHĀB AL-DĪN AḤMAD B. ᶜUMAR B. ISMĀᶜĪL AL-MAQSĪ

Suter, no. 383; *GAL*, I, p. 626, and SI, p. 869.

3.1.2 *K. al-Dāᵓir = K. Ghāyat al-intifāᶜ fī maᶜrifat al-dāᵓir min qibal al-irtifāᶜ*, tables of time since sunrise as a function of solar altitude and longitude for latitude 30°, Cairo, with an introduction: DM 444 (37 fols., 858H); DM 776,1 (fols. 1v + 10r–43r, 981H). These tables are contained in al-Bakhāniqī's edition of the Cairo corpus (see no. C28). See further *SATMI*.

3.1.2 *K. al-Dāᵓir wa-faḍlihi* tables of time since sunrise and the hour-angle for latitude 30°, Cairo: DM 777 (42 fols., *ca.* 900H, anonymous). The hour-angle tables were probably computed by Ibn al-Kattānī (no. C32) using al-Maqsī's tables of time since sunrise. They are also copied separately as *K. Faḍl al-dāᵓir* in DM 778 (33 fols., *ca.* 1050H, anonymous, disordered and incomplete); TR 191 (90 pp., 1215H, attributed to Ibn Yūnus (no. B59)); K 4044,1 (fols. 1r–41r, 1308H, attributed to al-Maqsī). See further *SATMI* on all these tables.

4.7.3 *Shifāᵓ al-asqām fī waḍᶜ al-sāᶜāt ᶜala l-rukhām*, a treatise on sundial theory with an introduction in 15 *faṣl*s and extensive tables, compiled in 675H: DM 103,1 (fols. 1r–28r + 57r–130v, partly 719H, defective and disordered, see also DM 517); DM 955 (72 fols., *ca.* 750H, defective); DM 517 (68 fols., partly [719H], defective, contains some tables missing from DM 103,1); *etc.* (3 fragments). Several complete copies of this work survive. Merits detailed analysis.

C16 NAJM AL-DĪN MUḤAMMAD B. MUḤAMMAD AL-MIṢRĪ

Suter, no. 460 (confused).

3.1.3 The second half of his *Jadwal al-dāᵓir al-āfāqī*: MM 132 (234 fols., *ca.* 800H, anonymous). The first half of the same manuscript is contained in MS Oxford Bodleian Marsh 672 and is there attributed to him. The work consists almost entirely of tables, the major one displaying the time since rising of the sun or any star as a function of its instantaneous altitude for any latitude. See further *SATMI*, I. An extract is shown in Pl. XXIX.

C17 SHARAF AL-DĪN ABŪ ᶜALĪ AL-ḤASAN B. ᶜALĪ B. ᶜUMAR AL-MARRĀKUSHĪ

Suter, no. 363; *Krause*, no. 363; *GAL*, I, p. 625, and SII, p. 866; *Mayer*, p. 46; omitted from *Azzawi* (!).

2.4.1 Solar declination tables for each minute of argument to 3 digits (obliquity 23;35°) are attributed to al-Marrākushī in DM 136,3 (fols. 53r–66r, 1147H). Two earlier anonymous copies of the same tables are in ṬM 80,3 (fols. 31v–46r, *ca.* 850H) and ṬM 235,3 (fols. 43r–58r, 866H).

3.2.5 Anonymous abridgement, without title, of the first part of *K. al-Mabādiᵓ wa-l-ghāyāt* (see below) dealing with spherical astronomy: DM 787,1 (fols. 1r–23v, *ca.* 750H, clear *naskhī*), unique. There is no specific indication of that the work is such an abridgment, and Abū ᶜAlī is mentioned only once (fol. 22r). See also no. C140.

4.1.1 *K. Jāmiᶜ al-mabādiᵓ wa-l-ghāyāt*, an extensive compendium of spherical astronomy and instruments arranged in 4 *fann*s of 80 *faṣl*s + 7 *qism*s + 14 *bāb*s + 4 *bāb*s, mostly subdivided further, compiled in Cairo *ca.* 680H: DM 935 (photos of MSS Paris B.N. ar. 2507 and 2508, 154 + 195 pls., *ca.* 750H); K 3821 (346 fols., *ca.* 900H in different hands); ṬH 48,2 (fol. 62v, *ca.* 700H, first page only!). See Pls. XXX and LIX–LXI.

Numerous fragments and extracts, mainly anonymous: Part I = I (on spherical astronomy) + II.1–3 (on sundials): DM 1208 (143 fols., *ca.* 1200H in different hands).

II.4–6 (on sinical, spherical, and astrolabic instruments): DM 115 (32 fols., *ca* 1200H).

I.11 (on trepidation): MM 213,2 (fols. 27r–28r, *ca.* 1150H); MM 9,2 (fols. 28v–29r, *ca.* 1200H).

I.87 (on formulae of spherical astronomy): DM 291,3 (fols. 4r–6r, *ca.* 900H); DM 125,3 (fols. 3r–4r, *ca.* 1000H). See Pl. XXX.

II.2.1 (on the *ḥāfir*): DM 597,2 (fols. 2r–40r, *ca.* 1050H).

II.6.4 (on the projection of the *Zarqāllīya* and *Shakkāzīya ṣafīḥas*): ṬM 155,4 (fols. 21v–29r, *ca.* 1050H); DM 194,2 (fols. 53r–65v, *ca.* 1100H).

III.12 (on the use of the *Zarqāllīya ṣafīḥa*): DM 194,1 (fols. 1v–53v, *ca.* 1100H).

Notes: TR 140,14 (pp. 74–82 + 88, *ca.* 1150H).

4.2.4 III.7 of the *K. al-Mabādiᵓ* on the use of the armillary sphere in 90 *faṣls*: K 4050 (47 fols., *ca.* 1150H), *etc.* (3 copies). Shorter anonymous version (in 88 *faṣls*, lacking 81 and 88 of the longer version): MM 175,1 (fols. 1v–30r, *ca.* 1100H), *etc.* (2 copies). *Cf. Krause*, p. 525, no. 17, and *GAS*, 149 for an Istanbul copy in 88 *faṣls*. It seems probable that al-Marrākushī lifted the treatise in 88 *faṣls* from one of his predecessors and added 2 *faṣls* of his own. The introduction mentions treatises on the armillary sphere by Eutocios "al-ᶜAsqalānī," Philon, Qusṭā b. Lūqā, Heron, and Theon (see *GAS*, V, pp. 185–186 and 401 and also *Krause*, p. 526, no. 16).

n. b. al-Marrākushī apparently wrote his compendium in Cairo not in Morocco as is usually stated. His work was known in Egypt, Syria, the Yemen, and Turkey. I know of no Maghribi copies.

It should also be stressed that only the first half of al-Marrākushī's work, namely the parts dealing with spherical astronomy and sundials, is translated in J. J. Sédillot, *Traité des Instruments Astronomiques des Arabes composé au treizième siècle par Aboul Hhassan Ali de Maroc . . .*, 2 vols., Paris: Imprimerie Royale, 1834–1835. The second half dealing with instruments was summarized in a very confused fashion in L. A. Sédillot, "Mémoire sur les Instruments Astronomiques des Arabes," *Mémoires de l'Academie Royale des Inscriptions et Belles-lettres de l'Institut de France*, 1 (1844), pp. 1–229. However, Sédellot-fils ignored most of the significant historical details included by al-Marrākushī. A table of contents of MSS Paris B.N. ar. 2507 + 2508, which are the copies used by the Sédillots, is included in the Arabic part of the catalogue (II, Section 4.1.1). This work merits publication and detailed analysis. Krause lists eight copies, of which MS Nurosmaniye 2902 (266 fols., 695H!) is apparently complete. The Cairo copies listed above can be ignored by anyone who wishes to work on the text alone, but they do indicate the popularity of the work in Egypt in later centuries.

See also the next entry.

C18 ANONYMOUS (AL-MARRĀKUSHĪ?)

4.3.15 A treatise on the construction of the astrolabe in 8 *bābs*: MM 211,1 (fols. 1v–9r, *ca.* 850H, lacks tables promised in text). The author mentions star coordinates for 681H, which is the epoch of al-Marrākushī, and Ibn Yūnus' value of precession, namely, 1° in 70¼ years. (There is no reference to Ibn Yūnus (no. B59) in al-Marrākushī's *Jāmiᶜ al-mabādiᵓ wa-l-ghāyāt*.) Another copy of the treatise in 8 *bābs* is probably MS Damascus Ẓāhiriya 9239 (*Cat.*, p. 156).

C19 NAJM AL-DĪN ABU L-ᶜABBĀS AḤMAD B. MUḤAMMAD B. ᶜALĪ B. AL-RIFᶜA AL-ANṢĀRĪ

Suter, no. 386; *GAL*, II, pp. 165–166, and SII, p. 164.

6.8.2 *K. al-Īḍāḥ wa-l-tibyān fī maᶜrifat al-mikyāl wa-l-mīzān*, a treatise on weights and measures, compiled in 703H: J 3964,12 (fols. 51r–57v, 862H); TR 359 (16 fols., *ca.* 1050H); ṬR 145 (7 fols., 1196H); *etc.* (2 late copies).

C20 ABŪ ᶜALĪ AL-FĀRISĪ[1]

(see also no. C21)

Awwad, no. 175.

4.3.19 ***Maqāṣid dhawi l-albāb fi-l-ᶜamal bi-l-asṭurlāb***, a treatise on the use of the astrolabe in 4 *maqālas* of 8 + 9 + 4 + 2 *faṣls*, compiled in Hama *ca.* 700H: QM 2,1 (fols. 1r–57v, *ca.* 800H), unique! See Pl. LVIII.

1. In the old Qawala catlog Abū ᶜAlī al-Fārisī is falsely identified with Abū ᶜAlī al-Ḥasan b. Aḥmad b. ᶜAbd al-Ghaffār al-Fārisī, who died in 377/987 (!) (*cf. GAL*, SI, pp. 175–176).

C21 ʿUMAR B. DĀʾŪD B. SULAYMĀN AL-FĀRISĪ[1]

(see also no. C20)

GAL, I, p. 346 and SI, p. 509.

1.2.10 ***(Takmilat al-Tadhkira)***, a complement to al-Ṭūsī's ***Tadhkira*** (see no. G23), compiled *ca.* 710H for Abu l-Fidāʾ (*cf. EI*$_2$, s.v.): TR 128 (99 fols., *ca.* 850H) and photos thereof in K 7193, unique. The work is arranged in 4 *bāb*s of 2 + 17 + 12 + 7 *faṣl*s, and the title mentioned in the introduction is illegible. Merits investigation in any study of the ***Tadhkira***. See Pl. VIIb.

C22 TĀJ AL-DĪN ABU L-ḤASAN ʿALĪ B. ABĪ MUḤAMMAD ʿABD ALLĀH B. AL-ḤASAN B. ABĪ BAKR AL- TABRĪZĪ, a teacher at the Ṭaranṭāy *madrasa* in Cairo.

Not mentioned in the modern sources.

1.2.13 ***K. al-Nuzha al-ʿAlāʾīya***, a treatise in four *qism*s dealing with arithmetic, geometry, astronomy, and astrology: ṬR 104 (39 fols., *ca.* 800H, lacks most of the 3rd *qism* and all of the 4th), unique. See Pl. CVIc.

C23 BADR AL-DĪN MUḤAMMAD B. IBRĀHĪM B. SAʿD ALLĀH B. JAMĀʿA AL-KINĀNĪ AL-ḤAMAWĪ

GAL, II, pp. 89–90 and SII, pp. 80–81; *Awwad*, no. 179.

4.3.18 ***Mulakhkhaṣ al-albāb fi l-ʿamal bi-l-asṭurlāb***, an Arabic treatise on the astrolabe in 60 *bāb*s, compiled 1597 Alexander (= 1285/86): MMT 6,1 (fols. 1v–20r, *ca.* 1150H), unique. This treatise is not listed by Brockelmann but is listed by Awwad, who refers to this unique manuscript.

C24 NĀṢIR AL-DĪN MUḤAMMAD B. SIMʿŪN

Suter, no. 398; *GAL*, II, p. 155.

3.2.2 ***al-Tuḥfa al-malakīya fi l-asʾila wa-l-ajwiba al-falakīya***, 21 questions and answers on timekeeping: DM 25 (7 fols., 1105H), unique.

4.6.16 ***Nihāyat al-musāmara fi l-ʿamal bi-l-musātara***, on the use of the instrument called *al-musātara* in 18 *bāb*s: QM 2,4 (fols. 86v–92r, 706H, anonymous). This treatise is probably by Ibn Simʿūn. Another anonymous copy of this work is MS Dublin Chester Beatty 4833,2 (fols. 9–11, *ca.* 750H). See also no. C25.

4.6.19 ***al-Uṣūl al-thāmira fi l-ʿamal bi-rubʿ al-musātara***, a treatise on the *musātara* quadrant in 18 *bāb*s: QM 2,3 (fols. 82r–85r, *ca.* 700H), unique.

4.7.4 ***Shifāʾ al-ṣudūr fi l-ʿamal bi-l-mīzān al-maḥfūr***, a treatise on the instrument called *al-mīzān al-Fazārī* in 20 *bāb*s: QM 2,5 (fols. 92r–98r, 706H, anonymous), unique. See Pl. CIIIb. The style indicates that the anonymous author is the author of QM 2,4 (see above). Merits detailed study.

C25 ANONYMOUS (treatise on the *musātāra*)

4.6.16 ***Nihāyat al-musāmara fi l-ʿamal bi-l-musātara***, on the use of the horizontal dial in 18 *bāb*s: QM 2,4 (fols. 86r–92r, 706H). Perhaps the author is Ibn Simʿūn (no. C24).

C26 SHIHĀB AL-DĪN AḤMAD B. ABĪ BAKR B. ʿALĪ B. AL-SARRĀJ

Suter, no. 508 (confused); *GAL*, II, p. 155, and SII, p. 156.

4.6.4 ***R. al-Āla al-Sarrājīya fi stikhrāj al-aʿmāl al-āfāqīya***, a treatise on the use of the universal astrolabe which the author invented: DM 291,1 (fols. 1r–2r, *ca.* 900H), unique. See Pl. LXV. Analyzed together with a description of the universal astrolabe made by Ibn al-Sarrāj preserved in the Benaki Museum, Athens, in D. A. King, *The Astronomical Instruments of Ibn al-Sarrāj*, to be published by the Benaki Museum.

4.6.13 ***R. fi l-ʿAmal bi-l-rubʿ***, a treatise on the universal (*jāmiʿ*) quadrant, arranged in 13 *bāb*s, DM 138,8 (fols. 14r–18r, 840H, anonymous), unique. The author is clearly the inventor of the

---

1. An abridgement by Zayn al-Dīn Abū Hafṣ ʿUmar al-Fārisī, known as al-ʿAjamī, of the Qurʾān commentary entitled ***al-Kashshāf*** by al-Zamakhsharī is in MS Cairo Dār al-Kutub *tafsīr* 81, copied in 722H.

instrument, and in his treatise on the *jayb ghāʾib* (see below) Ibn al-Sarrāj states that he had considered such an instrument in the form of a quadrant.

4.6.26 ***R. fi l-Rub*ᶜ *al-mujannaḥ***, a treatise on the winged quadrant in 25 *bābs*: DM 64,5 (fols. 74v–76v, 803H); DM 138,7 (fols. 11v–14r, 840H).

4.6.26 ***R. laṭīfa fi l-ᶜamal bi-ṣadr al-awizza***, a treatise without divisions on the instrument called the "goose's chest": ṬM 242,10 (fol. 116v, 1180H), unique.

6.3.21 ***Masāʾil handasīya***, geometrical problems: DR 694 (9 fols., 748H by author, lacks first page), unique. See Pl. CIIIc.

6.3.21 ***(Tisᶜ) masāʾil handasīya***, geometrical problems: MR 41,27 (fols. 159v–161r, 1149H), unique.

6.8.5 A short treatise on a balance for measuring gold: MR 30,6 (fols. 31v–32r, *ca.* 1100H), unique.

C27 TAQI L-DĪN ᶜALĪ B. ᶜABD AL-KĀFĪ AL-SUBKĪ

GAL, II, pp. 106–107, and SII, pp. 102–104.

6.9.2 ***Kashf al-ghumma fī mīrāth ahl al-dhimma*** a treatise in 15 *bābs* on the inheritance laws relating to Jews and Christians: B 23317,1 (fols. 1r–54v, 789H), unique. See Pl. LXXXVIIIb. This title is not listed by Brockelmann, and the work has been studied by Dr. Seth Ward of Yale University.

Three other works of al-Subkī relating to scientific topics listed by Brockelmann (*GAL*, SII, p. 103) are ***al-Adilla fī ithbāt al-ahilla*** in MS Jerusalem Khalidiya 71,1; ***Bayān al-adilla fī*** *ithbāt al-ahilla* in MS Jerusalem Khalidiya 71,2; and *al-ᶜAlam al-manshūr fī ithbāt al-shuhūr* in MS Cairo Dar al-Kutub *fiqh Shāfiᶜī* 1414 (12 fols., 1322H), printed in Cairo in 1328H. These titles may relate to one and the same work.

C28 SHAMS AL-DĪN AḤMAD B. MUḤAMMAD B. AḤMAD AL-AZHARĪ AL-BAKHĀNIQĪ

GAL, SII, pp. 158 and 1019; *MAY*, pp. 34–35.

3.1.10 His edition of the main corpus of tables for timekeeping for the latitude of Cairo is available in several manuscripts, no two of which contain the same tables and not one of which contains a reliable statement concerning the authorship of the tables—see Pls. XXVI and XXVII. The tables are variously titles ***K. al-Dāʾir wa-faḍluhu wa-l-samt***, ***K. Ghāyat al-intifāᶜ*** . . . or *K. Kifāyat al-waqt* . . . In some of the copies there is an introduction by al-Bakhāniqī giving some indication of the way in which he put together the corpus—see Plate XXVIII. The most complete copy is DM 108 (77 fols., 1218H, attributed to Ibn Yūnus (no. B59)). On the tables in this manuscript see *AHES*, 10 (1973), pp. 342–394. Other copies are DM 690 (58 fols., *ca.* 900H, defective); DM 53 (77 fols., 1117H); DM 616 (74 fols., *ca.* 900H); DM 739 (64 fols., *ca.* 1050H); *etc.* (5 defective copies). These manuscripts are investigated in *SATMI*.

3.1.10 ***Jadwal maᶜrifat faḍl dāʾir al-ᶜaṣr li-ruʾūs al-burūj***, tables of the hour-angle at the *ᶜaṣr* for different latitudes: MM 33,2 (fols. 6v–7r, *ca.* 1000), unique.

See also no. C116.

C29 KHALĪL B. ISḤĀQ AL-JUNDĪ

GAL, II, pp. 101–103, and SII, pp. 96–99, esp. SII, pp. 98–99, no. aa.

6.9.11 On a commentary by al-Qalaṣādī on the section on inheritance in his ***al-Mukhtaṣar***, a book of Mālikī law, see no. F34.

C30 ᶜALĀʾ AL-DĪN ᶜALĪ B. IBRĀHĪM, IBN AL-SHĀṬIR

Suter, no. 416; *GAL*, II, p. 156, and SII, p. 157; *Kennedy*, no. 11; E. S. Kennedy and I. Ghanem, eds., *The Life and Work of Ibn al-Shāṭir*, (Aleppo, 1976), contains photo-offset reproductions of all articles on Ibn al-Shāṭir up to 1975; article "Ibn al-Shāṭir" by D. A. King in *DSB*.

1.2.15 *Nihāyat al-suʾl fī taṣḥīḥ al-uṣūl*, a treatise on planetary astronomy in 30 *bābs* in which the author gives the reasons for a modification to the Ptolemaic models: TR 154 (74 pp., *ca.* 1325H, incomplete, wretched). There are several good copies of this important work. A preliminary translation of this treatise has been prepared by V. Roberts. An edition of the text is being prepared by Prof. G. Saliba of Columbia University.

2.1.19 ***al-Zīj al-Jadīd***: inconsequential fragments in DM 64,6 (fols. 77r–79r, star catalog only, 803H); ṬFF 13,2 (fols. 118v–121v, *ca* 1000H); DM 940,2 (photos of MS Paris B.N. ar. 2541,5, 2 pls., *ca.* 1200H); ṬM 73,4 (fol. 6v, 1235H); ṬJ 811,4 (fols. 34v–35v, *ca.* 1150H); Sh 76,5 (fols. 8v–10r, *ca.* 1300H); MMF 5,2 (fols. 3v–5r, *ca.* 1175H). Complete copies of this *Zīj* are rather rare: a good example is MS Oxford Seld. A. inf. 30. For a summary of the contents see *Kennedy*, pp. 162–164.

2.1.19 *Syrian recensions*: (a) ***al-Rawḍ al-ᶜāṭir fī talkhīṣ Zīj Ibn al-Shāṭir*** by Ibn Zurayq (no. C116): TR 235,1 (pp. 2–143, *ca.* 1150H, incomplete, no tables, Maghribi script).
(b) ***Nuzhat al-nāẓir bi-khtiṣār Zīj Ibn al-Shāṭir*** by al-Mukhallalātī (no. D90): MJ 7,2 (fols. 44r–73r, *ca.* 1150H, no tables); ṬJ 515,4 (fols. 58r–63v + 55r–57v + 64r–80r, *ca.* 1200, incomplete, contains some tables).

2.1.19 ***Egyptian recension*** entitled ***al-Lumᶜa*** by al-Kawm al-Rīshī (no. C41): TR 275 (103 pp., 858H); DM 637,1 (fols. 1r–45r, 868H); *etc.* (numerous copies, and various commentaries—see no. C41).

2.1.19 ***Algerian recension*** entitled ***Rāᶜi l-kawākib*** by an anonymous author (no. F66), with tables for the longitude of Algiers 33;30° based on the parameters of Ibn al-Shāṭir: DM 533 (43 fols., *ca.* 1150H), unique?

3.1.8 / 3.1.17 The prayer-tables for latitude 34° which accompany a star catalog for 1033H by al-Ḥulwānī (no. D22) in DM 1170,2 (fols. 11r–22v, *ca.* 1050H) may be due to Ibn al-Shāṭir. They correspond to the tables described by Ibn al-Shāṭir in MS Leiden Or. 1001. See further *SATMI*, II.

3.5.4 A note by ᶜIzz al-Dīn al-Wafāʾī (see no. C61) attributes a theory of refraction at the horizon to Ibn al-Shāṭir. See Pl. XXXIII.

4.1.3 *al-Ishārāt*, a treatise in 3 *faṣl*s on the astrolabe, almucantar, and sine quadrant: DM 476 (6 fols., 1st. *faṣl* only, *ca.* 1100H); DM 926,1 (photos of MS London B.L. Add. 9599, 5 pls., 1st. *faṣl* only, late); MM 144,1 (fols. 1r–10r, *ca.* 850H). No other known copies.

4.3.22 (***R. fi l-Uṣūl min ᶜilm al-asṭurlāb***), on the use of the astrolabe in 10 *bāb*s: DM 479 (5 fols., *ca.* 1100H); K 18359 (5 fols., *ca.* 1250H).

4.4.2 Three copies of *Hidāyat al-ᶜāmil* on the use of the almucantar quadrnat by Sibṭ al-Māridīni (no. C97) are attributed to Ibn al-Shāṭir, namely: DM 639,7; DR 363,2; DM 1063,5.

4.5.5 ***Īḍāḥ al-mughayyab fi l-ᶜamal bi-l-rubᶜ al-mujayyab***, a treatise on the use of the sine quadrant in 205 *bāb*s: DM 64,1 (fols. 1v–35v, 803H); DM 1212 (36 fols., *ca.* 1300H, purportedly 1199H); *etc.* (1 fragment). Another copy exists in Bursa (*GAL*).

4.5.5 ***Kashf al-mughayyab fi l-ḥisāb bi-l-rubᶜ al-mujayyab***, in 54 *bāb*s: DM 64,2 (fols. 36v–56v, 803H). No other known copies.

4.6.5 ***al-Ashiᶜᶜa al-lāmiᶜa fi l-ᶜamal bi-l-jāmiᶜa***, on a universal astrolabe of his own invention: arranged in 60 *bāb*s: TR 169,1 (pp. 2–73, 1320H from a copy in Dar al-Quṭayna (in Jerusalem?)). Two other copies are in Leningrad and Aleppo. An anonymous commentary, probably due to Taqi l-Dīn (no. H12), entitled ***al-Thimār al-yāniᶜa min quṭūf al-āla al-jāmiᶜa***: DM 557,2 (fols. 35v–43v, 1058H). Two other copies of this commentary are MS Manchester 361,5 (anonymous) and Oxford (attributed to Taqi l-Dīn).

4.6.13+ ***Nuzhat al-sāmiᶜ fi l-ᶜamal bi-l-rubᶜ al-jāmiᶜ***, a treatise on the universal (*jāmiᶜ*) quadrant, arranged in 41 *bāb*s: DM 199 (31 fols., 789H); DM 138,11 (fols. 22r–34v, 840H).

4.6.13+ ***R. fi l-ᶜAmal bi-l-rubᶜ al-jāmiᶜ***, a treatise on the same instrument, arranged in 19 *bāb*s: DM 64,3 (fols. 57v–63r, 803H).

4.6.13+ ***al-Barq al-lāmiᶜ fi l-ᶜamal bi-l-rubᶜ al-jāmiᶜ***, a treatise on the same instrument, arranged in 16 *bāb*s, the first four of which correspond to those in the treatise listed above: DM 17 (8 fols., *ca.* 850H, anonymous), unique.

4.6.22 ***al-Nafᶜ al-ᶜāmm fi l-ᶜamal bi-l-rubᶜ al-tāmm***, on the use of the quadrant in 200 *bāb*s followed by a quiz of 100 *masʾala*s: DM 138,12 (fols. 34v–110v, 840H); plus one late fragment. Abridgement in 46 *bāb*s by Ibn al-Shāṭir: QM 2,2 (fols. 59v–80v, 739H). See also nos. C31 and H35.

4.6.25 ***R. fi l-ᶜAmal bi-l-murabbaᶜa*** on the use of a square grid in 8 *bāb*s: DM 138,5 (fols. 9r–10r, 840H, anonymous). Another anonymous copy of this treatise is contained in MS Damascus Ẓāhiriya 3098. Ibn al-ᶜAṭṭār in his *Kashf al-qināᶜ* (see no. C66) states that Ibn al-Shāṭir invented the *murabbaᶜa*. This instrument has not been described in the modern literature.

8.8 A treatise on chess: DM 560 (4 fols., 1128H), unique?

C31 ANONYMOUS (treatise on the "complete" quadrant of Ibn al-Shāṭir)

4.6.24+ *R. mukhtaṣara fī l-Āla al-musammā bi-l-rubᶜ al-muthallath aw al-jayb al-tāmm*: K17289,2 (fols. 5v–7r, *ca.* 1250H). Another anonymous treatise on the *muthallath* is in MS London B.L. Add. 9600,7 (*cat.*, pp. 194–195).

C32 MUḤAMMAD B. MUḤAMMAD B. MUḤAMMAD B. ᶜABD AL-QAWĪ AL-QURASHĪ, IBN AL-KATTĀNĪ

Suter, no. 411.

3.1.7 *Jadwal irtifāᶜ al-shams*, tables of solar altitude as a function of time, for latitude 30°, Cairo: MM 72 (32 fols., 747H by Ibn al-Kattānī, missing the beginning of the brief introduction), unique. These tables were probably compiled by Ibn al-Kattānī himself. See *SATMI*, I and also Pl. XXXIc.

C33 ABŪ ᶜABD ALLĀH MUḤAMMAD B. MUḤAMMAD, IBN AL-GHAZŪLĪ

Suter, no. 412; *GAL*, II, pp. 331–332 and SII, p. 364.

4.3.21 The *R. fī l-Aṣṭurlāb . . . al-maᶜrūfa bi-l-ᶜashara fuṣūl*, a treatise on the use of the astrolabe in 10 *faṣl*s attributed to Ibn al-Ghazūlī in DM 594 (9 fols., 1077H) is elsewhere attributed to al-Mizzī (see no. C34).

4.5.4 (*R. fī Ṣifat al-ᶜamal bi-l-rubᶜ al-mujayyab*), a treatise on the use of the sine quadrant in 10 unnumbered *faṣl*s: DM 443 (3 fols., *ca.* 900H).

4.6.20 *R. fī Rubᶜ al-musātara*, on a special variety of quadrant in 15 *bāb*s: MM 118 (9 fols., *ca.* 800H), unique.

4.6.34 (*R. fī l-ᶜAmal bi-thumn al-dāʾira*), a treatise on the octant in unnumbered *faṣl*s, compiled *ca.* 745H[1]: DM 138,3 (fols. 5v–8r, 840H); DM 170,2 (fols. 8r–10v, *ca.* 880H); ṬM 79,3 (fols. 29r–31v, incomplete, deals only with *jayb al-thumn, ca.* 950H); *etc.* (1 copy and 1 fragment).

4.6.39 *R. fī l-ᶜAmal bi-l-jayb al-ghāʾib*, a treatise on the use of the "concealed" trigonometric grid in 18 *bāb*s: ZK 782,7 (fols. 35v–40r, 1116H).

4.7.5 *Faṣl fī ᶜamal basīṭa^tin munḥarifa^tan bi-l-handasa*, on constructing a vertical inclined sundial from a horizontal sundial: DM 600,4 (fols. 6r–6v, *ca.* 1100H), unique.

C34 SHAMS AL-DĪN MUḤAMMAD B. AḤMAD B. ᶜABD AL-RAḤĪM AL-MIZZĪ

Suter, no. 406; *GAL*, II, pp. 155–156 and SII, pp. 156 and 1018 (no. 15).

3.1.4 Prayer-tables for latitude 33;27°, Damascus: MM 62 (10 fols., *ca.* 800H), unique. Analyzed in *SATMI*.

4.3.21 *Nakhab/Nakhabat al-albāb fī l-ᶜamal bi-l-aṣṭurlāb* = *R. fī l-ᶜAmal bi-l-aṣṭurlāb*, on the use of the astrolabe in 10 *faṣl*s: DM 771,1 (fols. 1r–10v, *ca.* 875H); DM 827,2 (fols. 8v–12v, 980H); DM 594 (9 fols., 1077H, attributed to Ibn al-Ghazūlī (no. C33)); *etc.* (4 copies). An anonymous fragment of this treatise exists in MS Damascus Ẓāhiriya 10163.

4.4.1 *al-Rawḍāt al-muzhirāt fī l-ᶜamal bi-rubᶜ al-muqanṭarāt* (in 35 *bāb*s): TJ 391,2 (pp. 41–78, 745H by the author, complete!); DM 763 (13 fols., 799H); DM 765 (20 fols., *ca.* 750H); DM 620,4A (fols. 19v–34v, 840H); MM 218 (17 fols., *ca.* 900H); *etc.* (6 copies). Only the first of these contains the tables of solar longitude for each day over a period of four Syrian years which al-Mizzī appended at the end of his treatise. See Pl. XIXc.

4.5.1 *Kashf al-rayb fī l-ᶜamal bi-l-jayb*, a treatise on the use of the sine quadrant in various recensions.

95 *bāb*s: DM 113 (21 fols., 787H, lacks first page, disordered).

---

1. DM 138,3 has 747H, DM 170,2 has 744H, MM 91,1 has 746H.

67 *bābs*: ṬM 79,1 (fols. 1v–23v, *ca.* 950H, lacks end); DM 116 (23 fols., 1027H); MM 195,2 (fols. 14v–37v, *ca.* 1150H).

43 *bābs*: MM 154 (18 fols., *ca.* 1100H).

24 *bābs*: TR 139,5 (pp. 100–110, *ca.* 1100); MM 187,2 (fols. 4v–8r, 1126H).

4.5.1 *Naẓm al-luᵓluᵓ al-muhadhdhab fi l-ᶜamal bi-rubᶜ al-mujayyab*, a poem on the sine quadrant in 20 *bābs*: TJ 367,1 (fols. 2v–10v, *ca.* 1100H). Another copy exists in MS Rabat 452,5.

4.6.27 *R. fi l-ᶜAmal bi-l-āla al-maᶜrūfa bi-l-mujannaḥa*, a treatise on the winged quadrant in 40 *bābs*: DM 1093,11 (fols. 51v–68r, 1088H); DM 771,6 (fols. 23r–27v, *ca.* 900H, anon. fragment); *etc.* (1 defective copy).

4.6.31 *R. fi l-ᶜAmal bi-rubᶜ al-dāᵓira al-mawḍūᶜ ᶜalayhi l-muqanṭarāt al-maṭwīya*, on the use of a special kind of almucantar quadrant, in 12 *bābs*: DM 138,2 (fols. 3v–5v, 840H, anon.). The author mentions his treatise on the almucantar quadrant in 35 *bābs* (see above).

C35 ZAYN AL-DĪN ABŪ BAKR MUḤAMMAD B. AYYŪB AL-ṢŪFĪ AL-KARAKĪ, a *muwaqqit* in Jerusalem

GAL, SII, p. 156.

4.4.1 MS ṬJ 391,2 of the treatise on the almucantar quadrant by al-Mizzī (no. C34) contains an *ijāza* in the author's hand to his student Zayn al-Dīn. . . . al-Ṣūfī—see Pl. CIVa. See also no. D222.

C36 ABU L-MAᶜĀLĪ MUḤYI L-DĪN MURTAFAᶜ B. ḤASAN B. MURTAFAᶜ AL-SĀᶜĀTĪ

Awwad, no. 119.

4.3.26 *Iẓhār al-ᶜajāᵓib min al-asṭurlāb al-ghāᵓib*, a treatise on the use of the astrolabe in 30 *bābs*: MM 11 (6 fols., anonymous, *ca.* 1100H); ṬM 242,3 (fols. 33v–40r, anonymous, *ca.* 1250H); plus on anonymous fragment in MM 43,4 (fols. 50r–50v, *ca.* 850H).

Ḥājjī Khalīfa (II, 1496) mentions a treatise on the sine quadrant by this author in 50 *bābs* entitled *Kashf al-mughayyab fi l-ᶜamal bi-l-rubᶜ al-mujayyab*: no known copies.

C37 SHAMS AL-DĪN ABŪ ᶜABD ALLĀH MUḤAMMAD B. MUḤAMMAD AL-KHALĪLĪ

Suter, no. 418; *GAL*, II, pp. 156–157, and SII, p. 157; article "al-Khalīlī" by D. A. King in *DSB*, supplement.

3.1.9 *al-Jadwal āfāqī*, universal auxiliary tables for solving problems in spherical astronomy, with a short introduction in 8 *bābs*: ṬM 228,1 (fols. 1r–3r + 4r–7v + 61r–68r, *ca.* 900H); MM 43,1 (fols. 1v–24r, *ca.* 850H); *etc.* (2 copies). See Pl. XXXII also nos. C145 and D152 below. These tables are analysed in *JHA*, 4 (1973), pp. 99–110. See also *SATMI*.

On al-Khalīlī's other auxiliary tables, see no. F30.

3.1.9 *Jadwal faḍl al-dāᵓir li-ᶜarḍ Dimashq*, tables of the hour-angle for latitude 33;30°, Damascus: K 8525,3 (fols. 20v–56r, *ca.* 850H by al-Ḥalabī (no. C69)); ṬM 228,5 (fols. 21v–60v, *ca.* 900H). See Pl. XXXIa.

3.1.9 *Jadwal al-dāᵓir wa-faḍlihi li-ᶜarḍ Dimashq*, tables of the time since sunrise and hour-angle for latitude 33;30°, Damascus: MM 71,1 (fols. 1r–66r + 67r–67v, *ca.* 1000H). A late copy of the hour-angle tables is ṬM 81,2 (fols. 85r–118r, *ca.* 1300H).

3.1.9 *Jadwal aᶜmāl al-layl wa-l-nahār li-ᶜarḍ Dimashq*, prayer-tables for latitude 33;30°, Damascus: K 8525,1 (fols. 1r–13v, *ca.* 850H by al-Ḥalabī (no. C69)); ṬM 218 (45 fols., *ca.* 1150H, preceded by an introduction by Ibn al-Kayyāl—see no. D60). Some of these tables are also contained in ṬM 255,7 (fols. 61v–80v, *ca.* 1100H). See also al-Manāshīrī (no. D8).

3.1.9 A table of geographical coordinates and qibla values stated to be taken from the handwriting of Shams al-Din al-Mizzī (no. C34) with qibla values computed by al-Khalīlī: ṬM 226,2 (fols. 95v–97r, 894H); MM 167,11 (fols. 203v–204r, 989H). See Pl. LII.

4.5.2 *al-Nujūm al-ẓāhira fi aᶜmāl al-jayb bi-ghayr murī wa-lā-dāᵓira*, a treatise on non-standard procedures with the sine quadrant, in 25 *bābs*: MM 201,3 (fols. 10r–16r, *ca.* 1250H); K 4045 (15 fols. 1315H). See also no. C149 below.

4.7.6 *Fāᵓida fī khaṭṭ al-ẓill alladhī fī maḥall al-ādhān fī maᵓdhanat al-ᶜarūs bi-l-jāmiᶜ al-umawī* *bi-Dimashq*, a note on the sundial (?) on the main minaret of the Umayyad Mosque in Damascus: TR 161,2 (p. 10, *ca.* 1100H from al-Khalīlī's handwriting), unique. It is not completely clear that the note refers to Ibn al-Shāṭir's sundial described in L. Janin, "Le Cadran Solaire de la Mosquée Umayyade de Damas," *Centaurus*, 16 (1971), pp. 285–298.

6.4.2 Notes on interpolation attributed to al-Khalīlī: ZK 917,4 (pp. 50–54, 1163H).

C38 SHARAF AL-DĪN ABŪ ᶜIMRĀN MŪSĀ B. MUḤAMMAD AL-KHALĪLĪ, a nephew of SHAMS AL-DĪN AL-KHALĪLĪ (no. C37)

Suter, no. 427; *GAL*, II, p. 157 and SII, p. 158.

3.2.8 *Masāᵓil mulakhkhaṣāt fi maᶜrifat awqāt al-ṣalawāt wa-jihat al-qibla ᶜinda ᶜadam al-ālāt*, a treatise on the prayer-times and qibla in 7 *masᵓala*s: KhM 10,5 (fols. 50v–61r, 1095H).

4.4.5 (*R. fi l-ᶜAmal bi-rubᶜayi l-jayb wa-l-muqanṭarāt*), a treatise on the sine quadrant and almuqantar quadrant in three *risāla*s, the first two comprising 5 *maqāla*s and dealing with the two quadrants together and the third comprising 7 faṣls and dealing with the determination of the prayer-times and qibla with a sine quadrant: DM 832,3 + 4 (fols. 50r–60v + 61v–68v, 832H). The third treatise alone is contained in DM 454 (10 fols., 1076H; attributed to "a student of (Shams al-Dīn) al-Khalīlī"). See Pl. CIVa.

C39 SHAMS AL-DĪN ABŪ ᶜABD ALLĀH MUḤAMMAD B. BURHĀN AL-DĪN IBRĀHĪM AL-RASHĪDĪ

Not mentioned in the modern sources.

3.1.5 *ᶜUmdat al-muᵓtamm wa-l-imām fī maᶜrifat awqāt al-ṣalawāt bi-l-balad al-ḥarām*, an introduction in 8 *bāb*s and prayer-tables for latitude 21°, Mecca: Sh 89,4 (fols. 29v–32v, 1025H, lacks tables!). The entries in the tables mentioned in the numerical examples presented in the introduction do not correspond to those in MM 68 (see no. C40).

3.1.5 Two anonymous fragments from a set of hour-angle tables for latitude 32°, Jerusalem: DM 153,3 (fols. 11 and 13, *ca.* 1050H) and DM 45,5 (fol. 27, *ca.* 1050H), unique. The attribution to Ibn al-Rashīdī is suggested in MS Leipzig Universitätsbibliothek 808, in the introduction to the timekeeping tables for Jerusalem by al-Karakī (no. C35). On these tables, see *SATMI*, I. See also no. D222.

C40 ANONYMOUS (prayer tables for Mecca)

3.1.6 Prayer-tables for latitude 21°, Mecca: MM 68 (8 fols., *ca.* 900H), unique. See further *SATMI*, II.

C41 SHIHĀB AL-DĪN AḤMAD B. GHULĀM ALLĀH AL-KAWM AL-RĪSHĪ

Suter, no. 428; *GAL*, II, p. 157.

2.1.19 *K. al-Lumᶜa fī ḥall al-sabᶜa*, a *zīj* for Cairo with an introduction in 12 *bāb*s and tables for Cairo, based on the *Zīj* of Ibn al-Shāṭir (no. C30): TR 275 (103 pp., 858H) and photos thereof in K 7197; DM 637,1 (fols. 1r–45r, 868H); TR 207 88 pp., 873H); DM 404 (45 fols., 926H); *etc.* (10 copies and 7 fragments). Commentaries by Ibn Abi l-Khayr al-Ḥusnī (no. C124), entitled *al-Durra al-muḍīya* . . . : K 4009 (72 fols., 1325H), *etc.* (1 fragment); and by Muḥammad al-Khuḍrī (no. D111): DM 94 (92 fols., 1239H), KhM 3 (111 fols., 1240H), *etc.* (10 copies, some incomplete), with an abridgement. See no. J4 and Pl. CVII.

A late collection of tables from the *zīj*es of Ibn al-Shāṭir (no. C30) and Riḍwān Efendī (no. D58): K 3978 (22 fols., 1302H).

2.3.5 *Kifāyat al-taᶜlīm fī waḍᶜ al-taqwīm*, a treatise on the compilation of ephemerides in 8 *faṣl*s: DM 117 (17 fols., 1182H); DM 977,1 (fols. 1r–19r, 1300H); *etc.* (1 copy).

al-Kawm al-Rīshī copied K 3823,2.

C42 AL-QAYMARĪ

Not mentioned in the modern sources.

3.1.12 *Jadwal al-bāqī li-l-ᶜaṣr li-ᶜarḍ Miṣr*, tables of the time remaining to the *ᶜaṣr* as a function of solar altitude: DM 620,8 (fols.

58r–59r + 64r–72v, *ca.* 800H), unique. See further *SATMI*, II.

C43 ABŪ BAKR B. ISMĀᶜĪL, IBN AL-MUSHRIF[1]

GAL, SI, p. 869.

3.1.14 ***Nūr al-aḥdhāq li-maᶜrifat aᶜmāl al-falak fī sāʾir al-āfāq***, auxiliary tables for timekeeping: MM 241 (18 fols., *ca.* 900H); DM 468,3 (fols. 5r–17v, *ca.* 900H, defective); plus three abridgements, the first by Sūdūn al-Bashtakī (no. C81) in DM 512 (16 fols., *ca.* 1080H); the second anonymous in K4054 (9 fols., 1157H); and the third by Ibn Abi l-Fatḥ al-Ṣūfī (no. C98) in K 4024,2 (fols. 2v–9v, 1293H). Another copy of Ibn al-Mushrif's tables may be in MS Istanbul Nurosmaniye 2461 (not examined).

3.1.14 / 3.1.24 Tables of normed right ascensions for each minute of ecliptic longitude: MM 209,1 (fols. 1r–37r, 873H by Ibn Abi l-Fatḥ al-Ṣūfī (no. C98) from author's copy dated 848H), unique.

3.1.24 Introduction to the ascension tables of Baylak al-Qipjāqī (no. C9) in MM 41 (41 fols., 1191H).

See also no. C13.

C44 MUḤAMMAD B. IDRĪS

Not mentioned in the modern sources.

3.2.6 ***Muqaddima fi l-mīqāt***, a treatise on timekeeping in 20 *bābs*, compiled in 790H: DM 894 (photos of MS Paris B.N. ar. 2548, 29 pls., late, unique); ṬM 117 (copied from DM 894).

C45 AḤMAD AL-ḤARĪRĪ

Mayer, p. 35 (lists a sundial in Cairo made by al-Ḥarīrī in 785/1383).

2.4.3 Some planetary equation tables for Jupiter are attributed to al-Ḥarīrī in DM 909,8 (photos. of MS Paris B.N. ar. 2496, fols. 131r–131v, *ca.* 1000H).

1. His *kunya* is given as Zayn al-Dīn in K4054 and Sayf al-Dīn in MM 41 (*q.v.*).

C46 ᶜIZZ AL-DĪN ᶜABD AL-ᶜAZĪZ B. SAᶜD AL-DĪN MASᶜŪD B. ᶜABD AL-ᶜAZĪZ

Not mentioned by Suter, Brockelmann, or Renaud!

See no. F49.

4.5.10 His ***R. Kāfiya fi l-ᶜamal bi-l-jayb*** on the sine quadrant in 4 *maqālas* of 7 + 50 + 10 + 10 *bābs*, compiled in Cairo in 795H, is in MS Escorial ar. 918, 14 (fols. 84v–106r, 888H), unique. Merits detailed study.

C47 JAMĀL AL-DĪN ᶜABD ALLĀH B. KHALĪL AL-MĀRIDĪNĪ

Suter, no. 421; *GAL*, II, p. 218, and SII, p. 218.

3.1.11 ***al-Shabaka***, auxiliary tables with an introduction: K 4026,1 (fols. 1v–2v + 6v–12r, 1308H); plus a commentary on the use of the tables by Aḥmad al-Basaṭī (no. C48): K 4026,2 (fols. 3r–6r, 1308H). The introduction to al-Māridīnī's tables is published and analyzed in *AIHS*, 24 (1974), pp. 231–240, using the older copy MS Paris B.N. ar. 2525,1, fols. 1v–16v. See also *SATMI*, I.

4.4.7 ***Waraqāt fi l-ᶜamal bi-rubᶜ al-dāʾira alladhī ᶜalayhi l-muqanṭarāt***, a short treatise on the almucantar quadrant: DM 620,10 (fols. 73r–78r, 861H); K 4061 (11 fols., 848H); *etc.* (6 copies); plus a commentary by Sibṭ al-Māridīnī (no. C97): MM 97 (8 fols. *ca.* 1150H)—*cf.* MS Berlin Ahlwardt 5843; plus an anonymous commentary (perhaps by the copyist ᶜAlī b. Muḥammad al-Dāghistānī, student of ᶜAbd al-Qadir al-Awfī): ṬJ 811,16 (fols. 151v–155v, 1135H).

4.4.7 On MM 167,1, fol. 76v, there is a statement that al-Māridīnī died in 896H, according to Ibn al-ᶜAṭṭār (no. C66). This date cannot be correct, not least because Ibn al-ᶜAṭṭār (*fl. ca.* 830H) was probably dead by 896H.

4.5.6 ***al-Durr al-manthūr fi l-ᶜamal bi-rubᶜ al-dustūr***, an extensive treatise on the sine quadrant also dealing with sundials, arranged in 60 *bābs*: DM 181,6 (fols. 105r–163v, in two hands one dated 1178H); MM 167,1 (fols. 1r–76v, 988H, lacks first page); DM 174,2 (fols. 21r–80r, 1115H); *etc.* (1 copy and 1 fragment).

Commentary by Ibn al-Majdī (no. C62) entitled ***Irshād al-sā'il ilā uṣūl al-masā'il***: DM 1096 (192 fols., 999H); DM 1057 (101 fols., *ca.* 1100H, incomplete); *etc.* (2 copies).

4.5.6 ***Mujmalat al-maṭlūb fī l-ᶜamal bi-rubᶜ al-juyūb***, a treatise on the sine quadrant in 20 *bābs*: ṬM 79,5 (fols. 40v–47v, *ca.* 1000H); DM 183,3 (fols. 11v–20v, 823H); ṬJ 442,2 (fols. 22v–27r, *ca.* 1050H; TR 283 (14 pp., *ca.* 950H); *etc.* (5 copies).

Also ***R. fī l-ᶜAmal bi-rubᶜ al-dā'ira al-mawḍūᶜ fīhi l-juyub***, in 20 *bābs*: K 6704 (11 pp., *ca.* 900H, incomplete); *etc.* (1 fragment). A poem based on one of these treatises by ᶜAlī b. Muḥammad al-Haythamī (no. C49) entitled ***Rāḥat al-qulūb*** . . . : DM 87 (9 fols., 871H by author), *etc.* (1 fragment).

4.5.6 ***Ghāyat al-intifāᶜ fī l-ᶜamal bi-l-bakhsh alladhī fī ākhir qaws al-irtifāᶜ***, a treatise on the sine quadrant in 11 *bābs*: DM 138,6 (fols. 10v–11v, 840H); DM 1093,12 (fols. 68r–70r, 1088H).

4.5.6 ***R. Mulakhkhaṣa fī l-ᶜAmal bi-l-rubᶜ al-mujayyab***, in 30 *bābs*: MM 245,2 (fols. 6r–9r, 1125H).

4.5.6 An anonymous treatise based on one of al-Māridīnī's and entitled *Takhlīṣ al-rubᶜ al-mujayyab* and arranged in 17 *bābs*: K 3773,2 (fols. 69v–75v, 1330H).

4.6.11 ***Muqaddima fī maᶜrifat al-ᶜAmal bi-rubᶜ al-shakkāzīya***, on the use of a double *shakkāzī* quadrant, without divisions: DM 138,4 (fols. 8r–9r, 840H, anonymous). Other copies of this treatise are in MSS Damascus Ẓāhirīya 7463 and 3098, and Istanbul Fatih 5397,20. Only MS Ẓāhirīya 7463 is attributed to al-Māridīnī (not Sibṭ al-Māridīnī as stated on p. 234 of the catalogue). Published with translation and commentary in *AIHS*, 24 (1974), pp. 219–242.

C48 SHIHĀB AL-DĪN ABU L-ᶜABBĀS AḤMAD AL-BASĀṬĪ

Not listed in the modern sources.

3.1.11 Notes on al-Māridīnī's auxiliary tables entitled ***al-Shabaka*** (see no. C47): K 4026,2 (fols. 3r–6r, 1308H), unique?

C49 ᶜALĪ B. MUḤAMMAD B. AḤMAD AL-HAYTHAMĪ AL-ṬUBNAWĪ AL-MĀLIKĪ AL-ASHᶜARĪ

GAL, II, pp. 92–93, and SII, p. 84.

4.5.6 ***Rāḥat al-qulūb min taᶜb al-aṭnāb fī l-juyūb***, a a poem based on the treatise on the sine quadrant in 20 *bābs* by Jamāl al-Dīn al-Māridīnī (no. C47): DM 87 (9 fols., 871H by the author); *etc.* (1 late fragment). DM 87 was copied by al-Haythamī in the Ḥusaynīya in Cairo in 854H and concludes with an *ijāza* to Abu l-Faḍl Naṣīr al-Dīn Muḥammad b. Muḥammad al-Ghazūlī of Damietta dated 857H.

C50 MAḤMŪD B. AḤMAD B. MAḤMŪD AL-SHĀFIᶜĪ AL-ṢĀLIḤĪ, AL-MURSHIDĪYA (?)[1]

Suter, no. 500; *GAL*, SII, p. 1022, no. 53.

4.6.28 ***al-Lafẓ al-muṣarraḥ fī l-ᶜamal bil-l-rubᶜ al-mujannaḥ***, a treatise on the winged quadrant in 35 *bābs*: DM 142 (8 fols., 794H), unique.

C51 ABU L-BAQĀ' ᶜALĪ B. ᶜUTHMĀN B. MUḤAMMAD B. AḤMAD B. AL-QĀṢIḤ

Suter no. 419; *GAL*, II, p. 214 & SII, p. 212, *etc.*; *Azzawi*, p. 100.

4.4.4 ***Tuḥfat al-ṭullāb fī l-ᶜamal bi-rubᶜ al-asṭurlāb***, a treatise on the almucantar quadrant in 90 *bābs*: DM 26 (29 fols., *ca.* 800H), unique. The other copy (MS Princeton Garrett 1024) listed by Brockelmann is unrelated to this work, being a treatise in 15 chapters on the astrolabe. An anonymous treatise in 18 *bābs* bearing the same title is in MS Berlin Ahlwardt 5808. However, Brockelmann also lists a work *al-Manhal al-ᶜadhb al-musayyab fī sharḥ al-ᶜamal bi-l-rubᶜ al-mujayyab*, extant in MS Vatican ar. 317, 4 (fols. 95r–113v, *ca.* 800H) which *is* by Ibn al-Qāṣiḥ. See *JAOS*, 104 (1984), pp. 111.

---

1. His name is given thus in MS DM 142. Suter and Brockelmann have incorrectly Muḥammad b. Aḥmad . . . .

C52 ISMĀᶜĪL B. HIBAT ALLĀH AL-ḤAMAWĪ

Not listed in the modern sources.

4.8.4 *R. fi l-ᶜAmal bi-dhāt al-shuᶜbatayn*, on the use of the parallactic ruler in 7 *bābs*: ṬM 102,8 (fols. 80v–83v, *ca.* 1150H). Another copy is in MS Escorial ar. 961,1 (fols. 1r–8v, 863H). Merits study.

C53 ᶜALĀᵓ AL-DĪN ṬĪBUGHĀ AL-DAWĀRDĀR AL-BAKLAMĪSHĪ

GAL, II, p. 135, and SII, p. 167; *Azzawi*, pp. 171–172.

4.6.7 (*R. fi l-ᶜAmal bi-rubᶜ al-shakkāzīya*), on the use of the single *shakkāzī* quadrant in 16 *bābs*: DM 774 (14 fols., 864H by Abd al-ᶜAzīz al-Wafāᵓī (no. C61)). Another copy is in MS Princeton Mach 4912 = Yehuda 373 (fols. 149v–157v, 1060H).

C54 ᶜALĀᵓ AL-DĪN ABU L-ḤASAN ᶜALĪ B. ṬĪBUGHĀ, *muwaqqit* at the Umayyad Mosque in Aleppo.

Not mentioned in the modern sources.

4.4.6 (*R. fi l-ᶜAmal bi-rubᶜ al-muqanṭarāt*), a treatise on the use of the almucantar quadrant in 10 *bābs*: DM 832,2 (fols. 48r–49v, 832H), unique?

4.5.7 *Ghāyat al-maṭlab fi l-ᶜamal bi-l-rubᶜ al-āfāqī al-mujayyab*, a treatise on the use of the sine quadrant arranged in 3 *fanns* of 60 + 60 + 15 *bābs*, including various tables such as a qibla table: DM 832,1 (fols. 1r–47v, 831H); MM 176,2 (fols. 33r–124v, anonymous, *ca.* 1100H, defective). Another anonymous copy is MS Paris B.N. ar. 2519,5. Merits study. See Pls. LIa and CIVb.

4.6.10 (*R. fi l-ᶜAmal bi-rubᶜ al-shakkāzīya*), on the use of the single *shakkāzī* quadrant in 20 *bābs*: DM 64,4 (fols. 63v–73v, 803H). This treatise is attributed to "Ibn Ṭībughā" in the margin in a late hand. *Cf.* J. Samsó Moya and M. A. Catalá in *Memorias de la Real Academia de Buenas Letras de Barcelona*, 13:1 (1971), pp. 5–31, where this individual is confused with Ibn al-Majdī (no. C62), who also had Ṭibughā as an ancestor.

C55 AḤMAD B. ᶜUMAR AL-SHĀDHILĪ

Azzawi, p. 203.

4.6.8 *R. fi l-ᶜAmal bi-rubᶜ al-shakkāzīya*, on the use of the single *shakkāzī* quadrant in 14 *bābs*: DM 988 (2 fols., 1102H). Another copy of this treatise is MS Oxford arab. 93 (fols. 47–49).

C56 TAQI L-DĪN ABŪ ṬĀHIR

Mayer, p. 33.

4.6 *R. wajīza mufīda fi l-ᶜAmal bi-rubᶜ al-shakkāzīya*, on the use of the single *shakkāzī* quadrant in 10 *bābs*: TR 169,2 (pp. 74–79, 1320H from a copy in Dar al-Quṭayna (in Jerusalem?), anonymous). Another copy of this treatise exists in MS Manchester 361,22 attributed to Taqi l-Dīn Abū Ṭāhir. The work is also mentioned by Ḥājjī Khalīfa who refered to the author as Taqi l-Dīn, and this misled Suter to attribute the work to Taqi l-Dīn b. Maᶜrūf (no. H12).

C57 ABŪ ᶜABD ALLĀH MUḤAMMAD B. .... B. JAMĀᶜA AL-KINĀNĪ

GAL, SII, pp. 111–112.

6.3.23 *Khulāṣat al-qawāᶜid wa-ghāyat al-maqāṣid*, a short treatise on 14 branches of knowledge, including arithmetic and algebra: ZK 9 (15 fols., *ca.* 1200H, anonymous). Another Cairo copy which I have not seen is mentioned by Brockelmann.

C58 SHIHĀB AL-DĪN ABU L-ᶜABBĀS AḤMAD B. MUḤAMMAD, IBN AL-HĀᵓIM

Suter, no. 423; *GAL*, II, pp. 153–155, and SII, pp. 154–155.

6.3.8 *K. al-Durr al-thamīn fī sharḥ Urjūzat Ibn al-Yasamīn*, a commentary on Ibn al-Yasamīn's poem on algebra (see no. F12): K 4313,1 (fols. 1r–81v, 814H); DR 1 (114 fols., 857H); *etc.* (2 copies and 2 fragments).

6.3.19 *al-Ḥāwi fi l-ḥisāb*, an abridgement of the *Talkhīṣ* of Ibn al-Bannāᵓ (no. F23): J 3964,1 (fols. 1r–7v, 862H); TJ 86,12 (pp. 301–319, 1177H); plus commentaries by Ibrāhīm al-Ḥalabī (no. C113): DR 667 (172 pp., 1192H); by ᶜAbd al-Qādir al-Ṣūfī (no. C93): K 7229 (37 fols., 915H); and by Sibṭ al-Māridīnī (no. C97): K 4392 (58 fols., *ca.* 1000H).

6.4.1 ***al-Maᶜūna***, a treatise on arithmetic in 3 *qism*s of 2 + 10 + 3 *bāb*s, compiled in 791H: DR 256,2 (fols., 35r–153r, 850H); MR 32 (182 fols., *ca.* 1150H); DR 68 (190 fols., 950H); *etc.* (4 copies). Commentary by Ibn al-Bilbaysī (no. C60): DR 11 (83 fols., *ca.* 1150H).

6.4.1 ***al-Wasīla fī ṣināᶜat al-hawāʾī***, a treatise on arithmetic in 3 *qism*s and a *khātima* of 2 *bāb*s and 8 + ? + 5 *faṣl*s, compiled in 772 or 792H: DR 256,1 (fols. 1r–34v, 850H); TR 1 (74 pp., 883H); TR 105,2 (fols. 16v–57v, 951H); *etc.* (5 copies); plus commentaries by Nūr al-Dīn al-Faraḍī (no. C77), entitled ***al-Fawāʾid al-jalīla fi ḥall majhūlāt al-Wasīla***: K 4023,1 (fols. 1r–37r, 1060H); *etc.* (3 copies); by Sibṭ al-Māridīnī (no. C97), entitled ***Irshād al-ṭullāb ilā Wasīlat al-ḥussāb***, compiled 902H: DR 2 (62 fols., 903H, with an *ijāza* by the author); K 3815,3 (fols. 21v–101r, 985H), *etc.* (1 copy and 2 fragments); by (Ibn) al-Bilbaysī (no. C60), entitled ***al-Zahra al-jalīla fī ḥall alfāẓ al-Wasīla***: K 4325 (105 fols., 1109H); by Ibn al-Muqriʾ (no. D187) with an anonymous note: K 4023,2 (fols. 37v–38v, 1060H); by Muḥammad b. Ibrāhīm al-Ḥalabī (no. C112), entitled ***Rafᶜ al-ḥijāb ᶜan qawāᶜid al-ḥisāb***: DR 557,4 (fols. 132r–160v, 999H).

6.4.1 ***Murshidat al-ṭālib ilā asna l-maṭālib***, a treatise on arithmetic in 2 *qism*s and a *takmila* of 6 + 10 + 3 *bābs*/*faṣl*s: K 3815,4 (fols. 103r–165r, 987H); K 4290 (37 fols., 1260H); plus commentary by al-Shinshawrī (no. C111), entitled ***Bughyat al-rāghib fī sharḥ Murshidat al-ṭālib***: DR 318 (300 fols., 997H); DR 13 (191 fols., 997H); *etc.* (6 copies).

6.4.1 ***Mukhtaṣar wajīz fī ᶜilm al-ḥisāb al-maftūḥ al-hawāʾī*** a treatise on arithmatic in 5 *bāb*s subdivided in *faṣl*s: DR 112,2 (fols. 3v–20v, *ca.* 800H, incomplete).

6.4.1 ***al-Lumaᶜ fī ᶜilm al-ḥisāb*** a treatise on arithmetic in unnumbered *bāb*s: K3815,1 (fols. 1v–11r, 985H); DR 64 (28 fols., *ca.* 1150H); *etc.* (13 copies). Commentaries by Sibṭ al-Māridīnī (no. C97): DR 315 (34 fols., 902H); DR 344 (26 fols., 1003H); *etc.* (14 copies); by Akovali Zāde Khātim (no. H56): DR 1027 (29 fols., 993H by the author); *etc.* (1 copy); by Zayn al-ᶜĀbidīn al-Durrī (no. D13): K 17290 (34 fols., 1062H); K 4305 (98 fols., 1086H); *etc.* (3 copies).

6.4.1 ***Nuzhat al-nuẓẓār fī ṣinaᶜat al-ghubar = Nuzhat al-ḥisāb = Mukhtaṣar al-Murshida***: DR 297 (10 fols., *ca.* 900H, incomplete, + DR 308 (11 fols., *ca.* 900H, completes DR 297, with an *ijāza* by Sibṭ al-Māridīnī (no. C97)); DJ 226,3 (fols. 90v–101v, *ca.* 1000H); *etc.* (16 copies). Commentary by Aḥmad al-Ghazzī (no. C115): DJ 861,7 (fols. 91r–138v, 966H); K 6830 (93 fols., 1024H); *etc.* (5 copies); an abridgement by Yaḥyā al-Khaṭṭāb (no. C119): TR 152, (27 pp., *ca.* 1150H); *etc.* (1 copy); a commentary by Muḥammad b. Ibrāhīm al-Ḥalabī (no. C112) entitled ***ᶜIddat al-ḥāsib wa-ᶜumdat al-muḥāsib***: K 4301,1 (fols. 1r–56r, 973H); DR 557,1 (fols. 1r–80r, 999H); a commentary by Yaḥyā b. Taqi l-Dīn al-Ḥalabī (no. D14): DR 1100 (65 fols., 1027H by author); an anonymous commentary: DR 82 (38 fols., 1132H), MR 11 (100 fols., 1183H); a commentary by ᶜAbd al-Qādir al-Fayyūmī (no. D12): DR 558 (82 fols., *ca.* 1100H); *etc.* (1 copy); a commentary by Jamāl al-Dīn b. Aᶜazz al-Dimashqī (no. C59), entitled ***Tuḥfat al-ṭullāb*** . . . : DR 662 (86 fols., *ca.* 1000H, anonymous); TR 124 (344 pp., 1092H, anonymous); a commentary by Ḥusayn b. Muhammad al-Maḥallī (no. D85), entitled ***Kashf al-astār*** . . . : MR 24 (135 fols., 1187H); a commentary by ᶜAbd Allāh al-Kāzūnī (no. X17), entitled ***Rafᶜ al-niqāb*** . . . : K 3953 (95 fols., *ca.* 1100H, incomp.): and a poem based on the *Nuzha* by Mūḥammad b. ᶜĀbidīn (no. D180), entitled *Manāhil al-surūr* . . . : DR 817,1 (fols. 1v–5r, 1282H); *etc.* (2 copies).

6.4.1 ***al-Muqniᶜ***, a poem on algebra: TJ 289,3 (pp. 38–41, 876H); *etc.* (3 copies).

al-Mumtiᶜ fī sharḥ al-Muqniᶜ: ṬR 127 (55 fols., *ca.* 1150H); DR 920 (85 fols., lacks first page, *ca.*, 1150H).

al-Musriᶜ, an abridgement of ***al-Mumtiᶜ***, compiled in 810H): TR 17 (16 pp., 845H); TJ 289,4 (pp. 42–57, 867H); K 3815,2 (fols. 11v–20r, 985H); *etc.* (2 copies). Commentaries on ***al-Muqniᶜ*** by Sibṭ al-Māridīnī (no. C97), entitled *al-Qawl al-Mubdi*ᶜ . . . : DJ 144,2 (fols. 46r–57r, 880H); DJ 861,6 (fols. 71r–90r, 903H); *etc.* (9 copies); by Abu Yaḥyā al-Anṣārī (no. C78), entitled *Fatḥ al-mubdi*ᶜ . . . : DR 307 (23 fols., 1091H); *etc.* (7 copies), and a *ḥāshiya* thereon by al-Janājī (no. D172): DR 1101 (169 fols., *ca.* 1275H); *etc.* (3 copies); by Qāsim al-Khānī (no. D43): TJ 196,15 (pp. 440–465, *ca.* 1000H); and an anonymous commentary: DR 357 (16 fols., 1171H, lacks beginning).

6.9.7 A commentary on his ***al-Fuṣūl al-muhimma fī mawārīth al-umma*** on inheritance, by Sibṭ al-Māridīnī (no. C97): B 19141 (84 fols., 908H).

6.9.7 ***Ghāyat al-su***ᵓ***l fi l-iqrār bi-l-dayn al-majhūl***, on inheritance, compiled in 797H: MR 21 (62 fols., partly 858H); DJ 33,2 (fols. 25r–52v, 997H); *etc.* (1 copy).

6.9.7 ***Faṣl fī*** ᶜ***Ilm al-munāsakhāt bi-l-jadwal*** in some copies stated to be taken from Ibn al-Hāᵓim's ***Sharḥ al-Kifāya***: TR 138,2 (fols. 20r–35r, *ca.* 850H); *etc.* (6 copies). See Pl. XCIIa.

6.9.7 ***Ithnā*** ᶜ***ashar mas***ᵓ***ala jāmi***ᶜ***a li-uṣūl masā***ᵓ***il al-iqrār bi-l-dayn al-majhūl al-dawrī***: DJ 33,3 (fols. 55v–62v, 997H); DR 1090 (18 fols., *ca.* 1150H).

6.9.7 A commentary on his *Kifāyat al-ḥuffāẓ*, a poem on inheritance, by Sibṭ al-Māridīnī (no. C97): B 22292 (176 fols., 996H).

6.9.7 ***al-Tuḥfa al-Qudsīya fī*** ᶜ***ilm al-farā***ᵓ***iḍ***, a poem on inheritance abridged from ***al-Raḥbīya*** by Ibn al-Mutaqqina (no. C171): J 3964,8 (fols. 32v–35v, 862H). Commentary by Sibṭ al-Māridīnī (no. C97) entitled *al-Lum*ᶜ*a al-shamsīya*: DJ 144,6 (fols. 104v–111r, 884H).

6.9.7 An anonymous commentary on his *R. fi* ᶜ*Amal al-qīrāṭ*: B 23837,3 (fols. 30v–75v, 1179H).

C59 JAMĀL AL-DĪN B. AᶜAZZ AL-DIMASHQĪ

GAL, SII, p. 154.

6.4.1 *Tuḥfat al-tullāb fī sharḥ Nuzhat al-ḥisāb* (see no. C58): DR 662 (86 fols., *ca.* 1000H); TR 124 (344 pp., 1092H), both anonymous (*cf. GAL*, SII, p. 154).

C60 MUḤAMMAD B. MUḤAMMAD B. ABĪ BAKR AL-AZHARĪ, IBN AL-BILBAYSĪ

Suter, no. 506.

6.4.1 A commentary on *K. al-Ma*ᶜ*ūna* by Ibn al-Hāᵓim: see no. C58.

6.4.1 ***al-Zahra al-jalīla fī ḥall alfāẓ al-Wasīla***, a commentary on the *Wasīla* of Ibn al-Hāᵓim: see also no. C58.

C61 ᶜIZZ AL-DĪN ᶜABD AL-ᶜAZĪZ B. MUḤAMMAD AL-WAFĀᵓĪ

Suter, no. 437 and 437N; *GAL*, II, pp. 159–160 and SII, p. 160. See Pl. CVIIIa.

2.1.19 Notes on lunar crescent visibility by the method of *al-Lum*ᶜ*a* (see no. C41) and the method of Ibn al-Majdī (no. C62), which is that outlined in the *Zīj* of Ibn al-Shāṭir (no. C30): MM 93,1 (fols. 1v–2v, *ca.* 1100H).

2.2.3 ***al-Ḥabṭaq al-aṣghar***, lunar tables based on the parameters of Ibn al-Shāṭir (no. C30): see no. C132.

2.3.6 A table of the lunar mean motion according to the method of Ibn al-Majdī's *al-Durr al-yatīm* (see no. C62) by al-Wafāᵓī is in DM 448,3 (fols. 5v–6v, *ca.* 900H).

2.3.6 ***R. fī Ḥall al-Durr al-yatīm li-muddat sana kāmila***, on the use of Ibn al-Majdī's auxiliary tables for compiling ephemerides: K 3833,1 (fol. 1r, *ca.* 1150H).

3.1.12 ***Natījat al-afkār fī a***ᶜ***māl al-layl wa-l-nahār***, an introduction on determining the solar longitude with a table for that purpose, followed by an incomplete set of prayer-tables from the main Cairo corpus: ṬJ 367,5 (fols. 35r–40r, *ca.* 850H perhaps by author). No other copies, though *cf.* MS Princeton Yahuda 861.

3.1.24 Tables of the ascensions at the time of the *salām* for each 0;6° of solar longitude, for latitude 30° (Cairo): DM 157,2 (fols. 1v–7r, *ca.* 850H).

3.2.10 ᶜ***Umdat al-ṭullāb wa-kayfīyat istikhrāj al-***ᶜ***amal bi-l-ḥisāb***, a treatise on timekeeping: DM 620,8 (fol. 60r, *ca.* 900H, title page only!). Abridgement by the author entitled *Tuḥfat al-ṭullāb fī kayfīyat istikhrāj al-a*ᶜ*māl al-falakīya bi-l-ḥisāb*, not subdivided: DM 620,5A (fols. 37r–41v, *ca.* 900H, defective); DM 588,3 (fols. 2v–4v, *ca.* 1100H); *etc.* (1 copy).

3.4.3 Notes on the lunar crescent visibility theory of Ibn al-Majdī (no. C62): MM 93,1 (fols. 1v–2v, *ca.* 1100H).

3.4.5 ***Nafāʾis al-durar fī maᶜrifat al-waqt bi-l-qamar***, treatise in 2 *faṣl*s with 10 tables: DM 588,4 (fols. 5r–7v, *ca.* 1100H, anonymous), unique. al-Wafāʾī mentions this treatise in his *Nuzhat al-naẓar* . . . . Abridgement by the author entitled ***Zād al-safar fi l-ᶜamal bi-l-qamar***: ṬM 165,2 (fols. 18r–20v, *ca.* 1150H, anonymous), also unique. DM 588,3 contains only the first two tables entitled "velocities" (*abhāt*) with no entries. The other tables are described in the introduction. Another short treatise by al-Wafāʾī entitled ***Khulāṣat al-durar fi l-ᶜamal bi-l-qamar***, extant in MS Manchester Rylands 361L (fols. 44r–45v, 1154H), contains no tables but refers the reader to "the treatise which I wrote on the tables for the moon by Shaykh Abū ᶜAbd Allāh al-Khalīlī (= no. C37)."

3.5.4 ***Nubdha fi l-ᶜamal bi-daqāʾiq ikhtilāf al-āfāq al-marʾiya wa-hiya llatī ḥasabahā . . . ᶜAlāʾ al-Dīn Ibn al-Shāṭir***, a note on the effect of refraction at the horizon with a table: TR 303,2 (fol. 33v, 1266H); MM 181,2 (fols. 4v–5r, 1030H); MM 155,2 (fol. 23v, *ca.* 1100H, lacks table). See Pl. XXXIII. The attribution of this theory to Ibn al-Shāṭir (no. C30) is doubtful. See further *SATMI*, II.

3.5.4 A set of values of the arc Sine function arranged in the form of a poem: DM 771,3 (fol. 10v, *ca.* 850H); DM 905,2 (photos. of MS Paris B.N. ar. 2486, fol. 255r, *ca.* 850H). In the latter manuscript the poem is written by the colophon of the *Zīj* of al-Baghdādī (no. G25).

4.3.29 Tables for marking azimuth curves on astrolabes: MM 27,4 (fol. 26r, 1118H).

4.4.11 ***al-Nujūm al-zāhirāt fi l-ᶜamal bi-rubᶜ al-muqanṭarāt***, a treatise on the use of the almuqantar quadrant in 25 *bāb*s: MM 229 (18 fols., *ca.* 840H in the author's hand, incomplete); DM 775,1 (fols. 1r–21v, 847H); DM 197 (18 fols., 1059H); *etc.* (6 copies); plus commentary by al-Ujhūrī (no. D148): DM 606 (3 fols., *ca.* 1200H). Two abridgements by al-Wafāʾī, the first entitled ***Qaṭf al-zāhirāt fi l-ᶜamal bi-rubᶜ al-muqanṭarāt*** (in 15 *bāb*s): DM 771,4 (fols. 11r–16r, *ca.* 850H); MM 150 (22 fols., *ca.* 850H), *etc.* (3 copies), and the second entitled ***al-Durar al-muntatharāt fi l-ᶜamal bi-rubᶜ al-muqanṭarāt*** in unnumbered *faṣl*s: DM 135,1 (fols. 1r–4r, 847H); DM 1093,3 (fols. 7r–10v, 1088H).

4.5.9 ***Nuzhat al-naẓar fi-l-ᶜamal bi-l-shams wa-l-qamar***, a treatise on the use of the sine quadrant in 25 *bāb*s with a *khātima* on the prayer-times and crescent visibility: DM 489 (18 fols., 1019H); DM 617,1 (fols. 1r–28v, *ca.* 1150H); ṬM 254,5 (fols. 83v–117r, *ca.* 1150H); *etc.* (1 fragment). In the text al-Wafāʾī mentions another treatise of his entitled ***Nafāʾis al-durar*** . . . which identifies DM 588,3, *etc.*, (see above). Also ***R. laṭīfa badīᶜa fi l-ᶜAmal bi-l-rubᶜ al-mujayyab***, an abridgement of ***Nuzhat al-naẓar*** in 10 *bāb*s: ṬM 233,2 (fols. 89r–93r, 1681A); TJ 196,12 (pp. 374–379, *ca.* 1000H); *etc.* (3 copies). See also Berlin Ahlwardt 5824. Finally, ***R. mukhtaṣara fi-l-ᶜAmal bi-l-rubᶜ al-mujayyab*** in 10 *bāb*s and differing from the above abridgement: DM 66 (5 fols., *ca.* 1100H).

4.6.11+ ***Kalām ᶜan Muqanṭarāt khaṭṭ al-istiwāʾ***, on the *shakkāzī* grid: DM 1093,2B (fols. 5v–7r, 1088H).

4.6.23 An edition by Ibn Abi l-Fatḥ al-Ṣūfī (no. C98) of his ***R. fi l-ᶜAmal bi-l-muthallath*** on the complete (*tāmm*) quadrant of Ibn al-Shāṭir (no. C30) in 15 *bāb*s: ṬM 242,9 (fols. 113r–116v, 1180H).

4.6.35 ***R. fi l-ᶜAmal bi-jayb al-thumn***, on the octant in 10 *faṣl*s: DM 1093,2A (fols. 2v–5v, 1088H).

4.6.43 (***R. ᶜala l-Āla al-musammā bi-dāʾirat al-muᶜaddil***), on the "equatorial semicircle" in 15 *bāb*s: ṬM 83,4 (fols. 35r–42v, *ca.* 900H); DM 911 (photos of MS Paris B.N. ar. 2532, 18 pls., 911H); MM 79 (10 fols., 1125H); ZK 706,5 (fols. 15v–19v, *ca.* 1100H); *etc.* (12 copies, including an anonymous treatise in 15 *bāb*s based on al-Wafāʾī's treatise). Published in S. Tekeli, "Equatorial Armilla of ᶜIz al-Dīn b. Muḥammad al-Wafāī and Torquetum," Ankara, 1962.

4.6.47 ***R. fi l-ᶜAmal bi-l-muqawwar***, on the instrument called *al-muqawwar*, arranged in 10 *bāb*s and compiled for the Sultan Īnāl: DM 451 (11 fols. *ca.* 850H in the author's hand);

DM 504,1 (fols. 1v–9r, *ca.* 1050H). See also the anonymous notes on this instrument in MM 96,2 (fol. 6v, *ca.* 1100H) and the treatise by al-Ḥusnī (no. C124).

4.7.9 ***Naẓm al-ᶜuqūd fī ᶜamal al-sāᶜāt ᶜala l-ᶜumūd***, on the construction of sundials on columns: MM 93,4 (fols. 6v–9v, *ca.* 1100H), unique. See also the later treatise of al-Ḥamzāwī (no. C91) on this subject.

5.5.3 A treatise on the level of the Nile and its astrological significance: DM 632,2 (fols. 6v-7r, *ca.* 1150H).

6.5.2 *Nuzhat al-ṭullāb fī maᶜrifat al-ḥisāb*, a treatise on sexagesimal arithmetic in 5 *bābs*, compiled in 851H: DR 354 (25 fols., *ca.* 900H), plus an abridgement entitled *al-Luᵓluᵓa al-muḍiyᵓa fi l-ᶜamal bi-l-nisba al-sittīnīya*: DM 170,1 (fols. 1v–7v, *ca.* 880H), *etc.* (2 copies).

C62 SHIHĀB AL-DĪN AḤMAD B. RAJAB B. ṬĪBUGHĀ, IBN AL-MAJDĪ

Suter, no. 432; *GAL*, II, pp. 158–159, and SII, pp. 158–159. See also *JHAS*, 4 (1980), pp. 48–49.

2.3.6 A fragment of his ***K. al-Jāmiᶜ al-mufīd fī bayān uṣūl al-taqwīm wa-l-mawālīd***, an extensive treatise on ephemerides, calendars, and astrology: DM 499,3 (fols. 17r–19v, *ca.* 1150H). This fragment deals with the Syrian and Coptic calendars. Complete copies of this work are available. Merits investigation.

2.3.6 ***Ghunyat al-fahīm wa-l-ṭarīq ilā ḥall al-taqwīm***, a treatise on the compilation of ephemerides: ṬM 82 (37 fols., 870H); ṬM 140 (67 fols., *ca.* 1000H); *etc.* (3 copies). See Pl. CIXa.

2.3.6 ***al-Manhal al-ᶜadhb al-zalāl fī taqwīm al-kawākib wa-ruᵓyat al-hilāl***, a treatise on the determination of planetary positions and lunar crescent visibility in 5 unnumbered *faṣl*s with tables for lunar crescent visibility: MM 183,1 (fols. 1r–12v, *ca.* 1000H), unique? See Pl. XLVb.

2.3.6 ***K. al-Tashīl wa-l-taqrīb fī bayān ṭuruq al-ḥall wa-l-tarkīb***, a treatise on the compilation of planetary ephemerides: K 8524 (68 fols., *ca.* 1000H); ḤM 15 (36 fols., *ca.* 1100H); ḤM 16,1 (fols. 1r–32v, *ca.* 1100H); *etc.* (3 copies). Merits investigation.

2.3.6 ***K. al-Durr al-yatīm fī tashīl ṣanāᶜat al-taqwīm***, auxiliary tables for compiling annual ephemerides for the sun and moon: DM 405 (39 fols., *ca.* 850H); MM 44,2 (fols. 22r-27v, *ca.* 850H); DM 391 (41 fols., *ca.* 1000H); DM 93 (38 fols., *ca.* 1050H); MM 44,1 (fols. 1r–21v, *ca.* 1050H); no other known copies. See Pls. XXII and XXVa. Solar tables only in MM 85,1 (fols. 1r–2v, *ca.* 850H). Lunar equation tables only in MM 25 (30 fols., *ca.* 825H); and MM 26 (30 fols., *ca.* 850H); *etc.* (1 fragment). These tables are stated to be based on the parameters of Ibn Yūnus (no. B59).

2.3.6 Treatise on the use of these tables by Ibn al-Majdī: DM 141,1 (fols. 1v–8v, *ca.* 900H); DM 1017,3 (fols. 7v–12r, *ca.* 1100H); *etc.* (5 copies); an example on finding the solar longitude: DM 1017,1 (fols. 1v–3r, *ca.* 1100H); *etc.* (1 copy); and a *faṣl* on the use of the solar tables: DM 448,2 (fols. 3v–5r, *ca.* 900H).

2.3.6 ***R. fi Kayfīyat rasm al-dastūr wa-waḍᶜ mā yahtāj ilayhi li-muqawwam al-qamar sana^{tan} kāmila^{tan}***, on the compilation of lunar ephemerides: DM 448,1 (fols. 1v–3r, *ca.* 900H); DM 990,2 (fols. 7r–10v, *ca.* 1150H from a copy dated 835H); DM 56 (7 fols., 850H, with some tables from *al-Durr al-yatīm* copied *ca.* 975H); *etc.* (1 copy).

2.3.6 An anonymous ***R. fi Maᶜrifat istikhrāj muqawwamat al-qamar min al-Durr al-yatīm wa-ᶜamal al-dastūr li-sana kāmila***, on the compilation of lunar ephemerides: DM 990,1 (fols. 1v–6v, *ca.* 1150H); DM 1017,2 (fols. 3v–7r, *ca.* 1100H); *etc.* (1 copy).

2.3.6 An anonymous ***R. fi l-ᶜAmal bi-l-Durr al-yatīm wa-baqiyat al-kawākib al-sabᶜa wa-l-jawzahar wa-ṭariq ᶜamal dastūrihā li-sana kāmila***, on the compilation of planetary ephemerides: DM 591,1 (fols. 1v–8v, *ca.* 1100H).

2.3.6 An anonymous ***Mithāl fī maᶜrifat muqawwam al-qamar li-raᵓs al-sana al-mafrūḍa***: DM 448,3 (fols. 5v–6v, *ca.* 900H). This also contains two tables for the lunar mean motion according to the opinions of ᶜIzz al-Dīn al-Wafāᵓī (no. C61) and Nūr al-Dīn al-Naqqāsh (no. C74).

2.3.6 Planetary tables based on the method of ***K. al-Durr al-yatīm***, some of wich were computed by ᶜAbd al-Raḥīm b. al-Bannāᵓ (no. C135), Shams al-Dīn al-Hunayd (no. C134), and Muḥammad b. al-Qalaᶜī (no. C136): ṬM 113,1 (fols. 1r–128r, 1053H); DM 1200 (83 fols., *ca.* 1050H); plus related tables for specific planets in MM 24, MM 238, DM 43, K 4022, DM 441, DM 619,1, DM 813.

Ibn al-Majdī's tables were extremely popular amongst later Egyptian astronomers, several of whom wrote commentaries on their use or developed more extensive tables. Treatises on the use of the tables in *al-Durr al-yatīm* by Ibn Abi l-Fatḥ al-Ṣūfī (no. C98), al-Karādīsī (no. C90), Ibn ᶜAbd al-Ghaffār (no. C106), al-Khaṭṭāb (no. C119), al-Khurfānī (no. D55), Ibn Bakhshīsh (no. C117), Ḥusayn the *muwaqqit* (no. D146), al-Wardānī (no. D107). Tables based on the method of *al-Durr al-yatīm* by Riḍwān Efendī (no. D58) and al-Khawānikī (no. D78). See also DM 678 (71 fols., *ca.* 1150H), an anonymous collection of tables, including some due to Ibn al-Majdī. His tables have been analyzed in E. S. Kennedy and D. A. King, "Ibn al-Majdī's Tables for Calculating Ephemerides," *JHAS*, 4 (1980), pp. 46–68.

3.2.12 Notes on lunar crescent visibility dictated by Ibn al-Majdī in 830H and recorded by Nūr al-Dīn ᶜAlī b. Muḥammad al-Naqqāsh (no. C74) and miscellaneous topics of *mīqāt*: DM 681,3 (fols. 5r–7v, *ca.* 875H), unique.

3.3.7 ***Tuḥfat al-aḥbāb fī naṣb al-bādhāhanj wa-l-miḥrāb***, a short treatise on the orientation of ventilators and the direction of the qibla: J 3964,17 (fols. 77r–80v, 862H); DM 183,2 (fols. 7r–11r, *ca.* 850H); *etc.* (4 copies). See *JAOS*, 104 (1984), p. 112.

3.4.3 ***ᶜIqd al-durar fi l-ᶜamal bi-l-qamar***, tables with no introduction: MM 193,1 (fols. 1r–6r, 870H), unique?

3.4.3 A treatise on the use of a table for determining crescent visibility: MM 58,5 (fols. 15r–15v, *ca.* 850H).

3.4.3 Notes on Ibn al-Majdī's crescent visibility theory by al-Wafāʾī (no. C61): MM 93,1 (fols. 1v–2v, *ca.* 1100H).

4.3.23 See no. C63.

4.4.10 ***R. fi l-ᶜAmal bi-l-rubᶜ al-mawsūm/al-marsūm bi-l-muqanṭarāt***, on the use of the almucantar quadrant in 10 *faṣl*s: DM 783 (6 fols., in two hands one *ca.* 900H, the other *ca.* 1250H); DM 738,1 (fols. 1r–4r, 982H); MM 112 (9 fols., *ca.* 900H), *etc.* (41 copies!); plus two commentaries by Yūnus b. ᶜAbd al-Qādir al-Rashīdī (no. D137), the first entitled ***Ghāyat al-suʾl fī sharḥ al-ᶜashara fuṣūl***: DM 544 (18 fols., 1083H), *etc.* (7 copies), the second without title: DM 642 (41 fols., 1076H); plus three anonymous commentaries, the first entitled ***Fayḍ al-āla al-mubdī***, in MM 146,2 (fols. 1v–24v, 1177H); DM 713,2 (fols. 8r–14r, *ca.* 1150H) and DM 710,2 (fols. 10r–13v, 1141H). See also MS Berlin Ahlwardt 5847.

4.5.6 ***Irshād al-sāʾil ilā uṣūl al-masāʾil***, an extensive commentary on ***al-Durr al-manthūr*** by al-Māridīnī (no. C47): DM 1096 (192 fols., 999H); DM 1057 (101 fols., *ca.* 1100H, incomplete); *etc.* (2 copies).

4.5.8 ***Khulāṣat al-aqwāl fī maᶜrifat al-waqt wa-ruʾyat al-hilāl***, a treatise on the use of the sine quadrant, with a section on lunar crescent visibility: DM 183,1 (fols. 1r–6v, *ca.* 850H); MM 193,2 (fols. 6v–8v, *ca.* 1000H, defective).

4.6.32 ***al-Rawḍ al-azhar fi l-ᶜamal bi-l-rubᶜ al-musattar***, on a special variety of almucantar quadrant, compiled in 812H and arranged in 10 *bāb*s: DM 751,1 (fols. 1r–5r, 812H). See Pl. CIXa. Brockelmann (*GAL*, SII, p. 159, no. 26) mentions another copy, MS Berlin 1023,3; yet others are in Cairo (Azhar) and Istanbul.

4.7.7 ***Irshād al-ḥāʾir ilā takhṭīṭ faḍl al-dāʾir***, a treatise on sundial theory with tables for latitude 30°, Cairo: DM 2 (30 fols., *ca.* 1100H); DM 685 (31 fols., *ca.* 850H); DM 684 (36 fols., *ca.* 1100H); *etc.* (5 copies and 1 fragment).

Abridgement by the author entitled ***Zād al-musāfir li-maᶜrifat rasm faḍl al-dāʾir***: DM 91 (17 fols., 876H); DM 534,1 (fols. 1r–19r, 899H); K 3823,2 (fols. 6r–13v, *ca.* 815H); *etc.* (12 copies). Commentaries by al-Ḥusnī (no. C124), the first entitled *Rāḥat al-fuʾād taysīr al-Zād*: ZK 917,1 (pp. 1–25, 1163H), *etc.* (2 copies), and the second entitled *Nuzhat al-khāṭir fī waḍᶜ ḥudūd ᶜalā Zād al-musāfir*: DM 175,2 (fols. 11r–18v, *ca.* 1200H); DM 1005 (108 fols., *ca.* 1275, with additional tables for latitude 31;25°, Damietta), *etc.* (3 copies).

Qurrat al-nāẓir fī maᶜrifat waḍᶜ faḍl al-dāʾir treatise on sundial theory in 8 *bāb*s, is attributed to Ibn al-Majdī in MM 148, elsewhere to Sibṭ al-Māridinī (no. C97).

5.5.1 A horoscope of the *amīr* Nāṣir al-Dīn Abu l-Fatḥ Muḥammad, born 802H: MM 141,2 (fols. 16v–20r, *ca.* 1175H).

6.3.19 *Ḥāwi l-lubāb wa-sharḥ Talkhīṣ al-ḥisāb*, a commentary on the *Talkhīṣ* of Ibn al-Bannāʾ: see no. F23.

6.4.2 ***al-Mubtakirāt al-ḥisābiya***, a short treatise on arithmetic: J 3964,10 (fols. 46v–48r, 862H); J 3964,2 (fols. 8r–9v, 862H). Brockelmann lists another copy in the Escorial.

6.4.2 Solutions to some simple problems of *Unterhaltungsmathematik*: DM 447,2 (fols. 11v–12r, *ca.* 900H), unique?

6.4.2 Notes on interpolation attributed to Ibn al-Majdī: ZK 917,4 (pp. 50–54, 1163H).

6.5.1 Two fragments of ***Kashf al-ḥaqāʾiq fī ḥisāb al-daraj wa-l-daqāʾiq***: DM 751,2 (fol. 5v, 812H); DM 775,2 (fols. 22r–24r, 847H). This treatise is arranged in 2 *bābs* of 8 + 7 *faṣls* dealing with sexagesimal arithmetic and sexagesimal multiplication tables, respectively, and a *khātima* on linear interpolation. Abridgements by Sibṭ al-Māridīnī (see no. C97).

6.8.7 Notes on *dirhams* and Ashrafī *dīnārs*, *etc.*: ZK 490,2 (pp. 73–84, 1002H).

6.9.8 ***Masāʾil fi l-murakkabat***, on inheritance: J 3964,9 (fols. 36r–44r, 862H).

6.9.8 ***al-Iqrār al-dawrī idhā kān li-thnayn***: J 3964,4 (fols. 22r–24v, 862H); J 3964,11 (fols. 48r–50v, 862H).

C63 ANONYMOUS (= IBN AL-MAJDĪ (?))

4.3.23 ***al-Ḍawʾ al-lāʾiḥ fī uṣūl al-tasṭīḥ wa-rasm al-ṣafāʾiḥ***, a treatise on the theory of stereographic projection in 4 *faṣls* with tables: DM 620,1 (fols. 1r–12v, *ca.* 1000H, incomplete, anonymous), and a smaller fragment from another copy of the same work, DM 620,1B (fols. 14r–14v, *ca.* 1000H), unique. *Azzawi*, p. 184 attributes this title to Ibn al-Majdī (no. C62).

C64 ZAKARĪYĀ B. YAḤYĀ AL-BILBAYSĪ (?)[1]

Suter, no. 522; *GAL*, SII, p. 1025, no. 87.

4.4.9 ***R. fi l-ʿAmal bi-rubʿ al-muqanṭarāt al-shamālīya***, on the use of the almucantar quadrant in 30 *bābs*: ṬJ 367,3 (fols. 13r–25v, 995H). Other copies are MSS Berlin 5864 and Istanbul Hamidiye 1453,11. This treatise is an abridgement of the author's ***Bughyat al-ṭullāb fi l-ʿamal bi-rubʿ al-asṭurlāb***, no copies of which are known to me.

1. MS Berlin 5864 has al-Talbīsī; MS Cairo ṬJ 367,2 has al-Blsy.

C65 ʿABD AL-GHANĪ B. ḤUSĀM AL-DĪN AḤMAD B. AL-ʿARABĀNĪ AL-MIṢRĪ

Suter, no. N433; *GAL*, II, p. 159 and SII, pp. 159–160.

5.3.13 ***Gharāʾib al-funūn wa-mulaḥ al-ʿuyūn wa-nuzhat al-ʿushshāq li-l-ṭālib al-mushtāq***, a treatise on astrology arranged in unnumbered *bābs*: DM 876,1 (fols. 1r–18v, 1051H, anonymous). Brockelmann lists several other copies. al-ʿArabānī mentions in his introduction that he used the works of several early Muslim astrologers. Merits investigation.

C66 MUḤIBB AL-DĪN ABŪ ʿABD ALLĀH MUḤAMMAD B. MUḤAMMAD, IBN AL-ʿAṬṬĀR

Suter, no. 431; *GAL*, II, pp. 157–158 and SII, p. 158; *Azzawi*, pp. 202–203.

4.1.4 ***Kashf al-qināʿ fī rasm al-arbāʿ***, a treatise on different kinds of quadrants in 2 *qisms* of 10 + 9 *faṣls*, mentioning those who invented them: DM 515 (20 fols., mostly *ca.* 900H); DM 491 (12 fols., *ca.* 1050H, with additional information about Ibn al-ʿAṭṭār); MM 153 (15 fols., *ca.* 1000); *etc.* (7 copies). Commentary entitled ***al-Rayy wa-l-ishbāʿ fī sharḥ Kashf al-qināʿ*** by Ibn Abi l-Khayr al-Ḥusnī (no. C124): MM 192,2 (fols. 4r–57v, *ca.* 1100H); MM 129 (80 fols., *ca.* 1200H). Ibn al-ʿAṭṭār's work merits publication and analysis. The only work on the quadrant in Islam available at present is P. Schmalzl, *Zur Geschichte des Quadranten bei den Arabern*, Munich: Salesianische Offizin, 1929. This work, despite its excellence, is now inadequate in view of the material now available, namely, the anonymous treatises from the ninth or tenth century (no. B105); the compendium of al-Marrākushī (no. C17); and the treatise of Ibn al-ʿAṭṭār.

4.3.24 ***Fuṣūl fī maᶜrifat al-mawqiᶜ wa-niṣf al-quṭr wa-buᶜd al-markaz li-l-muqanṭarāt*** on the construction of curves on astrolabe plates: ṬM 185,2 (fols. 15v–20r, *ca.* 1200H, anonymous). The anonymous author mentions his treatise *Kashf al-qināᶜ* on fol. 18r and al-Ḥalabī (= no. C69?).

(3.1.28) Ibn Abi l-Khayr al-Ḥusnī (no. C124) mentions a work entitled ***Jawharat al-yawāqīt*** by Ibn al-ᶜAṭṭār. No copies of this are known to me. The same treatise by Ibn al-ᶜAṭṭār is mentioned in the anonymous commentary to ***al-Risāla al-Fatḥīya*** of Sibṭ al-Māridīnī (no. C97) in MM 139.

C67 ᶜABD AL-RAḤMĀN B. MUḤAMMAD B. MUḤAMMAD, IBN AL-MUHALLABĪ

Not mentioned in the modern sources.

4.7.8 *ᶜUmdat al-dhākir li-waḍᶜ khuṭūṭ faḍl al-dāʾir*, a treatise on sundial theory with extensive tables for Cairo, compiled in 829H: MM 220 (6 fols., *ca.* 1250H, introduction only). A fine complete copy of this work is MS Dublin Chester Beatty 3641 (57 fols., copied in 858H). Merits investigation.

C68 ᶜABD AL-RAḤMĀN B. ᶜALĪ B. MUḤAMMAD AL-AQFAHSĪ

Suter, no. 440; *GAL*, II, p. 159; *Azzawi*, pp. 184–185.

4.3.25 A fragment of the text and tables of his *al-Jawhar al-maknūn fi l-ḥisāb al-maṣūn* dealing with the construction of astrolabes is MM 246 (5 fols., *ca.* 1100H). A complete copy is MS Berlin Ahlwardt 5692. Merits investigation.

C69 SHIHĀB AL-DĪN AḤMAD B. IBRĀHĪM B. KHALĪL AL-ḤALABĪ

Suter, no. 434.

2.1.12 *al-ᶜIqd al-yamānī fī ḥall Zīj al-Īlkhānī*, an Arabic recension of the *Zīj* of al-Ṭūsī (no. G23) with an introduction in 40 *bāb*s and with additional tables for Aleppo: ṬM 226,1 (fols. 1r–95r and 97v (list of mean motion parameters), 894H in Aleppo, lacks some pages of the introduction). Another copy (complete?) is MS Istanbul M. Hafid Ef. 181,1. This *zīj* is not listed in *Kennedy*.

3.1.13 ***Jadwal al-samt li-ᶜarḍ Dimashq***, solar azimuth tables for Damascus: K 8525,3 (fols. 20v–56r, *ca.* 850H by author); MM 71,1 (fols. 1r–66r + 67r–68v, *ca.* 1000H). Another copy is in MS Damascus Ẓāhirīya 9227.

3.1.13 ***Jadwal irtifāᶜ al-kawākib al-thābita ᶜinda ṭulūᶜ al-fajr***, a table of the altitude of various fixed stars at daybreak throughout the year: K 8525,2 (fols. 14r–20r, *ca.* 850H in author's hand), unique.

These tables are analyzed in *SATMI*.

C70 BADR AL-DĪN ABŪ ᶜABD ALLĀH MUḤAMMAD B. QĀḌĪ SHUHBA AL-ASADĪ AL-SHĀFIᶜĪ

GAL, II, p. 37, and SII, pp. 25–26 (no scientific works mentioned). Article "Ibn Ḳāḍī Shuhba" by J. Schacht in EI_2.

6.4.6 Three anonymous copies of a treatise on multiplication without title in 11 unnumbered *bābs*: ṬJ 825,6 (fols. 138v–140v, *ca.* 1100H); MR 15,1 (fols. 1v–3r, *ca.* 1200H); ṬJ 688,2 (fols. 30r–32r, 1045H); *etc.* (1 copy); plus an anonymous commentary in which the original is attributed to Ibn Qāḍī Shuhba in MJ 56,2 (fols. 16v–24v, *ca.* 900H). Not listed by Brockelmann. Other anonymous copies are MSS Princeton Mach 4844 = Yehuda 3021, fols. 150v–151v, *ca.* 1150H, and Selly Oak 1900 (Mach). See also an anonymous ***Mukhtaṣar fi l-ḥisāb*** also in 11 unnumbered *faṣl*s and dealing with multiplication: ṬJ 492,2 (fol. 184r, *ca.* 1000H).

C71 ABŪ ᶜABD ALLĀH MUḤAMMAD B. SULAYMĀN AL-KĀFIYAJĪ

GAL, II, pp. 138–140, and SII, pp. 140–141.

3.3.10 *K. al-Kāfī fī bayān al-ṣaff al-ṭawīl* = ***R. fi Bayān istiqbāl al-qibla***, a treatise on the religious aspects of facing the *qibla*: DJ 273,1 (fols. 4v–8r, 871H by author): DM 530 (5 fols., *ca.* 1150H). Brockelmann (*GAL*, SII, 141, no. 50) mentions one other copy of this treatise, MS Hyderabad Asaf. 455.

6.7.6 A fragment of a commentary (?) on *Ḥall al-ishāl fī mabāḥith al-ashkāl* dealing with geometry and surveying: DJ 394,4 (fols. 100v–104r, *ca.* 900H, anonymous, difficult *naskhī*), unique. The title is attributed to al-Kāfiyajī by Bagdatli (I, p. 416) and listed by Brockelmann (*GAL*, II, 139, no. 21).

C72 SAYF AL-DĪN SĀṬILMISH

Not identified.

2.4.1 The trigonometric tables in ṬM 80 (69 fols., *ca.* 850H) are copied in his elegant hand and it may be that he computed them. They consist of tables of the Sine and Cotangent and Versed Sine functions and the solar declination (for obliquity 23;35°) with entries to three digits for each minute of argument. See Pls. XXIVa and CVIa and also no. C137.

C73 MUḤAMMAD AL-GHAZĀWĪ (?), *muwaqqit* at the Muqassam Mosque in Cairo

3.1.24 Tables of the ascensions at nightfall for each 0;6° of solar longitude, for latitude 30° (Cairo): DM 1218 (14 fols., *ca.* 900H).

C74 NŪR AL-DĪN ᶜALĪ B. MUḤAMMAD AL-NAQQĀSH

Not mentioned in the modern sources.

2.3.6 A table of the lunar mean motion by al-Naqqāsh according to the method *al-Durr al-yatīm* by Ibn al-Majdī (no. C62) is in DM 448,3 (fols. 5v–6v, *ca.* 900H).

3.2.12 *ᶜUmdat al-ḥudhdhāq fi l-ᶜamal bi-hā fī sāʾir al-āfāq*, a treatise on timekeeping without subdivisions: ZK 706,2 (fols. 9r–10r, *ca.* 1100H, anonymous). Another anonymous copy of the treatise *ᶜUmdat al-aḥdhāq* is MS Paris B.N. ar. 2560,9, but in MS Princeton Mach 4927 = Yahuda 4103, fols. 18a–19v, *ca.* 1150H, the work is attributed to Abu l-Ḥasan Nūr al-Dīn al-Naqqāsh. The author states that the treatise is an abridgement of his ***Nūr al-aḥdhāq***, which is unrelated to the tables entitled ***Nūr al-aḥdhāq*** by Ibn al-Mushrif (see no. C43).

3.2.12 Notes on lunar crescent visibility (dictated by Ibn al-Majdī (no. C62) in 830H and recorded by Nūr al-Din ᶜAlī b. Muḥammad al-Naqqāsh) and miscellaneous topics of *mīqāt*: DM 681,3 (fols. 5r–7v, *ca.* 875H), unique. Merits study.

5.5.3 Notes on the computation of the astrological houses: DM 771,2 (fols. 9v–10v, *ca.* 850H).

C75 NŪR AL-DĪN ᶜALĪ B. MUḤAMMAD AL-ZAMZAMĪ

Suter, no. 446; *GAL*, II, p. 229 and SII, p. 230.

6.4.12 ***Fatḥ al-Wahhāb***, a poem on arithmetic, compiled in 878H: TR 138,1 (pp. 1–16, 889H), plus an anonymous commentary compiled in 895H): DR 1099 40 fols., 1147H). Another commentary by Zayn al-Dīn ᶜArafa al-Faraḍī (no. C76): DR 56 (32 fols., in two hands, one dated 919H); *etc.* (1 fragment).

C76 ZAYN AL-DĪN ᶜARAFA AL-FARAḌĪ

Not identified.

6.4.12 Commentary on the poem *Fatḥ al-Wahhāb* by al-Zamzamī: see no. C75 (manuscript dated 919H).

C77 NŪR AL-DĪN ᶜALĪ B. ᶜABD AL-QĀDIR AL-FARAḌĪ AL-SHARĪF AL-ḤUSNI AL-SHĀFIᶜĪ

6.4.1 ***al-Fawāʾid al-Jalīla fī ḥall majhūlāt al-Wasīla***, a commentary on the ***Wasīla*** of Ibn al-Hāʾim: see no. C58.

6.9.9 ***Fāʾida fi stikhrāj al-aqārīr bi-l-khaṭaʾayn***, on inheritance: J 3964,13 (fols. 58r–60v, 862H by author).

6.9.9 ***Fāʾida fī maᶜrifat al-dīnār wa-l-dirham wa-naṣb al-zakāt min al-dhahab***: J 3964,14 (fols. 61r–65r, 862H by author).

C78 ABŪ YAḤYĀ ZAKARĪYĀ B. MUḤAMMAD B. AḤMAD AL-ANṢĀRĪ

GAL, II, pp. 122–124, and SII, pp. 117–118, *etc.*

6.4.1 ***Fatḥ al-mubdiᶜ fī sharḥ al-Muqniᶜ*** (see no. C58): DR 307 (23 fols., 1091H); *etc.* (7 copies), plus *ḥāshiya* by Aḥmad b. Muḥammad al-Janājī (no. D172): DR 1101 (169 fols., *ca.* 1275H); *etc.* (3 copies).

C79 ZAYN AL-DĪN ᶜUMAR AL-ZUHRĪ

Not identified.

3.3.8 Notes on the determination of the qibla: ṬM 228,2 (fols. 3r–3v + 8v, *ca.* 900H).

C80 MUḤAMMAD B. AḤMAD AL-BILBAYSĪ AL-SHĀFIᶜĪ, IBN AL-ᶜAṬṬAR

Not identified.

3.2.11 *Kifāyat al-mushtāq li-maᶜrifat faḍl al-dāʾir fī sāʾir al-āfāq*, a treatise on timekeeping in 10 *bāb*s: DM 442 (8 fols., 1078H), unique. The treatise describes the use of al-Khalīlī's auxiliary tables (see no. C37), and is the only medieval Arabic work known to me which specifically mentions these tables.

C81 SŪDŪN AL-BASHTAKĪ, a *muʾadhdhin* in Cairo

(see no. C82)

3.1.14 An abridged version of the auxiliary tables entitled *Nūr al-aḥdhāq* by Ibn al-Mushrif (see no. C43): DM 512 (16 fols., *ca.* 950H), unique. See *SATMI* for further details.

C82 IBN SŪDŪN[1]

(see no. C81)

Not identified.

4.7.10 *Jadāwil al-munḥarifāt al-maḥsūba ilā ṣād daraja*, tables for marking vertical sundials for the latitude of Cairo for each degree of inclination to the meridian: DM 1170,1 (fols. 1r–10v, 893H); DM 534,3 (fols. 20r–31r, *ca.* 900H, anonymous) *etc.* (2 anonymous copies). No other copies have come to my attention.

C83 ABU L-BAQĀʾ YAḤYĀ B. AL-JĪᶜĀN

GAL, II, p. 38 and SII, p. 26. See also *GAL*, II, p. 163, and SII, p. 163 (Sharaf al-Dīn Yaḥyā b. al-Jīᶜān).

2.5.6 *Ṭawāliᶜ al-budūr fī taḥwīl al-sinīn wa-l-shuhūr*, a treatise in 3 *bāb*s on the *kharājīya* (fiscal) years, their relation to the lunar calendar, and their use in official documents: DM 214 (29 fols., 8 lines on each page, *ca.* 850H, beautiful *thulth*, *waqf* of Sultan Qaytbāy 890H); DM 105 (29 fols., 968H, fine *thulth*); *etc.* (1 late copy). See Pl. CV. Another fine copy of this treatise in MS Paris B.N. ar. 2557. Brockelmann lists also MSS Istanbul Aya Sofia 2665 and Nurosmaniye 4919.

C84 IBRĀHĪM B. ᶜUMAR B. ḤASAN AL-RIBĀṬ AL-BUQĀʾĪ AL-SHĀFIᶜĪ

Suter no. 441; *GAL*, II, pp. 179–180, *etc.*

6.4.3 *Ibāḥat al-Bāḥa fī ᶜilmayi l-ḥisāb wa-l-misāḥa*, a commentary on the author's poem on arithmetic and surveying entitled *al-Bāḥa*: DR 3 (99 fols., *ca.* 850H). Another copy of both poem and commentary is in the British Library.

C85 ḤASAN B. MUḤAMMAD B. AL-MAḤALLĪ

Not identified.

1.4.22 *al-Jawhara al-muḍīʾa*, a poem on the stars and lunar mansions: DM 772,1 (fols. 1r–3v, *ca.* 900H), unique?

C86 TAQI L-DĪN AḤMAD B. ᶜALĪ B. ᶜABD AL-QĀDIR B. MUḤAMMAD AL-MAQRĪZĪ, the historian

GAL, II, pp. 47–50 and SII, pp. 36–38.

6.8.3 *R. fī Asmāʾ al-awzān wa-l-akyāl al-sharᶜīya*, a treatise on weights and measures in 7 unnumbered *faṣl*s: ṬR 144,1 (fols. 1r–12v, 1196H); *etc.* (1 late copy). Brockelmann mentions MS Leiden 1014 as the only other copy (*GAL*, SI, p. 44 and SII, p. 37, no. 8b), which was published in 1800.

C87 ABŪ HURAYRA ᶜABD AL-RAḤMĀN B. BANAFSHĀ AL-JAWHARĪ AL-ṢĀLIḤĪ

Suter, no. 454; *GAL*, II, pp. 160–161 and SII, p. 161.

---

1. The name is virtually illegible in MS DM 1170,1: all I can make of it is Sayyidī . . . b. . . . b. Sūdūn. The other copies are anonymous.

2.2.4 *al-Durr al-naẓīm fī tashīl al-taqwīm*, planetary tables for the longitude of Damascus, with an introduction in unnumbered *bābs*: DM 1006,1 (fols. 1r–137r, *ca.* 1000H); DM 140 (33 fols., *ca.* 870H by author, lacks most of introduction and some tables); K 8526 (114 fols., in two hands 1290H and *ca.* 1150H); DM 20 (123 fols., *ca.* 1150H, attributed to Muḥammad al-Miṣrī (no. C98), lacks introduction); *etc.* (4 fragments).

C88 GHARS AL-DĪN KHALĪL B. AḤMAD AL-NAQĪB AL-ḤALABĪ

Suter, no. 465.

3.3.13 A treatise on the determination of the qibla in 3 subdivided *bābs*: MM 114 (9 fols. *ca.* 1000H from author's copy dated 969H) *etc.* (copy of MM 114), unique? al-Ḥalabī mentions the problem of the qiblas of the mosques of Cairo. See Pl. Lb. On this subject, see now *JAOS*, 104 (1984), pp. 97–133, esp. pp. 112–115.

4.4.16 *Qawāᶜid fī maᶜrifat samt al-qibla wa-l-awqāt bi-aqrab al-ṭuruq wa-ashal al-ālāt*, on the use of the almucantar quadrant in 8 *bābs*: MMT 7,6 (fols. 122v–124r, 1002H); MJ 180,19 (fols. 149v–151r, *ca.* 1100H, anonymous).

4.5.17 (*R. fi l-ᶜAmal bi-l-rubᶜ al-mujayyab*), on the use of the sine quadrant in 20 *bābs*: ṬM 255,4 (fols. 40v–50r, 1060H); MJ 180,16 (fols. 143v–146r, *ca.* 1150H).

C89 MUḤAMMAD B. AḤMAD AL-ḤAFRĪ

Not mentioned in the modern sources.

1.2.10 *al-Takmila fī sharḥ al-Tadhkira*, a commentary on the *Tadhkira* of al-Ṭūsī (no. G23) compiled in 932H: TR 230 (388 pp., 1053H). Unique.

C90 BADR AL-DĪN AL-ḤASAN B. KHALĪL AL-ṬUBNĪ AL-KARĀDĪSĪ

Suter, no. 180; *GAL*, II, p. 160 and SII, p. 160.

2.3.6 *al-Sirr al-maṣūn fi l-Durr al-maknūn*, on the use of the auxiliary tables of Ibn al-Majdī (no. C62) for compiling ephemerides: DM 1107,2 (fols. 51r–59v, 1172H), unique?

3.2.13 *Kifāyat al-muḥtāj min al-ṭullāb ilā maᶜrifat al-masāʾil al-falakīya bi-l-ḥisāb*, a treatise on timekeeping in 19 *bābs*: DM 761 (46 fols., 902H); DM 121 (30 fols., *ca.* 900H); *etc.* (3 copies).

3.4.4 *Nūr al-baṣar fi l-ᶜamal bi-l-qamar*, on lunar crescent visibility: DM 550 (5 fols., 887H, lacks the tables promised in the text).

3.4.4 *R. fī ᶜAmal al-ahilla bi-ṭarīq al-jadāwil*, without tables: DM 523 (3 fols., 1148H).

3.4.4 *R. fī ᶜAmal al-ahilla bi-l-ḥisāb*, without tables: K 4027,2 (fols. 5r–6v, *ca.* 1250H).

3.4.4 *Faṣl fī ᶜAmal al-ahilla*, with tables: MM 214 (7 fols., *ca.* 1050H).

3.5.5 *R. fī Ṭulūᶜ al-shams wa-l-ghurūb wa-l-tawassuṭ fī balad bi-l-nisba ilā balad ukhrā*, on the difference in the rising and setting times of the sun in different localities: DM 620,6 (fols. 44v–45v, *ca.* 900H, lacks end); K 4027,1 (fols. 1v–4v, *ca.* 1300H).

4.7.12 *Ashkāl al-wasāʾiṭ fi l-munḥarifāt wa-l-basāʾiṭ*, a treatise on sundial theory with tables for latitude 30° (Cairo), arranged in 2 *qisms* of 7 *bābs* each, compiled in 882H: DR 892 (55 fols., 885H); DM 6 (52 fols., *ca.* 1175H); *etc.* (7 copies); plus an anonymous recension in DM 175,7 (fols. 66r–87v, *ca.* 1200H). MS Paris B.N. ar. 2543 of this work is in the author's hand, dated 882H. Merits investigation. See Pl. LXVIII.

4.7.16+ *R. fī Maᶜrifat waḍᶜ khayṭ al-musātara wa-waḍᶜ khuṭūṭ faḍl al-dāʾir taḥtahu*, on a special variety of sundial: MM 181,1 (fols. 1r–4r, 1030H); TR 343,2 (pp. 61–68, *ca.* 1250H, anonymous). See also an anonymous *Fāʾida fī naṣb khayṭ al-musātara*: Sh 89,3 (fol. 29v, 1025H), published in *JHAS*, 1 (1977), p. 256.

C91 AL-JAMĀLĪ YŪSUF B. QURQMĀS AL-ḤAMZĀWĪ, AMĪR AL-ḤĀJJ AL-ḤALABĪ

Not listed in the modern sources.

2.4.5 Tables of the ascensions of the astrological houses: K 4003,4 (fols. 30v–31r, *ca.* 1100H). No other known copies.

4.7.11 *R. fi Maᶜrifat waḍᶜ khuṭūṭ faḍl al-dāʾir wa-qusiy al-ᶜaṣr wa-l-bāqī minhu li-l-ghurūb wa-l-sāᶜāt al-mustawiya wa-l-zamanīya bi-l-aᶜmida al-thābita*, on the construction of sundials on columns: MM 93,3 (fols. 5r–6v, *ca.* 1100H); ṬM 102,3 (fols. 36v–38v, *ca.* 1150H). No other known copies. See also the earlier treatise of al-Wafāʾī (no. C61) on this subject.

5.3.16 An astrological history of events relating to the campaign between Yashbak and Shahsiwār around Adana from 875H to 877H: Sh 74 (13 fols., 1318H); and fragments in MM 13,3 (fols. 4r + 49r–49v + 99v, 894H by Aḥmad b. Aḥmad Timurbāy—see no. C94). Neither author or work is mentioned by Suter or Brockelmann, and I know of no other copies. There is no indication which *zīj* was used to compute the several horoscopes in the text. Merits detailed investigation. See Pl. LXXXIII.

See also his treatise ***Durr al-maṭlūb fī sirr al-ghālib wa-l-maghlūb*** extant in MS Istanbul Nurosmaniye 4901, mentioned in F. Rosenthal, *Ibn Khaldūn: The Muqaddimah*, Princeton: 1967 ed., I, p. 235, note 359.

6.6.2 *K. al-Nisba al-kubrā*, extensive sexagesimal multiplication tables: TR 119 (614 pp., *ca.* 900H, incomplete: first half only), unique. See Pl. XCIa and *HM*, 6 (1979), p. 409.

C92 JAMĀL AL-DĪN YŪSUF B. ṬŪGHĀN AL-QIṬJĪ (?) AL-MĪQĀTĪ

Not mentioned in the modern sources.

3.5.5+ *Faṣl fi Maʿrifat mughīb al-qamar wa-ṭulūʿihi taqrīban fī kull yawm*, a short treatise accompanied by a table for finding the times of moon-rise and -set: MM 167,4 (fols. 158v–160r, 989H), unique. See Pl. XXXVIb.

5.3.21 *Nuzhat al-afkār fī maʿrifat li-maʿrifat al-asʿār*, an astrological treatise dealing with prices: DM 74 (10 fols., *ca.* 1150H); DM 1210,2 (fols. 14v–29r, *ca.* 1100H); *etc.* (3 copies). This treatise is compiled from the works of various early Muslim astrologers.

C93 MUḤYI L-DĪN ABU L-JŪD ʿABD AL-QĀDIR B. ʿALĪ B. SHAʿBĀN AL-DANJĀWĪ / AL-DANJĪHĪ (?) AL-FARAḌĪ AL-SHĀFIʿĪ

See *Suter*, no. 528, where the footnote is to be deleted.

6.3.10 A commentary on Ibn al-Maghribī's poem on finger arithmetic (no. F36): DR 674 (8 fols., *ca.* 1150H). Other copies include MSS Gotha Forschungsbibliothek A1495 and Princeton Mach 4837 = Yehuda 1028, fols. 110v–113v, *ca.* 1050H.

6.3.19 A commentary on *al-Ḥāwī fi l-ḥisāb* by Ibn al-Hāʾim, compiled in 854H: see no. C58.

6.4.9 *Mukhtaṣar fī ʿilm al-ḥisāb*, a treatise on arithmetic in 11 *bābs*: K 4614 (6 fols., 1093H); DR 888 (6 fols., 1156H). Another copy is MS Berlin Ahlwardt 6001, which is not the same as the treatise of ʿAbd al-Qādir b. ʿAlī al-Sakhkhāwī (no. C131), as stated by Ahlwardt (*cf.* also *Suter* nos. 476 and 528).

6.9.10 Notes on *al-nisāb al-sharʿī*, the minimum amount of property liable to payment of the *zakāt* tax: J 3964,15 (fols. 65v–70r, *ca.* 850H).

C94 AL-SHIHĀB AḤMAD B. AḤMAD TIMURBĀY

GAS, VII, p. 187. See Pls. XIIIb and LXXXIII.

2.2.3 Tables for the sun and moon for longitude 54;55° (Cairo) based on the parameters of Ulugh Beg (no. G49): DM 639,27 (fols. 242r–245r, *ca.* 900H by author), unique.

5.2.18 *al-Barq al-sāṭiʿ fī mukhtaṣar al-Bāriʿ*, an abridgement of the astrological treatise of Ibn Abi l-Rijāl (no. B76): MM 13,2 (fols. 4v–99v, 894H in author's hand); ṬM 112 (159 fols., 1089H); *etc.* (2 copies). Other copies are listed in *GAS*, VII, p. 187.

5.3.15 (*R. fī Maʿrifat al-ḥawādith al-suflīya min dalālāt al-ashkhāṣ al-ʿulwīya*), an astrological treatise in 4 *qisms*: Sh 89,1 (fols. 1r–9v, *ca.* 1000H, attributed to Sayyidī Aḥmad Timurbāy); DM 1006,2 (fols. 139r–143r, *ca.* 1100H, anonymous); DM 180,4 (fols. 73v–84v, 1238H, anonymous.).

He copied MM 13, 3 *ca.* 1480: see Pl. LXXXIII.

C95 ABŪ ʿABD ALLĀH MUḤAMMAD B. MUḤAMMAD B. ABĪ BAKR AL-ḤALABĪ AL-TĪZĪNĪ

Suter, no. 450; *GAL*, II, p. 160 and SII, p. 484.

4.4.14 *R. fī Rubʿ al-muqanṭarāt*, on the use of the almucantar quadrant in 12 *bābs*: TR 167 (18 pp., *ca.* 950H); *etc.* (1 incomplete copy).

4.4.14 *R. mulakhkhaṣa fi l-ʿAmal bi-rubʿ al-dāʾira al-mawḍūʿa ʿalayhi l-muqanṭarāt al-shamālīya*, in 16 *bāb*s: MM 187,1 (fols. 1r–4r, 1125H).

4.4.14 *Malḥa fi l-ʿAmal bi-l-rubʿ al-mawḍūʿ ʿalayhi l-muqanṭarāt al-shamālīya*, in 15 *bāb*s: DM 162 (10 fols., 1054H, anonymous). The attribution is suggested by the related treatise in 12 *bāb*s in MS Berlin Ahlwardt 5804.

4.5.13 *R. fi l-ʿAmal bi-l-rubʿ al-mujayyab*, on the use of the sine quadrant in 12 *faṣl*s: MM 245,3 (fols. 9v–10v, 1125H). Note that MS Paris B.N. ar. 2547,22 contains another treatise on the sine quadrant in 20 *bāb*s attributed to al-Tīzīnī.

4.5.13 *Fāʾida fi l-ʿAmal bi-qaws al-ʿaṣr al-mawḍūʿ ʿala l-jayb*: TR 161,3 (fol. 11r, *ca.* 1100H from a copy by al-Tīzīnī).

C96 ABŪ BAKR B. AL-IMĀM[1]

Not mentioned in the modern sources.

3.2.16 *R. Āfāqīya fi l-ʿamal bi-l-nisba al-sittīnīya*, a treatise on calculations in spherical astronomy in 12 *bāb*s, sometimes 13: DM 495 (4 fols., 1077H); DM 128,1 (fols. 1r–3r, *ca.* 1100H, in 13 *bāb*s); DM 127,1 (fols. 1r–2v, *ca.* 1100H); *etc.* (5 copies).

C97 BADR AL-DĪN ABŪ ʿABD ALLĀH MUḤAMMAD B. MUḤAMMAD B. AḤMAD, SIBṬ AL-MĀRIDĪNĪ

Suter, no. 445 and p. 222 (note 90a) and 445N; *GAL*, II, pp. 216 and 468, and SII, pp. 215–217 and 484 (confused). See Pl. CIXc.

3.2.14 *Laqṭ al-jawāhir fī maʿrifat al-dawāʾir* and variations on this title, a collection of definitions of terms relating to spherical astronomy: DM 997 (8 fols., 873H); MM 163 (15 fols., 881H); *etc.* (25 copies); plus commentary by al-Sajjāʿī (no. D86) entitled *Fatḥ al-ʿalīm al-qādir . . .* : DM 645 (37 fols., 1330H), *etc.* (2 copies).

3.2.14 *Wasīlat al-ṭullāb wa-nuzhat al-albāb ilā maʿrifat al-awqāt bi-l-ḥisab*, a treatise on timekeeping on 25 *bāb*s: DM 596,1 (fols. 1r–6r, 1030H; DM 449 (10 fols., 1179H); *etc.* (1 fragment).

3.3.9 *al-R. li-Khurūj (?) al-qibla min ghayr āla*, a treatise on the computation of the qibla in 2 *faṣl*s: DM 639,21 (fols. 180v–185v, 1132H). No other known copies.

4.4.7 A commentary on ***al-Waraqāt*** by his grand-father al-Māridīnī: see no. C47.

4.4.12 *Iẓhār al-sirr al-mawḍūʿ fi l-ʿamal bi-l-rubʿ al-maqṭūʿ*, on the use of almucantar quadrant in 20 *bāb*s: ZK 782,15 (fols. 100v–114r, 1118H); KhM 10,1 (fols. 1r–8v, 1095H).

4.4.12 Abridgement by Sibṭ al-Māridīnī, *Kifāyat al-qanūʿ fi l-ʿamal bi-l-rubʿ al-maqṭūʿ*, in 15 *bāb*s: ṬM 83,2 (fols. 15v–26r, *ca.* 900H); ṬJ 179,5 (fols. 105r–110v, *ca.* 1100H); *etc.* (22 copies); plus an anonymous Turkish translation (no. H86) in QM 6,3 (fols. 16v–23r, *ca.* 1200H).

4.4.12 *Ḥāwi l-mukhtaṣarāt fi l-ʿamal bi-rubʿ al-muqanṭarāt*, in 30 *bāb*s: DM 686 (46 fols., *ca.* 1100H); MM 75 (50 fols., *ca.* 1000H); ṬM 242,8 (fols. 78v–113r, 1108H in the hand of the copyist of MM 38); DM 651,2 (fols. 12r–36r, *ca.* 900H, incomplete); *etc.* (11 copies); plus an abridgement in 30 *bāb*s by Sibṭ al-Māridīnī entitled *Naẓm al-jawhar al-ghālī* in MM 234 (21 fols., 1004H) and *takmila* in 14 *bāb*s by him in MM 170,2 (fols. 22r–24r, 1072H).

4.4.12 *Tadrīb al-ʿāmil fi l-ʿamal bi-l-rubʿ al-kāmil*, in 25 *bāb*s: TR 64,2 (pp. 7–22, 971H); ṬJ 179,4 (fols. 91r–105r, *ca.* 1000H); *etc.* (1 copy); plus abridgement in 15 *bāb*s by author: DM 492 (13 fols., *ca.* 1200H).

4.4.12 *Hidāyat al-sāʾil fi l-ʿamal bil-rubʿ al-kāmil = R. fi l-Rubʿ al-kāmil*, in 20 *bāb*s: ṬM 83,1 (fols. 1v–14v, *ca.* 900H); DM 453,2 (fols. 1v–8r, *ca.* 1100H); *etc.* (3 copies). Another version (in 15 *bāb*s) also entitled *Hidāyat al-sāʾil fi l-ʿamal bi-l-rubʿ al-kāmil* or *Hidāyat al-ʿāmil* and in some copies incorrectly attributed to Ibn al-Shāṭir (no. C30): ṬM 255,5 (fols. 52v–62r, 1060H); ṬM 254,2 (fols. 8v–18v, *ca.* 1100H); *etc.* (8 copies); plus commentary ***Khulāṣat al-masāʾil fī sharḥ Hidāyat al-sāʾil*** in 15 *bāb*s by ʿAbd Allāh b. ʿAbd al-Raḥmān al-Ḥanbalī al-Ḥalabī (no. D128) in ṬM 121,1 (fols. 1v–47r, *ca.* 1325H).

4.4.12 Also, an anonymous treatise in 15 *bāb*s on the almucantar quadrant, very similar to

1. This name is given only in DM 495; the other copies are all anonymous. Another anonymous copy is MS Princeton Mach 5010.

Kifāyat al-qanūᶜ: ṬM 101,2 (fols. 38v–51v, *ca.* 1250H).

4.4.12 ***R. fi l-ᶜAmal bi-rubᶜ al-muqanṭarāt***, in 14 *faṣls*: DM 717,1 (fols. 1r–5v, 1120H); *etc.* (5 copies).

4.4.12 Other such treatises (in 20 *faṣls*): DM 804,4 (fols. 9v–11v, 1231H) and DM 768 (6 fols., *ca.* 1250H); and (in unnumbered *faṣls*): DM 992 (10 fols., 1117H).

4.5.12 ***al-Maṭlab fi l-ᶜamal bi-l-rubᶜ al-mujayyab***, a treatise on the use of the sine quadrant in 100 *bābs*, compiled in 844 or 849H: TR 102 (166 pp., 889H); MM 167,2 (fols. 77r–154r, 988H); *etc.* (7 copies and 2 fragments).

4.5.12 ***Tuḥfat al-mukhtaṣarāt fī maᶜrifat al-qibla wa-awqāt al-ṣalawāt***, a treatise on the use of the sine quadrant in 25 *bābs*: K 4028 (7 fols., *ca.* 1150H); ṬM 77,1 (fols. 1v–12v, 1124H).

4.5.12 ***Qaṭf al-zuhūr fi l-ᶜamal bi-rubᶜ al-dastūr***, a treatise on the use of the sine quadrant in 20 *bābs*; K 4047 (8 fols., *ca.* 950H).

4.5.12 (***R. fi l-ᶜAmal bi-l-rubᶜ al-mujayyab***), in 15 *bābs*: DM 582 (8 fols., 1187H).

4.5.12 (***R. ᶜala l-Rubᶜ al-mujayyab***), in 20 *bābs*: DM 651,3 (fol. 36v, *ca.* 900H, first page only).

4.5.12 ***Laṭāʾif al-ikhtirāᶜ fi l-ᶜamal bi-l-rubᶜ alladhī quṭbuhu fī ṭaraf qaws al-irtifāᶜ***, in 15 *bābs*: DM 537 (4 fols., 1082H); DM 596,2 (fols. 6v–8r, 1031H).

4.5.12 ***al-R. al-Fatḥīya fi l-aᶜmāl al-jaybīya = R. fi l-ᶜAmal bi-l-rubᶜ al-mujayyab***, in 20 *bābs*: ṬM 83,3 (fols. 26v–34v, *ca.* 900H); TR 64,4 (fols. 33v–44r, 971H); DJ 273,2 (fols. 9v–12r, 867H, abridged); *etc.* (47 copies!).

Commentaries by al-Tājūrī (no. C110): MM 17 (42 fols., *ca.* 1050H), DM 959 (20 fols., *ca.* 1000H), *etc.* (4 copies); by al-Sunbāṭī (no. C126): DM 587 (20 fols., 1053H), DM 99 (22 fols., *ca.* 1050H), *etc.* (7 copies); by Yūsuf al-Maḥallī (no. D52), entitled ***al-Shams al-muḍīya*** . . . : DM 485 (23 fols., 1146H), MM 140 (19 fols., 1146H), *etc.* (2 copies); by ᶜAlī al-Nabtītī (no. D26), entitled ***al-Futūḥāt al-Wahbīya*** . . . : MM 171,1 (fols. 1r–34r, 1145H), MJ 42,3 (fols. 39v–76v, *ca.* 1100;H), *etc.* (4 copies); by Daʾūd al-Qāriḥ (no. D149): ṬJ 366,1 (fols. 1v–6v, *ca.* 1250H); by Ḥasan al-Jabartī (no. D91), the first entitled ***al-Thamarāt al-majnīya*** . . . : DM 503 (5 fols., *ca.* 1150H), DJ 11,4 (fols. 58r–60v, 1160H), and the second entitled ***al-Nasamāt al-Fatḥīya*** . . . : DM 259 (8 fols., *ca.* 1100H), MM 232 (10 fols., *ca.* 1200H); a commentary by ᶜAlī al-Bāṣo (?) al-Tūnisī (no. F63): DM 521,14 (fols. 248r–294r, *ca.* 1175H).

Two poems based on ***al-Fatḥīya***, the first by Ibn Ghānim (no. C127) entitled ***al-Nasama al-nafḥīya***: ṬM 242,1 (fols. 1v–8r, 1284H), ṬM 156 (5 fols., 1029H), plus several commentaries thereon, and the second by Shams al-Dīn al-Qaṭarī (no. D31) in a commentary by ᶜAlī al-Rashīdī (no. D32): DM 439 (26 fols., 1064H by the author).

Anonymous commentaries on ***al-Fatḥīya*** in (a) ṬJ 366,6 (fols. 72v–91r, 1267H), *etc.* (3 copies); (b) DM 168,2 (fols. 25v–50v, 1237H), (c) DJ 844,6 (fols. 139v–187r, *ca.* 1150H), *etc.*

4.6.29 ***al-Istīᶜāb li-l-ᶜamal bi-ṣadr al-wizza wa-jināḥ al-ghurāb***, a treatise on the winged quadrant in 10 *maqālas*: DM 135,3 (fols. 5r–9v, *ca.* 900H); TR 287 (10 pp., *ca.* 1300H).

4.7.13 ***Qurrat al-nāẓir fī maᶜrifat waḍᶜ faḍl al-dāʾir***, a treatise on sundial theory in 8 *bābs*: MM 148 (7 fols., *ca.* 1100H, attributed to Ibn al-Majdī (no. C62)); DM 762 (5 fols., *ca.* 1100H, anonymous); MM 147 (9 fols., 1214H, attributed to Sibṭ al-Māridīnī).

4.7.13 ***Jādawil fī rasm al-munḥarifāt ᶜala l-ḥayṭān***, tables for marking vertical sundials, with a brief introduction in 3 *bābs*: MM 53 (12 fols., *ca.* 1000H); MM 180,2 (fols. 3r–15r, *ca.* 1100H); ṬM 174 (11 fols., *ca.* 1150H); *etc.* (15 copies, mostly defective). On these tables, see C. Schoy, "Sonnenuhren der spätarabischen Astronomie," *Isis*, 6 (1924), pp. 332–360. At the end of the 3rd *bāb* of the introduction the author mentions a work of his entitled ***Jāmiᶜ al-mubtakarāt fī aᶜmāl al-waqt wa-waḍᶜ al-ālāt***, no copies of which are known.

6.3.8 Four commentaries on the poem on algebra by Ibn al-Yasamīn (no. F12), as follows:

Sharḥ ᶜala l-Arjūza al-Yasamīnīya: DR 314 (19 fols., *ca.* 1150H); DR 361 (20 fols., 1180H), *etc.* (3 copies).

Sharḥ al-Yasamīnīya: DR 181,14 (fols. 483r–494v, *ca.* 1150H).

al-Tuḥfa al-Māridīnīya fī sharḥ al-Yasamīnīya: DR 362 (10 fols., 1109H); DJ 39,16 (fols.

168v–173v, 1132H); *etc.* (1 copy).

al-Lumᶜa al-Māridīnīya fī sharḥ al-Yasamīnīya: DR 656 (9 fols., 1111H); DR 89,1 (fols. 1v–15v, 1172H); *etc.* (18 copies); plus a commentary by Muḥammad al-Ḥifnī (no. D104), entitled *Fawāʾid ᶜawāʾid jabrīya . . .* : DR 922 (23 fols., *ca.* 1200H), *etc.* (8 copies); and a *ḥāshiya* by Aḥmad al-Janājī (no. D203): DR 627 (31 fols., 1237H); and an anonymous commentary: MR 27,2 (fols. 15r–20v, *ca.* 1200H). This treatise has been published by M. Souissi (Kuwait, 1983).

6.3.19 A commentary on ***al-Ḥāwī fi l-ḥisāb*** by Ibn al-Hāʾim, compiled in 895H: see no. C58.

6.4.1 ***Irshād al-ṭullāb ilā Wasīlat al-ḥisāb***, a commentary on the ***Wasīla*** of Ibn al-Hāʾim (no. C58) compiled in 902H: DR 2 (62 fols., 903H, contains *ijāza* in the author's hand); *etc.* (2 copies).

6.4.1 A commentary on the *Lumᶜa* of Ibn al-Hāʾim, compiled in 873H or 893H in Abū Tīj in Upper Egypt: see also no. C58.

6.4.1 ***al-Qawl al-mubdiᶜ fī sharḥ al-Muqniᶜ***, a commentary on the ***Muqniᶜ*** of Ibn al-Hāʾim (no. C58), compiled in 876H: see no. C58. DJ 861,6 (fols. 71v–90r, 903H) of this work contains information about the various compilations of Sibṭ al-Māridīnī.

6.4.8 ***Tuḥfat al-aḥbāb fī ᶜamal al-ḥisāb***, a treatise on arithmetic in 3 *bābs*, compiled in 895H in Maḥalla al-Kubra: ṬR 105,1 (fols. 1v–15v, 901H); K23015 (14 fols., *ca.* 1050H); *etc.* (7 copies). Commentary by al-Shinshawrī (no. C111): DJ 861,3 (fols. 20v–70v, *ca.* 960H), *etc.* (7 copies), with *ḥāshiya* by al-Adfīnī (no. D49).

6.5.1 ***Raqāʾiq al-ḥaqāʾiq fī maᶜrifat al-daraj wa-l-daqāʾiq***, an abridgement of the *Kashf al-ḥaqāʾiq* of Ibn al-Majdī (no. C62) in 10 *bābs*: DM 51,1 (fols. 1r–26v, *ca.* 1000H), KhM 7,2 (fols. 9v–28r, *ca.* 1050H); TR 296,4 (pp. 167–208, 1123H); *etc.* (18 copies). Turkish translation by Yūnus Ef. (no. H67); ṬFT 51,1 (fols. 1v–46r, *ca.* 1300H). Abridgements by the author entitled ***al-Ṭuruq al-sanīya fi l-ᶜamal bi-l-nisba al-sittīnīya*** in 15 *bābs*: DM 186,1 (fols. 1v–10r, 1181H), and another version with the same title in 7 *bābs*: DJ 323,3 (fols. 15r–20v, *ca.* 1300H), *etc.* (1 fragment), and ***Zubd al-raqāʾiq fī ḥisāb al-daraj wa-l-daqāʾiq*** in 8 *faṣls*: DM 131,5 (fols. 188v–205v, 1146H).

Abridgement by Ibn Abi l-Fatḥ al-Ṣūfī (no. C98), entitled ***Nihāyat al-rutba bi-jadwal al-nisba*** in 5 *bābs*: MM 240 (11 fols., 1033H). Commentary by Ḥasan al-Jabartī (no. D98), entitled *Ḥaqāʾiq al-raqāʾiq ᶜala Raqāʾiq al-ḥaqāʾiq*: DM 186,3 (fols. 14v–49v, 1177H), *etc.* (4 copies). *Ḥawāshī* by Ibrāhīm al-Ḥalabī (no. C113): DM 877 (22 fols., *ca.* 1100H by the author).

6.9.6 An abridgement of al-Killāʾī's (no. G37) ***Majmūᶜ*** on inheritance: DJ 144,3 (fols. 58v–67v, 884H), and a commentary on the ***Majmūᶜ***: DJ 144,4 (fols. 67v–84v, 884H).

6.9.7 commentaries on ***al-Fuṣūl al-muhimma*** and ***Kifāyat al-huffāẓ*** and ***al-Tuḥfa al-Qudsīya*** by Ibn al-Hāʾim: see no. C58.

6.9.13 ***al-Mawāhib al-sanīya fī aḥkām al-waṣīya***, on inheritance: DJ 144,1 (fols. 1v–45v, 884H).

6.9.13 ***Kashf al-ghawāmiḍ fī ᶜilm al-farāʾiḍ***: DJ 144,5 (fols. 85v–103r, 884H).

C98 SHAMS AL-DĪN MUḤAMMAD B. ABI L-FATḤ AL-ṢŪFĪ AL-MIṢRĪ

Suter, nos. 447 and 460 (confused); *GAL*, SII, p. 159 (omitted from *GAL*, II); *Kennedy*, no. 37. See Pls. I and CIXb.

2.2.3 ***Zīj al-Ṣūfī*** is a recension of the *Zīj* of Ulugh Beg (no. G49) for Cairo, using the translation of the introduction prepared by al-Rifāᶜī (no. G99). A complete copy of al-Ṣūfī's *Zīj* in his own hand in MS Tehran Milli 768 (microfilm available from the Arab League Institute of Arabic Manuscripts, Cairo). This work merits detailed study. According to al-Rifaᶜi, al-Ṣūfī died before he could prepare the fourth *maqāla*, and al-Rifāᶜī's translation of the text of this *maqāla*, from the *Zīj* of Ulugh Beg, is appended at the end of the introduction in the Tehran copy (fols. 67v–70r, in a different hand!).

There are no complete copies in the Egyptian National Library. Fragments of little consequence, all of them late, are MJ 7,3 (fols. 73v–198r, *ca.* 1150H), *etc.* (over 10 copies).

2.2.3 ***Sullam al-manāra fī muqawwamāt al-kawākib al-sayyāra***, a small set of planetary tables: MM 136 (12 fols., 1150H); *etc.* (4 copies).

2.3.6 ***al-Ṣirāṭ al-mustaqīm fī ḥall muqawwamāt al-qamar min al-Durr al-yatīm***, on the use of the auxiliary tables of Ibn al-Majdi (no. C62) for compiling ephemerides for the moon: DM 468,2 (fols. 3r–4v, *ca.* 1050H); *etc.* (2 copies).

2.4.7 Miscellaneous astrological tables: DM 571 (46 fols., *ca.* 1200H).

2.5.7 ***Nubdhat al-isᶜāf fī maᶜrifat qaws al-khilāf***, a short treatise on a technique used in the compilation of ephemerides: K 4031,1 (fols. 1r–5v, *ca.* 1100H), plus an abridgement by ᶜAbd al-Qādir al-Minūfī (no. C121): DM 577,3 (fols. 31v–38r, 1200H), and a commentary by al-Khawānikī (no. D78) entitled *al-Itḥāf ᶜalā Nubdhat al-isᶜāf*: DM 525 (7 fols., *ca.* 1150H). Merits investigation.

3.1.14 Auxiliary tables related to those in the set entitled *Nūr al-aḥdhāq* by Ibn al-Mushrif (see no. C43), with an introduction in 7 *bābs*: K 4024,2 (fols. 2v–9v, 1293H from author's copy 916H), unique. See *SATMI* for further details.

3.2.15 ***Munyat al-ṭullāb fī taḥsīl ghālib al-qawāᶜid al-falakīya bi-l-ḥisāb***, a treatise on timekeeping in unnumbered sections: ṬM 91,3 (fols. 17v–21r, *ca.* 1300H), unique?

3.5.6+ ***Natāʾij al-fikar fi l-mubāshara bi-l-qamar***, a short treatise on timekeeping by the moon: DM 201 (4 fols.,, 1088H); *etc.* (1 copy).

3.5.6+ ***Bulūgh al-waṭar fi l-ᶜamal bi-l-qamar***, another treatise on the same subject in 5 *bābs*: DM 19 (4 fols., 1092H, anonymous). al-Ṣūfī mentions another treatise of his entitled ***Taḥrīr al-naẓar***.

4.3.29 A table displaying the hour-angle as a function of the solar azimuth for marking astrolabe plates (?): MM 27,1 (fols. 1r–5v, 1118H).

4.5.14 ***al-Shamsīya fi l-aᶜmāl al-jaybīya***, a treatise on the use of the sine quadrant in 16 *bābs*: ṬM 254,6 (fols. 118v–123r, *ca.* 1050H); DM 617,2 (fols. 102r–108v, *ca.* 1150H); *etc.* (3 copies).

4.6.23 A recension of the treatise by al-Wafāʾī on the complete (*tāmm*) quadrant of Ibn al-Shāṭir (no. C30): see no. C61.

4.6.44 ***al-Mufaṣṣal bi-l-ᶜamal bi-niṣf dāʾirat al-muᶜaddil***, a treatise on the "equatorial (semi-) circle" in 5 *faṣls*: DM 181,8 (fols. 174v–181v, 1178H); DM 879,2 (fols. 4v–7r, 1090H); *etc.* (2 fragments).

4.7.14 ***Nuzhat al-nāẓir fī waḍᶜ khuṭūṭ faḍl al-dāʾir***, a treatise on sundial theory in 2 *bābs* containing an extensive table entitled *al-jadwal al-mushtarak*, compiled in 878H: MM 231 (11 fols., *ca.* 1100H); DM 196 (10 fols., *ca.* 1100H); *etc.* (5 incomplete copies). On the table, see *SATMI*, I.

4.7.14 ***(R. fī) Ṭarīqat ḥisāb al-māʾila wa-rasmihā bi-samt al-iᶜtidāl***, on sundial theory: MM 178,1 (fols. 1v–2v, 1183H; DM 5,2 (fols. 39r–40v, *ca.* 1150H); *etc.* (2 copies).

4.7.14 ***M. ᶜala ᶜAmal al-basīṭa al-musammā bi-l-rukhāma bi-ṭarīq al-handasa***, on sundial theory: ṬM 103,1 (fols. 1r–3v, *ca.* 1100H); DM 588,1 (fols. 1r–1v, *ca.* 1100H), *etc.* (1 copy).

4.7.14 ***Faṣl fi l-munḥarifa allatī waḍaᶜahā bi-l-qubba al-Muʾayyadīya ᶜām 824H***, on a sundial constructed in Cairo: DM 588,2 (fols. 1v–2r, *ca.* 1100H). By virtue of the date of construction of the sundial, this short note is either garbled or not by al-Ṣūfī.

4.7.14 Tables for drawing vertical sundials with inclination 59;9° and 60;27° for an unspecified latitude: MM 191,1 (fols. 2r–4r, *ca.* 1150H).

6.4.7 ***Irshād al-ᶜajam li-aᶜmāl al-judhūr al-aṣamm***, a treatise on the extraction of roots in 2 *fanns* of 4 + 5 *faṣls*: DR 663 (55 fols., 897H in author's hand, lacks first page), unique. This title is not listed in previous bibliographical sources.

6.5.1 ***Nihāyat al-rutba bi-jadwal al-nisba***, an abridgement of Sibṭ al-Māridīnī's *Raqāʾiq al-ḥaqāʾiq* (see no. C97): MM 240 (11 fols., 1033H).

6.6.3 ***Fāʾida fī sharḥ qiṭaᶜa fī jins khārij al-qisma***, a note on a section of a treatise by an unidentified author (possibly Ibn al-Majdī—no. C62) on sexagesimal products: MM 209,3 (fol. 38v, 876H in author's hand).

6.8.8 ***Irshād al-wazzān li-maᶜrifat al-awzān bi-l-qabbān***, on the balance, in 5 *faṣls*: DR 748,1 (fols. 1r–5v, *ca.* 1100H); K3831,1 (fols. 1r–6r, 1097H); *etc.* (1 copy).

6.8.8 ***R. fī Iṣlāḥ fasād al-qabbān***, on the balance: MR 14 (10 fols., partly 993H); MR 28,2 (fols. 2v–10r, 1166H); *etc.* (4 copies).

6.8.8 ***R. fi Qismat al-qabbān bi-ṭarīq al-handasa wa-l-misāḥa wa-l-ḥisāb bi-l-nisab al-arbaᶜ***, on the balance: MR 30,3 (fols. 25v–29v, *ca.* 1100H).

6.8.8 ***Faṣl fī maᶜrifat qismat al-qabbān bi-ṭarīq al-ḥisāb***, on the balance: MR 30,5 (fols. 30v–31v, *ca.* 1100H).

6.8.8 ***Tuḥfat al-nuẓẓār fī inshāʾ al-ᶜiyār min aṣl al-miᶜyār***, on weights and measures: DJ 286,3 (fols. 12r–14v, *ca.* 1300H).

6.8.8 An anonymous *R. fi l-Usṭuwāna* . . . , on the balance, in MR 30,4 (fols. 29v–30v, *ca.* 1100H) may also be due to al-Ṣūfī.

7.1.5 ***al-Iᶜlām bi-shadd al-minkām/al-binkām***, a treatise on clocks in 5 *bābs*: MM 204,1 (fols. 1r–8v, 1052H); DM 1169,7 (fols. 48r–52v, 1158H, Maghribi script); DM 521,4 (fols. 30r–51v, 1159H). Other copies in MSS Istanbul Fatih 5397,4 (fols. 55–65, 1113H) and Hamidiye 874,6 (fols. 86–98, n.d.). Merits study.

C99 YAḤYĀ B. ᶜALĪ AL-RIFĀᶜĪ

GAL, II, p. 276, and SII, p. 298; *Storey*, p. 71.

2.2.3 An Arabic version of the introduction of the *Zīj* of Ulugh Beg (no. G49) commissioned by Ibn Abi l-Fatḥ al-Ṣūfī (no. C98): QM 1,1 (fols. 1r–29v, 1135H by Yūsuf b. Yūsuf Kalārjī (no. D61)); *etc.* (2 late copies, and 1 fragment). Other copies listed by Brockelmann.

C100 ANONYMOUS

2.2.7 DM 744 (21 fols., *ca.* 875) is the author's copy of the introduction to a *zīj* compiled in Cairo in 878H for Qāytbāy (see fol. 11v). The folios are in disorder and the handwriting is difficult to read. The first few folios (fols. 1v then 3r–7v then 2r–2v) contain calculations of *tasyīr* for various dates in the mid 9/15th century: it is not clear whether these belong to the introduction (fol. 1v) which begins *al-faṣl r* (perhaps read *z*) *fī tasyīrat al-Mirrīkh*, the 200th (or perhaps 7th) section on the *tasyīr* of Mars. These calculations are followed (fols. 8r–9v) by tables of the *ḥudūd* relating to each degree of the zodiac and a series of undated horoscopes (fol. 10r). The introduction (fol. 11v) breaks off shortly after the dedication to Qāytbāy, and the author mentions two scholars Shihāb al-Dīn b. Subḥ (?) and Tāj al-Dīn ᶜAbd al-Wahhāb. The chapters which follow (fols. 14r–18v then 12r–13v) are 10 *bābs* from the third *faṣl* dealing with planetary theory. The remaining pages (fols. 19r–21v) contain calculations of the projections of the rays for various stars.

DM 615 (10 fols., *ca.* 875H) copied in the same hand contains similar but unnumbered chapters from an introduction to an Egyptian *zīj*, dealing with the equation of time, mean motions, lunar latitude, syzygies, parallax, crescent visibility, and lunar eclipses. The writer mentions (fol. 3r) Ulugh Beg (no. G49) and the recension of his star tables by Ibn Abi-l-Fatḥ al-Ṣūfī (no. C98).

DM 718 (6 fols., 917H (?)) contains notes and calculations on the distance of the sun and moon according to Ibn Yūnus (no. B59) and *al-Zīj al-Jadīd* (of Ulugh Beg), and on solar and lunar parallax, as well as (fols. 3v–4r) Chapters 50–51 of *al-Zīj al-Mustawfī* by Ibn al-Raqqām (see no. F22).

C101 MUḤAMMAD AL-WAFĀʾĪ, a student of IBN ABI L-FATḤ AL-ṢŪFĪ (no. C89)

Suter, no. 459.

3.1.18 ***K. Nuzhat al-abṣār fī aᶜmāl al-layl wa-l-nahār***, prayer-tables for latitude 27° (Assiut): DM 188,1 (fols. 1r–10r, *ca.* 1100H), unique? Analyzed in *SATMI*, II.

4.7.15 ***al-Jawāhir al-nayyirāt fī rasm al-basāʾiṭ wa-l-munḥarifāt***, a treatise on sundial theory: no longer extant. Abridgement (?) by al-Asyūṭī's pupil ᶜAbī al-Andalusī entitled ***al-Waḍᶜ*** . . . : see no. C102.

C102 ᶜALĪ AL-MĀLAQI AL-ANDALUSĪ

Suter, no. 459.

4.7.15 ***al-Waḍᶜ ᶜala l-jihāt fi l-basāʾiṭ wa-l-munḥarifāt***, a treatise on sundial theory based on the no longer extant *Jawāhir al-nayyirāt* . . . by his teacher Muḥammad b. Dallāl al-Wafāʾī al-Asyūṭī (no. C101): DM 452,1 (fols. 1r–6r, 1106H); DM 166,2 (fols. 6r–10v, 1120H); *etc.* (7 copies).

C103 JALĀL AL-DĪN ᶜABD AL-RAḤMĀN B. ABĪ BAKR AL-SUYŪṬĪ, the celebrated polymath

Suter, no. 449; *GAL*, II, pp. 180–204, and SII, pp. 178–198.

1.5.1 ***al-Hīʾa al-sanīya fi l-hayʾa al-sunnīya***, a treatise on early traditions dealing with astronomy: ṬH 37 (22 fols., *ca.* 1100H); ZK 594 (19 pp., 1230H); plus a commentary by al-Qaramānī (no. H22). *Cf. GAL*, II, p. 187, no. 66, and SII, p. 185, no. 66. This treatise has recently been published and analyzed in detail: see A. Heinen, *Islamic Cosmology: A Study of al-Suyūṭī's al-Hayʾa as-sanīya . . .*, Beirut (for Franz Steiner Verlag, Wiesbaden), 1982.

1.5.1 Two short treatises on the prediction of eclipses and on where the sun goes when it sets: QJ 25,114 (fols. 363r–363v, 1115H) and QJ 25,115 (fols. 364v–365r, 1115H), unique?

3.5.6 A poem on the time of the *ẓuhr*: MJ 39,3 (fol. 58r, 1069H).

C104 ZAYN AL-ᶜĀBIDĪN ᶜUMAR B. IBRĀHĪM B. NUJAYM AL-MIṢRĪ AL-ḤANAFĪ

GAL, II, p. 401, and SII, p. 426 (esp. no. 5).

3.5.7 ***Rafᶜ al-ghishāʾ ᶜan waqtayi l-ᶜaṣr wa-l-ᶜishāʾ***, a treatise without subdivisions on the times of the afternoon and nightfall prayers: ṬJ 200,5 (fols. 27r–30r, 1146H); *etc.* (1 copy). Other copies are MSS Berlin Ahlwardt 3592 and 3593. This treatise has been investigated by my student Ms. Diana Kimiatek.

C105 ṢĀRIM AL-DĪN IBRĀHĪM B. ᶜUMAR AL-BAJALĪ (?) (*GAL*: AL-NAJALĪ)

GAL, SII, p. 94, and p. 1021, no. 43.

6.9.12 ***Burhān al-burhān al-rāʾid fi l-jabr wa-l-ḥisāb wa-l-khaṭaʾayn wa-l-aqdār wa-l-farāʾiḍ***: DO 470 (23 fols., *ca.* 1100H, incomplete). Another copy is MS Berlin Fol. 3029. This short treatise on the arithmetic of inheritance is, as Brockelmann says, arranged "in komplizierter, akrostichischer Anordnung." According to Brockelmann, the work was compiled in 909/1503 for Nūr al-Dīn ᶜAlī b. ᶜUmar al-Ḥamawī. The Cairo manuscript appears to be of Yemeni provenance, and the author is named Ṣārim al-Dīn not Ḥāzim al-Dīn (*cf. GAL*, SII, p. 1021).

C106 SHIHĀB AL-DĪN AḤMAD B. MŪSĀ B. ᶜABD AL-GHAFFĀR

Suter, no. 461; *GAL*, II, p. 154, and SII, pp. 155 and 536.

2.3.6 ***Silk al-durrayn fī ḥall al-nayyirayn***, a commentary on *al-Durr al-yatīm* by Ibn al-Majdī (see no. C62), compiled in 922H: DM 131,1 (fols. 1v–90r, 1146H); MM 135 (36 fols., *ca.* 1050H); *etc.* (2 copies); plus abridgement in MM 31,1 (fols. 1r–26v, *ca.* 1150H) and TR 317,1 (pp. 1–56, *ca.* 1200H), *etc.* (2 copies); plus abridgement entitled *Jawāhir al-silk* in MM 134 (21 fols., 1250H) and DM 509 (21 fols., *ca.* 1200H). Another abridgement by Yaḥyā al-Khaṭṭāb (no. C119): DM 167,1 (fols. 1r–58r, 1259H); *etc.* (1 copy).

C107 AḤMAD B. ᶜALĪ ZUNBUL AL-MAḤALLĪ AL-MUNAJJIM

GAL, II, pp. 54 and 384–385, and SII, pp. 409–410.

5.3.19 ***K. al-Maqālāt wa-ḥall al-mushkilāt***: a treatise on astrology and magic in 30 *maqālas*: MḤ 86 (211 fols., *ca.* 1100H). Another copy (wretched) is in MS Damascus Ẓāhiriya 3582 (*cf. Cat.*, p. 294).

The various works of this author merit investigation.

C108 SHAMS AL-DĪN MUḤAMMAD B. ᶜALĪ B. MUḤAMMAD B. ṬŪLŪN AL-DIMASHQĪ

GAL, II, pp. 481–483, and SII, p. 494; *Azzawi*, pp. 310–311. Brockelmann lists no scientific works; Azzawi lists four titles of works on arithmetic, astronomy, and magic, taken from Ibn Ṭūlūn's biography, but no manuscripts thereof.

6.4.16 ***al-Ajwiba al-muᶜallala fi l-masāʾil al-mujahhala***, on selected problems in simple applied mathematics: DR 839 (4 plates, photos from an unidentified manuscript, *ca.* 1000H). Not listed by Azzawi.

C109 ANONYMOUS

4.4.27 ***Munyat al-muwaqqit***, a poem on the use of the almucantar quadrant in 12 *faṣl*s, compiled in Mecca in 945H: Sh 89,16 (fols. 323v–326r, *ca.* 1100H), no other copies known.

C110 ʿABD AL-RAḤMĀN B. MUḤAMMAD (sometimes ʿABD ALLĀH) B. AḤMAD AL-TĀJŪRĪ

Suter, no. 512; *GAL*, II, pp. 217 and 469, SII, p. 485.

3.1.28 ***Muqaddima fī ʿilm al-falak yuʿrafu minhā awāʾil al-layl wa-l-nahār***, a treatise in which the durations of the prayer-times are given in words for each month of the year: ṬM 210 (6 fols., 1161H); TR 333,1 (pp. 1–7, 1146H); *etc.* (1 fragment).

3.3.11 A *fatwa* by the *ʿulamāʾ* of Egypt on the *miḥrāb*s of the Maghrib, recorded in 934H by al-Tājūrī: DM 540 (12 fols., 1014H from al-Tājūrī's copy), unique. al-Tājūrī's name is given as ʿAbd al-Raḥmān b. Muḥammad b. Aḥmad.

3.2.18 ***R. fi l-Fuṣūl al-arbaʿa wa-awqāt al-ṣalawāt wa-ajzāʾ al-layl wa-jihat al-qibla bi-ghayr āla***, a treatise on timekeeping in 20 *bāb*s: DM 176,4 (fols. 33r–53v, *ca.* 1150H); DM 463 (17 fols., 1150H); *etc.* (4 copies), plus MM 216,1 (attributed to al-Qalyūbī (no. D30)).

4.4.21 ***R. ʿalā Rubʿ al-muqanṭarāt*** on the almucantar quadrant in 16 *bāb*s: DM 174,1 (fols. 1r–20r, 1166H); MM 115 (20 fols., *ca.* 1100H); *etc.* (2 copies).

4.5.12 A commentary on *al-R. al-Fatḥīya* by Sibṭ al-Māridīnī, compiled in 944H: see no. C97. This commentary ends with a poem on the prayer-times attributed to the Imām al-Shāfiʿī (see no. B2).

4.8.6 (***R. fī Bayt al-ibra***), a treatise on the magnetic compass: MM 181,4 (fols. 6v–8v, 1030H); DM 628,2 (fols. 3v–6r, *ca.* 1150H); DM 779,2 (fols. 13v–21r, *ca.* 1150H), *etc.* (2 copies).

C111 ʿABD ALLĀH B. BAHĀʾ AL-DĪN MUḤAMMAD AL-SHINSHAWRĪ

GAL, SIII, p. 515 (s.v.).

6.4.1 ***Bughyat al-rāghib fī sharḥ Murshidat al-ṭālib***, a commentary on the ***Murshida*** of Ibn al-Hāʾim, compiled in 995H: see no. C58.

6.4.8 A commentary on Sibṭ al-Māridīnī's ***Tuḥfat al-aḥbāb*** (see no. C97): DJ 861,3 (fols. 20v–70v, *ca.* 960H, with a comment in the author's hand at the end); *etc.* (7 copies); plus a *ḥāshiya* by Muḥammad b. ʿAlī b. Muḥammad al-Adfīnī (no. D49): DR 286 (52 fols., *ca.* 1150H), *etc.* (1 copy).

C112 MUḤAMMAD B. IBRĀHĪM AL-ḤANBALĪ AL-ḤALABĪ AL-RUBʿĪ (?)

Suter, no. 464; *GAL*, II, pp. 483–484, and SII, pp. 495–496.

6.4.1 ***Rafʿ al-ḥijāb ʿan qawāʿid al-ḥisāb***, a commentary on the ***Wasīla*** of Ibn al-Hāʾim: see no. C58.

6.4.1 ***ʿIddat al-ḥāsib wa-ʿumdat al-muḥāsib***, a commentary on ***al-Nuzha*** by Ibn al-Hāʾim, compiled in 954H: see also no. C58. This commentary begins with some remarks about various contemporary scholars in Aleppo.

6.7.7 ***Makhāyil al-malāḥa fī masāʾil al-misāḥa***, a treatise on surveying based on the ***Ghunyat al-ḥussāb*** of Jamāl al-Dīn Aḥmad b. Thibāt (no. C11): K 4301,2 (fols. 56v–73v, 972H).

See also no. C113.

C113 IBRĀHĪM AL-ḤALABĪ

Not identified. See the previous entry.

1.2.7 A treatise on the problem concerning the altitude of mountains as discussed in the commentary by Qāḍī Zāde (no. G50) on the *Mulakhkhaṣ* of al-Jaghmīnī (no. G17): MH 4,2 (fols. 39r–46v, *ca.* 1200H). Another copy is MS Istanbul Laleli 2126,3 (fols. 52–60, 1119H).

6.3.19 A commentary on ***al-Ḥāwī fi l-ḥisāb*** by Ibn al-Hāʾim: see no. C58. The attributions to Ibrāhīm al-Ḥalabī may be an error for Muḥammad b. Ibrāhīm al-Ḥalabī, no. C112.

6.5.1 *Ḥawāshī* on the ***Raqāʾiq al-ḥaqāʾiq*** of Sibṭ al-Māridīnī (no. C97): DM 877 (22 fols., *ca.* 1100H by the author).

C114 SHIHĀB AL-DĪN AḤMAD B. MUḤAMMAD B. MUḤAMMAD . . . B. JIBRĪL AL-GHAZZĪ (?)

GAL, SNII, p. 154. See next entry.

6.4.1 A commentary on the *Nuzha* of Ibn al-Hā⁾im, compiled in 947H: see no. C58.

C115 MUḤAMMAD B. AḤMAD B. ḤASAN AL-GHAZZĪ

See previous entry?

6.3.19 *Takhṣīṣ ūli l-albāb fī sharḥ Talkhīṣ a*ᶜ*māl al-ḥisāb*, a commentary on the *Talkhīṣ* of Ibn al-Bannā⁾: see no. F23.

C116 MUḤAMMAD B. ᶜALĪ B. IBRĀHĪM, IBN ZURAYQ AL-KHAYRĪ

Suter, no. 426; *GAL*, SII, p. 1023, no. 61.

2.1.19 A fragment from the introduction (in unnumbered *faṣl*s) to his ***al-Rawḍ al-*ᶜ*āṭir fī talkhīṣ Zīj Ibn al-Shāṭir***, an abridgement of the *Zīj* of Ibn al-Shāṭir (no. C30): TR 235,1 pp. 2–143, *ca.* 1150H, Maghribi script).

4.4.13 ***al-Lafẓ al-muḥarrar/al-mu*ᶜ*aṭṭar fi l-*ᶜ*amal bi-l-rub*ᶜ *al-muqanṭar***, a treatise on the use of the almucantar quadrant in 14 *bāb*s: TR 64,1 (pp. 1–7, 971H); MM 187,3 (fols. 8r–10v, 1128H). The Damascus manuscripts bearing these titles listed in the Ẓāhirīya catalog (p. 223) are attributed to Ibn Zurayq, not to al-Bakhāniqī (see no. C28) as stated. On these titles, see also the notes on TJ 391,2 of the treatise by al-Mizzī (no. C34).

4.5.18 Two versions of ***al-Nashr al-muṭayyab fi l-*ᶜ*amal bi-l-rub*ᶜ *al-mujayyab***, on the use of the sine quadrant, the first in 15 *bāb*s: DM 559 (10 fols., *ca.* 1150H), *etc.* (2 copies); and the second in 20 *bāb*s; TR 64,3 (pp. 23–33, 971H), DR 363,4 (fols. 80v–92r, 1115H), *etc.* (3 copies), plus an anonymous commentary in 20 *bāb*s; entitled ***Iẓhār al-mughayyab min al-Nashr al-muṭayyab***: MM 10,1 (fols. 1r–7r, 1122H).

C117 SULAYMĀN B. ḤAMZA B. BAKHSHĪSH (?) AL-RŪMĪ AL-ᶜUTHMĀNĪ

Azzawi, p. 181; *GAL*, II, p. 469, and SII, p. 484. See Pl. CXIa.

2.3.6 ***Tarẓ al-ghurar fī ḥall al-durar***, a treatise on the use of ***al-Durr al-yatīm*** by Ibn al-Majdī (no. C62): DM 791 (9 fols., 1072H); DJ 323,6 (fols. 33v–34r, *ca.* 1250H); *etc.* (2 copies). Another copy is MS Yale Nemoy 1453.

5.3.23 ***al-Fayḍ al-*ᶜ*amīm fī ma*ᶜ*rifat aḥkām ṣadr al-taqwīm***, an astrological treatise in 3 *bāb*s: DM 885 (128 fols., *ca.* 1100H); DM 886 (145 fols., 1199H); *etc.* (6 copies).

5.3.23 ***al-Durr al-thamīn fi l-ḥukm*** ᶜ***alā taḥāwīl al-sinīn***, an astrological treatise in 3 *bāb*s: DM 633 (19 fols., *ca.* 975H in author's hand); *etc.* (6 copies).

5.3.23 ***Ẓuhūr al-Thurayyā wa-khafā*⁾ *ma kāna wabiy*ᵃⁿ**, a treatise on the influence of the Pleiades on the plague which struck Egypt in 987H: DJ 215,2 (fols. 20r–33v, *ca.* 1150H); DJ 48,1 (fols. 1r–11v, 1172H); *etc.* (1 copy). Merits study.

5.3.23 ***al-Aḥkām fī fuṣūl al-aḥkām li-taḥwīl al-sinīn wa-l-ayyām***, an astrological treatise: ṬM 163 (128 fols., *ca.* 1200H); MM 1,1 (fols. 1r–142v, 1184H).

5.3.23 An answer to a question from Damietta on the treatment of the astrological concept of *sahm* by Ibn al-Shāṭir (no. C30): DM 131,2 (fols. 90r–93r, 1146H).

C118 MUḤAMMAD B. MUḤAMMAD AL-ḤAṬṬĀB (or AL-KHAṬṬAB?)

GAL, II, p. 508, and SII, p. 526.

3.2.17 ***R. mukhtaṣara fī Ma*ᶜ*rifat istikhrāj awqāt al-ṣalāt wa-shay*⁾ *min al-tawārīkh wa-l-a*ᶜ*māl al-falakīya min ghayr āla***, a treatise on timekeeping in 10 *bāb*s with some simple tables: ṬM 145 (42 fols., 1110H); TR 107 (22 pp., *ca.* 1150H, defective); *etc.* (2 copies), plus abridgement by author's son (no. C119).

C119 YAḤYĀ B. MUḤAMMAD AL-ḤAṬṬĀB / AL-KHAṬṬĀB

Suter, no. N474a; *GAL*, II, pp. 515–516, and SII, p. 537.

2.3.6 ***Mukhtaṣar Silk al-durrayn***, an abridgement of the treatise by Ibn ᶜAbd al-Ghaffār: see no. C106.

3.2.17 ***Wasīlat al-ṭullāb fī maᶜrifat aᶜmāl al-layl wa-l-nahār bi-ṭarīq al-ḥisāb***, an abridgement in 7 *bāb*s of the treatise of his father (see no. C118): MM 243 (13 fols., 1026H); *etc.* (6 copies).

4.5.21 *Muqaddima mukhtaṣara fī maᶜrifat istikhrāj aᶜmāl al-layl wa-l-nahār . . . bi-rubᶜ al-mujayyab*, a treatise on the use of the sine quadrant in 17 *bāb*s: MM 217 (11 fols., 1076H); KhM 10,4 (fols. 40v–49v, 1096H); DM 599 (13 fols., 1138H); *etc.* (8 copies); plus a commentary by al-Jālī (no. D54): DM 1082,1 (fols. 1v–32v, 1174H); ṬJ 366,6 (fols. 48v–71r, 1257H).

6.4.1 *Muqaddima fi l-ḥisāb*, an abridgement of *al-Nuzha* by Ibn al-Hāᵓim: see no. C58.

C120 SHAMS AL-DĪN MUḤAMMAD AL-MINŪFĪ

Not listed in the modern sources.

3.1.23 ***ᶜUnwān al-muhimmāt fī taḥrīr al-awqāt***, prayer-tables for latitude 30° (Cairo) with an introduction: DM 470 (50 fols., *ca.* 975H, by ᶜAbd al-Qādir al-Minūfī (no. C121)); DM 467 (47 fols., 1144H); *etc.* (7 defective copies). al-Minūfī's tables are those of the main Cairo corpus (see no. C28) modified for the effect of refraction at the horizon. See further *SATMI*, II.

3.2.20 ***Naẓm al-jawāhir wa-l-yawāqīt fī taḥrīr aᶜmāl al-mawāqīt***, a treatise in 5 *faṣl*s on problems relating to horizon phenomena with some tables, compiled in 981H: MM 235,1 (fols. 1r–7r, *ca.* 1000H by the author's son ᶜAbd al-Qādir (no. C121)); DM 547 (8 fols., *ca.* 1100H). Merits study.

C121 ᶜABD AL-QĀDIR B. MUḤAMMAD AL-MINŪFĪ, son of no. C120

Suter, no. 479 (confused); *GAL*, SII, p. 486 (confused).

2.4.6 A table of lunar parallax: DM 13,1 (fols. 1v–2r, *ca.* 1100H).

2.4.7 Miscellaneous astrological tables: DM 571 (46 fols., *ca.* 1200H).

2.5.7 An abridgement of Ibn Abi l-Fatḥ al-Ṣūfī's *Nubdhat al-isᶜāf*: see no. C98.

3.1.24 Tables of normed right ascensions (3 sexagesimal digits) for each minute of argument: MM 66 (10 fols., 973H by author); DM 755,1 (fols. 1v–2v, *ca.* 975H by author). Tables of the same functions (to 5 sexagesimal digits) for each minute (parameters of Ulugh Beg): MM 45 (38 fols., *ca.* 1025H).

3.1.24 Tables of the half diurnal arc (to visible horizon), hours of daylight, time from midday to *ᶜaṣr*, duration of evening twilight, with entries for each 0;3° of solar longitude, for latitude 30° (Cairo): DM 1100 (29 fols., 974H by author).

On these tables see *SATMI*, II.

3.5.9 ***Rafᶜ al-khilāf fī ᶜamal daqāᵓiq al-ikhtilāf***, a treatise on the time taken by the sun to move between the true and visible horizons: MM 123 (44 pp., *ca.* 1000H); K 4048 (19 fols., *ca.* 1150H); DM 182,1 (fols. 1r–24v, *ca.* 1100H). Merits study. See Pl. XXXIV.

3.5.9 Two tables for correcting the arc of depression of the sun at daybreak and nightfall: MM 235,2 (fol. 8r, *ca.* 975H), unique. Most of the entries in the tables are blank.

3.5.9 A treatise on problems relating to shadows, in response to questions posed by Taqi l-Dīn b. Maᶜrūf (no. H12) in the presence of (*bi-majlis*) Muḥammad b. Īlyās: DM 792 (5 fols., 978H by the author); DM 577,2 (fols. 12v–30v, *ca.* 1200H).

5.5.1 A horoscope of the *qāḍī* Muḥammad al-ᶜIbādī, born 917H: MM 141,1 (fols. 1r–16v, *ca.* 1175H). Another anonymous horoscope for the same individual is in DM 945,1 (fols. 1v–8r, *ca.* 1300H).

C122 IBRĀHĪM B. QĀYTBĀY AL-ASHRAFĪ AL-ḤANAFĪ, a student of MUḤAMMAD AL-MINŪFĪ (no. C120)

Not listed in the modern sources.

3.1.24 Tables for timekeeping with values for each minute of solar longitude, all copied *ca.* 1025H in the author's hand, as follows:

(1) time between midday and the *ᶜaṣr*: DM 34 (8 fols.) + DM 152,1 (fols. 1r–29r), originally the same manuscript;

(2) time between the ^{c}*aṣr* and the setting of the center of the sun over the visible horizon: DM 152,2 (fols. 29v–31v);

(3) half diurnal arc (to visible horizon): DM 33 (37 fols.);

(4) the time taken by the sun to cross between the true horizon and the visible horizon: DM 682 (37 fols.);

(5) the ascensions at nightfall: DM 153,1 (fols. 1v–9v) + DM 740,2 (fols. 4r–58r), originally the same manuscript.

See further *SATMI*, II.

C123 MISCELLANEOUS ANONYMI

Tables of functions for timekeeping for latitude 30° (Cairo), other than those attributed to ^{c}Abd al-Qādir al-Minūfī (no. C121) and Ibrāhīm b. Qāytbāy (no. C122), as follows:

(1) time from nightfall to the *salām* for each 0;6° of solar longitude: DM 57,1 (fols. 1r–4v, *ca.* 1000H, incomplete) and DM 151,2 (fols. 9r–11r, *ca.* 1000H).

(2) the ascensions at the *salām* for each 0;3° of solar longitude: DM 545 (23 fols., *ca.* 1000H, incomplete); DM 158 (13 fols., *ca.* 1000H); DM 423 (25 fols., *ca.* 1100H).

(3) the oblique ascensions for latitude 30° for each 0;6° of ecliptic longitude: DM 42 (7 fols., *ca.* 1200H).

See further *SATMI*, II.

C124 MUḤAMMAD B. MUḤAMMAD (?) B. ABI L-KHAYR AL-ḤUSNĪ AL-URMAYŪNĪ

Suter, no. 511; *GAL*, SII, p. 485.

1.4.19 ***Qiṭ^{c}a fī taḥrīr al-manāzil al-qamarīya wa-anwā^{ɔ}ihā wa-ṭulū^{c} al-kawākib al-thābita bi-l-fajr***, on folk astronomy: DJ 323,9 (fols. 38r–39r, *ca.* 1275H).

2.1.19 ***al-Durra al-muḍīya fī sharḥ al-Lum^{c}a al-bahīya***, a commentary on the *Lum^{c}a* of al-Kawm al-Rīshī (no. C41) followed by the tables of the *Lum^{c}a*: K4009 (72 fols., 1325H), *etc.* (1 fragment).

3.1.28 ***Itḥāf al-ḥabīb bi-ma^{c}rifat al-tawqī^{c}at wa-l-awqāt wa-l-qibla bi-l-taqrīb***, a treatise on timekeeping arranged in 7 *faṣls*: TR 114 (74 pp., *ca.* 1100H); ZK 154 (91 fols., *ca.* 1100H), *etc.* (4 copies). About one half of the treatise is devoted to the lengths of the intervals for the prayers, written out in words for each day of the Coptic year. al-Ḥusnī mentions that he has used a work entitled ***Jawharat al-yawāqīt*** by Shams al-Dīn Muḥammad b. al-^{c}Aṭṭar al-Bakrī al-Shāfi^{c}ī (no. C168) for his *faṣl* on the determination of the qibla by practical means. See further *SATMI*, II.

3.2.22 ***Kashf al-karubāt fī taḥqīq masā^{ɔ}il yaḥtāj ilayhā ṭālib ^{c}ilm al-awqāt/al-mīqāt***, a treatise on timekeeping without subdivisions: DM 1108,3 (fols. 5v–10v, 1052H); DM 511,2 (fols. 6r–8v, *ca.* 1100H, defective). This work includes tables and a discussion of the effect of refraction on horizon phenomena.

3.3.12 An answer to a question about the *qibla* at Manfalūṭ: DM 1093,9 (fols. 40r–43r, 1088H), unique.

3.5.8 ***al-Manhal al-sākib fī taḥrīk al-kawākib***, a treatise on timekeeping by the stars: DM 961 (10 fols., 1158H from author's copy); DM 741 (10 fols., *ca.* 1150H, lacks last page); *etc.* (2 copies).

3.5.8 ***al-Nujūm al-shāriqāt fī dhikr ba^{c}ḍ al-ṣanā^{ɔ}i^{c} al-muḥtāj ilayhā fī ^{c}ilm al-mīqāt***, a treatise on the materials used in the construction of astronomical instruments: DJ 208,1 (fols. 1v–21r, 1161H); ṬJ 791,2 (fols. 83r–120v, 1849 A.D.); *etc.* (5 copies). A copy of the published text which appeared in Aleppo in the late 1930s exists in the E.N.L. A study of this work based on the published text was made by A. Siggel in *Quellen und Studien zur Geschichte der Naturwissenschaften und der Medizin*, 8 (1941–42), pp. 435–457.

3.5.8 A treatise on the duration of morning and evening twilight: MM 167,5 (fols. 160v–162v, 989H), unique.

4.1.4 ***al-Rayy wa-l-ishbā^{c} fī sharḥ Kashf al-qinā^{c}***, a commmentary on the treatise of Ibn al-^{c}Aṭṭār: see no. C66.

4.6.47 ***Qiṭ^{c}a fī ^{c}amal al-muqawwar wa-naṣbihā (!) bi-ṭarīq al-handasa wa-l-ḥisāb***, on the construction of the instrument called *al-muqawwar*: DM 597,1 (fols. 1r–1v, *ca.* 1050H).

4.7.7 Two commentaries on Ibn al-Majdi's *Zād al-musāfir* entitled ***Rāḥat al-fu^{ɔ}ād fī taysīr al-Zād*** (compiled 941H), and ***Nuzhat al-khāṭir fī waḍ^{c}***

ḥudūd ᶜalā Zād al-musāfir: see no. C62.

4.7.19 *al-Uṣūl al-rawāsikh fī maᶜrifat al-buᶜd wa-jihatihi*, on sundial theory, without tables, compiled in 977H: ZK 917,2 (pp. 26–37, 1163H from author's copy).

A treatise on the technical terms of sundial construction: ZK 917,3 (pp. 38–48, 1163H, anonymous). The author mentions his treatise *Kashf al-qināᶜ*.

Tables for marking vertical sundials (inclination 53° for latitude 30°): DM 746,2 (fols. 7r–7v, *ca.* 1150H).

6.4.1 Perhaps al-Ḥusnī is the author of the anonymous commentary on the *Nuzha* of Ibn al-Hāʾim (no. C58) in DR 82 (38 fols., 1136H); MR 11 (100 fols., 1183H).

C124X IBRĀHĪM B. MUḤAMMAD AL-ANDALUSĪ, a student of AL-ḤUSNĪ

3.2.23 *R. fī Masāʾil ᶜilm al-waqt bi-ghayr āla*, a treatise on timekeeping in 11 *bāb*s: ḤM 13 (22 fols., 986H). The author mentions al-Ḥaṭṭāb (no. C118), al-Tājūrī (no. C110), Ibn Abi l-Qāsim al-Andalusī (no. C169), and his own teacher Muḥammad b. Abi l-Khayr al-Ḥusnī (no. C124).

C125 DĀʾŪD B. ᶜUMAR AL-ANṬĀKĪ

GAL, II, p. 478 and SII, pp. 491–492; *Azzawi*, p. 312.

5.3.20 *R. fī ᶜIlm aḥkām al-nujūm*, an astrological treatise stated to be taken from his book on medicine entitled ***al-Tadhkira***: ṬM 227,6 (fols. 124v–133v, *ca.* 1145H); DM 931 (photos of MS London B.L. add. 9599, 16 pls., 1222H). An extract from the ***Tadhkira*** on the same subject prepared by Muḥammad b. ᶜAbd al-ᶜAzīz b. Muḥammad al-Ḥalabī is in MS Princeton Mach 5064 = Yehuda 246, fols. 30–50v, 1114H.

6.4.15 A short treatise on algebra in 5 *bāb*s: DR 649,2 (fols. 95v–99v, *ca.* 1050H, anonymous); TR 142 (15 pp., 1153H). No copies listed by Brockelmann.

C126 SHIHĀB AL-DĪN AḤMAD B. AḤMAD B. ᶜABD AL-ḤAQQ AL-SUNBĀṬĪ

Suter, no. 470; *GAL*, II, pp. 217 and 484, and SII, p. 496.

4.5.12 A commentary on *al-R. al-Fatḥīya* by Sibṭ al-Māridīnī: see no. C97.

C127 ᶜALĪ B. MUḤAMMAD B. ᶜALĪ, IBN GHĀNIM

GAL, II, pp. 404–405, and SII, p. 429; *Azzawi*, p. 191.

4.5.12 *al-Nasama al-nafḥīya*, a poem based on Sibṭ al-Māridīnī's ***al-R. al-Fatḥīya*** (see no. C97): ṬM 242,1 (fols. 1v–8r, 1284H); ṬM 156 (5 fols., 1029H); *etc.* (1 copy); plus a commentary by the author: DM 102 (19 fols., *ca.* 1100H); DM 1098 (20 fols., *ca.* 1200H); ṬM 242,2 (fols. 9r–32v, 1284H), *etc.* (3 copies); plus an anonymous commentary entitled ***al-Maᶜālim*** . . . : K 6863 (42 fols., *ca.* 1150H), plus commentaries by Aḥmad al-Sharafī (no. D48), entitled ***Fatḥ Rabb al-barīya*** . . . : MJ 39,1 (fols. 1r–49v, 1069H); and by Saᶜīd al-Tanbuktī (no. D25), entitled ***al-Maẓāhir al-Aḥmadīya*** . . . : Sh 89,7 (fols. 53r–84r, 1071H).

C128 ANONYMOUS

5.3.25 *R. fī ᶜIlm aḥkām al-nujūm*, a treatise on nativities dedicated to Awīs Bāshā and compiled in 995H: DM 477 (6 fols., *ca.* 1000H).

C129 SHAMS AL-DĪN ABŪ ᶜABD ALLĀH MUḤAMMAD AL-SHĀFIᶜĪ

Not identified.

5.3.24 *K. al-ᶜIqd al-farīd fī ṭāliᶜ al-mawla l-saᶜīd*, an astrological treatise with some tables, compiled in 995H: DM 945,2 (fols. 9r–42r, *ca.* 1300H).

C130 KHIḌR B. ᶜABD AL-QĀDIR AL-BIRLISĪ AL-QABBĀNĪ

Not listed in the modern sources.

3.4.9 *Ijābat al-suʾāl fī maᶜrifat ᶜamal al-hilāl ᶜalā ṭarīq al-jadāwil*, a treatise on the determination of lunar crescent visibility, without tables: DM 513,3 (fols. 4r–5r, *ca.* 1100H); DJ 323,10 (fols. 39r–40v, *ca.* 1300H).

3.4.9 ***Qaṭf al-zalāl fī ma*^c*rifat* ^c*amal al-hilāl***, another treatise without tables, compiled in 1002H: DM 1108,1 (fols. 1v–4v, 1052H).

6.8.12 ***al-Jawāhir al-ḥisān (wa-shams* ^c*ayn al-zamān) fī* ^c*ilm al-qabbān***, a treatise on the balance in 10 *bāb*s: MR 30,1 (fols. 1v–15r, *ca.* 1100H); *etc.* (3 copies).

6.8.12 ***R. fī Ma*^c*rifat al-wizn al-qā*^ɔ*im wa-tarḥ al-*^c*iyār minhu***, a treatise on the balance in 3 *faṣl*s: MR 30,8 (fols. 44v–47v, *ca.* 1100H).

See also no. D64.

C131 ^cABD AL-QĀDIR B. ^cALĪ AL-SAKHKHĀWĪ

Suter, no. 476; *GAL*, II, p. 468, and SII, p. 483.

6.4.24 ***al-R. al-Sakhkhāwīya = Mukhtaṣar fī* ^c*ilm al-ḥisāb***, a treatise on arithmetic in 11 *bāb*s: MR 29,1 (fols. 1v–16r, *ca.* 1050H); TR 299 (24 pp., 1197H); *etc.* (14 copies). Commentaries by Ḥusayn al-Maḥallī (no. D85), entitled ***Fatḥ Rabb al-barīya*** . . . : DR 616 (33 fols., 1296H); *etc.* (9 copies), with a commentary by al-Manfalūṭī (no. D84), entitled ***al-Durra al-saniya*** . . . : DR 350 (74 fols., 1161H by the author); by ^cAbd al-Fattāḥ al-Daystī (no. D62), entitled ***al-Nubdha al-wāfiya*** . . . : DR 900 (110 pp., 1171H), *etc.* (1 copy); by al-Manfalūṭi (no. D84), entitled *al-Durar al-bahiya* . . . : MR 12 (69 fols., *ca.* 1200H); and a *ḥāshiya* by Aḥmad al-Maktabī (?) (no. D200): K 3943 (71 fols., 1319H).

C132 SHAMS AL-DĪN MUḤAMMAD B. MUḤAMMAD AL-LĀDHIQĪ

GAL, SII, p. 1023.

2.2.3 ***Jadwal ḥabṭaq al-qamar***, introduction and lunar mean motion tables beginning 990H and lunar *ḥabṭaq* tables: DM 1108,5 (fols. 24v–26v, 1053H); DM 639,26 (fols. 237r–241v, *ca.* 1100H, incomplete). No other copies known to me. In the brief introduction the author explains that the tables entitled *al-ḥabṭaq al-aṣghar* by ^cIzz al-Dīn al-Wafā^ɔī (no. C61) were the most useful that he knew, but since they were based on the parameters of Ibn al-Shāṭir (no. C30) he decided to recompute them using the parameters of Ulugh Beg (no. G49).

3.1.25 ***Natījat al-afkār fī a*^c*māl al-layl wa-l-nahār = Bughyat al-nafs fī ḥall al-shams***, prayer-tables for Cairo taken from the main corpus (see no. C28) preceded by a short introduction in 8 or 10 *faṣl*s, simple calendrical tables, and a solar longitude table, and followed by a star catalog: K 6700 (22 fols., *ca.* 1100H); DM 190,1 (fols. 1r–24v, *ca.* 1150H in different hands); DM 307 (30 fols., *ca.* 1050H in different hands, the prayer-tables are arranged in columns); MM 226 (23 fols., in two hands *ca.* 1100H); MM 222 (20 fols., 1038H); *etc.* (22 copies). Most of these copies are wretched. The different star catalogs which occur are attributed to ^cAbd al-Raḥmān al-Ṭūlūnī (no. D16); Aḥmad al-Sharafī (no. D48); Muṣṭafā al-Khayyāṭ (no. D101); ^cAbd al-Raḥmān al-Ṭanānī (no. D106); Muḥammad al-Ikhṣāṣī (no. D3); and ^cAbd al-Ṣamad (no. D95). See also the similar work attributed to al-Ikhṣāṣī (no. D3). Other copies of al-Lādhiqī's work are preserved in Europe and Istanbul.

I have the impression that al-Lādhiqī in his innocence compiled a short treatise entitled *Bughyat al-nafs fī ḥall al-shams* followed by some simple calendrical tables and a *shabaka* table displaying the solar longitude for each day of the Coptic year. Various prayer-tables from the main Cairo corpus were then added to this, not necessarily by al-Lādhiqī because there is no mention of them in the introduction, and the resulting text and tables were given the title *Natījat al-afkār fī a*^c*māl al-layl wa-l-nahār*. Analysed in detail in *SATMI*, II.

C133 MUḤAMMAD AL-HUNAYDĪ

See next entry.

3.1.27 A table for the solar longitude, compiled 1007H: DM 746,5 (fol. 19r, *ca.* 1100H).

C134 SHAMS AL-DĪN MUḤAMMAD AL-HUNAYD

See previous entry.

2.3.6 Planetary tables attributed to him based on the method of ***al-Durr al-yatīm*** of Ibn al-Majdī (no. C62): ṬM 113,1 (fols. 1r–128r, 1053H).

C135 ᶜABD AL-RAḤĪM B. AL-BANNĀᵓ

2.3.6 Planetary tables attributed to him based on the method of ***al-Durr al-yatīm*** of Ibn al-Majdī (no. C62): ṬM 113,1 (fols. 1r–128r, 1053H).

C136 MUḤAMMAD B. AL-QALAᶜĪ

2.3.6 Planetary tables attributed to him based on the method of ***al-Durr al-yatīm*** of Ibn al-Majdī (no. C62): ṬM 113,1 (fols. 1r–128r, 1053H).

C137 MISCELLANEOUS ANONYMI (trigonometric tables)

2.4.1 Trigonometric tables computed for each minute of arc (the tables listed below are all of Egyptian provenance). Note that the tables in ṬM 80 (69 fols., *ca.* 850H) are copied in the elegant hand of Sayf al-Dīn Sāṭilmish (no. C72) and may have been computed by him—see Pls. XXIVa and CVIa. The tables in ṬM 235 (74 fols., 866H) are copied in the elegant hand of al-Dalāmī. The tables in DM 136 (98 fols., 1147H), which were copied in Fayyum, are attributed to various authors. Thus the tables in DM 136,1 are attributed to Shams al-Dīn Muḥammad b. Muḥammad al-Baghdādī (no. X7), on whom I have very little further information. The Cotangent tables in DM 136,4 are attributed to Abu l-Fatḥ al-Samarqandī, a scholar of the fourth century Hijra (no. B58). Likewise the solar declination tables in DM 136,3 are attributed to Abū ᶜAlī al-Marrākushī (no. C17). The only other tables of this kind which I have come across are in MS Berlin Ahlwardt 5742,2 which contain the same Cotangent tables as K 4017 (see below), and which are attributed to Ibn Yūnus (no. B59), and in MSS Berlin Ahlwardt 5752 and Damascus Ẓāhirīya 3109 which contain Sine tables to five digits also attributed to Ibn Yūnus.

The Cairo collection contains the following sets of tables:

Sine tables, 3 sexagesimal digits: ṬM 80,1 (fols. 1r–16r, *ca.* 850H by Sayf al-Dīn Sāṭilmish (no. C72)); DM 514 (16 fols., *ca.* 900H); *etc.* (2 copies).

Sine tables, 4 digits: ṬM 235,2 (fols. 27v–42r, 866H); DM 136,1 (fols. 1r–24r, 1147H, attributed to Shams al-Dīn Muḥammad b. Muḥammad al-Baghdādī (no. X7)).

Sine tables, 3 digits: DM 993 (25 fols., mainly 829H); *etc.* (4 copies plus a fragment in the hand of Ibn Abi l-Fatḥ al-Ṣūfī (no. C98)).

Sine tables, 3 digits: MM 202,2 (fols. 17r–26r, 1063H by Quṭb al-Dīn al-Maḥallī (no. D37)).

Sine tables, 4 digits: DM 469 (24 fols., *ca.* 1250H).

Versed Sine tables, 3 digits with entries up to argument 140°: ṬM 80,4 (fols. 46v–69v, *ca.* 850H by Sayf al-Dīn Sāṭilmish (no. C72)), MM 56 (15 fols., *ca.* 1150H); *etc.* (1 copy). The same tables with entries up to argument 180°: ṬM 235,1 (fols. 1r–26r, 866H, with the additional tables in a later hand); DM 577,8 (fols. 78r–96r, *ca.* 1150H); *etc.* (1 copy).

Versed Sine tables, 3 digits: MM 57,1 (fols. 1v–20r, *ca.* 850H); *etc.* (2 copies).

Cotangent tables to base 12, 3 digits: ṬM 80,2 (fols. 16v–31r, *ca.* 850H by Sayf al-Dīn Sāṭilmish (no. C72)); ṬM 235,4 (fols. 59r–74r, 866H); *etc.* (3 copies).

Cotangent tables to base 12, 3 digits: DM 715 (16 fols., *ca.* 1050H); DM 136,4 (fols. 67r–82r, 1147H, attributed to Abu l-Fatḥ al-Samarqandī (no. B58)).

Cotangent tables to base 12, 2 digits: MM 172,2 (fols. 11r–26r, *ca.* 900H) ≈ DM 1222 (10 fols., *ca.* 1150H); *etc.* (see the Arabic catalog).

Cotangent tables to base 12, 3 digits: MM 202,1 (fols. 1v–16v, 1063H by Quṭb al-Dīn al-Maḥallī (no. D37) who claims to be the compiler).

Tangent tables to base 60, 3 digits: DM 631 (11 fols., *ca.* 900H by Ibn Abi l-Fatḥ al-Ṣūfī (no. C98)); K 4017 (10 fols., *ca.* 1000H, with a short introduction as contained in MS Berlin Ahlwardt 5742,2); *etc.* (4 copies). One other late set (see the Arabic catalog).

Cotangent tables to base 5, called *ẓill al-faḍla*, 2 digits: DM 32 (11 fols., *ca.* 1100H).

Solar declination tables, obliquity 23;35°, 3 digits: ṬM 80,3 (fols. 31v–46r, *ca.* 850H by Sayf al-Dīn Sāṭilmish (no. C72)); ṬM 235,3 (fols. 43r–58r, 866H); DM 136,3 (fols. 53r–66r, 1147H, attributed to Sharaf al-Dīn Abū ᶜAlī al-Marrākushī (no. C17)); *etc.* (1 copy). One other incomplete set (see the Arabic catalog).

C138 ANONYMOUS

2.3.12 An incomplete ephemeris for an unspecified year, probably of Egyptian origin: DM 909,2 (photos of MS Paris B.N. 2496, fols. 3r–8v, *ca.* 1000H?).

C139 ANONYMOUS

2.5.4 A fragment from an anonymous treatise entitled ***K. al-Durar wa-l-yawāqīt fī uṣūl al-mawāqīt*** (the 1st *bāb* of the 2nd *qism* dealing with the Sine) occurs together with a fragment from a treatise by Shams al-Dīn al-Baghdādī (see no. X7) in K 4031,2 (fols. 5v–9v, *ca.* 1150H). The above mentioned treatise is distinct from the ***K. al-Durar wa-l-yawāqīt fī ᶜilm al-raṣad wa-l-mawāqīt*** preserved in MS Oxford Bodleian 133 (fols. 94v–130r, 734H). The Cairo fragment merits investigation.

C140 ANONYMOUS

2.5.5 A short treatise on computing planetary longitudes and stations: DM 787,2 (fols. 23r–26r, *ca.* 750H). This may be due to the author of the treatise in DM 787,1 (see no. C17).

C141 MISCELLANEOUS ANONYMI (Egyptian tables for timekeeping)

3.1.12 Miscellaneous anonymous tables for timekeeping mainly belonging to the main Cairo corpus: MM 204,10 (fols. 80v–81r, 1052H); DM 954,2 (fols. 2r–10v, *ca.* 875H); MM 58,2 (fols. 2v–5v + 6v–12v, *ca.* 850H); DM 190,2 (fols. 25r–33r, *ca.* 950H); MM 43,2 (fols. 24v–30v + 37r–49v, *ca.* 850H); MM 184,4 (fols. 7v–10r, *ca.* 1100H); DM 651,6 (fols. 59r–64v, *ca.* 1100H); DM 45,1 (fols. 1r + 28r, *ca.* 1050H); *etc.* (see Arabic catalog).

C142 MISCELLANEOUS ANONYMI (Syrian tables for timekeeping)

3.1.17 Prayer-tables for latitudes 41;15° (Istanbul) and 36° (Aleppo) together with some of the prayer-tables of al-Khalīlī (no. C37) for latitude 33;30° (Damascus): ṬM 255,6 (fols. 61v–80v, *ca.* 1100H).

3.1.17 Prayer tables for latitude 34° (Tripoli): ṬM 228,3 (fols. 9r–9v + 16r–21r, *ca.* 900H), unique. See Pl. XXXIb.

3.1.17 Table of oblique ascensions for latitude 35;50° (Aleppo): ṬM 177,2 (fol. 15v, *ca.* 1325H).

All of these tables are analyzed in *SATMI*, II.

C143 MISCELLANEOUS ANONYMI (Egyptian tables for timekeeping)

3.1.21 Tables of the time since sunrise, hour-angle, and azimuth as functions of solar altitude and solar longitude for latitude 31° (Alexandria and al-Maḥalla): TR 354 (142 pp., *ca.* 1100H, falsely attributed to Ibn Yūnus al-Mawṣilī—(see no. B59); ṬM 216,3 (fols. 14r–87v excluding fols. 19r–22v, *ca.* 1200H, lacks azimuth tables). No other known copies. See Pl. XXXV.

3.1.21 Prayer-tables for latitude 31° probably by the same astronomer with an introduction in 4 *faṣl*s: DM 1211 (28 fols., 1101H, in the hand of Riḍwān Efendī (no. D58)); DM 1207,1 (fols. 1r–57v, *ca.* 1175H); *etc.* (1 copy and 1 fragment).

All of these tables are analyzed in *SATMI.*

C144 MISCELLANEOUS ANONYMI (auxiliary tables for timekeeping)

3.1.22 ***Fatḥ al-karīm al-bāqi fī maᶜrifat al-dāʾir wa-faḍlihi āfāqī***, auxiliary tables for finding the hour-angle, with a short introduction: DM 644,1 (fols. 1r–16v, *ca.* 1100H). Another copy is MS Istanbul S. Esad Efendi Madresesi 119. The tables were lifted from the minor auxiliary tables of al-Khalīlī (no. C37), extant in the unique MS Dublin Chester Beatty 4091.

3.1.22 ***Daqāʾiq al-raqāʾiq fī maᶜrifat faḍl al-dāʾir li-sāʾir al-āfāq***, auxiliary tables for timekeeping with a short introduction in 5 *bāb*s: MM 203,3 (fols. 153r–160v, *ca.* 1100H), unique.

Each of these tables is analyzed in *SATMI.*

C145 MISCELLANEOUS ANONYMI (treatises on timekeeping)

3.2.9 ***Risāla/Muqaddima fī Ḥisāb al-masāʾil al-jaybīya wa-l-aʿmāl al-falakīya***, a treatise on timekeeping in 10 *bāb*s: MM 177,2 (fols. 3v–4v, *ca.* 1100H); ṬM 230,4 (pp. 88–90, *ca.* 1300H), *etc.* (1 copy). The procedures outlined involve the use of the first and second functions underlying al-Khalīlī's universal auxiliary tables (see no. C37).

3.2.9 The end of an anonymous treatise on timekeeping in 12 *bāb*s: DM 620,7 (fols. 60v–63v, *ca.* 900H).

C146 ANONYMOUS

3.5.6 A note on finding the local latitude from a circumpolar star: MM 144,2 (fol. 10v, *ca.* 850H).

3.5.6 Two notes on finding the meridian by a geometrical construction and finding the solar altitude at the seasonal hours in all latitudes: MM 167,7 (fols. 165r–166v, 989H).

3.5.6 Notes on the planets and the hours of daylight in Cairo for each zodiacal sign: MM 204,8 (fols. 82r–83r, 1052H).

3.5.6 A note on the time of the prayer of the *ʿīd*: DM 534,3 (fol. 19v, *ca.* 900H).

C147 ANONYMOUS

4.1.5 List of treatises on quadrants, of which it is stated that they were available in the handwriting of al-Sanhūrī and that most of them were short: DM 504,2 (fol. 9v, *ca.* 950H), unique. (See Pl. LXVI and the Arabic catalog for details.)

C148 ANONYMOUS

4.4.8 ***R. fi l-ʿAmal bi-rubʿ al-muqanṭarāt***, on the use of the almuqantar quadrant in 20 *bāb*s: ṬM 79,4 (fols. 31v–40v, *ca.* 1000H); plus a fragment in MM 144,3 (fols. 11r–16r, *ca.* 850H).

C149 ANONYMOUS

4.5.3 ***R. mukhtaṣara fi l-ʿAmal bi-l-jayb al-sittīnī bi-lā murī wa-la-madār***, a treatise on non-standard procedures with the sine quadrant, in unnumbered sections: DM 138,9 (fols. 18r–18v, 840H). See also no. C37.

C150 MISCELLANEOUS ANONYMI (treatises on universal astrolabes and *shakkāzīya* plates)

4.6.6 ***R. fi l-ʿAmal bi-l-ṣafīḥa al-āfāqīya al-musammā al-jāmiʿa***, in 15 *bāb*s: DM 1001 (20 fols., 911H). Another copy of this anonymous treatise exists in MS Berlin 5813 (copied *ca.* 1150H).

4.6.6 (***R. laṭīfa fi l-ʿAmal bi-l-āla al-musammā bi-l-shakkāzīya***), arranged in unnumbered *faṣl*s: DM 88 (12 fols., 1060H), unique.

4.6.6 *R. fi l-ʿAmal bi-l-safiḥa al-sakkajīya (!)*, in 40 *bāb*s: ZK 706,1 (fols. 1v–8v, *ca.* 1100H). Another copy of this treatise exists in Alexandria Municipal Library. In both copies the instrument is called *al-sakkājīya*, but is simply the single *shakkāzīya* plate.

4.6.6 ***R. mukhtaṣara fī Kayfīyat al-ʿamal bi-l-ṣafīḥa al-jāmiʿa***, in 23 *bāb*s: TR 131,3 (pp. 2–13, 1186H in Maghribī script). Another anonymous copy exists in the Ẓāhirīya Library in Damascus.

4.6.6 ***R. al-Ṣafīḥa al-shakkāzīya*** in 23 *bāb*s: TR 159,4 (pp. 98–120, 1320H from a copy in the Khālidīya Library in Jerusalem dated 1139H). Another copy exists in Tunis (*cf.* J. Samsó Moya in *al-Andalus*, 31 (1966), pp. 391-392).

4.6.6 ***R. mujīza ʿala l-Āla al-musammā bi-l-shakkāzīya***, in 10 *bāb*s: DM 129,1 (fols. 1v–3r, *ca.* 1150H); ṬM 102,10 (fols. 86r–89r, *ca.* 1150H). No other known copies.

C151 MISCELLANEOUS ANONYMI (treatises on the *musātara*)

4.6.18 ***R. al-Musātara***, in 10 *bāb*s: DM 138,13 (fols. 110v–111r, *ca.* 1150H in a wretched *naskhī* hand).

4.6.18 (***Qiṭʿa fī ʿAmal al-musātara***): MM 211,2 (fol. 9v, *ca.* 850H).

4.6.18 A treatise on the construction of a kind of *musātara*: ṬM 73,2 (fols. 3r–4r, 1235H).

C152 MISCELLANEOUS ANONYMI (treatises on the "winged" quadrant)

4.6.30 (*R. fi l-ᶜAmal bi-l-rubᶜ al-mujannaḥ*), in 17 *bābs*: DM 728,2 (fols. 4v–8r, 1143H), unique.

4.6.30 (*R. mukhtaṣara fi l-āla al-musammā bi-janāḥ al-ghurāb*), not subdivided: K 17289,1 (fols. 1v–5r, *ca.* 1250H).

C153 MISCELLANEOUS ANONYMI (treatises on the "crescent" quadrant)

4.6.33 *R. fi l-ᶜAmal bi-l-rubᶜ al-hilālī*, in 10 *bābs*: DM 138,1 (fols. 1v–3v, 840H).

4.6.33 *Naẓm al-laᵓāliᵓ fi l-ᶜamal bi-l-rubᶜ al-hilālī*, not subdivided: DM 138,10 (fols. 18v–22r, 840H); DM 1093,15 (fols. 74r–79r, 1088H. This treatise differs from Sibṭ al-Māridīnī's treatise entitled *ᶜUqūd al-laᵓāliᵓ fi l-ᶜamal bi-l-rubᶜ al-hilālī*, preserved in MS Manchester Rylands 361.

C154 MISCELLANEOUS ANONYMI (treatises on the octant)

4.6.36 (*R. fi l-ᶜAmal bi-l-thumn*), in 10 *bābs*: DM 781,1 (fols. 1v–5r, *ca.* 900H), unique.

4.6.36 *R. fī maᶜrifat ᶜAmal al-jayb bi-l-thumn . . .*, in 8 *bābs*: DM 781,2 (fols. 5v–7v, *ca.* 900H), unique.

C155 MISCELLANEOUS ANONYMI (treatises on a semi-circular grid)

4.6.38 *R. al-Ṣafīḥa al-āfāqīya (?)*[1] = *R. fi l-ᶜAmal bi-l-juyūb al-sittīnīya*, in 11 *bābs*: MJ 180,2 (fol. 31r, *ca.* 1000H), unique.

4.6.38 *Natījat al-afkār fi l-ᶜamal bi-jayb al-awtār*, in 9 *bābs*: DM 1093,13 (fols. 70r–72v, 1088H); DM 138,14 (fol. 112r, *ca.* 1150H, lacks beginning).

C156 ANONYMOUS (treatise on the concealed trigonometric grid)

4.6.40 (*R. fi l-ᶜAmal bi-l-jayb al-ghāᵓib*), a treatise on the use of the "concealed" trigonometric grid in 16 *bābs*: DM 568,2 (fols. 5v–8v, *ca.* 1050H), unique.

1. MJ 180,2 has the title *R. al-Ṣaḥfa al-āfāqī* (*sic*).

C157 MISCELLANEOUS ANONYMI (treatises on the instrument called *dāᵓirat al-muᶜaddil*, the equatorial (semi-)circle)

4.6.46 Anonymous treatise, not subdivided: DM 781,4 (fols. 8v–10r, *ca.* 900H).

4.6.46 Anonymous treatise, divided into unnumbered *faṣls*: MM 201,1 (fols. 1v–2v, *ca.* 1100H).

4.6.46 Anonymous treatise in 10 *faṣls*: K3993,2 (fols. 5v–6r, *ca.* 1100H).

4.6.46 Anonymous treatise, in 13 *faṣls*: DM 793, (4 fols., *ca.* 1050H); ṬM 73,1 (fols. 1v–2v, 1235H).

4.6.46 Anonymous *faṣl* on the markings of the diameter of the instrument: DM 720 (5 fols., *ca.* 1200H).

C158 MISCELLANEOUS ANONYMI (treatises on sundial theory)

4.7.16 A short treatise on *al-radāda*, a gnomon serving two sundials simultaneously: MM 96,3 (fols. 6v–10r, *ca.* 1100H).

4.7.16 Tables for marking horizontal sundials for the latitudes of Mecca, Medina, Jerusalem, Damascus: MM 57,2 (fol. 20v, *ca.* 850H); *etc.* (1 late copy). These tables differ from those in the treatise by al-Maqsī (no. C15).

4.7.16 Sundial tables for latitude 30°: MM 58,1 (fols. 1v–2r, *ca.* 850H).

4.7.16 A note on the construction of a sundial in the plane of the celestial equator: DM 597,3 (fols. 5r–7r, *ca.* 1100H). This note differs from the discussion of such sundials in the treatise of al-Maqsī (no. C15).

C159 ANONYMOUS

4.7.23 *al-Durr al-fākhir fī waḍᶜ al-sāᶜāt wa-khuṭūṭ faḍl al-dāᵓir*, on sundial theory, not extant? Abridgement by the author entitled *Naẓm al-jawāhir min al-Durr al-fākhir*, in 5 *bābs* with tables for latitude 30°: DM 185,1 (fols. 1r–30r, *ca.* 1150H).

C160 ANONYMOUS

4.8.5 *(Fāʾida) fi l-Qiyās bi-masṭaratayn*, a note on the use of the parallactic ruler: MM 144,4 (fols. 16v–17r, *ca.* 850H. Merits study.

C161 MISCELLANEOUS ANONYMI (treatises on mathematics)

6.3.23 *R. mukhtaṣara fī Maᶜrifat al-aᶜdād bi-l-aṣābiᶜ*, on finger arithmetic: J 3964,6 (fols. 30v–31r, 862H).

6.3.23 Geometrical problems introduced by the phrase *Masāʾil handasīya mutarjama bi-l-maqāla al-muhdā wa-hiya muqaddima li-masāʾil . . . (?) ustikhrijat bi-l-handasa* . . . : DR 702,5 (fols. 63r–66r, *ca.* 1000H).

6.3.23 A note from a work entitled *K. al-Muᶜtamad fī funūn al-ᶜadad*: DR 1066,1 (fol. 1r, *ca.* 1100H).

6.3.23 Notes on the area of a crescent: see TJ 258,2 (pp. 39–43, *ca.* 1150H).

6.3.23 Notes on the determination of the Sine of 1° and on the theory of parallels: DR 703,5 (fols. 130v–134r, 1006H).

6.4.26 *R. fī ᶜIlm al-jabr wa-l-muqābala wa-l-taᶜdīl wa-l-takmīl wa-l-radd*, on algebraic methods: KhR 2 (107 fols., *ca.* 1200H, purportedly 948H).

6.4.26 *R. fi l-ḥisāb wa-l-mīrāth*, on arithmetic, with particular regard to inheritance, in 4 *faṣl*s; ṬJ 688,5 (fols. 80v–148r, 1045H).

C162 MISCELLANEOUS ANONYMI (multiplication tables)

On these tables, see D. A. King, "On Medieval Islamic Multiplication Tables," *HM*, 1 (1974), pp. 317–323, and a supplement in *HM*, 6 (1979), pp. 405–417.

6.6.3 *Jadwal sittīnī = Jadwal al-nisba al-sittīnīya*, standard 60 × 60 sexagesimal multiplication tables: ZK 740,2 (pp. 3–16, *ca.* 850H); DM 64,7 (fols. 79v–94r, 803H); DM 797 (26 fols., *ca.* 1000H); *etc.* (20 copies).

6.6.3 *Jadwal jins khārij al-ḍarb wa-l-qisma*, as appended to many sexagesimal multiplication tables: DM 620,3 (fols. 18r–18v, 913H). See also MM 209,3.

6.6.4 Tables for multiplying by 60, 3600, and 216000: MM 58,4 (fols. 13r–14r, *ca.* 850H by Ibn Abi l-Fatḥ al-Ṣūfī (no. C98)).

Tables for multiplying by 44: TR 321,2 (fols. 2v–6r, *ca.* 1250H).

Tables for decimal multiplication: DM 900,1 (fols. 3v–5r, *ca.* 1200H). These tables may be due to Taqi l-Dīn b. Maᶜrūf (no. H12).

A curious alphabetical rule for multiplication: DM 26,2 (fol. 29v, *ca.* 800H).

See also D96.

C163 ANONYMOUS

6.8.11 Names of units of weights and measures in Arabic and Coptic, written in Arabic: DM 291,2 (fols. 3r–3v, *ca.* 900H). See Pl. XXc.

C164 ANONYMOUS (treatises on inheritance)

6.9.17 *Muqaddima fī ḥaṣr masāʾil al-jadd*, a treatise on the division of the inheritance of a person who is a grandfather: J 3964,16 (fols. 71v–75v, 862H).

6.9.17 Notes on inheritance arranged in 11 *wajh*s: DR 302,2 (fols. 8r–9v, 1135H).

6.9.17 See also B 23837,2 (fols. 28v–30r, *ca.* 1150H); B 25293,2 (fols. 59r–60v, *ca.* 1100H); TR 222,2 (pp. 70–75, 1099H).

See also Sections X and Z.

ADDENDA:

C165 ABU NAᶜĪM MUSLIM B. ISMĀᶜĪL AL-FĀRIQĪ

Not identified.

5.5.3 A treatise on the level of the Nile and its astrological significance, stated to be taken from the *Mathalāt* (?) of al-Fāriqī: DM 1210,1 (fols. 1v–3v, *ca.* 1200H).

C166 YŪSUF B. AL-DAMĪRĪ

3.3.6 The qibla table in DM 1030,3 (fols. 3r–4r, *ca.* 1200H) is attributed to Yūsuf b. al-Damīrī (see no. Z25). This copy contains a short introduction on the use of the table, mentioning Ibn Yūnus (no. B59).

C167 AḤMAD ᶜUBAYD ALLĀH, a *muwaqqit*

See no. C13.

C168 SHAMS AL-DĪN MUḤAMMAD B. AL-ᶜAṬṬĀR AL-BADRĪ AL-SHĀFIᶜĪ

See no. C124.

C169 IBN ABI L-QĀSIM AL-ANDALUSĪ

See no. C124X

C170 MUᵓAYYAD AL-DĪN AL-ᶜURḌĪ

Suter, pp. 147 and 154.

He or his son copied ṬH 43 of the recension of the ***Almagest*** by al-Ṭūsī (no. G23) *ca.* 650H. See Pl. I.

al-ᶜUrḍī's treatise on planetary theory is currently being investigated by Prof. George Saliba, Columbia University.

C171 ABŪ ᶜABD ALLĀH MUḤAMMAD B. ᶜALĪ AL-RAḤBĪ, IBN AL-MUTAQQINA

GAL, I, pp. 490–491, and SI, pp. 675–676.

6.9.7 An abridgement of his poem on inheritance known as ***al-Raḥbīya***, written by Ibn al-Hāᵓim: see no. C58.

C172 YAᶜĪSH B. IBRĀHĪM B. YŪSUF AL-UMAWĪ

Suter, no. 453; *Krause*, no. 453 (where a manuscript dated 774H is listed); *GAL*, II (1st ed. only!), p. 126, and SII, pp. 155 and 379; and the article "al-Umawī" by A. S. Saidan in *DSB.*

6.8.6 *R. fī ᶜIlm al-qabbān*, on the balance: DR 86,3 (pp. 25–29, 1256H).

C173 Al-SAḤTĪTĪ

See Pl. XIXb.

D. Egyptian and Syrian Authors (after *ca.* 1600)

References to scientists from this period are to be found in *Suter*, Brockelmann's *GAL*, and Azzawi. No survey of the activities of the astronomers of this period is available.

D1 ᶜUTHMĀN B. ᶜALĀʾ L-DĪN B. YŪNUS B. MUḤAMMAD, IBN AL-MALIK AL-DIMASHQĪ

GAL, II, p. 468.

6.4.20 *al-Isᶜāf al-atamm bi-ḥisāb al-qalam*, a treatise on arithmetic in 7 *maqāla*s, compiled in Cairo in 1002H: ṬR 140 (83 fols., 1198H); ṬR 117 (49 fols., *ca.* 1150H, incomplete); *etc.* (2 copies). No other copies listed by Brockelmann.

6.8.9 *Nukhbat al-zamān fī ṣināᶜat al-qabbān*, a poem on the balance, compiled in 997H: DR 562 (3 fols., 997H in author's hand).

D2 ᶜUMAR B. MUḤAMMAD B. ABĪ BAKR AL-FĀRISKŪRĪ

Suter, no. 478; *GAL*, II, p. 419, and SII, pp. 443 & 484.

4.7.20 *Nafḥ al-fuyūḥ bi-sharḥ Rayḥānat al-rūḥ*, a commentary on Taqi l-Dīn's treatise on sundials: see no. H12.

D3 MUḤAMMAD B. HAYKAL AL-IKHṢĀṢĪ, *muwaqqit* in the Azhar Mosque, Cairo.

Not mentioned in the modern sources.

3.1.25 *al-Durra al-muḍīya fi l-aᶜmāl al-shamsīya*, a short introduction followed by calendrical tables and solar longitude table, prayer-tables for Cairo lifted from the main corpus (see no. C28), and a star catalog: MM 81 (22 fols., *ca.* 1100H); DM 60 (22 fols., *ca.* 1100H); *etc.* (2 copies). A similar work is attributed to al-Lādhiqī (no. C132). See further *SATMI*, II.

3.1.25 Star table appended to al-Lādhiqī's *Natījat al-afkār* (see no. C132) in MM 221 (*ca.* 1100H), MM 52 (*ca.* 1100H), *etc.*

D4 MUḤAMMAD B. AḤMAD B. MUḤAMMAD B. BĪRĪ(?) AL-ḤANAFĪ AL-MAKKĪ AL-ᶜALWĀNĪ

Not mentioned in the modern sources.

6.4.28 *K. al-Yawāqīt al-mufaṣṣalāt bi-l-laʾāliʾ al-nayyirāt fī aᶜmāl dhawāt al-asmāʾ wa-l-munfaṣalāt*, a commentary compiled in 1017H on his poem *al-Laʾālīʾ al-nayyirāt* dealing with number theory: DR 99 (50 fols., *ca.* 1100H); K 4300 (54 fols., 1181H).

D5 ᶜABD AL-RAḤMĀN B. ᶜĪSĀ B. MURSHID AL-ᶜUMRĪ AL-ḤANAFĪ

GAL, II, p. 499, and SII, p. 513; *Azzawi*, p. 339.

3.4.8 *Riyāḍ al-faḍāʾil = Barāᶜat al-istihlāl wa-mā yataᶜallaq bi-l-shahr wa-l-hilāl*, a treatise in 3 *bāb*s on crescent visibility, mainly from a religious point of view: KhM 4 (77 fols., *ca.* 1150H, purportedly 1043H); DM 15 (152 pp., 1066H). Brockelmann lists 3 other copies (*GAL*, SII, p. 513), to which add MSS Istanbul Yeni Cami 989, Nurosmaniye 3152, perhaps also Bagdatli Vehbi Ef. 901. Merits study.

D6 SĀLIM B. AḤMAD B. SHAYKHĀN

GAL, II, p. 537, *etc.*

5.3.27 *al-ᶜArḍ al-kāfī li-l-ᶜarḍ al-shāfī wa-huwa l-bayān ᶜan ᶜumr al-zamān*, a treatise on the length of time the earth will last, compiled in 1036H: TJ 250,11 (pp. 175–199, 1214H).

Brockelmann lists another copy in MS Berlin Ahlwardt 2764.

D7 MUḤAMMAD B. ᶜALĪ AL-SHABRĀMALLISĪ

GAL, II, p. 480, SII, p. 493.

6.4.27 *Bughyat al-ḥāsib wa-bulghat al-kātib*, a treatise on arithmetic in 2 *bābs*: DR 1065 (21 fols., 1027H). No other known copies.

6.4.27 *Īḍāḥ al-muktatam fī ḥisāb al-raqam*, a treatise on arithmetic in 2 *qisms* of 5 + 5 *bābs*: MR 3 (140 fols., *ca.* 1250H). No other known copies.

6.4.27 *Mabāhij al-taysīr bi-manāhij al-taksīr*, a treatise on geometry in 2 *bābs* and a *khātima* in 3 + 3 + 3 *faṣls*: DR 299,1 (fols. 1r–18r, *ca.* 1200H). No other known copies.

6.4.27 *al-Irshād li-l-ᶜilm bi-khawaṣṣ al-aᶜdād*, on number theory in 2 *qisms* of 4 + 3 *bābs*: see MS Berlin Ahlwardt 5997 (34 fols., *ca.* 1150H), also unique.

4.7.21 *al-Durra al-bahīya fī waḍᶜ basāʾiṭ faḍl al-dāʾir bi-l-ṭuruq al-handasīya*, a treatise on sundial theory in 3 *qisms*, without tables: MM 80 (14 fols., 1021H in author's hand).

5.3.26 *Bahjat al-muḥādith fī aḥkām jumla min al-ḥawādith*, a treatise on astrology arranged in 2 *bābs*: TR 118,1 (pp. 2–56, 1068H); DM 972 (43 fols., *ca.* 1100H); *etc.* (2 copies).

See also no. Z31.

D8 MUḤAMMAD B. MAḤMŪD AL-MANĀSHĪRĪ

GAL, II, p. 427.

2.5.13 *K. al-Falak al-dawwār li-l-shams al-munayyira wa-l-qamar al-sayyār*, a treatise on the motions of the sun and moon, arranged in 5 *bābs*, with tables: DM 184,2 (fols. 61v–75v, 1171H). See also no. D9.

D9 ANONYMOUS

3.1.17 *Nafaḥat al-misk al-khitām wa-minḥat al-mutanassik min al-anām*, a treatise on timekeeping with prayer-tables for latitudes 33;30° (Damascus) computed by al-Khalīlī (no. C37), 30;0° (Cairo), 21;30° (Mecca), and 41;15° (Istanbul): DM 184,1 (fols. 1v–61r, 1171H). This work is anonymous but is followed by a treatise by al-Mahāshīrī (no. D8) in the same hand. These tables are analyzed in *SATMI*, II.

D10 MUḤAMMAD B. ᶜUMAR B. ṢIDDĪQ B. ᶜUMAR AL-BAKRĪ AL-FAWĀNĪSĪ

Suter, no. 475; *GAL*, II, p. 469, and SII, p. 485.

3.1.26 *K. Natījat al-afkār fī ᶜamal al-layl wa-l-nahār*, extensive prayer-tables for latitude 30° (Cairo) with an introduction: DM 950 (photos of MS Paris B.N. ar. 2545, 69 pls., *ca.* 1000H). See *SATMI*, II for an analysis.

D11 ᶜABD AL-QĀDIR B. AḤMAD B. ḤASAN AL-ᶜAJMĀWĪ AL-AZHARĪ, *muwaqqit* at the *madrasa* of Sultan Ḥasan in Cairo.

GAL, SII, p. 1018.

3.1.28 *Nuzhat al-nāẓir fī maᶜrifat mā bayn al-awqāt mina l-dawāʾir*, a treatise giving the durations of the intervals for the prayers in words for each day of the Coptic year: ZK 287 (203 fols., 1056H); ṬM 223 (82 fols., 1027H, lacks beginning and end). See further *SATMI*, II.

D12 ᶜABD AL-QĀDIR B. MUḤAMMAD AL-FAYYŪMĪ

GAL, SII, p. 486 (confused).

6.4.1 A commentary on *al-Nuzha* by Ibn al-Hāʾim: see no. C58.

D13 ZAYN AL-DĪN B. SURA L-DĪN . . . AL-DURRĪ AL-MĀLIKĪ

GAL, SII, p. 154.

6.4.1 A commentary on *al-Lumaᶜ* by Ibn al-Hāʾim, compiled in 1030H: see no. C58.

D14 YAḤYĀ B. TAQI L-DĪN B. ISMĀᶜĪL AL-ḤALABĪ

Not mentioned in the modern sources.

6.4.1 **A commentary on *al-Nuzha*** by Ibn al-Hāᵓim (see no. C58): DR 1100 (65 fols., 1027H by author).

D15 ANONYMOUS (treatises on timekeeping)

3.2.21 The last page of a treatise on timekeeping compiled in 1038H in which the author mentions another work of his entitled *Naẓm al-yawāqīt fī ᶜilm al-mawāqīt*: DM 172,4 (fols. 24r–24v, 1038H by author?).

D16 ᶜABD AL-RAḤMĀN B. ᶜABD ALLĀH AL-ISKIRĪ AL-ṬŪLŪNĪ, *muwaqqit* and *imām* at the Mosque of Ibn Ṭūlūn in Cairo

GAL, II, p. 480, and SII, p. 493. See also Pl. CXb.

3.1.25 A star catalog appended to al-Lādhiqī's *Natījat al-afkār* (see no. C132) in MM 15 (1234H).

4.1.10 ***Taḥsīl al-intifāᶜ wa-ghāyat al-irtifāᶜ fī waḍᶜ al-maqāyīs wa-waḍᶜ al-arbāᶜ***, a treatise on different kinds of instruments in 2 *qism*s of 74 + 15 *bāb*s with numerous diagrams, compiled in 1034H: MM 21 (71 fols., 1126H); DM 445 (69 fols., *ca.* 1150H, lacks diagrams); DM 21 (59 fols., 1245H); *etc.* (1 fragment).

4.7.16+ ***al-Kawākib al-ẓāhira fī waḍᶜ khayṭ al-musātara***, on the construction of a special kind of sundial: DM 982 (6 fols., *ca.* 1100H). Another copy of this treatise is MS Princeton Mach 5012.

4.7.22 ***R. fī Kayfīyat ᶜamal al-basīṭa***, a treatise on sundial theory in 12 *bāb*s, with tables for latitude 30°, Cairo: MM 95 (9 fols., *ca.* 1150H); DM 421 (8 fols., 1151H); *etc.* (3 copies).

D17 ᶜABD ALLĀH B. ᶜABD AL-RAḤMĀN AL-ṬŪLŪNĪ, son of no. D16

4.8.8 ***Kashf al-rayb wa-bayān al-sirr al-maghmūḍ fi l-ᶜamal bi-dāᵓirat rijāl al-ghayb wa-bi-l-basīṭa dhāt al-ᶜurūḍ***, on an instrument comprising four horizontal sundials for different latitudes and a circle for determining the position of the *rijāl al-ghayb*: ṬJ 811,6 (fols. 48v–57r, *ca.* 1150H); *etc.* (1 copy). The scientific part is published in *JHAS*, 1 (1977), pp. 251–255. The section on the mysterious *rijāl al-ghayb* has been studied by my student Ms. Zeina Matar.

D18 ᶜABD AL-RAḤMĀN B. ᶜABD ALLĀH AL-KĀTIB AL-AZHARĪ

Not identified.

4.8.8 ***Bahjat al-nāẓir fīmā yataᶜallaqu bi-maᶜrifat waḍᶜ dāᵓirat al-buldān wa-faḍl al-dāᵓir***, on the use of the qibla indicator: DM 173,8 (fols. 148v–150r, *ca.* 1100H).

D19 ABU L-KHAYR ᶜABD AL-RAḤMĀN AL-WAFĀᵓĪ AL-KĀTIB, a *muwaqqit* at the Ghawrīya *madrasa* in Cairo.

3.5.13 ***al-R. al-Juzᵓīya fī taᶜdīl al-daraja al-shamsīya***, a treatise on finding the solar longitude, arranged in 2 *bāb*s and a *khātima*, without tables: MM 184,2 (fols. 3r–4r, *ca.* 1100H); MM 180,1 (fols. 1v–3r, *ca.* 1100H).

4.8.8 ***Bayān al-sirr al-gh?m?ḍ fī rasm dāᵓirat al-maḥārīb***, on the use of the qibla indicator: DM 760,2 (fols. 5r–10r, *ca.* 1100H); *etc.* (1 copy).

D20 ABŪ BAKR MUḤAMMAD AL-KUTĀMĪ[1]

Not listed in the modern sources.

2.5.15 ***K. al-Isᶜāf ᶜala l-ikhtilāf fī ḥisāb al-izdilāf***, a treatise in 2 *faṣl*s on the *izdilāf*, on which, for want of a better reference, see *Azzawi*, pp. 85–87: K4006,1 (fols. 1r–15r, 1117H); TR 172 (4 fols., *ca.* 1100H).

He copied MM 181 in 1030H.

---

1. Brockelmann (*GAL*, II, pp. 432–433, and SII, p. 457) lists Abū Bakr b. Ṣāliḥ al-Kutāmī, who wrote on religious matters and died in Cairo of the plague in 1051/1641. The title *al-Isᶜāf* is not mentioned by Brockelmann.

D21 MUḤAMMAD AL-ṢAḤRĪ/AL-ṢAKHRĪ AL-ḤARRĪ (?)[1]

Not listed in the modern biographical sources.

4.6.21 *R. fī l-ᶜAmal bi-l-ṣafīḥa al-ᶜajamīya*, a treatise in 15 *bāb*s, compiled in 1038H: DM 912 (photos of MS Paris B.N. ar. 2532,2, 22 pls., *ca.* 1100H). The name of the author occurs at the end of the treatise and was overlooked by De Slane. The treatise deals with an instrument fitted with a *kursī*, *ᶜalāqa*, four *ṣafīḥa*s, and a cone bearing an *ᶜiḍāda*, for finding the solar and lunar equations and solving problems of spherical astronomy (!). Merits study.

D22 MUḤAMMAD B. IBRĀHĪM AL-ḤULWĀNĪ

Not listed in the modern sources.

3.1.17 Prayer-tables for latitude 34° and a table of positions of fixed stars for 1033H: DM 1170,2 (fols. 11r–22v, *ca.* 1050H). These prayer-tables are probably due to Ibn al-Shāṭir (no. C30).

3.1.17 Table for finding the ascendant at daybreak from the lunar mansions and the ascensions of the lunar mansions: MM 71,2 (fols. 66v–67r, *ca.* 1000H).

D23 MUṢṬAFĀ B. SHAMS AL-DĪN B. AḤMAD B. KHIḌRBĀY AL-JARKASHĪ AL-ṬĀHIRĪ

GAL, SII, p. 486.

4.5.30 *al-Durr fī l-jayb al-nafīs*, a poem on the sine quadrant in 15 *bāb*s, compiled in 1038H: MJ 39,2 (fols. 50r–57v, 1069H); DM 161,2 (fols. 6v–12v, *ca.* 1200H).

D24 MUṢṬAFĀ AL-MUQRIᵓ AL-KHALWATĪ AL-FALAKĪ AL-DIMYĀṬĪ

Not identified.

4.4.27 *Kifāyat al-mubtadiᵓ*, a poem on the use of the almucantar quadrant in 11 *bāb*s: DM 161,1 (fols. 1r–5v, *ca.* 1200H).

1. MS Berlin Ahlwardt 5852 contains a treatise on the almucantar quadrant in 15 chapters attributed to Sayyidī Muḥammad al-Ṣakhrī.

D25 SAᶜĪD B. ᶜABD ALLĀH AL-TANBAKTĪ AL-JANAWĪ

Not identified.

4.5.12 *al-Maẓāhir al-Aḥmadīya bi-sharḥ al-Nasama al-nafḥīya*, a commentary on the poem by Ibn Ghānim (no. C127): Sh 89,7 (fols. 53r–84r, 1071H).

See also no. D42.

D26 ᶜALĪ B. ᶜABD AL-QĀDIR AL-NABTĪTĪ

GAL, II, p. 217, *etc.*

4.5.12 *K. al-Futūḥāt al-wahbīya bi-sharḥ al-R. al-Fatḥīya*, a commentary on the treatise by Sibṭ al-Māridīnī: see no. C97.

D27 MUḤAMMAD AL-KUTUBĪ

Not identified.

3.1.27 An almanac for Cairo, 1044H, with prayer-times for each 15 days: ṬM 87 (23 fols., *ca.* 1044H).

D28 ᶜABD AL-QĀDIR B. MAḤMŪD AL-NABTĪTĪ AL-QĀDIRĪ

GAL, II, p. 477 (has ᶜAq. b. Muḥammad incorrectly); *etc.*

1.4.16 A treatise on the attributes of the days of the Syrian months: DM 108,1 (fols. 1v–23r, *ca.* 1150H by the author), unique? The treatise contains the text of a poem attributed to al-Fazārī (no. B3).

An almanac containing quotations from al-Fazārī and also Ibn Muqaffaᶜ is in Rabat Khazāna ᶜĀmma 29J (112 fols., *ca.* 1050H?). Quotations of the same kind occur in the almanac entitled *ᶜĀdāt al-nujūm* by al-Shayzarī (no. C5).

D29 ḤUSAYN B. MUḤAMMAD AL-KILĀBĪ

Not listed in the modern sources.

4.6.37 *R. mukhtaṣara fi-l-ᶜAmal bi-l-thumn al-dāᵓira (sic) al-mawḍūᶜ ᶜalayhi l-muqanṭarāt*, on the use of the octant in 19 *bāb*s: DM 1093,10 (fols. 44v–51r, 1088H), unique.

4.6.41 *R. mukhtaṣara* ᶜ*ala l-Jayb al-ghā*ᵓ*ib*, on the use of the "concealed" trigonometric grid in 22 *bābs*: DM 1093,4 (fols. 11r–17r, 1088H), unique.

D30 SHIHĀB AL-DĪN AḤMAD B. AḤMAD AL-QALYŪBĪ

GAL, II, pp. 478–479, and SII, pp. 492–493.

3.2.24 *K. al-Hidāya min al-dalāla fī ma*ᶜ*rifat al-waqt wa-l-qibla min ghayr āla*, a treatise on timekeeping in 12 *bābs*: DM 203 (16 fols., 1094H); DM 494 (18 fols., 1113H); DM 984 (9 fols., 1071H); *etc.* (9 copies), plus an anonymous commentary in MM 185,2 (14 fols., 1201H).

3.2.24 Treatise on the seasons, the lunar mansions, timekeeping, and the *qibla*, arranged in 6 *bābs* and numerous *fā*ᵓ*idas*: K 4235 (35 fols., *ca.* 1150H, lacks beginning), unique? This work includes some simple prayer-tables for Damietta and contains considerable historical information, unique. Merits study.

3.2.24 *R. fi l-Awqāt wa-l-mawāsim wa-l-tawqī*ᶜ*āt*, an almanac giving the times of prayer: ṬM 132 (26 fols., 1108H).

D31 SHAMS AL-DĪN ABU L-ṢALĀḤ MUḤAMMAD AL-QAṬARĪ thumma AL-JAWLĀNĪ

Not identified.

4.5.12 A commentary on his poem *al-R. al-Shihābīya fi l-a*ᶜ*māl al-jaybīya* by ᶜAlī al-Rashīdī (no. D32) is in DM 439 (26 fols., 1064H).

D32 ᶜALĪ B. RAJAB MUSHĀQ(?) AL-ḤANAFĪ AL-RASHĪDĪ

Not identified.

4.5.12 *Sharḥ manẓūmat Shams al-Dīn Abi l-Ṣalāḥ Muḥammad al-Qaṭarī* ᶜ*ala l-Risāla al-Shihābīya (??) li-Sibṭ al-Māridīnī* (see no. D31): DM 439 (26 fols., 1064H in author's hand).

D33 MŪSĀ B. SHĀHĪR (?) AL-ABSHĀDĪ (?) AL-MUSLIMĪ (?) AL-ḤUSAYNĪ

Not listed in *GAL*.

2.5.9 *Ighāthat al-malhūf fī* ᶜ*amal al-khusūf wa-l-kusūf*, a short treatise on eclipse computations: DM 12 (4 fols., 1091H by author). Another copy is MS Istanbul Nurosmaniye 2484,3 (fols. 17v–21v, n.d.). al-Abshādī copied DM 88 of an anonymous treatise on the *shakkāzīya* in 1060H.

D34 ᶜALĪ (B. MUḤAMMAD ZAYN AL-ᶜĀBIDĪN) AL-UJHŪRĪ

GAL, II, pp. 413–414 and SII, p. 437 (does not mention any scientific works).

6.4.31 A commentary on his own poem on arithmetic: DR 1066,2 (fols. 1r–18v, *ca.* 1100H).

D35 ᶜABD ALLĀH B. AḤMAD AL-MUNĀWĪ

GAL, SII, p. 972.

6.9.16 *al-Kawākib al-bahīya fī qismat al-mīrāth li-l-barīya*, a poem on inheritance compiled in 1048H: DR 181,6 (fols. 206r–210v, *ca.* 1150H). Brockelmann lists another copy in Alexandria.

D36 MUṢṬAFĀ B. SAHRĀB

Not identified.

2.4.2 A star catalog for 1060H: DM 644,2 (fols. 17r–18v, *ca.* 1100H); MM 54,1 (fols. 1r–5r, *ca.* 1200H). This was modified by Yaḥyā al-Birshamsī (see no. D70). See Pl. XIXb.

D37 QUṬB AL-DĪN MAḤMŪD AL-MAḤALLĪ

Not listed in the modern sources.

2.4.1 Cotangent (base 12) and Sine tables for each minute of argument to 3 digits in MM 202,1 + 2 (26 fols., 1063H by al-Maḥallī). al-Maḥallī purports to be the compiler of at least the Cotangent tables.

3.1.29 Prayer-tables for latitude 31;25° (Damietta), with an introduction, compiled in 1088H: ṬM 241,1 (fols. 1r–33r, 1245H); DM 1014 (29 fols., 1287H, lacks introduction); DM 889,1 (fols. 1r–27r, 1296H); *etc.* (4 defective copies); plus commentaries by Muḥammad al-Nimra (no. D51): DM 991 (2 fols., 1089H by author); and by Shihāb al-Dīn Ibrāhīm b. al-Ḥasan (no. D112): ṬM 225 (28 fols., *ca.* 1300H); DM 889,3 (fols. 30r–75v, 1296H). The tables are analyzed in *SATMI*, II.

3.4.12 A short treatise on the possibility of crescent visibility in different localities: DM 159,1 (fols. 1r–3r, 1096H); *etc.* (1 copy).

3.5.13 A treatise on the determination of the times of rising of the fixed stars, compiled in 1077H: DM 159,2 (fols. 3v–4r, 1096H).

6.9.16 ***Muqaddima fi khtiṣār al-kusūr fī jadāwil qismat al-tarikāt***, on inheritance, with tables: DR 619 (14 fols., 1325H).

D38 ᶜABD AL-RAḤĪM AL-ᶜAJAMĪ, *muwaqqit* at the Umayyad Mosque in Damascus

See *JHAS*, 1 (1977), pp. 24–32.

2.4.2 A star catalog for 1070H: DM 184,3 (fols. 76r–88r, 1171H).

D39 MUḤAMMAD B. MUḤAMMAD B. SULAYMĀN AL-RŪDĀNĪ AL-MAGHRIBĪ

Suter, no. 527; *Renaud*, no. 527; *GAL*, II, pp. 610–611 and SII, pp. 691 and 709.

3.2.26 ***Qalāᵓid al-laᵓāliᵓ fī ᶜamal al-ayyām wa-l-layālī***, a poem on timekeeping in unnumbered *faṣl*s: DM 1063,6 (fols. 75r–94v, 1224H). Another copy is MS Istanbul Laleli 2756 (19 pp., n.d.). Extended commentary with some tables, entitled *Maqāṣid al-ᶜawāli bi-qalāᵓid al-laᵓālīᵓ*: K 8523,2 (fols. 21v–89v, 1118H). Brockelmann (*GAL*, SII, p. 709) lists another copy in Tunis.

4.2.6 ***al-Nāfiᶜa fī aᶜmāl al-jāmiᶜa***, a treatise on a variety of spherical astrolabe, compiled in Medina in 1072H and divided into 45 unnumbered *bāb*s: ṬM 94,2 (fols. 2v–52r, *ca.* 1100H); DH 28 (38 fols., *ca.* 1150H). Both of these copies are anonymous. This treatise has been published (C. Pellat) and translated (L. Janin) in *Bulletin d'Etudes Orientales*, 26 (1973), pp. 7–83, and 28 (1975), pp. 83–165, using a single copy apparently of Maghribi provenance. Also, an anonymous ***R. fī Asmāᵓ tajzīᵓat al-kura al-musammā bi-l-jāmiᶜa***..., probably by al-Rūdānī himself, in ṬM 94,3 (fols. 53v–59v, *ca.* 1100H); and a poem on the merits of the *jāmiᶜa* by Aḥmad b. Tāj al-Dīn (no. D40): ṬM 94,1 (fols. 1v–2r, *ca.* 1100H); as well as a ***Faṣl fī Rasm al-kura***, actually attributed to al-Rūdānī: DM 639,9A (fols. 82v–83r, 1133H).

4.3.35 ***Bahjat al-ṭullāb fi l-ᶜamal bi-l-asṭurlāb***: Sh 89,8 (fols. 85v–98r, *ca.* 1100H); KA 2,1 (fols. 1r–19v, *ca.* 1200H); *etc.* (1 copy).

4.3.35 (***Risāla fī Rasm al-asṭurlāb bi-l-handasa***): DM 639,9 (fols. 80v–82v, 1133H); ZK 782,10 (fols. 57v–58v, 1126H); *etc.* (1 fragment); plus a commentary by Aḥmad b. Muḥammad ᶜAbbād (?) al-Maḥallī *al-muwaqqit* (no. D67): K 4007,1 (fols. 1r–18r, anonymous, *ca.* 1300H); DM 41,1 (fols. 1r–1v, *ca.* 1100H, first page only). *Tarjamat al-Rūdānī* by Muḥammad al-Qaṭṭān al-Makkī: KA 2,2 (fols. 19v–20r, *ca.* 1200H).

D40 AḤMAD B. TĀJ AL-DĪN

Not identified.

4.2.6 A poem on the merits of the universal instrument of al-Rūdānī (no. D39): ṬM 94,1 (fols. 1v–2r, *ca.* 1100H).

D41 ᶜABD AL-MAJĪD AL-SĀMŪLĪ AL-HINDĪ

GAL, SII, p. 1018. See also II, p. 29.

6.4.19 A treatise on mathematics without title, in 3 *maqāla*s of 2 *qism*s, 2 *bāb*s, and 7 *faṣl*s: TR 314 (117 pp., *ca.* 1200H, only first *maqāla* and part of second); ṬR 113 (134 fols., *ca.* 1100H, second and third *maqāla*s only); K3831,4 (fols. 9r–20v, 1096H, third *maqāla* only); *etc.* (2 fragments). See also no. D42.

6.8.10 ***R. fī ᶜIlm al-qabbān*** on the balance, not subdivided: MR 13 (13 fols., *ca.* 1000H); *etc.* (2 copies).

D42 BADR AL-DĪN ḤUSAYN B. SHĀMĪ AL-HATTĀRĪ AL-FARAḌĪ (?)

GAL, II, p. 523 and SII, pp. 543, 1039 (no scientific works mentioned).

2.5.11 *al-Sirāj al-wahhāj fī aᶜmāl al-azyāj*, a treatise in 40 *bāb*s on the standard topics of *zīj*es, with some prayer-tables for latitude 25°, Medina: Sh 89,12 (fols. 175r–243r, 1071H). Another copy is MS Leiden Or. 2538 (*Cat.*, p. 339).

4.3.36 *ᶜUmdat al-ṭālib fī ᶜamal al-asṭurlāb*, a treatise on the use of the astrolabe in 20 *bāb*s containing some prayer-tables for Medina: Sh 89,11 (fols. 157r–174r, 1070H in the author's hand), unique.

4.4.22 *al-Ayyāt al-bayyināt fi l-ᶜamal bi-rubᶜ al-muqanṭarāt*, a treatise in 17 *bāb*s on the use of the almucantar quadrant, containing some prayer-tables for latitude 25°, Medina: Sh 89,14 (fols. 252r–284r, 1070H by author), unique.

4.5.24 *al-Manhaj al-muqarrab fi l-ᶜamal bi-l-rubᶜ al-mujayyab*, a treatise on the sine quadrant in 25 *bāb*s, containing some prayer-tables for latitude 25°, Medina: Sh 89,15 (fols. 285r–322v, 1070H in author's hand), unique—see Pl. XLIIb. al-Hattārī mentions Saᶜīd b. ᶜAbd Allāh al-Tanbaktī (no. D25) at the beginning of this treatise, and the star tables of al-Minūfī (no. C120), al-Ṭūlūnī (no. D16), and Sulaymān al-ᶜUthmānī (no. C117) at the end of his treatise.

6.4.35 *Miᶜrāj al-albāb ilā ᶜilm al-ḥisāb*, a short treatise on arithmetic in 2 *bāb*s of 5 + 3 *faṣl*s: Sh 89,13 (fols. 244r–251v, 1070H by author), unique? Another treatise by al-Hattārī on algebra compiled from works by Ibn al-Hāʾim (no. C58) and al-Sāmūlī (no. D41) is in MS Princeton Garrett 1051 (44 fols., 1088H by author).

D43 QĀSIM B. ṢALĀḤ AL-DĪN AL-KHĀNĪ

GAL, II, p. 452, *etc.*

1.2.21 A short treatise on astronomical terminology: TJ 196,18 (pp. 498–513, *ca.* 1100H), unique, not listed by Brockelmann.

6.4.1 A commentary on Ibn al-Hāʾim's poem *al-Muqniᶜ* (see no. C58): TJ 196,15 (pp. 440–465, *ca.* 1100H).

Note: TJ 196 is a compendium of treatises by al-Khānī.

D44 ᶜALĪ B. ᶜABD AL-QĀDIR AL-NABTĪTĪ AL-ḤANAFĪ

GAL, SII, p. 558 (*s.v.*).

6.9.15 *al-Budūr al-shāriqāt fī aᶜmāl al-munasikhāt*, on inheritance: B 22621 (46 fols., 1069H).

D45 ᶜABD ALLĀH B. AḤMAD AL-MAQDISĪ AL-ḤANBALĪ

GAL, II, p. 470, and SII, p. 486.

4.5.25 *Tuḥfat al-labīb wa-bughyat al-arīb*, a treatise on the use of the sine quadrant in 5 *bāb*s: DM 639,10 (fols. 83r–83v, *ca.* 1150H, incomplete). Another copy is in MS Berlin Ahlwardt 5856.

5.3.33 *Tuḥfat al-albāb fī bayān ḥukm al-adhnāb*, a treatise on the astrological significance of various kinds of comets and pseudo-comets, with some illustrations: DM 729 (24 fols., 1156H); DM 24 (18 fols., 1157H); TR 236 (98 pp., *ca.* 1250H, the treatise is followed by some Hermetic material); *etc.* (5 copies). Another copy is MS Princeton Mach 5065 = Yehuda 368 (37 fols., *ca.* 1150H). Merits study. See Pl. LXXXIV.

D46 ᶜALĪ AL-ZAᶜTARĪ AL-MIṢRĪ

I have no information on al-Zaᶜtarī other than that he made an astrolabe in 1082/1671, now preserved in the Topkapi Seray, Istanbul.

4.6.46 *Fawāʾid fī maᶜrifat al-ᶜamal bi-niṣf dāʾirat al-muᶜaddil*, seven notes on the "equatorial (semi-)circle": DM 879,1(fols. 1v–3v, 1090H); TR 176 (7 pp., *ca.* 1300H).

D47 MUḤAMMAD B. ABĪ BAKR B. AḤMAD SHILLĪ BĀᶜALAWĪ

GAL, II, p. 502 and SII, p. 516.

3.5.12 A short treatise on the midday shadows at latitude 21°, Mecca: DM 130,2 (fols. 14v–15v, *ca.* 1200H), unique. See *SATMI*, III.

D48 AḤMAD B. ᶜABD AL-ᶜAZĪZ AL-SHARAFĪ AL-ṢAFĀQUSĪ AL-AZHARĪ

GAL, II, p. 612, and SII, pp. 486 and 694.

2.4.2 A star catalog for *ca.* 1100H: DM 740,1 (fols. 1v–3v, *ca.* 1150H).

3.1.25 A star catalog for 1096H appended to al-Lādhiqī's *Natījat al-afkār* (see no. C132) in MM 226 (cop. *ca.* 1110); K6700 (cop. *ca.* 1100H); *etc.*

4.4.25 ***al-Durar al-fākhirat fī l-ᶜamal bi-rubᶜ al-muqanṭarāt***, a treatise on the use of the almucantar quadrant in 20 *bābs*: DM 58 (21 fols., *ca.* 1150H); DM 737 (24 fols., 1145H); *etc.* (2 copies).

4.5.12 ***Fatḥ Rabb al-barīya li-ḥall alfāẓ al-Nasama al-nafaḥīya***, a commentary on the poem by Ibn Ghānim (no. C127): MJ 39,1 (fols. 1r–49v, *ca.* 1100H).

D49 MUḤAMMAD B. ᶜALĪ B. MUḤAMMAD AL-ADFĪNĪ AL-SHĀFIᶜĪ AL-FARAḌĪ

GAL, II, p. 418, and SII, p. 442.

6.4.8 A *ḥāshiya* on al-Shinshawrī's commentary on Sibṭ al-Māridīnī's *Tuḥfat al-aḥbāb*: see nos. C97 and C111.

D50 ᶜABD AL-WAHHĀB AL-MUQRIᵓ AL-SARĀJĪ (?)

GAL, II, pp. 470–471.

2.5.9 ***Taḥbīr inkishāf al-labs fī taḥrīr inkisāf al-shams***, a treatise on the computation of solar eclipses: MM 19 (9 fols., 1092H); MM 18 (8 fols., 1146H). The first of these manuscripts concludes with an anonymous *jadwal al-samt li-maᶜrifat kusūf al-shams* which is ultimately of Abbasid origin: see no. B8. No other known copies.

D51 MUḤAMMAD B. MUḤYI L-DĪN AL-DIMYĀṬĪ, AL-NIMRA (?)

Not identified.

3.1.29 A commentary on the prayer-tables of Quṭb al-Maḥallī (no. D37): DM 991 (2 fols., 1089H by author).

D52 YŪSUF B. MUḤAMMAD AL-MAḤALLĪ AL-MASDĪ

Suter, no. 510; *GAL*, II, p. 217, and SII, p. 1025, no. 83.

2.3.8 ***Naẓm al-durar fī taqwīm al-shams wa-l-qamar***, an introduction in 6 *bābs* and tables for compiling solar and lunar ephemerides: K 4041 (20 fols., *ca.* 1100H); DM 458 (20 fols., *ca.* 1200H).

2.4.6 Tables of the hour-angle (*faḍl al-dāᵓir*) of the sun, moon and planets for computing conjunctions (?): MM 20,2 (fols. 5v–12r, 1154H), unique. Merits investigation.

2.4.9 Tables for calendar conversion: DM 127,2 (fols. 6v–7r, *ca.* 1100H).

4.4.17 ***Sharḥ al-Nujūm al-shāriqāt***, a commentary on the author's treatise on the almucantar quadrant, in 9 *bābs*: DM 536 (21 fols., *ca.* 1150H).

4.5.12 ***al-Shams al-muḍīᵓa ᶜala l-R. al-Fatḥīya***, a commentary on the treatise by Sibṭ al-Māridīnī: see no. C97.

D53 ᶜABD AL-QĀDIR B. IBRĀHĪM AL-MAḤALLĪ

Not listed in the modern sources.

6.4.34 ***Ḍābiṭ fī stikhrāj al-majhūlāt bi-l-aᶜdād al-mutanāsaba***, a short poem on arithmetical proportions: DR 899,2 (fols. 23v, *ca.* 1150H); plus a commentary by the author: ṬJ 635,1 (fols. 1v–2v, 1170H).

D54 ᶜUMAR B. AḤMAD AL-MĀᵓĪ WA-L-JALĪ

GAL, SIII, p. 735, *s.v.*

4.5.21 A commentary on the treatise on the sine quadrant by Yaḥyā al-Ḥaṭṭāb: see no. C119.

6.4.23 A commentary on al-ᶜĀmilī's *Khulāṣat al-ḥisāb*, compiled in 1089H: see no. G68.

D55 SHIHĀB AL-DĪN AḤMAD AL-KUTUBĪ AL-KHURFĀNĪ (?)

Not listed in the modern sources.

2.3.6 ***Faṣl fī maᶜrifat kayfīyat taqwīm ᶜUṭārid min al-Durr al-yatīm***, on the computation of the position of Mercury using tables based on the method of Ibn al-Majdī (see no. C62): ḤM 16,2 (fols. 33v–37v, *ca.* 1100H); DM 131,3 (fols. 93v–105r, 1146H); *etc.* (2 copies).

D56 ḤASAN AL-ṬABLĀWĪ

Not listed in the modern sources.

4.7.30 Some tables for marking sundials for latitudes 31° and 31;25°, Alexandria and Damietta: MM 184,3 (fols. 4v–7r, *ca.* 1100H by author).

D57 MUḤAMMAD B. ᶜALĪ AL-ḤAMĪDĪ

GAL, II, p. 472, and SII, p. 487.

4.2.9 ***R. fī Dhāt al-kursī***, a treatise on the armillary sphere in 18 *bābs*: ṬJ 524,2 (fols. 86v–97r, 1140H); MM 165,2 (fols. 58v–69r, *ca.* 1200H); *etc.* (7 copies).

4.3.39 ***Naḍrat al-lubāb fī sharḥ Bahjat al-albāb***: see no. H25.

D58 RIḌWĀN EFENDĪ B. ᶜABD ALLĀH AL-RAZZĀZ

GAL, II, p. 471 and SII, p. 487.

2.2.3 ***al-Zīj al-Mufīd ᶜalā uṣūl al-rasad al-jadīd***, a *zīj* for Cairo based on the parameters of Ulugh Beg (no. G49), with an introduction in 4 *maqālas* and tables: DM 1205,1 (fols. 1r–215r, 1250H); *etc.* (3 fragments).

2.2.3 Tables for the sun and moon based on the parameters of Ulugh Beg: K 4012 (22 fols., *ca.* 1200H).

2.2.3 ***al-Durr al-naẓīm fī ṣināᶜat al-taqwīm = al-Durr al-farīd ᶜala l-raṣad al-jadīd***, a *zīj* for Cairo: TR 188 (822 pp. !, *ca.* 1100H, tables in the author's hand, introduction in the hand of his student ᶜUthmān (no. D58X)).

2.2.3 Miscellaneous tables taken from the *zīj*es of Ibn al-Shāṭir (no. C30) and Riḍwān: K 3978 (22 fols., 1302H).

2.2.3 Tables for the longitude of Medina: ZK 97,2 (fols. 32v–43v, *ca.* 1300H).

2.3.6 ***al-Durr al-naẓīm fī ṣināᶜat al-taqwīm = al-Durr al-farīd ᶜala l-raṣad al-jadīd***, extensive planetary tables in the tradition of Ibn al-Majdī's *al-Durr al-yatīm* (no. C62) and based on the parameters of Ulugh Beg (no. G49): TR 188 (822 pp. !!, *ca.* 1100H, partly in author's hand); MM 83 (259 fols., *ca.* 1250H).

2.3.6 ***K. Asna l-mawāhib fī taqwīm al-kawākib*** (ditto): DM 802 (723 fols. !!, *ca.* 1100H in author's hand).

2.4.2 A star catalog for 1090H: DM 107,2 (fols. 46r–56r, 1101H); DM 746,4 (fols. 8r–20v, *ca.* 1200H); K 4039 (12 fols., *ca.* 1100H); MM 73 (11 fols., *ca.* 1100H).

2.4.8 Tables of geographical coordinates: DM 806 (4 fols., *ca.* 1300H).

2.5.9 ***Kalām kāfⁱⁿ fī maᶜrifat al-ijtimāᶜ wa-l-istiqbāl wa-l-kusūf wa-l-khusūf***, on the determination of eclipses: K 4019,1 (fols. 1r–9r, *ca.* 1150H). The work contains an eclipse computation for 1087H.

3.1.30 ***Dastūr uṣūl ᶜilm al-mīqāt wa-natījat al-naẓar fī taḥrīr al-awqāt***, an extensive treatise on timekeeping consisting of an introduction and tables lifted from the main Cairo corpus: DM 189 (80 fols., 1113H); DM 403 (80 fols., *ca.* 1100H); *etc.* (9 copies, mostly defective).

3.1.30 ***Natījat al-afkār fī maᶜrifat al-ᶜamal bi-l-layl wa-l-nahār***, a treatise on timekeeping consisting of an introduction in 4 *bābs* and tables lifted from the main Cairo corpus: DM 1206,1 (fols. 1r–54r, *ca.* 1125H); MM 152 (41 fols., *ca.* 1100H); *etc.* (5 copies, mostly defective).

3.1.30 ***Dastūr natījat al-muqaddimāt fī aᶜmāl ᶜilm al-mīqāt***, introduction in 5 *bābs*: DM 639,3 (fols. 33v–41v, 1100H in author's hand, complete?), unique.

3.1.30 Prayer-tables for latitude 21° (Mecca): DM 44,1 (fols. 1v–17r, *ca.* 1100H, anonymous, author's copy), unique.

3.2.29 ***Bughyat al-ṭālib fi stikhrāj al-aᶜmāl al-falakīya bi-l-ḥisāb***, a treatise on timekeeping with examples for Mecca (latitude 21°): K 4024,3 (fols. 10r–11v, 1290H).

3.4.10 ***Ṭirāz al-durar fī ruʾyat al-ahilla wa-l-ᶜamal bi-l-qamar***, on the determination of lunar crescent visibility: DM 639,28 (fols. 246r–249v + 252r–253r, *ca.* 1150H); K 4021,1 (fols. 1r–15v, *ca.* 1150H).

3.4.10 Tables of the oblique ascensions of the moon, declination, and half arc of visibility (for each degree of lunar latitude) for each degree of lunar longitude: K 4021,2 (fols. 16r–53v, *ca.* 1150H).

4.7.24 ***Bughyat al-sā'il fī waḍc al-mazāwil***, on sundial theory in 2 *qisms*, with tables for latitude 30°: TR 259 (16 pp., *ca.* 1100H, with additional tables for latitude 31°); DM 695 (10 fols., *ca.* 1200H, incomplete); ṬM 102,1 (fols. 1r–31r, *ca.* 1150H, with tables for latitude 41° rather than 30° not by Riḍwān); *etc.* (1 fragment).

Tables for vertical sundials with inclination 27° S and horizontal sundials for latitude 31;25° (Damietta): MM 191,2 (fols. 5r–6r, 1135H).

Tables for computing the coordinates of the *ʿaṣr* curve on a sundial for latitude 21° (Mecca): DM 44,2 (fols. 18v–19r, *ca.* 1100H by author).

Tables for various inclinations on vertical sundials for latitudes 30° and 31° computed by Riḍwān and Muḥammad al-Najāmī (no. D63): MM 85,2 (fols. 3r–8v, *ca.* 1150H).

4.7.30 Some tables for marking sundials for latitudes 31° and 31;25°, Alexandria and Damietta: MM 184,3 (fols. 4v–7r, *ca.* 1100H).

D58X ʿUTHMĀN, a student of RIḌWĀN EFENDĪ (no. D58)

He copied TR 188 of Riḍwān's *zīj*, *al-Durr al-naẓīm*, *ca.* 1100H.

D59 ʿABD AL-RAḤMĀN AL-QADDŪSĪ AL-MANZILĀWĪ, a student of RIḌWĀN EFENDĪ (no. D58)

2.4.2 A star catalog for 1111H: DM 669 (5 fols., *ca.* 1125H).

D60 ʿABD AL-LAṬĪF B. IBRĀHĪM, IBN AL-KAYYĀL

Suter, no. 474; *GAL*, II, p. 469.

3.1.9 Introduction to the prayer-tables of al-Khalīlī (see no. C37) and star catalog for 1100H in ṬM 218 (45 fols., *ca.* 1100H).

D61 AL-JAMĀLĪ YŪSUF B. YŪSUF, KALARJĪ of ḤASAN ĀGHĀ, and student of RIḌWĀN EFENDĪ (no. D58)

Not listed in *Suter* or *GAL*.

1.4.17 ***Kunz al-durar fī aḥwāl manāzil al-qamar***, a sophisticated treatise on the lunar mansions in 3 *bābs* with tables, compiled in 1133H: DM 484 (19 fols., 1135H by the author). Merits study.

3.1.31 ***Multaqaṭ al-durar wa-l-yawāqīt fi stikhrāj aʿmāl al-mawāqīt***, prayer-tables for latitude 35;30° (Crete), with an introduction in 12 *faṣls*: DM 834 (70 fols., 1120H by author?), unique. See *SATMI*, II for a brief analysis. The introduction concludes with the poem attributed to the *imām* al-Shāfiʿī: see no. B2.

4.7.25 ***K. Īḍāḥ al-ashkāl al-iʿtidālīya fī rasm al-sāʿāt ʿala l-asṭiḥa al-mustawīya***, an extensive treatise on sundial theory in 2 *maqālas* of 4 *qisms* with numerous diagrams and tables, compiled in 1127H: DM 677 (122 fols., *ca.* 1200H, lacks some diagrams), unique? Plus one fragment: DM 586,3. Merits study. Note added in proof: see now the newly-acquired copy K 26114.

D62 ʿABD AL-FATTĀḤ B. IBRĀHĪM AL-DAYSTĪ AL-MĀLIKĪ, a student of RIḌWĀN EFENDĪ (no. D58)

GAL, SII, p. 1017.

4.7.26 ***K. Aqrab al-wasā'il fī ʿamal al-mazāwil***, a treatise on sundial theory arranged in a *muqaddima* and 2 *qisms* of 8 + 9 *faṣls* and a *khātima*: DM 175,6 (fols. 31r–65v, in two hands, *ca.* 1200H). The *khātima* is copied separately as ***R. fī Waḍʿ al-āla al-musammā bi-l-ḥalzūn*** in DM 89 (2 fols., *ca.* 1200H).

Also, ***Asna l-ghāyāt fī ʿilm al-mīqāt***, prayer-tables for Lattakia (latitude 34;30°): MS Aleppo Awqaf 911 (28 fols., *ca.* 1175H).

6.4.24 ***al-Nubdha al-wāfiya bi-Sharḥ al-Muqaddima al-Sakhkhāwīya***, a commentary on al-Sakhkhāwī's arithmetic: see no. C131.

6.4.42 ***Kashf al-ḥijāb ʿan wajh Bughyat al-ṭullāb***, a commentary on a poem on arithmetic by the author: K 4310 (33 fols., 1190H).

D63 MUḤAMMAD AL-NAJĀMĪ

4.7.24 Tables for various inclinations on vertical sundials for latitudes 30° and 31° computed by Muḥammad al-Najāmī and Riḍwān Efendī (no. D61): MM 85,2 (fols. 3r–8v, *ca.* 1150H).

D64 MUḤAMMAD B. ʿABD AL-RAḤMĀN B. AḤMAD, IBN AL-JUNDĪ (AL-JANADĪ?)

Not listed in the modern sources.

3.2.28 *Kifāyat al-aḥbāb fī maʿrifat al-awqāt bi-l-ḥisāb*, a treatise on timekeeping in 10 *bāb*s, compiled in 1091H: DM 629 (24 fols., 1093H), unique. The author mentions al-Farghānī (no. B18), al-Marrākushī (no. C17), Ibn al-Shāṭir (no. C30), al-Wafāʾī (no. C61), al-Tājūrī (no. C110), and Khiḍr al-Birlisī (no. C130).

D65 ʿABD AL-MUʿṬĪ B. SĀLIM B. ʿUMAR AL-SAMALĀWĪ AL-SHĀFIʿĪ AL-AZHARĪ

GAL, II, p. 420, and SII, p. 444.

5.3.35 *R. fī Khusūf al-qamar wa-kusūf al-shams wa-l-raʿd wa-l-zalzala wa-dāʾirat al-qamar wa-l-nayrūz*, an astrological treatise in the Hermetic tradition: TGh 133 (19 fols., *ca.* 1200H).

D66 MUḤAMMAD B. ʿABD AL-MAḤMŪD AL-ḤAKĪM AL-LĀDHIQĪ

Not listed in the modern sources.

2.2.13 *Nihāyat al-tashīl li-l-ʿibāra wa-l-ikhtiṣār li-l-ghāya li-taqwīm al-kawākib al-sayyāra*, a small *zīj* for Lattakia: ṬM 92 (32 fols., *ca.* 1150H), unique. Merits investigation.

al-Lādhiqī mentions in his introduction that he based his work on a *zīj* entitled *Tashīl al-ʿibāra fī taqwīm al-kawākib al-sayyāra* by Ibn Yūnus (no. B59), and on the tables of Riḍwān Efendī (no. D58). It is of interest that there is a copy of a Maghribī *zīj* with the above title, attributed to Ibn Yūnus on the title folio, and copied in the hand of al-Lādhiqī himself, preserved in MS Aleppo 947. This Maghribī *zīj* merits investigation.

2.3.9 The *Khātima fī ʿamal dustūr muqawwam kull kawkab li-sana kāmila*, on the compilation of ephemerides stated to be taken from the *Zīj* of Ibn Yūnus (no. B59) in MM 116,2 (fols. 8r–9r, *ca.* 1150H) is in fact taken from the small *zīj* entitled *Tashīl al-ʿibāra . . .* which is falsely attributed to Ibn Yūnus on the title folio in MS Aleppo Awqāf 947 (see above). Since al-Lādhiqī was the copyist, he clearly thought the work he was copying was indeed by Ibn Yūnus.

D67 AḤMAD B. MUḤAMMAD ʿAYYĀD AL-MAḤALLĪ, a *muwaqqit*

2.1.17 *Tashīl* or *Taysīr al-maṭālib fī taʿdīl al-kāwakib*, a small *zīj* for Cairo, compiled in 1110H: see no. F23. See also no. C13.

4.3.35 A commentary on the treatise on astrolabe construction by al-Hattārī: see no. D39.

D68 MISCELLANEOUS (almanacs)

5.5.2 An astrological almanac for 1113–14H: MM 29 (17 fols., *ca.* 1113H). See also MM 189,2 (fols. 35v–109v, *ca.* 1100H), which contains several almanacs of this kind, and TR 306, 192, and 307 for other later almanacs.

D69 SHAMS AL-DĪN MAḤMŪD B. AḤMAD AL-AWFĪ AL-ḤIJĀZĪ

Cf. GAL, II, p. 427.

2.3.7 *R. fī Sharḥ kayfīyat istikhrāj al-taqwīm*, a treatise on the compilation of ephemerides in a *muqaddima* of 3 *baḥth*s and a *maqāla* of 13 *mabḥath*s: ṬJ 582,1B (fols. 1r–28v, 1180H); ṬM 227,2 (fols. 41v–71v, 1145H); *etc.* (5 copies, including one dated 1130H).

4.6.42 *Mirʾat al-ʿajāʾib fi l-ʿamal bi-l-jayb al-ghāʾib*, on the use of the "concealed" trigonometric grid in 26 *bāb*s, compiled in 1110H: DM 1082,4 (fols. 52v–61v, 1174H). Another copy is MS Alexandria Municipal Library D 4865.

6.3.8 An extended commentary on Ibn al-Yasamīn's poem on algebra (see no. F12): DR 1038 (105 fols., 1250H, lacks beginning and end). This commentary is attributed to Shams al-Dīn al-Ḥijāzī in a late hand.

D70 YAḤYĀ AL-BIRSHAMSĪ

2.4.2 A star catalog for 1114H adapted from that of Ibn Sahrāb (no. D36): MM 36 (4 fols., 1114H).

D71 MŪSĀ B. MUḤAMMAD B. MŪSĀ AL-QULAYBĪ AL-MĀLIKĪ AL-GHAMRĪ

GAL, SII, p. 487. See Pl. CXIc.

5.3.35 *Qāᶜida yuᶜrafu minha l-ḥawādith fī mustaqbal al-zamān*, on the prediction of future events by the day of the week on which the *Nuzūl al-nuqṭa* occurs: DM 79,1 (fols. 1v–4r, *ca.* 1150H). Two copies of a closely related *R. laṭīfa fī Maᶜrifat al-ḥawādith allatī taqaᶜu fī kull sana ᶜinda nuzūl al-nuqṭa* are: DM 973,2 (fols. 9v–10v, 1285H); K 4051,2 (fols. 9r–12v, *ca.* 1300H).

Other treatises on astrology and magic are his *al-Zāᵓiraja al-ᶜajamīya wa-l-ṭarīqa al-ᶜadadīya*: Sh 60,1 (fols. 1r–4v, *ca.* 1150H) and *al-Zāᵓiraja al-tusāᶜīya al-ḥarfiya*: Sh 60,2 + 3 (fols. 5r–6v and 5v–6v, 1143H by author).

D72 ANONYMOUS

3.4.11 Tables for predicting the visibility of the lunar crescent for each month for 1125H to 1130H: DṢ 166,2 (fols. 40r–41r, 1125H, fine *naskhī*), unique. See Pl. XLVIa. Investigated by D. A. King in the Memorial Volume for Malcolm Kerr.

D73 ᶜALĪ B. MUḤAMMAD AL-DĀGHISTĀNĪ

4.4.7 al-Dāghistānī is perhaps the author of the anonymous commentary on *al-Warāqāt* by al-Māridīnī (no. C47), which he copied: ṬJ 811,17 (fols. 151v–155v, 1135H).

D74 MUḤAMMAD B. MUḤAMMAD AL-FALLĀTĪ AL-KISHNAWĪ AL-ASHᶜARĪ AL-MĀLIKĪ AL-SŪDĀNĪ

GAL, I, p. 669, and SI, p. 924.

5.3.4 *al-Durr al-manẓūm wa-khulāṣat al-Sirr al-maktūm*: a commentary on the treatise on astrology and magic by al-Rāzī (no. G12): DḤ 14 (241 fols., 1246H).

D75 MUḤAMMAD AL-GHAMRĪ AL-FALAKĪ AL-SHĀFIᶜĪ

GAL, II, p. 471, and SII, p. 487.

2.5.15 *al-Muqtaṭafāt al-fikrīya ᶜala l-dāᵓira al-taᵓrīkhīya*, on the use of a graduated circle for converting dates in various calendars: DM 156 (4 fols., 1161H by author; DM 539 (14 fols., 1161H by author); *etc.* (1 copy).

3.4.12 *Muᶜarrabat sanat 1154H*, an almanac giving information about the lunar crescents at the beginning of each month and corresponding dates in various calendars: K 4014 (7 fols., 1154H).

6.4.41 *R. fī Kayfīyat al-ᶜamal fi stikhrāj al-majhūl*, on problems in algebra: DR 1041 (4 fols., *ca.* 1125H).

6.5.3 *Raqāᵓiq al-asrār fī ḥisābay daraj wa-daqāᵓiq aᶜẓam dawwār*, a treatise on sexagesimal arithmetic in 4 *bābs*: MM 124 (15 fols., *ca.* 1100H).

6.8.12 *al-Qawāᶜid al-ḥisābīya fī taḥwīlāt al-akyās al-Rūmīya ila l-akyās al-Miṣrīya*, on converting Syrian and Egyptian weights and measures: MR 22 (5 fols., 1151H).

6.8.12 *al-Asrār al-ḥisābīya wa-l-qawāᶜid al-ilhāmīya fi stikhrāj fiḍḍat al-yawm*, on measures of silver, compiled in 1107H: DR 1043 (11 fols., 1110H); *etc.* (1 copy).

6.8.12 *al-Qawāᶜid al-muqniᶜa fī taḥwīlāt al-maqādīr al-arbaᶜa*, on weights and measures, compiled in 1124H: ṬR 144,3 (fols. 17v–19v, 1196H).

6.8.12 *Muzīlat jahl al-jahala bi-miqdār mā fi l-muthamman al-kullī min masᶜari (?) l-masᵓala*, on weights and measures: DR 1042 (4 fols., *ca.* 1100H).

D76 ABU L-RIḌĀ ᶜABD AL-LAṬĪF B. AḤMAD AL-KUTUBĪ AL-DIMASHQĪ

GAL, SI, p. 558; *Azzawi*, pp. 321–322.

3.1.32 *al-Manhaj al-aqrab li-taṣḥīḥ mawḍiᶜ al-ᶜaqrab*, time-keeping tables for Cairo, with entries expressed according to the Turkish convention, compiled in 1150H: DM 1104 (36 fols., 1211H); ṬJ 811,8 (fols. 61v–78r, 1198H); *etc.* (6 copies). Some of these manuscripts have the prayer-tables of al-Khawānikī (no. D78) appended.

3.5.13 ***al-Anfiḥa al-miskīya fī masʾalat al-Makkīya***, a response to a question from some scholars in Mekka about the time difference between sunrise in the furthest east and the furthest west, compiled in 1140H: DM 1146,1 (fols. 1r–2v, 1183H); *etc.* (1 copy).

4.7.28 ***al-Alfāẓ al-mūjaza fi l-waḍʿīyat bi-l-handasa***, a treatise on sundial theory in 2 *bābs*: DM 723,1 (fols. 1r–8v, *ca.* 1100H); DM 1146,2 (fols. 3r–8v, 1183H, incomplete); *etc.* (1 copy).

6.4.42 ***Sharḥ Urjūza fī ḥall al-aʿdād***, a commentary on the author's poem on arithmetic: MR 10 (5 fols., 1132H); MR 16 (5 fols., *ca.* 1200H); *etc.* (1 copy).

6.7.2 ***Nukhbat al-Tuffāḥa***, a poem based on ***al-Tuffāḥa*** by Abu l-Ḥasan al-Ashʿarī (no. E3): no copies. Commentary by al-Dimashqī: DR 310 (26 fols., *ca.* 1200H); *etc.* (3 copies).

D77 AḤMAD B. ʿUMAR AL-DAYRABĪ AL-SHĀFIʿĪ AL-GHUNAYMĪ

GAL, II, p. 421, SI, p. 677, and SII, p. 445.

6.9.16 ***Fatḥ al-malik al-jawād bi-tasḥīl qismat al-tarakāt ʿalā baʿḍ al-ʿibād***, on inheritance, compiled in 1139H: TR 291 (37 pp., 1166H); ZK 778,4 (pp. 84–127, 1243H). Brockelmann lists other copies in Cairo and Alexandria.

D78 RAMAḌĀN B. ṢĀLIḤ AL-KHAWĀNIKĪ

GAL, II, pp. 471–472, and SII, p. 48; *Azzawi*, pp. 320–321.

2.2.3 Planetary tables for longitude 65° and 64;55° (Cairo) based on the parameters of Ulugh Beg (no. G49): DM 1197 (141 fols., 1226H); DM 706 (13 fols., *ca.* 1125H by author, Mercury tables only).

2.2.3 ***Nuzhat al-nafs bi-taqwīm al-shams***, solar longitude tables: K 3984 (15 fols., 1231H); *etc.* (1 copy); and a modified version thereof by Ḥasan al-Jabartī (no. D91): TR 300,1 (pp. 1–32, *ca.* 1150H by author).

2.3.6 Tables of the lunar equation based on the method of Ibn al-Majdī's ***al-Durr al-yatīm*** (see no. C62): MM 133 (90 fols., *ca.* 1150H in the author's hand).

2.4.2 A star catalog for 1139H: DM 76 (12 fols., *ca.* 1150H by author); ṬJ 811,12 (fols. 124r–132v, 1198H).

2.5.7 ***al-Itḥāf ʿalā Nubdhat al-isʿāf***, a commentary on a treatise by Ibn Abi l-Fatḥ al-Ṣūfī: see no. C98.

2.5.10 ***al-Kalām al-maʿrūf fī aʿmal al-khusūf***, on the determination of lunar eclipses: MM 157 (9 fols., *ca.* 1125H); MM 186,1 (fols. 1r–12r, *ca.* 1150H).

2.5.10 ***al-Qawl al-muḥkam fī maʿrifat kusūf al-nayyir al-aʿẓam***, on the determination of solar eclipses, compiled in 1130H: MM 186,2 (fols. 12v–18r, *ca.* 1150H); DM 166,1 (fols. 1r–5v, *ca.* 1135H).

2.5.10 ***Kashf al-ghayāhib ʿan mushkilāt al-kawākib***, a treatise on the fixed stars, with some tables: DM 486,2 (fols. 1r–10v, *ca.* 1140H); DM 501 (10 fols., *ca.* 1250H, lacks tables).

3.1.32 Prayer-tables for Cairo: DM 812,2 (fols. 29r–33v, *ca.* 1150H); ṬJ 811,9 (fols. 78v–79v, *ca.* 1150H); *etc.* (1 copy). See also al-Dimashqī (no. D76).

3.2.31 ***Kifāyat al-ṭālib li-ʿilm al-waqt wa-bughyat al-rāghib fī maʿrifat al-dāʾir wa-faḍlihi wa-l-samt***, a treatise on timekeeping in 23 *bābs*: DM 574 (54 fols., *ca.* 1150H); DM 114 (63 fols., *ca.* 1150H).

3.2.31 ***Nubdha mufīda fī maʿrifat istikhrāj al-dāʾir wa-faḍlihi wa-taʿdīlihimā wa-taʿdīl al-irtifāʿ***, on some of the operations of spherical astronomy: K 3024,1 (fols. 1v–2r, 1290H).

3.4.12 ***Rashf al-zalāl fī maʿrifat istikhrāj qaws makth al-hila***, a treatise on lunar crescent visibility in 2 *nawʿ*s, without tables: DM 535 (11 fols., *ca.* 1150H).

4.7.30 Tables for marking sundials for latitude 30°, Cairo: DM 498 (4 fols., *ca.* 1100H by author).

6.5.3 ***Maṭāliʿ al-budūr fi l-ḍarb wa-l-qisma wa-l-judhūr***, a short treatise on sexagesimal arithmetic: MR 31 (9 fols., 1153H).

See also no. B22 (5.2.1).

D79 ANONYMOUS

2.5.8 A treatise on the computation of eclipses and lunar crescent visibility: DM 1147 (14 fols., *ca.* 1150H). The anonymous author mentions Ibn Yūnus (no. B59), Ibn al-Shāṭir (no. C30); Ibn al-Majdī (no. C62); Aḥmad al-Ḥalabī (no. C69); and Ulugh Beg (no. C49). See also al-Jurdī (no. D82). Another copy of this treatise, now followed by a set of planetary tables, is MS Dublin Chester Beatty 4088.

D80 ḤUSAYN AL-QABBĀNĪ

2.4.2 A star catalog for 1176H: K 4034 (6 fols., 1822 A.D.).

D81 MUṢṬAFĀ ṢIDQĪ B. ṢĀLIḤ

Not listed in the modern sources. He copied both of the precious mathematical compendia MR 40 and 41, as well as ṬJ 635. See Pls. LXVII and LXIX.

4.1.8 Drawings of various instruments: MR 40,2 (fols. 40v–64r, 1160H by Muṣṭafā Ṣidqī). Merits detailed study. See Pl. LXVII.

6.1.8 On his *iṣlāḥ* of *Thābit b. Qurra's* translation of the *K. ᶜAmal al-dāʾira* . . . attributed to Archimedes, see no. A7.

6.2.24 An edition of *K. Istikhrāj al-awtār* . . . by al-Bīrūnī (no. B78): MR 41,11 (fols. 74v–82r, 1153H by Muṣṭafā Ṣidqī). This copy should be compared with the other versions of this work.

D82 YŪSUF B. AḤMAD AL-JŪRDĪ (?) AL-WANĀ (?) AL-SHĀFIᶜĪ AL-AZHARĪ

Not identified.

2.5.8 *Kashūfāt al-adilla fī maᶜrifat al-khusūfāt wa-l-ahilla*, a treatise on the computation of eclipses and lunar crescent visibility in 5 *maqāla*s subdivided in *faṣl*s, compiled in 1162H: DM 641 (40 fols., *ca.* 1175H), unique. See also nos. D79 and D159.

4.4.24 See no. D230.

D83 MUṢṬAFĀ B. MUḤAMMAD AL-ṬĀʾĪ AL-ḤANAFĪ

GAL, SII, p. 267.

6.4.42 *al-Durra al-Ṭāʾīya fī l-uṣūl al-arithmāṭīqīya*, a treatise on arithmetic compiled in 1161H: K 9659,3 (fols. 39v–45v, 1198H). See also MS Princeton Garrett 1053.

D84 MUḤAMMAD ABŪ SHAHYA AL-MANFALŪṬĪ AL-SHĀFIᶜĪ AL-SHĀDHILĪ AL-AZHARĪ

6.4.24 *al-Durra al-bahīya bi-ḥall alfāẓ al-Sakhkhāwīya*, a commentary on al-Sakhkhāwī's arithmetic (see no. C131), compiled in 1163H: MR 12 (69 fols., *ca.* 1200H).

6.4.24 *al-Durra al-sanīya ᶜalā Fatḥ rabb al-barīya*, a commentary on al-Maḥallī's commentary on al-Sakhkhāwī's arithmetic (see nos. C131 and D85), compiled 1161H: DR 350 (74 fols., 1161H by the author).

D85 ḤUSAYN B. MUḤAMMAD AL-MAḤALLĪ

GAL, SII, p. 483.

6.4.1 *Kashf al-astār ᶜan Nuzhat al-ghubār*, a commentary on *al-Nuzha* by Ibn al-Hāʾim, compiled in 1163H: see no. C58.

6.4.24 *Fatḥ rabb al-barīya ᶜalā matn al-Sakhkhāwīya*, a commentary on al-Sakhkhāwī's treatise on arithmetic (see no. C131), compiled in 1138H: DR 616 (33 fols., 1296H); *etc.* (9 copies); plus a commentary by al-Manfalūṭī (no. D84), entitled *al-Durra al-sanīya* . . . : DR 350 (74 fols., 1161H by the author).

6.4.42 *Jadāwil al-ghurbāl fī bayān al-aᶜdād al-murakkaba*, tables of factors with a brief introduction: DR 316 (55 fols., *ca.* 1200H).

D86 AḤMAD B. AḤMAD AL-SAJJĀᶜĪ

GAL, II, pp. 422–423, and SII, pp. 445–446.

1.4.22 A commentary on his own poem on the lunar mansions: ṬM 213 (5 fols., 1285H).

3.2.14 ***Fatḥ al-ᶜalīm al-qādir bi-sharḥ Laqṭ al-jawāhir***, a commentary on ***Laqṭ al-jawāhir*** by Sibṭ al-Māridīnī compiled in 1183H: see no. C97.

3.2.33 ***Hidāyat ūli l-baṣāʾir wa-l-abṣār ilā maᶜrifat ajzāʾ al-layl wa-l-nahār***, a commentary on a poem on timekeeping by Aḥmad ᶜAyyād (no. D67): MM 242 (25 fols., 1185H); DM 181,5 (fols. 49v–104v, *ca.* 1200H); *etc.* (6 copies).

6.3.8 ***Fatḥ dhi l-ṣifāt al-sanīya bi-sharḥ matn al-Yasamīnīya***, a commentary on the poem of Ibn al-Yasamīn: see no. F12.

6.4.44 A poem on fractions: ZK 778,3 (p. 83, 1243H). See also no. D89.

D87 ᶜALĪ B. SAᶜD AL-BAYSŪSĪ (?), a student of AL-SAJJĀᶜĪ (no. D86).

4.5.30 ***al-Nafaḥa al-zakīya fī ᶜamal bi-l-jiha al-jaybīya***, a poem on the sine quadrant compiled in 1184H: MM 237 (8 fols., 1184H by author).

D88 ANONYMOUS, also a student of AL-SAJJĀᶜĪ (no. D86)

4.5.30 ***Manẓūma fi l-ᶜAmal bi-l-rubᶜ al-mujayyab***, a poem on the use of the sine quadrant: K 3980 (7 fols., *ca.* 1200H).

D89 ANONYMOUS

6.4.44 An anonymous commentary on an anonymous poem on fractions (not that of al-Sajjāᶜī (no. D86)): ZK 778,5 (pp. 128–130, 1243H).

D90 MUḤAMMAD B. ᶜABD AL-RAḤĪM . . . AL-MUKHALLALĀTĪ

Not mentioned in the modern sources.

2.1.19 ***Nuzhat al-nāẓir bi-khtiṣār Zīj Ibn al-Shāṭir***, an abridgement of the ***Zīj*** of Ibn al-Shāṭir (no. C30) with an introduction in 15 *bāb*s and tables: MJ 7,2 (fols. 44r–73r, *ca.* 1150H, no tables); ṬJ 515,4 (fols. 44r–73r, *ca.* 1150H, no tables); ṬJ 515,4 (fols. 55r–80r, *ca.* 1200H, disordered, incomplete).

D91 ḤASAN B. IBRĀHĪM AL-JABARTĪ

GAL, II, p. 472, and SII, p. 487.

2.2.3 A recension of some solar longitude tables by al-Khawānikī (no. D78): TR 300,1 (pp. 1–32, *ca.* 1150H by author).

3.1.32 An almanac and prayer-tables for Cairo: DM 789 (7 fols., *ca.* 1200H); TR 346 (28 pp., *ca.* 1275H).

3.3.14 ***Bulūgh al-amal fī kayfīyat al-istiqbāl***, a non-mathematical treatise about facing the qibla: TR 155 (22 pp., *ca.* 1200H).

4.4.25 ***Akhṣar al-mukhtaṣarāt ᶜalā rubᶜ al-muqanṭarāt***, a treatise on the use of the almucantar quadrant arranged in *muqaddima*, *maqṣad*, and *khātima*: DM 462 (4 fols., 1156H).

4.5.12 ***al-Thamarāt al-majnīya (?) min abwāb al-Fatḥīya***, and ***al-Nasamāt al-fayḥīya (?) ᶜala l-Risāla al-Fatḥīya***, abridgements of the treatise by Sibṭ al-Māridīnī: see no. C97.

4.6.46 ***ᶜUjāla ᶜalā aᶜdal āla***, a treatise in a *muqaddima* and *maqsad* on the "equatorial (semi-) circle": TR 256 (12 pp., 1174H); ZK 402 (10 fols., 1288H).

4.7.27 ***al-R. al-Mufaṣṣaḥa fīmā yataᶜallaqu bi-l-asṭiḥa***, on sundial theory in 3 *bāb*s: TR 343,1 (pp. 1–55, *ca.* 1250H); DH 97 (19 fols., *ca.* 1250H, incomplete).

4.7.27 ***Bāb fī ḥisāb sumūt munḥarifāt 53° janūb bi-ᶜarḍ 30°***, with tables: MM 178,2 (fols. 3r–4r, 1183H); *etc.* (2 deficient copies).

4.7.27 Tables for drawing horizontal sundials: DM 721 (7 fols., *ca.* 1125H perhaps by author).

4.7.27 Tables for marking vertical sundials by Muḥammad Chelebī (no. D93) and Ḥasan al-Jabartī: DM 612 (6 fols., *ca.* 1150H).

5.3.35 ***R. fī Maᶜna l-haylaj wa-l-katkhudāh wa-rabb al-sana wa-l-sālkhudah wa-ma ashbaha dhālika***, a short treatise on specific concepts of astrology: Sh 82 (4 fols., *ca.* 1300H, purportedly 1147H).

6.4.42 ***Rafᶜ al-ishkāl fī ghālib al-ashkāl***, a treatise on mensuration: K 10970 (15 fols., *ca.* 1300H).

6.5.1 ***Ḥaqāʾiq al-raqāʾiq***, a commentary on Sibṭ al-Māridīnī's ***Raqāʾiq al-ḥaqāʾiq***: see no. C97.

6.8.12 ***al-ᶜIqd al-thamīn fīmā yataᶜallaqu bi-l-mawāzīn***, a treatise on the balance arranged in a *muqaddima*, *maqṣad*, and *tatmīm*, compiled in 1180H: ṬR 121 (55 fols., 1181H); ṬR 344 (97 pp., 1207H); *etc.* (5 copies).

D92 ᶜABD AL-RAḤMĀN B. ḤASAN AL-JABARTĪ

GAL, II, p. 632, and SII, pp. 730–731; article "al-Djabartī" by D. Ayalon in EI_2

2.2.3 Solar longitude tables: ṬR 300,2 (pp. 35–73, 1198H by author).

2.3.12 *Dastūr taqwīm al-kawākib al-sabᶜa . . .*, an ephemeris for 1209H: MM 86 (7 fols., *ca.* 1208H, may be in author's hand).

D93 MUḤAMMAD CHELEBĪ AL-SHURAYBĪ

4.7.27 Tables for marking vertical sundials by Muḥammad Chelebī and Ḥasan al-Jabartī (no. D91): DM 612 (6 fols., *ca.* 1150H).

4.7.30 Tables for marking sundials for latitudes 30°, Cairo, and 24°, Medina: MM 64 (10 fols., 1173H by author) and MM 70 (5 fols., *ca.* 1175H).

D94 MUḤAMMAD AL-NAFRAWĪ, a student of ḤASAN AL-JABARTĪ (no. D91)

4.7.27 Tables for marking horizontal sundials for latitude 30°: DM 466 (7 fols., 1150H by author).

D95 ᶜABD AL-ṢAMAD, *muwaqqit* at the Muᵓayyad Mosque in Cairo

3.1.25 A star catalog appended to al-Lādhiqī's *Natījat al-afkār* (see no. C132) in MS DM 459 (*ca.* 1200H).

D96 MAḤMŪD EFENDĪ B. SULAYMĀN JUMLIYĀN (?)

Not mentioned in the modern sources. He was the owner of several astronomical manuscripts: see Pl. CXb.

6.6.5 *K. al-Nisba al-sittīniya maḥlūla daqīqa daqīqa*, extensive sexagesimal multiplication tables without introduction: K 3992 (120 fols., *ca.* 1150H, incomplete). See also no. C162.

D97 MUḤAMMAD B. ᶜALĪ AL-ṢABBĀN

GAL, II, pp. 371–372, and SII, pp. 399–400.

1.2.23 *K. fī Fann al-hayᵓa*, a treatise on theoretical astronomy in 3 *maqāla*s, based on al-Jaghmīnī (no. G17), Qāḍi Zāde (no. G50), and ᶜAlī Qūshjī (no. G51): DH 6 (60 fols., *ca.* 1250H, lacks 2nd and 3rd *maqāla*s), unique.

D98 AḤMAD B. ISMĀᶜĪL ĀGHĀ AL-ḤANAFĪ, BALGHIYĀR (?) AL-FALAKĪ

Not mentioned in the modern sources.

5.3.34 *al-Durr al-yatīm fī gharīb al-tanjīm*, a treatise on astrology in 110 *bāb*s: ṬM 109 (237 fols., *ca.* 1150H), unique. The author, on whom I have no further information, mentions the following authors in his introduction: Ibn Abi l-Rijāl (no. B76); al-Israᵓilī (no. B15); al-Ṭabari (no. B6); al-Qaṣrānī (no. B21); al-Ghaznawī (no. G33); *al-Dawr al-ᶜUṭaridī* by Ḥasan b. Jaᶜfar al-Anṣārī (no. B49); Sulaymān al-Falakī (no. C117); and al-Qummī (no. B60).

MM 200 was copied by a student (*tābiᶜ*) of Balghiyār *al-falakī* in 1187H.

D99 AḤMAD B. ᶜABD AL-MUNᶜIM AL-DAMANHŪRĪ

GAL, II, pp. 487–488, and SII, pp. 498–499.

5.3.35 *R. fī ᶜIlm aḥkām al-nujūm*, a treatise on astrology: DM 798 (8 fols., 1213H); DJ 155,2 (fols. 15v–28v, 1228H).

5.3.35 *al-Nujūm al-zāhira*, a treatise on astrology: DH 93 (8 fols., *ca.* 1300H).

6.4.42 *Iḥyāᵓ al-fuᵓād bi-maᶜrifat khawaṣṣ al-aᶜdād*, a treatise on number theory: TR 86 (51 pp., *ca.* 1150H).

6.7.9 *ᶜAyn al-ḥayā(t) fī ᶜilm istinbāṭ al-miyāh*, a treatise on surveying in 2 *bāb*s: DR 556 (19 fols., 1150H); *etc.* (1 fragment).

D100 IBRĀHĪM B. MUḤAMMAD AL-ZAMZAMĪ

GAL, II, p. 516, and SII, p. 538.

4.4.27 ***Wasīlat al-thiqāt bi-fahm ālāt al-muqanṭarāt***, a poem on the use of the almucantar quadrant: DM 154,2 (fols. 2v–5v, *ca.* 1200H).

4.5.30 ***Manẓūma fi l-ᶜamal bi-l-rubᶜ al-mujayyab***, a poem on the use of the sine quadrant: DM 154,1 (fols. 1v–2v, *ca.* 1200H).

D101 ABU L-ITQĀN MUṢṬAFĀ AL-WAFĀʾĪ AL-KHAYYĀṬ

Not listed in the modern sources.

2.2.3 Planetary tables for Cairo based on the parameters of Ulugh Beg: DM 726 (5 fols., *ca.* 1200H).

2.3.9 ***Fāʾida fī Maᶜrifat buᶜd awwal ᶜuqd min ᶜuqūd al-ibtidāʾ wa-man yaqūmu maqāmahu fi l-darari l-khamsa ghayr ᶜUṭārid wa-l-qamar***, notes on the compilation of ephemerides: MM 94,2 (fols. 6r–7v, 1210H); *etc.* (1 defective copy).

2.4.2 A star catalog for 1180H: K 4030,1 (fols. 1r–9r, 1222H); MM 60 (10 fols., *ca.* 1200H).

2.4.5 Tables for computing the descensions of the moon: DM 13,2 (fols. 2v–3r, *ca.* 1100H). See also DM 57,2.

3.1.25 A star catalog for 1180H appended to the ***Natījat al-afkār*** of al-Lādhiqī (no. C132) is in DM 1217 (1197H), and a different one in MM 14 (1250H).

3.3.14 Calculation of the qibla at Cairo by the method of Kūshyār (no. B70): DM 639,17 (fol. 158v, 1188H by the author for his student Ḥasan Chelebī (no. D103); MM 213,5 (fol. 30v, 1169H). Calculation of the qibla at Cairo by the method of Ulugh Beg (no. G49): MM 213,3 (fols. 28v + 30r, 1169H).

D102 MUḤAMMAD ᶜUTHMĀN, a student of MUṢṬAFĀ AL-KHAYYĀṬ (no. D101)

4.3.29 Tables for marking the curves on astrolabe plates: DM 640,3 (fols. 23v–35v, *ca.* 1200H by author); *etc.* (2 copies).

D103 ḤASAN CHELEBĪ, a student of MUṢṬAFĀ AL-KHAYYĀṬ (no. D101)

See no. D101.

D104 MUḤAMMAD AL-ḤIFNĪ

6.3.8 ***Fawāʾid ᶜawāʾid jabrīya ᶜalā sharḥ al-Sibṭ li-l-Yasamīnīya***, notes on Sibṭ al-Māridīnī's commentary to Ibn al-Yasamīn's poem on algebra, compiled in 1167H: see no. C97.

D105 MUḤAMMAD AL-MUNĪR AL-SAMĀNŪDĪ AL-AZHARĪ AL-SHĀFIᶜĪ

3.5.14 A poem on the zodiac, compiled in 1193H: ṬM 194 (5 fols., *ca.* 1200H); *etc.* (1 copy).

D106 ᶜABD AL-RAḤMĀN AL-ṬANĀNĪ

Not listed in the modern sources.

3.1.25 A star catalog for 1187H appended to the ***Natījat al-afkār*** of al-Lādhiqī (see no. C132) is in MM 222, DM 307, and MM 47, *etc.*

D107 ᶜUTHMĀN B. SĀLIM AL-WARDĀNĪ

GAL, II, p. 472.

2.3.6 ***al-Silk al-qawīm fī maᶜrifat al-taqwīm min al-Durr al-yatīm***, on the use of the auxiliary tables of Ibn al-Majdī (no. C62) for compiling ephemerides: DM 98 (6 fols., 1219H); *etc.* (3 copies).

2.5.9 A treatise on the lunar eclipse of Shaᶜbān 1182H and related astrological predictions: MM 212 (4 fols., *ca.* 1200H).

3.4.12 ***Tashīl al-maqāl fī maᶜrifat al-ᶜamal bi-l-qamar wa-ruʾyat al-hilāl***, a treatise on lunar crescent visibility in 3 *maqāla*s, without tables: DM 564 (4 fols., 1192H by the author).

D108 MUḤAMMAD SANNĀR, a *muwaqqit* in the Aḥmadī Mosque (in Tanta)

4.5.28 ***Ghāyat al-maṭlūb fī maᶜrifat al-aᶜmāl bi-l-juyūb***, a treatise on the sine quadrant in 50 *bāb*s, compiled in 1208H: ṬM 115 (20 fols., 1220H). See also no. D230.

D109 MUḤAMMAD B. ḤUSAYN AL-ᶜAṬṬĀR

Not listed in the modern sources.

4.7.28 *R. fī Fann al-mazāwil*, a treatise on sundial theory in 3 *bāb*s: DM 1026 (9 fols., 1299H), no other known copies. The treatise contains a table for marking prime vertical sundials for each ½° of latitude from 20° to 45° and information on some of the sundials in Damascus which the author had seen. Merits study.

6.7.9 *K. al-Nabāh fī ᶜilm al-miyāh*, a treatise on surveying in 1 *aṣl* and 2 *far*ᶜs, compiled in 1212H: K 4159,3 (fols. 64r–82v, 1212H); TR 156 (19 pp., *ca.* 1215H).

6.8.12 *K. Ẓarāʾif al-afnān fī maᶜrifat mā yakhuṣṣu l-qabbān*, a treatise on the balance in 2 *maqṣad*s, compiled in 1217H: K 4159,2 (fols. 55r–63v, 1212H); DR 645 (14 fols., 1287H).

D110 MUḤAMMAD AMĪN B. ᶜUMAR ... B. ᶜĀBIDĪN

GAL, II, p. 648, and SII, pp. 773–774 (esp. nos. 13 and 31).

3.4.12 *Tanbīh al-ghāfil wa-l-wasnān ᶜalā aḥkām hilāl Ramaḍān*, on the religious obligation of sighting the lunar crescent: B 22641 (16 fols., 1284H). Brockelmann lists another copy in MS Damascus Ẓāhirīya 1301, which is a collection of works by this author. Merits study.

D111 MUḤAMMAD AL-KHUḌRĪ

Not listed in the modern sources.

2.1.19 A commentary on *al-Lumᶜa* (without tables) by al-Kawm al-Rīshī (no. C41), compiled in 1239H: DM 94 (92 fols., 1239H); KhM 3 (111 fols., 1240H); *etc.* (10 copies, some incomplete); plus abridgements by Shihāb al-Dīn Ibrāhīm b. Ḥasan (no. D112): DM 638 (70 fols., 1254H, by the author); and by an anonymous author entitled *al-Rawḍa al-bahīya fī taqwīm al-kawākib al-sanīya* and containing some tables from Ibn al-Shāṭir's *Zīj*: DM 979,1 (fols. 1v–23r, *ca.* 1325H).

4.7.7 A commentary on Ibn al-Majdī's *Zād al-musāfir* (see no. C62), compiled in 1240H: DM 1005 (108 fols., *ca.* 1275H, with additional sundial tables for latitude 31;25°, Damietta); *etc.* (3 copies).

D112 SHIHĀB AL-DĪN IBRĀHĪM B. ḤASAN

Not listed in the modern sources.

2.1.19 An abridgement of al-Khuḍrī's commentary on al-Kawm al-Rīshī's *al-Lumᶜa* (see no. D111): DM 638 (70 fols., 1254H by the author).

3.1.29 A commentary on the introduction to the prayer-tables of Quṭb al-Dīn al-Maḥallī, compiled in 1284H: see no. D37.

D113 SHIHĀB AL-DĪN AḤMAD AL-MARZŪQĪ

Not listed in the modern sources.

3.2.33 *Natījat al-mīqāt fīmā li-ᶜarḍ al-lām min al-awqāt*, a poem on timekeeping compiled in 1235H: TR 56,2 (pp. 39–46, 1270H). Commentary on al-Ibrāshī (no. D114), entitled *al-Anwār al-sāṭiᶜāt* . . . : TR 76 (76 pp., 1298H), *etc.* (1 fragment).

D114 MUḤAMMAD B. IBRĀHĪM AL-IBRĀSHĪ

Not listed in the modern sources.

3.2.33 *al-Anwār al-sāṭiᶜāt ᶜalā natījat al-mīqāt* . . . , a commentary on a poem by al-Marzūqī: see no. D113.

D115 MUḤAMMAD ᶜABD AL-ḤAYY AL-SHĪBĪNĪ

6.9.16 *Manjiyāt al-afkār min dawrīyāt al-iqrār*, on inheritance: DR 1095 (10 fols., 1253H by author); *etc.* (1 copy).

D116 ᶜABD AL-GHANĪ, *khādim al-sāda al-Wafāᵓīya*

2.3.12 / 3.4.12 A *muᶜarraba* displaying information on the new moons of the solar year 1240–41 H: DM 824,2 (fols. 5v–28r, *ca.* 1240H); DM 1221 (23 fols., *ca.* 1240H).

D117 MUṢṬAFĀ AL-DHAHABĪ AL-SHĀFIᶜĪ

1.7.4 A treatise criticizing the prediction of Rudolf Walp (no. 16) that the comet of Nov. 13, 1899 would collide with the earth: TR 82 (72 fols., *ca.* 1325H).

6.8.12 A treatise on weights and measures used in Egypt, compiled in 1256H: TR 330 (9 fols., *ca.* 1300H).

D118 AḤMAD B. QĀSIM

Not identified.

3.2.33 *Wasīlat al-mubtadiᵓīn li-ᶜilm ghurrat al-shuhūr wa-l-sinīn*, a poem on timekeeping: DM 1016,2 (fols. 3v–4v, 1287); DM 893,1 (fols. 1v–2v, *ca.* 1300H); *etc.* (2 copies). Commentary by the author entitled *Tuḥfat al-murīdīn* . . . , compiled in 1273H: K 3865 (59 fols., *ca.* 1300H); DM 893,2 (fols. 3r–39r, *ca.* 1300H); *etc.* (2 copies).

3.2.33 *Tuḥfat al-ikhwān*, another poem on timekeeping DM 1016,1 (fols. 1r–3r, 1287H). Commentary by Muhammad al-Nābalī (?) (no. D119), entitled *Fatḥ al-manān*: ṬM 166 (32 fols., 1303H); *etc.* (1 copy).

D119 MUḤAMMAD B. ᶜABD AL-RAḤMĀN AL-NĀBALĪ (read AL-NĀBULSĪ?)

3.2.33 *Fatḥ al-manān bi-sharḥ Tuḥfat al-ikhwān*, a commentary on a poem by Aḥmad b. Qāsim, compiled in 1278H: see no. D118.

D120 ᶜALĪ AL-KHASHSHĀB

Not identified.

2.5.9 A short treatise on the equation of time and solar and lunar eclipses: DM 261 (6 fols., 1293H).

3.3.14 Calculation of the *qibla* at Cairo and London: K 4011,2 (fol. 2r, 1285H).

D121 ABŪ ᶜAYYĀSHA MUḤAMMAD AL-SHĀFIᶜĪ AL-DAMANHŪRĪ

4.1.11 *Fatḥ al-ᶜAlīm al-Basīṭ fī rasm al-arbāᶜ wa-l-basāᵓiṭ*, a treatise on different kinds of instruments in 12 *bāb*s with tables for latitude 30° (Cairo), compiled in 1274H: TR 166,1 (pp. 2–22, 1304H); DM 110 (18 fols., *ca.* 1300H, lacks tables).

D122 ᶜABD AL-RAZZĀQ AL-MUNBĀWĪ

2.4.2 A star catalog for 1291H: DM 889,2 (fols. 27v–39r, 1296H).

D123 MUḤAMMAD B. MUṢṬAFĀ AL-ṬANṬĀWĪ, a *muwaqqit* at the Umayyad Mosque in Damascus

3.1.34 *Kashf al-qināᶜ ᶜan maᶜrifat al-waqt min al-irtifāᶜ*, tables for timekeeping and prayer-tables for latitude 33;30° (Damascus): ṬM 173,2 (fols. 11v–50v, 1300H); TR 129,2 (pp. 26–107, 1301H); DM 1007 (55 fols., *ca.* 1299H).

3.1.34 *Dībāja laṭīfa* on two tables for finding the solar longitude for any date: ṬM 173,1 (fols. 1v–10r, 1300H); TR 129,1 (pp. 2–23, 1301H).

4.3.29 Tables for marking astrolabe plates and sundials for latitude 33;30°, Damascus: DM 1159,1 (fols. 1r–12r, 1301H).

4.4.25 Various treatises and tables relating to quadrants: DM 1161, 1162, 1163, 1164, and 1159,2 (mostly *ca.* 1300H in author's hand).

D124 ᶜALĪ PĀSHĀ MUBĀRAK

Article in EI_2 by K. Vollers.

6.8.12 *K. fi l-Mawāzīn wa-l-mikāyīl wa-l-maqāyīs*, on weights and measures: K 4524 (125 pp., 1325H); *etc.* (1 copy).

Note: The remaining authors are not listed in the modern sources and have not been securely identified.

D125 MUḤAMMAD ᶜABD AL-ᶜAZĪZ B. ᶜALĪ AL-SĀᶜĀTĪ AL-SHUBRĀWĪ

3.1.27 Prayer-tables for Cairo, 1315H: DM 393 (33 fols., *ca.* 1315H).

D126 ᶜABD AL-RAḤĪM MANṢŪR AL-ḤALABĪ

5.3.35 ***K. Bughyat al-ṭālib fī aḥkām al-kawākib*, a treatise on astrology in 48 *bābs*: ṬM 152 (156 pp., 1342H by author).**

5.3.35 ***K. Tāj al-Madkhal al-musammā bi-l-Durr al-manẓūm*, a treatise on astrology based on the *Tāj al-madākhil* (see no. G1) and Abū Maᶜshar (no. B22): ṬM 245 (458 pp., 1345H by author); ṬM 74 (387 pp., *ca.* 1340H by author).**

D127 ĪLYĀ AL-MAṬRĀN

6.8.12 ***M. fi l-Maqāyīl wa-l-mawāzīn*, on weights and measures, in 16 *bābs*: TR 341 (18 pp., *ca.* 1050H); DR 1046 (10 fols., *ca.* 1100H); *etc.* (2 copies). See Pl. CXb.**

D128 ᶜABD ALLĀH B. ᶜABD AL-RAḤMĀN B. AḤMAD AL-ḤANBALĪ AL-ḤALABĪ

4.4.12 ***Khulāṣat al-masāᵓil fī sharḥ Hidāyat al-sāᵓil*: see no. C97 (manuscript dated 1190H).**

D129 AWĪS AL-QWJWY (?)

1.2.24 *Maṭlab fī bayān istifādat al-qamar nūr*[an] *min nūr al-shams*, on the light received by the moon from the sun: MJ 180,18 (fol. 149v, *ca.* 1150H).

D130 MUḤAMMAD B. SHUᶜAYB AL-ḤALABĪ ZĀᵓIR AL-QARĀFATAYN AL-SHĀFIᶜĪ AL-IBSHĪHĪ[1]

1.4.13 Treatise in 14 *faṣl*s on the Arab months, seasons, days of the weeks, and Coptic months: DM 134,1 (fols. 1r–19v, *ca.* 1150H, contains 4 *faṣl*s only).

D131 AḤMAD B. MUḤAMMAD B. ᶜABDŪ AL-JADALĪ

1.4.18 *K. Tashīl al-maṭālib fī ᶜilm al-nujūm wa-l-kawākib*, a treatise on folk astronomy in two *juzᵓ*s with 18 + 8 *bābs*: ṬM 160,1 (fols. 1r–35v, 1162H), unique.

D132 ABU L-FATḤ B. ᶜABD AL-RAḤMĀN AL-DANŪSHĪRĪ

2.5.16 *Jawharat al-nafs fī maᶜrifat al-taᵓrīkh wa-ḥall darajat al-shams*, a short treatise on the calendar and determination of the solar longitude, arranged in a *muqaddima*, a *maqṣad* in 2 *bābs*, and a *khātima*, with tables: DM 627 (9 fols., *ca.* 1100H); DM 878 (25 fols., *ca.* 1250H, with additional tables).

D133 SHAMS AL-DĪN MUḤAMMAD B. MUḤAMMAD AL-SHAMSĪ AL-KHAWĀNIKĪ

4.7.28 *al-Rawḍ al-zāhir fī maᶜrifat qisi l-dāᵓir*, a treatise on sundial theory in 3 *bābs*: DM 790,1 (fols. 1r–19r, *ca.* 1150H).

4.7.28 An abridgement of his treatise entitled *al-Fawāᵓid al-muhimma fī maᶜrifat ma yakhruju min al-jayb bi-l-ḍarb wa-l-qisma*, on the sine quadrant, in 15 *bābs*: DM 185,2 (fols. 31v–48r, *ca.* 1150H); *etc.* (1 copy). The treatise deals mainly with sundial theory.

---

1. Brockelmann (*GAL*, SII, p. 99) mentions a Muḥammad b. Shuᶜayb who wrote on *farāᵓiḍ*.

D134 KHALĪL B. IBRĀHĪM B. ḤASAN AL-ᶜAZZĀZĪ

3.1.32 *Muqaddima fi ᶜAmal mawāqiᶜ ᶜaqārib al-sāᶜāt*, on timekeeping: TR 304,1 (pp. 1–6, 1299H).

4.7.28 A treatise on the construction of vertical sundials in 3 *bābs*: K4013,2 (fols. 10r–30r, 1308H); *etc.* (2 copies), and another also in 3 *bābs* entitled ***Nukhbat qawl al-sādāt fī maᶜrifat mā yataᶜallaqu bi-l-munḥarifāt***: K 3990 (26 fols., 1316H); *etc.* (1 fragment).

4.7.28 A treatise on the construction of the hour-angle lines and *ᶜaṣr* arcs on horizontal sundials: DM 304 (14 fols., 1284H); *etc.* (5 copies).

See also no. D230.

D135 MUḤAMMAD B. MUḤAMMAD AL-ḤANAFĪ, a *mawlā* in Damanhūr

4.5.22 *K. al-Zahra al-mthwr (!) fī sharḥ manẓūm mujayyab rubᶜ al-dastūr*, a commentary on a poem on the sine quadrant by Taqi l-Dīn (no. H12): DM 626 (10 fols., 1069H); MM 138 (9 fols., *ca.* 1100H). The only other known copy of this treatise, MS Berlin Ahlwardt 5834, is anonymous.

D136 AL-NAJDĪ

4.7.28 A treatise on horizontal and vertical sundials in two *wajhs*: DM 862,5 (fols. 24v–27r, 1293H).

D137 YŪNUS B. ᶜABD AL-QĀDIR AL-RASHĪDĪ

4.4.10 *Ghāyat al-suʾl fī sharḥ al-ᶜashara fuṣūl* and another commentary without title on Ibn al-Majdī's treatise on the almucantar quadrant: see no. C62.

D138 AL-FAWZĪ

4.7.28 *R. mukhtaṣara fī Rasm rusūm al-basīṭa wa-l-mizwala wa-l-munḥarifa*, on sundial theory in 3 *bābs*: ṬM 202,3 (fols. 76r–80r, *ca.* 1150H).

D139 ᶜABD AL-RAḤMĀN ᶜUTHMĀN

2.2.6 A Turkish translation of the ***Zīj*** of Ulugh Beg (see no. G49) commissioned by Ḥasan Efendī in Cairo, and based on a copy sent by Ulugh Beg to Sultan Bayezid II: ṬFT 33 (183 fols., *ca.* 1250H).

D140 MUḤAMMAD ᶜABD ALLĀH BEY

6.9.16 Tables for inheritance: B 23837,1 (fols. 1r–27v, *ca.* 1150H).

D141 ABU L-BARAKĀT MUḤAMMAD B. MUḤAMMAD B. ᶜĀMIR AL-DALJAMŪNĪ AL-WAFĀʾĪ

6.9.16 / 4.5.16 *al-Jawāhir al-muhimma fi l-munāsakhāt wa-l-qisma*, on inheritance: B 25293,1 (fols. 1r–59r, *ca.* 1100H).

D142 ABU L-KHAYR AL-SHUKR AL-SĀMĪ (?)

4.4.23 *al-Ṣilāt fi stikhrāj awqāt al-ṣalāt*, a treatise in two *qisms* on the sine and almucantar quadrants: DM 986 (10 fols., 1176H).

D143 YŪSUF B. ᶜALĪ ABI L-ᶜĪD

6.9.16 *al-Daqāʾiq al-muḥkamāt fī taḥqīq qismat al-muḥāṣāt*, on inheritance: ZK 778,6 (pp. 131–149, 1245H).

D144 ḤASAN B. MUḤAMMAD AL-ᶜAṬṬĀR

6.1.4 A *ḥāshiya* on Qāḍī Zāde's commentary on al-Samarqandī's ***Ashkāl al-taʾsīs***: see no. G26.

D145 MUḤAMMAD SURŪR

6.7.9 *K. Burhān al-misāḥa fī maᶜrifat sutūḥ al-ashkāl*, on surveying: ṬR 124 (63 fols., 1240H).

D146 ḤUSAYN, *muwwaqit* at the Zaghlūl Mosque in Rosetta

2.3.6 *K. al-Durr al-mukhtaṣar fī ḥall al-shams wa-l-qamar*, tables based on *al-Durr al-yatīm* by Ibn al-Majdī (no. C62): DM 985 (5 fols., *ca.* 1200H).

D147 SHIHĀB AL-DĪN AḤMAD B. MUḤAMMAD AL-UJHŪRĪ

3.1.27 Prayer-tables for latitude 30°, Cairo, with values for each day of the year: K4005 (11 fols., *ca.* 1100H).

D148 BADR AL-DĪN ḤASAN B. ᶜALĪ B. AḤMAD AL-UJHŪRĪ

4.4.11 An abridgement in 10 *bāb*s of *al-Nujūm al-zāhirāt* by al-Wafāʾī: see no. D61.

D149 DĀʾŪD B. MUḤAMMAD AL-QĀRIḤ (?)

4.5.12 *ᶜUjāla laṭīfa ᶜala l-R. al-Fatḥīya*, a commentary on the treatise by Sibṭ al-Māridīnī: see no. C97.

D150 ᶜABD AL-ṢAMAD AL-AZHARĪ, a muwaqqit at the Mosque of al-Muʾayyad in Cairo.

4.5.28 *R. fi l-Rubᶜ al-mujayyab*, on the sine quadrant also called *al-rubᶜ al-muqaffaḍ* and *al-rubᶜ al-mufaṣṣaṣ* and *rubᶜ al-dustūr*, arranged in unnumbered *faṣl*s: K 4020,1 (fols. 1r–19r, 1118H). This treatise differs slightly from the anonymous commentary on *al-R. al-Fatḥīya* in ṬJ 366,6, *etc.* (see no. C97).

D151 SHAMS AL-DĪN MUḤAMMAD AL-ḤANAFĪ

2.5.16 *Laṭāʾif al-ishāra fī maᶜrifat amākin al-kawākib al-sabᶜa al-sayyāra*, a treatise on finding planetary longitudes approximately, arranged in 7 unnumbered *faṣl*s: DM 628,3 (fols. 6r–8v, *ca.* 1150H).

D152 ḤUSAYN B. ᶜALĪ AL-SUYŪṬĪ, AL-DĪK

6.4.42 *ᶜIddat al-ḥāsib wa-ᶜumdat al-kātib fī fann al-muḥāsaba*, a treatise on arithmetic and accounting in 3 *maqāla*s: DR 565 (80 fols., 1285H); *etc.* (1 copy). No other copies known.

D153 ᶜUMAR B. MAḤMŪD B. AL-ṢĀRIM

2.4.9 Calendrical tables for computing the Muslim festivals: K 4038,1 (fols. 1r–2v, *ca.* 1050H).

D154 MUḤAMMAD B. ᶜABD AL-RAḤMĀN B. MUḤAMMAD AL-DĀDĪ (?)

4.7.28 *al-R. al-Abadīya ᶜan mā yataᶜallaqu bi-l-asṭiḥa al-mustawīya*, on sundial theory in 4 *bāb*s: ZK 195 (33 fols., 1260H).

D155 ḤASAN AL-GHAWRĪ

6.4.42 *Munyat al-ṭullāb fī ᶜilm al-ḥisāb*, a treatise on arithmetic in 2 *qism*s of 7 + 3 *bāb*s: KhR 4 (25 fols., *ca.* 1250H). This treatise was compiled by order of ᶜUthmān Bey. No other known copies.

D156 MUḤAMMAD B. AḤMAD B. MUḤAMMAD AL-QABBĀNĪ

6.4.42 *Ghayāt al-suʾūl fī kayfīyat istikhrāj al-majhūl*, a treatise on algebra in 5 *bāb*s: ṬR 132,2 (fols. 43r–48v, 1156H, 1st *bāb* only).

D157 MUṢṬAFĀ AL-MANZILĪ

3.5.13 *Faṣl fī Maᶜrifat al-ṭāliᶜ fi l-waqt wa-l-kawākib fī ayy al-burūj taqrīban wa-fi manāzil al-qamar*, on the determination of the instantaneous ascendant approximately: DM 1055 (18 fols., *ca.* 1250H).

D158 ᶜABD ALLĀH AL-MUWAQQIT

5.3.36 *al-Durr al-thamīn fi l-ḥukm ᶜalā taḥāwīl al-sinīn bi-mazj al-adilla wa-l-qawānīn*, a treatise on astrology in 3 *bābs*: DM 63,1 (fols. 1v–12v, *ca.* 1250H).

D159 MUṢṬAFĀ

3.5.14 *Bughyat al-maṭālib (?) liman yurīdu l-ᶜamal bi-l-kawākib*, a poem on timekeeping by the stars: K 4016 (7 fols., *ca.* 1100H). The author mentions Yūsuf al-Jurdī (no. D82).

D160 YŪSUF EFENDĪ AL-ᶜAQQĀD AL-MAKKĪ

2.5.15 Three poems on calendars and on a calendrical circle: ṬJ 811,1 (fols. 1v–4v, 1198H). Another copy is probably MS Dublin Chester Beatty 4124, entitled *Muntaqaḥāt al-durar*.

D161 AḤMAD AL-DRDYR (?)

3.5.13 (*R. mukhtaṣara fī Ziyādat al-nahār wa-naqṣānihi fi l-shuhūr al-Qibṭīya*), a treatise on the length of daylight throughout the Coptic year: DM 820 (4 fols., *ca.* 1200H).

D162 SHAMS AL-DĪN B. ᶜABD ALLĀH FATḤ AL-FARGHALĪ AL-SABRABĀWĪ

3.4.12 *Aqrab al-adilla fi stikhrāj al-ahilla*, a treatise on lunar crescent visibility in 4 *faṣl*s with tables: DM 1194 (34 fols., *ca.* 1150H).

4.4.24 *Ijtināᵓ al-thamarāt fī rasm jayb al-dustūr wa-waḍᶜ al-muqanṭarāt*, on the construction of quadrants in 10 *bāb*s with diagrams and tables: DM 90 (10 fols., lacks *bāb*s 9 and 10, *ca.* 1200H).

D163 ṢĀLIḤ MANṢŪR AL-YĀFĪ AL-ḤANAFĪ

3.5.13 *Natījat al-barakāt fī dhikr ṭuruq tataᶜallaqu bi-ᶜilm al-mīqāt*, a treatise on the calendar and astronomical timekeeping in 4 *bābs*: DM 830 (19 pp., 1257H).

D164 YŪSUF (??)

3.5.13 (*R. fī Ḥisāb al-aẓlāl wa-l-juyūb wa-l-ashām*), on the calculation of shadows, sines, and versed sines: DM 628,1 (fols. 1r–3r, *ca.* 1150H); DM 521,5 (fols. 52v–57r, *ca.* 1200H).

D165 AḤMAD B. ᶜABD AL-RAḤMĀN, a preacher (*wāᶜiẓ*) at the Umayyad Mosque in Aleppo

3.2.32 *Hidyat al-mushtāq li-maᶜrifat al-awqāt fī jamīᶜ al-āfāq*, a treatise on timekeeping in 11 *bābs*: TR 60 (90 pp., *ca.* 1300H).

D166 MUṢṬAFĀ B. MUḤAMMAD B. MUṢṬAFĀ AL-MUᶜTAMADĀWĪ

3.2.32 *Tuḥfat al-ishārāt fī maᶜrifat ghurar al-sinīn wa-l-awqāt*, a treatise on timekeeping in 13 *bābs*: DM 311 (60 fols., 1294H).

D167 MŪSĀ B. SHĀHĪN AL-ABSHĀDĪ

3.2.32 *Tanqīḥ al-afkār fī aᶜmāl al-layl wa-l-nahār = al-Minan al-irshādīya fi l-ᶜamal bi-l-natīja al-Abshādīya*, a treatise on timekeeping in 12 *bābs*: MM 32 (7 fols., *ca.* 1200H).

D168 MUḤAMMAD AL-SHAMSĪ AL-MUQRIᵓ

3.2.33 *Zād man qad qalla wa-dalīl man ᶜan al-ṣawāb dalla*, a poem on timekeeping in 29 *bābs*: TR 88 (11 pp., *ca.* 1200H).

D169 ᶜABD ALLĀH B. AḤMAD B. AL-MUNĀWĪ

3.2.33 *al-Durra al-yatīma*, a poem on timekeeping: MM 82 (22 fols., 1082H); DM 181,2 (fols. 13r–39v, 1179H); *etc.* (2 copies). See also al-Manbāwī (no. D178).

D170 MUḤAMMAD B. MUḤAMMAD AL-BADĪRĪ AL-DIMYĀṬĪ

3.2.32 *Fawāʾid jawharīya ḥusn (?) al-dalāla fī maᶜrifat al-waqt wa-l-qibla bi-ghayr āla*, a treatise on timekeeping in 6 *bābs*: ṬM 131 (36 fols., 1239H); *etc.* (1 copy).

D171 MUḤAMMAD EFENDĪ, *muftī* of the Shāfiᶜīya

3.2.25 A commentary on the *Muqniᶜ* of al-Sūsī: see no. F49.

D172 AḤMAD B. MUḤAMMAD AL-JANĀJĪ

6.4.1 A *ḥāshiya* on Abū Yaḥyā al-Anṣārī's commentary on Ibn al-Haʾim's poem *al-Muqniᶜ*: see nos. C58 and C78.

D173 AḤMAD AL-AṢBAḤĪ

6.4.44 A commentary on his poem on fractions by Muḥammad b. Ḥasan al-Bayṭār (no. D174): K 3955 (3 fols., *ca.* 1255H).

D174 MUḤAMMAD B. ḤASAN AL-BAYṬĀR

6.4.44 A commentary on a poem by Aḥmad al-Aṣbaḥī: see no. D173.

D175 SIRĀJ AL-DĪN ᶜUMAR AL-FĀRISKŪRĪ

4.4.27 An anonymous commentary on a poem by al-Fāriskūrī on the use of the almucantar quadrant is contained in: Sh 89,10 (fols. 133v–155v, *ca.* 1200H). A copy of the poem is in MS Cairo Azhar 7673 (fols. 35–42, 1106H) (*Cat.*, p. 319).

D176 MUḤAMMAD EFENDĪ DARANDAR

4.4.25 *R. fī l-Rubᶜ al-mashhūr fi-l-muqanṭarāt*, a treatise on the use of the almucantar quadrant in 6 *faṣls*: DM 748,3 (fols. 8r–9v, 1144H); TR 106,6 (pp. 136–141, *ca.* 1150H); *etc.* (4 copies).

D177 ṢADR AL-DĪN

3.5.14 Notes in verse on the risings of the mansions and on daybreak throughout the year, and on the amounts to be added to the solar longitude at midday to obtain the solar longitude at nightfall and at the time of the *salām*: DM 71,2 (fols. 5r–6r, *ca.* 1100H).

D178 AL-MANBĀWĪ (?)

3.5.15 A note on modifying the operations of timekeeping for the sun's motion during the day, stated to be taken from *al-Durr al-muntashar* by al-Manbāwī (?): DM 260,2 (fol. 32r, 1290H). A work entitled *al-Durra al-yatīma fi l-mīqāt* is attributed to ᶜAbd Allāh b. Aḥmad al-Mināwī (no. D169).

D179 IMĀM AL-QALᶜA

4.4.25 *al-Durr al-manẓūm fi l-ᶜamal bi-l-rubᶜ al-maqṭūᶜ*, a treatise on the use of the almucantar quadrant in 35 *bābs*: DM 551 (16 fols., *ca.* 1250H).

D180 MUḤAMMAD B. ᶜĀBIDĪN

6.4.1 *Manāhil al-surūr li-mubtaghi l-ḥisāb bi-l-kusūr*, a poem on arithmetic based on *al-Nuzha* by Ibn al-Hāʾim: see no. C58.

D181 MUḤAMMAD ḤATTĀTA (?) AL-FĀRISKŪRĪ[1]

4.4.25 *R. fi l-ᶜAmal bi-rubᶜ al-muqanṭarāt*, a treatise on the use of the almucantar quadrant in 11 *bābs*: DM 639,4 (fols. 42r–46v, *ca.* 1200H).

4.5.28 *R. fi l-ᶜAmal bi-l-rubᶜ al-mujayyab*, on the use of the sine quadrant in 13 *bābs*: DM 639,5 (fols. 46v–51r, *ca.* 1200H).

1. Perhaps to be identified with the Muḥammad al-Fāriskūrī (d. 1053H) mentioned in *GAL*, II, p. 377.

D182 MUḤAMMAD B. MUḤAMMAD B. MUḤAMMAD B. BAHĀDUR AL-MAWLAWĪ AL-SHĀFIᶜĪ

4.4.25 *Nuzhat al-ᶜāmil fī l-ᶜamal bi-l-rubᶜ al-kāmil*, a treatise on the use of the almucantar quadrant in 10 *bāb*s: DM 1050,2 (fols. 41v–45v, *ca.* 1300H).

D183 BAṬRUS ḤANNĀ ᶜANJŪRĪ AL-DIMASHQĪ

2.5.17 *al-Maᵓkhadh al-ḥadīth al-ṣifāt bi-ᶜilm ḥālāt al-kusūfāt*, a treatise in 25 *faṣl*s on the determination of solar eclipses, with tables: ṬM 169,1 (fols. 1v–82v, *ca.* 1275H); DM 915 (44 fols., *ca.* 1275H); plus an abridgement by the author, compiled in 1816 A.D.: K 21657 (60 pp., *ca.* 1275H).

D184 NŪR AL-DĪN B. ᶜABD ALLĀH AL-WĀᶜIẒ

6.4.23 A commentary on Chapter 3 of al-ᶜĀmilī's *Khulāṣat al-ḥisāb*, compiled in 1113H: see no. G68.

D185 YŪSUF AL-TALĪNĪ (?), a priest in the Ezbekiya district of Cairo

2.5.17 A treatise on the Coptic calendar: DM 868 (57 fols., 1216H by the author).

D186 MUḤAMMAD ṢĀLIḤ

2.5.15 Rules for finding the *notae* of the years arranged in verse: DM 1063,4 (fols. 55r–56v, 1215H by the author).

D187 ISMĀᶜĪL B. SHARAF, IBN AL-MUQRIᵓ

6.4.1 Anonymous remarks on his commentary on the *Wasīla* of Ibn al-Hāᵓim: see no. C58.

D188 YAḤYĀ AL-KHAZRAJĪ AL-ANṢĀRĪ

6.8.12 *Urjūza fī l-Qabbān*, a poem on the balance: MR 30,9 (fols. 47v–49r, *ca.* 1100H); MR 28,1 (fols. 1r–2v, 1166H); TR 279,2 (pp. 62–64, 1151H); *etc.* (1 copy).

D189 ABU L-ḤASAN ᶜALĪ B. ABĪ MAḤMŪD AL-MAGHRIBĪ AL-MANZILĪ[1]

GAL, SII, p. 1039, no. 20.

5.3.38 *K. Nuzhat al-nufūs wa-muzīl al-ᶜakūs*, a treatise on magic and astrology in 44 *bāb*s: DM 859,1 (fols. 1v–20v, *ca.* 1300H); *etc.* (2 copies). Another copy is MS Berlin Ahlwardt 4208.

D190 MUḤAMMAD B. MUḤAMMAD B. MUḤAMMAD AL-MUBASHSHIR

3.5.14 A poem on timekeeping: MM 185,1 (fols. 1r–2v, 1201H); with an anonymous commentary: MM 185,2 (fols. 3v–14v, 1201H).

D191 IBRĀHĪM AL-ḌĀBIṬ

3.5.14 A commentary on his poem on finding the time of the two ᶜ*aṣr* prayers by Muḥammad al-Jirijāwī: see no. D192.

D192 MUḤAMMAD B. MUḤAMMAD AL-MARĀGHĪ AL-JIRJĀWĪ

3.5.14 *Nihāyat al-iqdām fī maᶜrifat al-ẓuhr bi-l-aqdām*, a commentary on a poem on the time of the two ᶜ*aṣr* prayers by Ibrāhīm al-Ḍābiṭ (no. D191): DM 1112 (10 fols., *ca.* 1310H); DM 819 (6 fols., *ca.* 1310H).

1. Author's name in MS Berlin 4208 is ᶜAlī b. al-Shaykh al-Fāḍil al-Madanī al-Maghribī.

D193 ABŪ SAᶜĪD MUḤAMMAD AL-DARWĪSH

3.1.28 *al-Mubāshara al-muḥarrara al-maḥbūra ᶜalā ᶜarḍ Jirja l-maᶜmūra*, a treatise displaying the intervals for the prayer-times in words for each degree of solar longitude, stated to be for the latitude of Jirjā: DM 1111 (38 fols., *ca.* 1100H, lacks beginning).

D194 MUḤAMMAD EFENDĪ ḤIFẒĪ, an official of the Egyptian Customs

3.1.35 Prayer-tables for different latitudes in Egypt and the Hejaz: K 3792–3802 (11 volumes in the same hand, about 50 fols. each, *ca.* 1325H). See Pl. XLIIc.

D195 ᶜUMAR NIᶜMAT ALLĀH

4.3.29 Tables for marking the curves on astrolabe plates for different latitudes: MM 27,2 (fols. 6r–25r, *ca.* 1100H).

D196 AḤMAD AL-JĀLĪ

3.5.14 A poem on the shadows at midday and the *ᶜaṣr*: MM 171,6 (fol. 59v, 1145H).

D197 ᶜABD AL-GHANĪ B. MUḤAMMAD HILĀL

4.5.30 *Fāᵓida (manẓūma) fi stikhrāj al-qibla bi-l-rubᶜ al-mujayyab*, a poem on finding the qibla using a sine quadrant: DM 263,2 (fol. 38r, *ca.* 1300H).

D198 MAḤMŪD AL-ANṬĀKĪ

4.5.28 *al-R. al-Ṣāliḥīya li-taḥsīl al-aᶜmāl al-jaybīya*, a treatise on the use of the sine quadrant in 16 *bābs*: TJ 177,7 (pp. 109–115, *ca.* 1150H).

D199 IBRĀHĪM JAWDA AL-DAMANHŪRĪ

4.4.25 *al-Qawl al-muᶜaṭṭar fī ᶜamal al-murī ᶜala l-rubᶜ al-muqanṭar*, a treatise without divisions on the use of the almucantar quadrant: ṬM 195 (5 fols., 1345H).

D200 AḤMAD AL-MAKTABĪ (?)

6.4.24 A *ḥāshiya* on al-Sakhkhāwī's arithmetic (see no. C131): K 3943 (71 fols., 1319H).

D201 ZAYN AL-DĪN ᶜABD AL-RAḤMĀN AL-SHABRĪSĪ (?)

4.5.28 *Tashwīq al-ḥabīb li-l-ᶜamal bi-l-rubᶜ al-mujayyab*, on the use of the sine quadrant in 10 *bābs*: DM 804,1 (fols. 1r–3v, 1231H); ZK 706,3 (fols. 11r–12v, *ca.* 1100H, lacks end).

D202 ᶜABD AL-FATTĀḤ AL-BANNĀᵓ

6.4.48 An abridgement of the *Hidāyat al-muhtadī* of ᶜUthmān b. ᶜAbd al-Manān (no. H43): DR 628 (110 pp., 1311H by author).

D203 AḤMAD B. MUḤAMMAD AL-SHĀFIᶜĪ AL-JANĀJĪ

6.3.8 A *ḥāshiya* on Sibṭ al-Māridīnī's commentary on Ibn al-Yasamīn's poem on algebra: see no. C97.

D204 ABŪ ᶜABD ALLĀH AL-KAWMĪ

5.3.35 An anonymous commentary on a treatise by him on astrology and magic entitled *K. Tasyīr al-maṭālib ᶜala l-kawākib al-sabᶜa al-sayyāra*: K 3952 (2 fols., *ca.* 1200H). Note that this title is the same as the mysterious *Pseudo-Muṣṭalaḥ Zīj* introduction preserved in MM 38 (see no. C13).

D205 ANONYMOUS

2.1.21 Chapters from the introduction of an unidentified *zīj*: TR 235,2 (pp. 145–156, *ca.* 1150H, Maghribi script). The anonymous author mentions Damascus.

D206 ANONYMOUS

2.2.4 ***Tuḥfat al-ṭullāb fi l-ᶜamal bi-l-ḥisāb***, a treatise on planetary astronomy in 48 *bābs*: K 3771,2 (fols. 59v–130v, 1329H). No other known copies.

D207 ANONYMOUS

2.2.14 An anonymous *zīj* containing tables by Ibn Yūnus (no. B59), Ibn al-Majdī (no. C62), and al-Ṭabarī (no. B92), with a table of stellar coordinates for the year 1070H, and some planetary tables for longitude 99;16° (Samarqand) although the oblique ascensions tables are for latitude 30° (Cairo) and 33;30° (Damascus): DM 678 (71 fols., *ca.* 1150H). Merits investigation. See Pl. XV.

D208 ANONYMOUS

2.2.3 ***Bahjat al-kawākib al-nayyira fī ḥall al-kawākib al-mutaḥayyira***, a treatise on the determination of planetary positions, with tables: DM 210 (14 fols., *ca.* 1250H).

D209 MISCELLANEOUS (ephemerides)

2.3.12 Egyptian ephemerides giving solar, lunar, planetary positions for each day of the period 1126H–1158H: DM 680 (168 fols., *ca.* 1125H).

2.3.12 Egyptian ephemerides giving solar, lunar, planetary positions for the solar or Jalālī year beginning in 1076H: MM 30 (23 fols.); 1090H: DM 546 (14 fols.); *etc.* (5 similar copies).

2.3.12 Egyptian ephemerides giving solar, lunar, planetary positions for the Muslim year 1194H: MM 59,1 (7 fols.); 1199H: MM 121 (8 fols.); *etc.* (9 similar copies). See Pl. XXV.

D210 MISCELLANEOUS (star catalogs)

2.4.2 Late star catalogs: TR 163,2 (fols. 26v–30v, *ca.* 1100H); DM 37 (6 fols., *ca.* 1200H); DM 794,1 (fols. 1r–4v, *ca.* 1150H); DR 363,5 (fols. 93r–97v, *ca.* 1115H); ZK 782,14A (fols. 91v–93v, *ca.* 1120H).

D211 MISCELLANEOUS (eclipse computations)

2.5.9 Computations for the solar eclipse of Ramaḍān 29, 1042H: DM 173,7 (fols. 146v–148v, *ca.* 1150H).

2.5.9 Ṣafar, 1116H: K 8525,4 (fols. 57v–58r, 1116H). 1187H, 1173H, 1196H: K 4019,2 (fols. 10r–11r, *ca.* 1200H). 1173H, 1174H: ṬM 250,3 (fols. 28r–30r, 1277H).

2.5.9 A note on computing conjunctions and oppositions for predicting eclipses: DM 591,2 (fols. 9r–9v, *ca.* 1150H).

2.5.9 A note on the computation of the amount of lunar surface obscured in a lunar eclipse: MM 9,3 (fol. 29v, *ca.* 1200H).

D212 ANONYMOUS

2.5.16 A treatise in 10 *qāᶜidas* on the determination of solar longitude, with some tables: DM 109,3 (fols. 33v–39r, *ca.* 1200H, incomplete).

D213 ANONYMOUS (Christian)

2.5.17 ***K. al-Niḥla al-nafīsa fī lawāzim khuddām al-kanīsa***, a treatise in 15 bābs with tables for computing Christian feastdays: DM 1025 (7 fols., *ca.* 1250H).

D214 ANONYMOUS (Jewish)

2.5.18 ***K. fī maᶜrifat Kusūf al-shams wa-khusūf al-qamar***, a treatise on the computation of eclipses stated to be translated from Hebrew, with a short introduction and 22 pages of sophisticated tables: K 3770 (15 fols., 1316H). Merits investigation.

D215 ANONYMOUS (prayer-tables)

3.1.27 Miscellaneous prayer-tables for latitude 30°, Cairo, with entries for each day of the Coptic year: MM 34 (6 fols., *ca.* 1100H); DM 200 (7 fols., *ca.* 1150H); MM 142,1 (fols. 1r–7v, 1128H); *etc.*

3.1.27 The same with entries for each five days: MM 198,4 (fols. 46v–51v, *ca.* 1100H); DM 531,1 (fols. 1r–6v, 1187H), *etc.*

3.1.27 Miscellaneous prayer-tables with values written in words for each day: TR 94,1 (pp. 1–96, 1168H).

3.1.27 The same for each five days: MM 219 (23 fols., *ca.* 1250H) = TR 94,2 (pp. 98–108, 1168H)≠DM 542 (20 fols., 1175H), *etc.*

3.1.27 The same with values for each month: DM 754 (4 fols., *ca.* 1100H)≠MM 33,1 (fols. 1v–6r, *ca.* 1000H).

3.1.27 An almanac for Cairo, 1033H, giving prayer-times for each 15 days: DM 510 (10 fols., *ca.* 1033H).

D216 ANONYMOUS

3.1.28 A treatise giving prayer-times in words and astrological information for the Coptic year: MJ 183,7 (fols. 147r–149v, *ca.* 1150H).

D217 ANONYMOUS

3.1.29 Prayer-tables for latitude 31;20° (Rosetta): DM 1207,2 (fols. 57v–58v, *ca.* 1200H), unique. Analyzed in *SATMI*, II. See Pl. XLIIa.

D218 ANONYMOUS

3.1.29 Prayer-tables for latitude 31;25° (Damietta) arranged like those in the *Ruznāme* of Shaykh Vefā (no. H2), with an introduction in Turkish: MMT 1,2 (fols. 3v–11v, *ca.* 1100H), unique. Analyzed in *SATMI*, II.

D219 ANONYMOUS

3.1.29 Tables for timekeeping for latitude 32;10°, Nablus: ṬM 81,3 (fols. 119r–120v, *ca.* 1300H), unique. See further *SATMI*, II.

D220 MISCELLANEOUS (Egyptian prayer-tables)

3.1.32 Late prayer-tables for Cairo: ṬM 142 (7 fols., 1260H); ṬJ 811,3A (fol. 6r, *ca.* 1150H); DM 821 (13 fols., 1277H); ṬM 241,3 (fols. 36r–39r, 1245H); TR 261 (17 pp., *ca.* 1200H); ṬJ 811,4 (fols. 6r–25v, *ca.* 1100H). Almanacs giving *tawqīᶜāt* for each day of the Coptic year: K 4011,1 (fols. 1v–9v, 1287H); MM 117 (4 fols., *ca.* 1200H); and for each day of the lunar year: DM 184,2 (fols. 89r–117v, 1171H).

D221 MISCELLANEOUS (Syrian prayer-tables)

3.1.34 Prayer-tables for Damascus: DM 1160 (8 fols., *ca.* 1250H). These tables differ from those of al-Tanṭāwī (no. D123).

3.1.34 An almanac arranged according to the days of the Syrian year: ṬM 255,8 (fols. 86v–89v, *ca.* 1100H).

D222 ANONYMOUS

3.1.34 Tables for timekeeping for latitude 32° (Jerusalem): ṬM 81,1 (fols. 1r–84v, *ca.* 1300H), unique? These tables, which display the time of day in the Turkish convention as a function of solar longitude and altitude, are probably based on those of al-Karakī (no. C35), preserved in the unique MS Leipzig Universitätsbibliothek 808. See further *SATMI.*

D223 ANONYMOUS

3.2.32 An anonymous treatise on timekeeping: DM 728,3 (fols. 8r–11r, 1143H).

D224 ANONYMOUS

3.2.33 A poem on timekeeping in 26 *bābs*: DM 712 (4 fols., *ca.* 1200H, lacks first 8 *bābs*).

D225 ANONYMOUS

3.4.5 *Taḥrīr al-maqāl fī maᶜrifat ᶜamal al-hilāl*, a treatise on lunar crescent visibility, without tables: MM 20,1 (fols. 1v–4v, 1154H); DM 583 (4 fols., *ca.* 1150H).

D226 MISCELLANEOUS ANONYMI (lunar crescent visibility)

3.4.13 *Faṣl fī maᶜrifat ᶜamal al-ahilla*: DM 474,2 (fols. 5r–5v, *ca.* 1150H).

3.4.13 A *faṣl* on the technical terms used in crescent visibility calculations, without tables: DM 1108,8 (fols. 33v–34v, 1053H).

3.4.13 *Nubdha laṭīfa fī ᶜamal al-hilāl ᶜalā atamm minwāl wa-aḥsan mithāl*, with tables: DM 1195 (14 fols., 1008H).

3.4.13 A treatise on lunar crescent visibility with tables: DM 1108,4A (fols. 11v–23r without 20v, 1053H).

3.4.13 Tables for determining lunar crescent visibility: MM 42 (5 fols., *ca.* 1100H)≠MM 46 (5 fols., *ca.* 1050H)≠MM 99,1 (fols. 1v–4v, *ca.* 1175H).

D227 MISCELLANEOUS ANONYMI

3.5.5 *R. fī Ṭulūᶜ al-shams wa-tawassuṭihā wa-ghurūbihā fī ayy balad kānat bi-l-nisba ilā baladihi*, on the difference in the rising and setting times of the sun in different localities: DM 549 (4 fols., 1182H); MM 244,2 (fols. 7v–9v, *ca.* 1200H); *etc.* (1 copy).

3.5.5 (*R. fī Ikhtilāf ṭulūᶜ al-shams wa-ghurūbihā bi-mukhtalif al-buldān*), on the same topic, with some tables: MM 204,7 (fols. 76v–81r, 1052H).

D228 MISCELLANEOUS (timekeeping)

3.5.12 A note of telling time of day using fingers: DJ 323,7 (fol. 34r, *ca.* 1300H). See Pl. XLb.

3.5.12 A note on the shadows at each seasonal hour: DM 119,4 (fol. 9v, 1109H) (Egyptian provenance). See Pl. XLIII.

3.5.12 A poem on the midday shadows for each Coptic month: DM 440,2 (fol. 13v, *ca.* 1000H).

3.5.14 A poem on timekeeping arranged in *fāʾidas*: ZK 706,7 (fols. 27v–28v, *ca.* 1100H, incomplete).

3.5.14 A poem on the midday shadows of the Coptic months: DM 440,2 (fol. 13v, *ca.* 1000H).

3.5.14 A poem on timekeeping by the stars: DM 181,7 (fols. 166v–169r, 1178H).

3.5.15 Rules for finding the times of prayer (by approximate methods): DM 181,3 (fols. 43r–46v, *ca.* 1200H).

3.5.15 Sundry notes on timekeeping: DM 1046,7; DM 1030,1; DJ 323,8; *etc.* (see Arabic catalog).

D229 MISCELLANEOUS

4.3.29 Various tables for marking the curves on astrolabe plates and quadrants: DM 117,5 (fols. 81r–81v, *ca.* 850H); MM 43,3 (fols. 31r–36v, *ca.* 850H); MM 27,2 (fols. 6r–25r, *ca.* 1100H); DMT 13,2 (fols. 16r–23r, 1183H); DM 701,2 (fols. 8v–23v, *ca.* 1000H); and notes on the same subject in MM 167,6 (fols. 163r–164r, 989H).

D230 MISCELLANEOUS

4.4.24 *Ghunyat al-awḍāᶜ fī ḥisāb al-muqanṭarāt wa-rasm al-arbāᶜ*, a treatise on the construction of almucantar quadrant in 10 *bāb*s with several tables: TR 104 (30 pp., 1259H, attributed to Yūsuf b. Aḥmad al-Jundī (no. D82)); ḤM 8 (9 fols., *ca.* 1200H, attributed to Sannār, *muwaqqit* at the Sayyid Aḥmad al-Badawī mosque (in Tanta) (no. D108)); DM 436 (10 fols., *ca.* 1250H, attributed to al-ᶜAzzāzī (no. D134)); K 4007,5 (fols. 35v–59v, *ca.* 1300H, anonymous). The author mentions Ibn al-Majdī (no. C62) and his treatise *Kashf al-qināᶜ*, by which he probably meant the treatise with this title by Ibn al-ᶜAṭṭār (no. C66).

D231 MISCELLANEOUS (treatises on the almucantar quadrant)

4.4.26 Arabic treatise, not subdivided: ṬJ 179,2 (fols. 79v–84v, *ca.* 1100H).

4.4.26 *R. fī Maᶜrifat al-ᶜamal bi-l-rubᶜ al-marsūm bi-l-muqanṭarāt*, in 10 *bāb*s and differing from the treatise of Ibn al-Majdī (no. C62): MJ 180,14 (fols. 131r–134r, 1012H).

4.4.26 Arabic treatises without titles in 15 *bāb*s: ṬJ 410,5 (fols. 21v–28v, *ca.* 1200H); not subdivided: DM 760,1 (fols. 1r–5r, *ca.* 1100H); in 2 *faṣl*s: DM 701,1 (fols. 1v–7v, *ca.* 1200H); in 15 unnumbered *faṣl*s: K 17235 (12 fols., *ca.* 1200H); in 25 *bāb*s: MJ 180,12 (fols. 118v–126v, *ca.* 1100H); plus a note about the use of the almucantar quadrant in a latitude other than the one for which it was constructed (!): DM 781,3 (fol. 8r, *ca.* 1100H); *etc.* (see Arabic catalog).

D232 MISCELLANEOUS (treatises on the sine quadrant)

4.5.3 ***R. fī l-ʿAmal bi-l-rubʿ al-mujayyab***, in 10 *faṣl*s, based on the methods underlying the universal auxiliary tables of al-Khalīlī (no. C37), although this is not stated: DM 624,2 (fols. 5r–7v, *ca.* 1150H).

4.5.3 (***R. fī l-ʿAmal bi-l-rubʿ al-mujayyab***), in at least 28 unnumbered *bāb*s, using methods similar to the above: MM 167,8 (fols. 167r–199v, 989H, lacks first page and perhaps some chapters).

D233 ANONYMOUS

4.5.23 ***Rayḥānat ʿawāṭir al-juyūb fī aʿmāl al-nisab wa-l-suṭūḥ wa-l-juyūb***, an unusual treatise on the use of the sine quadrant in 2 *qism*s of 3 *bāb*s and 3 *marṣad*s: ZK 440 (71 pp., *ca.* 1300H).

D234 MISCELLANEOUS (treatises on the sine quadrant)

4.5.29 A fragment from the middle of a treatise compiled in 1090H: DM 713,1 (fols. 1r–7v, *ca.* 1150H).

4.5.29 ***R. fī l-ʿAmal bi-l-rubʿ al-mujayyab***, in 19 *bāb*s: K 3824,6 (fols. 34v–39r, *ca.* 1150H).

4.5.29 ***R. fī Rubʿ al-dustūr***, not subdivided: ṬJ 179,3 (fols. 85r–88r, *ca.* 1100H).

4.5.29 A treatise in 20 *bāb*s, which the author says he compiled in one hour: DM 478 (4 fols., *ca.* 1100H).

4.5.29 Fragments: MM 44,2; ṬM 136,1; DM 138,14. See also TR 140,17 (p. 85, *ca.* 1150H) for an illustration of the sine quadrant.

D235 MISCELLANEOUS (sundials)

4.7.29 Various treatises on the construction of sundials, arranged in 2 *bāb*s: DM 179,2 (fols. 3v–10v, 1118H); *etc.* (2 copies); without subdivisions: MM 180,3 (fols. 15v–17v, *ca.* 1100H), *etc.* (1 copy); in numerous *tanbīhāt*: DM 23 (10 fols., *ca.* 1200H); in 2 *bāb*s of 12 + 9 *faṣl*s: DM 178,3 (fols. 20r–29v, *ca.* 1150H); without subdivisions: MM 96,1 (fols. 1v–5v, *ca.* 1100H); in 2 *maqāla*s: MM 184,5 (fols. 10v–14r, *ca.* 1100H); *etc.* (1 copy); without subdivisions and containing a table for latitude 31°: DM 600,1 (fols. 1v–2v, *ca.* 1100H), in 5 *bāb*s: DM 179,3 (fols. 11r–16v, 1118H); in 2 *bāb*s, with some tables for latitude 30°: DM 173,3 (fols. 84r–87v, *ca.* 1150H); *etc.* (1 copy); in 2 *bāb*s: MM 201,4 (fols. 17r–26v, *ca.* 1100H, defective); without subdivisions: DM 513,4 (fols. 5v–6v, *ca.* 1100H); in 2 *bāb*s: DM 800,1 (fols. 1r–5v, *ca.* 1150H).

4.7.29 A treatise entitled ***Hidāyat al-ḥāʾir ilā maʿrifat waḍʿ faḍl al-dāʾir***: MM 191,3 (fols. 7r–11v, 1135H). This is not identical to MS Princeton Mach 5014.

D236 ANONYMOUS

4.7.30 Miscellaneous tables for constructing sundials for latitude 27°, Assiut: DM 179,4 (fols. 17r–18v, 1118H); for latitude 30°, Cairo: DM 974 (5 fols., *ca.* 1250H) ≠ K4060,1 (11 fols., *ca.* 1200H) ≠ DM 72 (5 fols., *ca.* 1150H); and for latitude 31;25° Damietta: DM 406 (12 fols., 1200H).

4.7.32 Miscellaneous notes on sundial construction: see the Arabic catalogue.

D237 MISCELLANEOUS (mathematics)

6.4.43 ***Nuzhat al-afkār fī fann al-ḥisāb***, a treatise on arithmetic without subdivisions: ḤR 12 (96 fols., *ca.* 1250, attributed to Ibn al-Hāʾim (no. C58)).

6.4.43 ***Fāʾidatān fī bayān al-aʿdād al-arbaʿa al-mutanāsaba wa-maʿrifat al-qisma bi-l-muḥāṣṣa (?)***, on proportions: ZK 188,2 (fols. 47r–57v, 1242H).

6.5.3 A short treatise on sexagesimal division and sexagesimal multiplication tables: DM 128,2 (fols. 3r–4v, *ca.* 1100H).

6.7.9 A collected work on surveying and geometry: DR 364 (10 fols., in two hands one *ca.* 1150H and the other 1245H). The author describes arithmetical operations in terms of equations, thus (fol. 2r): $17 + 10 = 27 \times 7 = 189 \div 21 = 9 - 12 = 12 = 6 \times 6 = 36 - 10 \times 10 = 100 = \sqrt{64} = 8$, *etc.*! See Pl. XCIIIb.

D238 MISCELLANEOUS (treatises on the balance)

6.8.13 *R. fī ᶜIlm al-qabbān*, not subdivided: K 3831,3 (fols. 7v–9r, 1096H); *etc.* (2 copies).

R. fi l-Qabbān, in 3 *maqṣad*s subdivided into 8 + 4 *bāb*s and unnumbered *faṣl*s: MR 30,2 (fols. 15v–25r, *ca.* 1100H).

R. fī Ṣināᶜat al-qabbān, in 2 *bāb*s: DR 339 (14 fols., *ca.* 1100H); *etc.* (2 copies).

R. fī Ṣināᶜat al-qabbān, not subdivided: DR 338,1 (fols. 1v–3r, *ca.* 1100H), *etc.* (2 copies).

K. al-Qawānīn fī ṣifat al-qabbān wa-l-mawāzīn, in 10 *bāb*s: TR 279,1 (pp. 1–62, 1151H, defective).

R. fī ᶜAmal al-qabbān, not subdivided, compiled 1096H: ṬR 108,5 (fols. 15r–18r, 1289H).

A note on the division of a balance: DR 86,2 (pp. 23–25, 1256H). See also MM 146,1 (fol. 1r, 1177H) and TJ 358,8 (fols. 199r–200r, 1153H).

ADDENDUM:

D239 TĀJ AL-DĪN, *muwaqqit* at the Prophet's Mosque in Medina

3.1.15 A commentary on the ***Ruznāme*** of Shaykh Vefā: see no. H2.

E. Yemeni Authors

Limited information on Yemeni astronomers and mathematicians is contained in *Suter*, Brockelmann's *GAL*, *Azzawi*, and *Hibshi*. More information is contained in my monograph *Mathematical Astronomy in Medieval Yemen* (hereafter *MAY*).

E1 AḤMAD (or ISḤĀQ?) B. YŪSUF AL-ṢARDAFĪ[1]

Suter, no. 260 (has Isḥāq b. Yūsuf); *GAL*, I, p. 620, and SI, p. 855; *MAY*, Appendix A, no. 1.

6.3.3 ***K. Ḍarb al-hindī = K. Mukhtaṣar al-hindī = K. al-Kāfī fī mukhtaṣar al-hindī***: TR 97 (84 pp., *ca.* 1100H); DJ 703,4 (fols. 23r–84r, 1133H); DJ 704,2 (fols. 89r–118v, 1111H); DR 84,1 (fols lr–27r, *ca.* 1200H); several other copies available. An anonymous commentary in DR 84,2 (fols. 27v–42r, 1205H), in which the author mentions various individuals such as Raḍi l-Dīn, Taqi l-Dīn ᶜUmar b. Muḥammad, Aḥmad b. Mūsā ᶜUjayl, *inter alios.*

E2 NASHWĀN B. SAᶜĪD AL-ḤIMYARĪ

GAL, SI, p. 528, no. 10; *MAY*, II, no. 3.

1.4.6 ***Urjūza fi l-shuhūr al-rūmīya***, a poem on the Syrian months: DJ 705,3 (fol. 19r, *ca.* 1300H).

E3 ABU L-ḤASAN AḤMAD B. MUḤAMMAD AL-ASHᶜARĪ AL-YAMANĪ

GAL, I, p. 558; *MAY*, Appendix A, no. 2.

6.7.2 For an abridgement of his ***al-Tuffāḥa fī ᶜilm al-misāḥa*** on surveying by ᶜAbd al-Laṭīf al-Dimashqī see no. D76.

1. DJ 703,4 has the name Abū ᶜAbd Allāh Aḥmad b. Yūsuf b. Yaᶜqūb b. ᶜAbd al-Ṣamad al-Ṣardafī al-Yamanī and DM 84,1 is anonymous. The other two copies have simply al-Ṣardafī. The Berlin MSS listed by Suter are anonymous. Ḥājjī Khalīfa has Isḥāq b. Yūsuf al-Ṣardafī.

E4 AL-FAḌL B. ABĪ SAᶜD AL-ᶜUṢAYFIRĪ

GAL, I, p. 510, and SI, p. 702; *MAY*, *addenda.*

6.7.3 A chapter on surveying from a work entitled *al-ᶜIqd*: TR 353,1 (pp. 1–40, 1319H from a copy dated 1082H). Brockelmann (*GAL*, SI, p. 702) lists a work by al-ᶜUṣayfirī entitled ***ᶜIqd al-aḥādīth fī ᶜilm al-mawārīth***, on inheritance.

6.9.1 ***Miftāḥ al-fāʾiḍ fī ᶜilm al-farāʾiḍ***, on the arithmetic of inheritance: see no. E16.

E5 IBRĀHĪM B. ᶜALĪ B. MUḤAMMAD B. MANṢŪR AL-AṢBAḤĪ

MAY, II, no. 5.

1.4.8+ ***K. al-Yawāqīt fī ᶜilm al-mawāqīt***, a treatise on folk astronomy and simple timekeeping: DM 948,1 (fols. 1v–53r, 1320H). Other copies in Baghdad, Oxford, and Cambridge; see *MAY* for details and for a brief description of this work.

E6 ABŪ ᶜABD ALLĀH MUḤAMMAD B. AḤMAD B. ᶜUMAR AL-KHUZĀᶜĪ

Suter, no. 367 (has al-Ḥaraᶜi following the Oxford catalogue); *GAS*, V, p. 240; *MAY*, Appendix A, no. 3.

6.2.1 ***Sharḥ Mukhtaṣar al-Khwārizmī***, a commentary on the chapter on inheritance in the treatise on algebra by al-Khwārizmī (no. B14): ṬJ 207,4 (fols. 52r–61v + 65r–69v, 734H), with additional notes by the copyist ᶜAlī b. Rāshid al-Nasrī (no. E12) in ṬJ 207,5 (fols. 62r–64v, 734H). Should be compared with MS Istanbul Şehit Ali

2706,5 (fols. 147v–217v, 878H). *Cf. GAS.* V, p. 240, no. 1. See Pl. LXXXVI.

6.3.16 ***al-Muqaddima fi-l-ḥisāb li-ᶜāmmat aḥdāth al-kuttāb***, a treatise on arithmetic: K18362,1 (fols. 1r–18v, 1001H). Other copies listed in *MAY*.

E7 MUḤAMMAD B. ABĪ BAKR AL-FĀRISĪ

Suter, no. 349; *GAL*, I, p. 625, SI, pp. 866–867; and *MAY*, II, no. 6.

2.1.14 ***K. Maᶜārij al-fikar al-wahīj fī ḥall mushkilāt al-zīj***, a treatise on mathematical astronomy in 15 *bāb*s, without tables: DM 817,1 (fols. 1r–51v, *ca.* 725H); ṬM 227,1 (fols. 1r–40r, 1145H); DM 145 (56 fols., *ca.* 1100H). For a brief summary of this work, see *MAY*.

5.3.8 ***Nihāyat al-idrāk fī asrār ᶜulūm al-aflāk***, a treatise on astrology arranged in 3 *maqṣad*s: DM 191 (40 fols., 1072H); Sh105 (45 fols., *ca.* 1200H); DM 983 (43 fols., 1181H); *etc.* (6 copies). In several of these copies the colophon states that the work was completed in 606H: see further *MAY*.

E8 ᶜUMAR B. YŪSUF B. ᶜUMAR, the Rasulid Sultan AL-ASHRAF

Suter, no. 394; *Mayer*, pp. 83–84; *MAY*, II, no. 8.

4.1.2 ***K. Muᶜīn al-ṭālib ᶜalā ᶜamal al-asṭurlāb (wa-l-rukhāma wa-l-ṭāsa)***, an extensive treatise on the construction of the astrolabe, sundial, and the use of the magnetic compass (!), inspired by the compendium of al-Marrākushī (no. C17): TR 105 (149 fols., *ca.* 690H). See Pls. LXII, LXIII, LXIV, CII and CIIIa. The last folios contain a series of *ijaza*s by the Sultan's teachers Ibrāhīm b. Mamdūd (no. E9) and Ḥasan b. ᶜAlī al-Fihrī (no. E10) on a series of astrolabes made by him, one of which survives in the Metropolitan Museum of Art in New York. See further *MAY* and *ZGAIW*, 2 (1986). This work merits detailed investigation.

E9 IBRĀHĪM B. MAMDŪD

See no. E8.

E10 ḤASAN B. ᶜALĪ AL-FIHRĪ

See no. E8.

E11 ANONYMOUS (annual ephemerides)

2.3.2 An ephemeris for the year 727 Hijra, dedicated to (or perhaps even compiled by) the Rasulid Sultan al-Muᵓayyad: DM 817,2 (fols. 55r–84v, 727H).

2.3.2 An ephemeris for the year 808H, containing over fifty pages of lists of historical events, mainly pertaining to the Yemen, in addition to the kind of astronomical and astrological tables contained in the above mentioned ephemeris: TR 274 (152 pp., 808H) and photos thereof in K 7198. See Pl. XXI.

Both of these ephemerides merit detailed investigation: for a brief account, see *MAY*.

E12 ᶜALĪ B. RĀSHID B. AḤMAD B. MŪSĀ B. YAḤYĀ B. NASR AL-NASRĪ

6.2.1 Additions to the commentary of al-Khuzāᶜī (no. E6) on the algebra of al-Khwārizmī (no. B14): ṬJ 207,5 (fols. 62r–64v, 734H by author).

E13 ABU L-QĀSIM ᶜIZZ AL-DĪN IBRĀHĪM B. MUḤAMMAD B. SULAYMĀN AL-BAWSĪ

GAL, SI, p. 702, and SII, p. 242; *MAY*, *Addenda*.

6.7.3 A fragment of a poem on surveying: DJ 703,2 (fols. 18v, 1117H, attributed to al-Bwsy). Brockelmann (*GAL*, SI, 702) mentions a poem on inheritance by al-Bawsī based on the ***Miftāḥ al-fāᵓiḍ*** of al-ᶜUṣayfirī (no. E4), extant in MS Vatican V, 1047_2.

E14 ABŪ MUḤAMMAD ᶜABD ALLĀH B. ASᶜAD B. ᶜALĪ AL-YĀFIᶜĪ

GAL, II, pp. 226–228, and SII, pp. 227–228; *MAY*, II, no. 15.

1.4.9 ***Sirāj al-tawḥīd al-bāhij al-nūr***..., a treatise on folk astronomy in 20 *faṣl*s: TR 322

(79 pp., 877H) and photos thereof in K 7194; ṬJ 179,1 (fols. 1v–79r, *ca.* 1100H).

1.4.9 A poem on the Syrian months: DJ 709,23 (fols. 182r–184r, 1049H); DM 949,1 (fols. 1r–6r, *ca.* 1100H); DJ 319,6 (fols. 78r–83r, *ca.* 1100H, Maghribī?); DJ 705,7 (fols. 45v–46v, *ca.* 1100H). See also the anonymous poem in MM 168,4 (fol. 16v, *ca.* 1100H).

E15 ABŪ BAKR B. ABI L-MAᶜĀLĪ

GAL, SII, p. 253; *MAY*; II, no. 19.

2.3.3 ***Madkhal al-taᶜlīm fī inshāʾ al-tasyīr wa-ᶜamal al-taqwīm***, a poem on the compilation of ephemerides, composed in 794H: DM 1015 (10 fols., *ca.* 1200H). Another copy is MS Manchester Rylands 361A (copied in Istanbul in 1741).

E16 AḤMAD B. MUḤAMMAD AL-KHĀLIDĪ

GAL, I, p. 510, and SI, p. 702; *MAY*, *Addenda*.

6.9.1 ***Īḍāḥ al-ghāmiḍ***, a commentary, compiled in 867H, on al-ᶜUṣayfirī's ***Miftāḥ al-fāʾiḍ fī ᶜilm al-farāʾiḍ***: for manuscripts see *GAL*. *Ḥawāshī* on this commentary by Ibrāhīm b. Khālid b. Aḥmad . . . al-ᶜAlfī (no. E17): DJ 703,8 (fols. 109r–136r, 1135H).

Also a commentary entitled *K. al-Īḍāḥ al-multaqaṭ limā abhama l-Miṣbāḥ* by ᶜImād al-Dīn Yaḥyā b. Muḥammad b. Ḥasan (no. E18) on the chapter on surveying in a work by al-Khālidī called *al-Miṣbāḥ* (?): DJ 703,9 (fols. 137r–157v, 1133H).

E17 IBRĀHĪM B. KHĀLID B. AḤMAD AL-ᶜALFĪ

6.9.1 *Ḥawāshī* on the commentary of al-Khālidī on al-ᶜUṣayfirī's *Miftāḥ*: see no. E16.

E18 ᶜIMĀD AL-DĪN YAḤYĀ B. MUḤAMMAD B. ḤASAN

6.9.1 ***K. al-Īḍāḥ al-multaqaṭ limā abhama l-Miṣbāḥ***, a commentary on the section on surveying in a work called ***al-Miṣbāḥ*** by al-Khālidī: see no. E16.

E19 ᶜAFĪF AL-DĪN ᶜABD ALLĀH B. ᶜUMAR B. ᶜABD ALLĀH B. AḤMAD BĀ MAKHRAMA

MAY, II, no. 26.

1.4.11 ***K. al-Shāmil fī dalāʾil al-qibla wa-ḥisāb al-Rūm wa-l-manāzil***, an extensive treatise on folk astronomy, compiled in 907H: DM 948,2 (fols. 53v–82r, *ca.* 1320H). A fragment from the beginning of this work in DM 899,1 (fols. 1r–7v, *ca.* 1100H) is attributed to Abū Ḥamad Maḥfūẓ b. ᶜAbd al-Raḥmān al-Ḥaḍramī (no. E20).

3.4.6 A list of places in Yemen and Hijaz with the same oblique ascensions for determining crescent visibility: DJ 713,15 (fol. 161v, 1013H).

E20 ABŪ ḤAMAD MAḤFŪẒ B. ᶜABD AL-RAḤMĀN AL-ḤAḌRAMĪ

1.4.11 The fragment of the ***K. al-Shāmil***. . . on folk astronomy in DM 899,1 (fols. 1r–7r, *ca.* 1100H) is attributed to al-Ḥaḍramī. See also no. E19.

E21 ṢĀRIM AL-DĪN IBRĀHĪM B. YAḤYĀ JAḤḤĀF[1]

GAL, SII, p. 567; *MAY*, Appendix A, no. 9.

6.4.29 ***al-Ṭarīqa fi l-ḥisāb = Ṭarīqat Jaḥḥāf***, a treatise on arithmetic in unnumbered *bābs*: K 4309,1 (fols. 1r–85v, 1222H); DJ 705,11 (fols. 67v–84v, 1311H).

E22 AḤMAD B. YAḤYĀ AL-ṢAᶜDĪ AL-DAWWĀRĪ

GAL, SII, p. 559; *MAY*, *Addenda*.

6.7.3 A fragment of his encyclopaedic treatise ***al-Maqṣad al-ḥasan*** dealing with surveying: TR 353,2 (pp. 41–45, 1319H, anonymous).

---

1. In K 4309 the name given is Ṣārim al-Dīn Ibrāhīm Jaḥḥāf, and in DJ 705 Shams al-Dīn Ibrāhīm b. Mahdī b. al-Hudā Jaḥḥāf. For other variants, see *MAY*.

E23 (NŪR AL-DĪN)[1] ᶜALĪ B. ᶜABD ALLĀH AL-ṬAWĀSHĪ

MAY, II, no. 27.

1.4.13 *K.* [*Miftāḥ*][1] *al-asrār fī ᶜilm al-falak al-dawwār*, a treatise in 10 *bābs* on folk astronomy, astrology, and magic: DJ 709,6 (fols. 29r–53v, *ca.* 1150H). Another copy is MS Princeton Garrett 1016. See further *MAY* for a brief description.

E24 ᶜABD ALLĀH B. ḤAMZA AL-QĀḌĪ AL-DAWWĀRĪ

MAY, II, no. 41 (12).

3.2.32 *Bulghat al-muqtāt fī maᶜrifat al-awqāt*, a treatise on timekeeping in 3 *bābs*: K 3769,2 (fols. 5r–11v, 1328H).

1. Supplied from MS Princeton Garrett 1016.

E25 ANONYMOUS

3.2.32 A treatise on timekeeping divided into unnumbered *faṣls*: K 3769,3 (fols. 12r–20r, 1328H). This treatise contains a poem on the standard Indian formula for reckoning time of day from shadow lengths.

See also no. C105.

F. Andalusian and the Maghribi Authors

Limited information on Andalusian and Maghribi astronomers and mathematicians is contained in *Suter*, *Renaud* (mainly adding notes on Andalusian and Maghribi authors to Suter), Brockelmann's *GAL*, and *Azzawi*. A brief overview of astronomy in medieval Maghrib (mainly after 1200) is found in *JHAS*, 6 (1982), pp. 5–9. On the Maghribi tradition in mathematics, see *Djebbar*.

F1 JĀBIR B. AFLAḤ

Article in *DSB* by R. Lorch.
See al-Abharī (no. G16).

F2 ABU L-WALĪD ISMĀᶜĪL B. AḤMAD AL-QURṬUBĪ

Not identified.

4.6.1 *R. fi l-ᶜAmal bi-wajh ṣafīḥat al-ustādh . . . al-Zarqāllu*, in 60 *bāb*s, based on al-Zarqāllu's treatise in 61 *bābs* (see no. B87): TR 131,6 (pp. 82–129, 1186H in Maghribi script). Another copy (dated 1301H) is preserved in MS Damascus Ẓāhirīya 9241 (*cf. Cat.*, pp. 227–228).

F3 [IBN MUᶜĀDH?]

Suter, no. 213 and 213N; article "al-Jayyānī" by Y. Dold-Samplonius and H. Hermelink in *DSB*; *GAS*, V, pp. 49, 109, 364.

4.3.10 (*R. fī ᶜAmal al-asṭurlāb*), compiled in Cordova: ḤM 19,2 (fols. 28v–45r, 1144H, anonymous). Another anonymous copy in MS Berlin Ahlwardt 5807 (fols. 70v–76r). Merits study. A fragment of a treatise on the construction of almucantars and azimuths on astrolabes in ṬM 155,6 (fols. 31r–33r and 34r–34v, *ca.* 1050H) is attributed to the *qāḍī* Abū ᶜAbd Allāh b. Muᶜādh. This was inserted in an anonymous treatise entitled *Tadhkirat al-albāb* (see no. F7) by an earlier copyist.

F4 ANONYMOUS

4.7.2 *R. fī Bayān takhṭīṭ al-ẓuhr wa-khaṭṭ al-zawāl wa-khaṭṭ al-ᶜaṣr wa-khaṭṭ ākhir waqt al-ᶜaṣr fi l-balāṭa al-mabsūṭa allatī laysat fihā sāᶜāt*: ḤM 19,3 (fols. 45v–53v, 1144H, anonymous). The treatise deals with the marking of times of religious significance on a sundial. The author mentions two works of his entitled *R. Takhṭīṭ al-asṭurlāb* and *K. Takhṭīṭ al-balāṭa*.

F5 ANONYMOUS

4.7.2 (*R. mukhtaṣara fī ᶜAmal rukhāma bi-l-ẓill al-mabsūṭ*), also on sundial theory, with a table for marking sundials (for the latitude of Fez): ḤM 19,4 (fols. 55v–58r, 1144H, anonymous). Some remarks at the end of the treatise are due to Ibrāhīm Khōja al-Muhandisī (no. H46).

F6 ANONYMOUS

4.6.12+ *Faṣl fī waḍᶜ al-sāᶜāt ᶜala l-musaṭṭaḥāt*: ṬM 155,3 (fols. 19r–21v, *ca.* 1050H). This short treatise or extract describes certain varieties of early Islamic quadrants not known from other Arabic sources. Analyzed in D. A. King, "On the Early History of the Quadrant in Islamic Astronomy" (forthcoming).

F7 ANONYMOUS

4.3.30 A treatise on the astrolabe entitled *Tadhkirat al-albāb fī ṣifat ᶜamal al-asṭurlāb*: TM 155,5 (fols. 29v–42r, *ca.* 1050H); DM 177,1 (fols 1v–12v, 1103H). See also no. F3.

F8 ANONYMOUS

4.3.30 ***Bāb fī Maᶜrifat rasm al-ᶜankabūt li-l-asṭurlāb***: ṬM 155,7 (fols. 42v–43r, *ca.* 1050H).

F9 MUḤAMMAD B. ᶜALĪ B. YAḤYĀ AL-NAṬṬĀḤ

Suter, no. 499.

4.3.12 A note in DH 10, fol. 39v states that the best treatises on the astrolabe are those of Ibn al-Naṭṭāḥ and Ibn al-Samḥ (no. B115). The former is extant in MS London B.L. 405,1.

F10 ABŪ ᶜABD ALLĀH MUḤAMMAD B. AḤMAD B. HISHĀM AL-LAKHMĪ

GAL, (I, p. 375, *etc.* and) SI, p. 854, no. 50.

1.4.7 A commentary on a poem attributed to Ibn al-Haytham (no. B77) on the entry (*tarḥīl*) of the sun into the lunar mansions: DM 1051 (23 fols., 1220H). A copy of the poem without the commentary is in MS Princeton Mach 4872 = Yehuda 1168 (fols. 62v–64a, *ca.* 1150H).

F11 AḤMAD B. YŪSUF, IBN AL-KAMMĀD

Suter, no. 487.

See no. B40.

F12 ABŪ MUḤAMMAD ᶜABD ALLĀH B. AL-ḤAJJĀJ, IBN AL-YASAMĪN

Suter, no. 320; *Renaud*, no. 320; *Krause*, no. 320; *GAL*, I, p. 621, and SI, p. 858.

6.3.8 ***al-Urjūza al-Yasamīnīya***, a poem on algebra: K 8522,3 (fols. 45r–48v, 836H); DR 112,4 (fols. 25r–27v, *ca.* 900H); TJ 289,1 (pp. 2–5, *ca.* 900H), *etc.* (9 copies). See Pl. LXXXVIIIa. Commentaries by Ibn al-Hāᵓim (no. C58) entitled ***al-Durr al-thamīn***: K 4313,1 (fols. 1r–81v, 814H), DR 1 (114 fols., 857H), *etc.* (4 copies); by an anonymous author, perhaps al-Qalaṣādī (no. F34), entitled ***Tuḥfat al-nāshiᵓīn***: TJ 289,2 (pp. 6–36, *ca.* 900H); plus four commentaries by Sibṭ al-Māridīnī (no. C97), the first without title: DR 314 (19 fols., *ca.* 1150H), DR 361 (20 fols., 1181H), *etc.* (4 copies); the second without title: DR 181,14 (fols. 483r–494v, *ca.* 1150H); the third entitled ***al-Tuḥfa al-Māridīnīya*** . . . : DR 362 (10 fols., 1109H), DJ 39,16 (fols. 168v–173v, 1132H), *etc.* (1 copy), and the fourth entitled ***al-Lumᶜa al-Māridīnīya***: DR 656 (9 fols., 1111H), DR 89,1 (fols. 1v–15v, 1172H), *etc*, (17 copies), and commentaries thereon by Muḥammad al-Ḥifnī (no. D104) and Aḥmad al-Janājī (no. D203); an extensive commentary attributed to Shams al-Dīn al-Ḥijāzī (no. D69): DR 1038 (105 fols., *ca.* 1250H, lacks beg. and end); plus a commentary by al-Sajjāᶜī (no. D86), entitled ***Fatḥ dhi l-ṣifāt al-sanīya*** . . . : DR 181,10 (fols. 410r–419r, 1180H); plus two anonymous commentaries: DR 55 (9 fols., *ca.* 1150H) and TJ 246,8 (pp. 64–69, 1140H, lacks end).

6.3.8 ***Urjūza fi l-Judhūr = al-Yasamīnīya fī aᶜmāl al-Judhūr***, a poem on root extraction: TR 138,3 (pp. 35–38, *ca.* 850H); TJ 86,9 (pp. 255–260, 1177H); DR 112,1 (fols. 1r–3r, *ca.* 800H). Commentary by al-Qalaṣādī (no. F34), entitled ***Bughyat al-ṭālibīn*** . . . : TJ 86,8 (pp. 233–254, 1177H).

Some biographical information on Ibn al-Yasamīn is contained at the end of K4313,1.

F13 AḤMAD B. ᶜALĪ B. YŪSUF AL-BŪNĪ

Suter, no. 341; *GAL*, I, pp. 655–656, and SI, pp. 910–911; *EI*$_2$, Supplement.

5.3.5 ***R. fī Aḥkām al-burūj wa-l-kawākib wa-maᶜrifat awāᵓil al-sinīn***, an astrological treatise stated to be taken from ***K. Shams al-maᶜārif al-wusṭā***: Sh 84 (17 fols., *ca.* 1150H). A magical treatise by al-Būnī is in MM 67,3, and a ***R. fī Sirr al-basmala*** by him is in TR 118,3.

F14 ABŪ ᶜABD ALLĀH MUḤAMMAD B. ᶜALĪ, IBN TŪMART AL-ANDALUSĪ[1]

GAL, I, pp. 506–507, and SI, p. 697, and I, p. 274 and SI, p. 424 (are these references to the same

---

1. In ḤJ 47,1 the author is named as Jamāl al-Dīn Muḥammad b. ᶜAlī b. Tūmart, perhaps a confusion with Jamāl al-Din (Abu l-ᶜAbbās Aḥmad b. ᶜAlī) b. Tamāt, a thirteenth-century mathematician (no. C11).

individual?); article "Ibn Tūmart" in EI_2 by J. F. P. Hopkins.

1.6.5 ***Kanz al-ʿulūm wa-l-durr al-manẓūm fī ḥaqāʾiq ʿilm al-sharīʿa wa-daqāʾiq ʿilm al-ṭabīʿa***, a treatise on philosophy and the sciences in 5 *bābs*: ḤJ 47,1 (fols. 1v–85r, 1158H); DṬ 124,1 (pp. 1–74, 1257H); *etc.* (1 fragment). A fine copy which came to my attention after the cataloguing was completed is MS Dār al-Kutub *falsafa* 411.[1]

On this work see G. Vajda, "Une synthèse peu connue de la révélation et de la philosophie: le *Kanz al-ʿulūm* de Muḥammad b. ʿAlī Ibn Tūmart al-Andalusī," in *Mélanges L. Massignon*, III (Damascus, 1957), pp. 359–374. Vajda (p. 361) has shown that the work cannot pre-date *ca.* 575/1175. The section on astronomy merits detailed investigation. The parts on timekeeping are discussed in *SATMI*. The text is currently being edited by ʿĀmir ʿAṭāʾ ʿAbd al-Ghanī al-Bayānī of the Center for the Revival of the Arab Scientific Heritage in Baghdad.

F15 ABŪ MUḤAMMAD ʿABD AL-ḤAQQ B. IBRĀHĪM, IBN SABʿĪN

GAL, I, p. 611, SI, p. 844, and SII, p. 1017, no. 6; article "Ibn Sabʿīn" in EI_2 by A. Faure.

5.3.6 *K. al-Daraj*, an astrological treatise: MJ 202,2 (fols. 14v–32r, 1092H in clear *naskhī*), unique—see Pl. CVIIIc. Another astrological work by Ibn Sabʿīn entitled ***Lisān al-falak al-nāṭiq ʿalā wajh al-ḥaqāʾiq*** is extant in MS Damascus Ẓāhiriya 9467, and consists entirely of tables of a kind not attested in any other source known to me. Brockelmann lists a copy of this work in MS Hyderabad Asafiya I, p. 802,109. An astronomical work entitled ***al-Tamkīn*** attributed to Ibn Sabʿīn is in MS Gotha 1379,3 (*cf. GAL*, SII, p. 1017, no. 6). These works deserve detailed investigation! See also the non-numerical astrological tables entitled *jadāwil baḥr al-ḥurūf* with a short introduction in MM 189,3 (fols. 110r–122r, *ca.* 1100H), listed in no. Z36 and extracted in Pl. CIb.

1. The other copies of this work listed in the 1890 catalogue appear to have been lost.

F16 ANONYMOUS

3.5.12 A note on finding the time of day in seasonal hours from the shadow lengths, appended to the treatise of Ibn Sabʿīn (no. F15) in MJ 202,3 (fol. 33v, 1092H). See further *SATMI*, III.

F17 ABŪ MIQRAʿ

Renaud, no. 531.

3.2.25 See ***al-Muqniʿ fī ʿilm Abī Miqraʿ, etc.***, by al-Sūsī: no. F49.

F18 ABŪ ʿABD ALLĀH MUḤAMMAD B. ʿUMAR, IBN BADR

Suter, no. 493; *GAL*, SI, p. 860.

6.3.14 A fragment of his ***K. al-Jabr wa-l-muqābala***: K6829,2 (fol. 53v, *ca.* 1050H), corresponding to pp. 18–19 of published text (Madrid, 1916). The author is named as Ibn Nadr.

F19 ABŪ ʿALĪ AL-ḤUSAYN B. AḤMAD B. YUSŪF B. BĀṢO AL-ASLAMĪ (?)

Suter, no. 381b; *Renaud*, no. 381b; *GAL*, SI, p. 869 and SII, p. 709; Renaud, "Les Ibn Baso," *Hespéris*, 24 (1937), pp. 1–12.

4.6.3 His *R. fi l-Ṣafīḥa al-jāmiʿa* on a universal astrolabe which he claims is superior to the *zarqāllīya* and *shakkāzīya* plates, in 161 *bābs*, compiled in 673H, is extant in MSS Escorial ar. 956,7 and Rabat 451,1. On this instrument, see Renaud's article cited above. An anonymous abridgement in 11 unnumbered *bābs*: TR 159,2 (pp. 62–80, 1320H from a copy in the Khālidīya Library, Jerusalem, dated 1129H). Another copy of this may be MS London B.L. ar. 408,9. *Cf. Renaud*, nos. 381b and 543.

F20 ʿABD ALLĀH B. MUḤAMMAD AL-TUJĪBĪ

Suter, no. 196; *GAL*, SI, 402.

4.6.2 ***R. fi l-Ṣafīḥa al-shakkāzīya***, a treatise on the use of the single *shakkāzīya* plate in 25 *bābs*:

TR 159,1 (pp. 1–61, 1320H from a copy in the Khālidīya Library in Jerusalem). No other known copies.

F21 ABU L-QĀSIM AL-ZUBAYR B. AḤMAD B. IBRĀHĪM B. AL-ZUBAYR AL-THAQAFĪ AL-QĀḌĪ

Suter, no. 513; *GAL*, SII, 1025, no. 88; *Awwad*, no. 33. On his father, see the article "Ibn al-Zubayr" by Ch. Pellat in EI_2.

4.3.16 ***Tadhkirat dhawi l-albāb fi stifāʾ al-ᶜamal bi-l-asṭurlāb***, on the use of the astrolabe in 3 *qism*s with 1 + 1 + 138 *bāb*s: DM 173,2 (fols. 12v–84r, *ca.* 1150H); K 8535,1 (fols. 1r–93r, *ca.* 1000H); disordered fragments in ZK 714,2 (fols. 15v–21v, *ca.* 1250H). al-Thaqafī quotes opinions of Ibn al-Samḥ (no. B115), Ibn al-Ṣaffār (no. B74), and Abu l-Ṣalt (no. B89), all of whose works on the astrolabe are extant. Other copies in London B.L. 407,1; Algiers 1466; Mosul 103/66/2; Rampur I 422,11.

F22 ABŪ ᶜABD ALLĀH MUḤAMMAD B. IBRĀHĪM AL-AWSĪ, IBN AL-RAQQĀM

Suter, nos. 388 and 417 (??); *Renaud*, no. 388; *GAL*, SII, p. 378; not listed in *Kennedy*.

2.1.18 A fragment from the introduction of his ***al-Zīj al-Mustawfī*** is in DM 718,2 (fols. 3v–4r, 917H). See no. C100.

4.3.20 ***Manẓūma fi l-ᶜamal bi-l-asṭurlāb***, a poem on the use of the astrolabe in 53 *bāb*s: K3982 (24 fols., *ca.* 1250H), unique?

F23 ABU L-ᶜABBĀS AḤMAD B. MUḤAMMAD, IBN AL-BANNĀʾ

Suter, no. 399; *Renaud*, no. 399; *GAL*, II, pp. 330–331, and SII, pp. 363–364; article by J. Vernet in *DSB*.

1.4.10 Chapters on determining the solar longitude and *nuḥūs al-qamar* (?) and a fragment from his work entitled ***al-Manākh***: DM 923 (photos of MS London B.L. add. 9599, fols. 41 + 52–54!, *ca.* 1200H).

1.4.10 ***Qānūn li-tarḥīl al-shams wa-l-qamar fi l-manāzil wa-maᶜrifat awqāt al-layl wa-l-nahār***, on folk astronomy: ṬM 160,2 (fols. 36r–39r, 1162H, anonymous). Brockelmann lists other copies attributed to Ibn al-Bannāʾ in London B.L. 407,2 and Tunis (*GAL*, II, p. 331 and SII, p. 364 (no. 6)).

2.1.17 His ***Minhāj al-ṭalib li-taᶜdīl al-kawākib***, extant in several copies, including MSS Escorial 904 and Chester Beatty 4087, is a *zīj* with extensive tables that merit detailed investigation. Only the introduction is published and translated in J. Vernet Gines, *Contribución al-Estudio de la Labor Astronómica de Ibn al-Bannāʾ*, (Tetuan, 1952), and the *zīj* is not listed in *Kennedy*.

2.1.17 ***al-Yasāra fī taqwīm al-kawākib al-sayyāra***, a short treatise on planetary astronomy in 11 *bāb*s: DM 924 (photos of MS London B.L. Add. 9599, fols. 38v–40r, *ca.* 1200H); plus an anonymous supplement ***Tashīl al-ᶜibāra fī takmīl mā naqaṣa min al-Yasāra***: DM 922 (photos of MS London B.L. Add. 9599, fols. 55v–79r, *ca.* 1200H), which includes some tables. No other known copies of either.

A small *zīj* for Tlemcen based on the ***Yasāra*** (although the London copy has no tables) is ***Tashīl al-maṭālib fī taᶜdīl al-kawākib*** by Ibn al-Qunfudh (no. F25), extant in several copies including one in the Madrid Museo Naval de Madrid (unnumbered). In Cairo we find two copies of another small *zīj* for Cairo entitled ***Tashīl*** or ***Taysīr al-maṭālib fī taᶜdīl al-kawākib*** by Aḥmad b. Muḥammad ᶜAyyād (no. D67) in 1110H: MM 23,1 (9 fols., 1153H); TR 49 (50 fols., 1305H). The author states that he used ***al-Yasāra*** by Ibn al-Bannāʾ but it appears that he actually simply plagiarized the work of Ibn Qunfudh. Both Cairo copies contain what I think is an early Andalusian crescent visibility table (see no. B99). See also the Maghribi *zīj* used by al-Lādhiqī (no. D66).

5.3.11 (***R. fī Aḥkām al-nujūm***), a treatise on astrology: MM 204,2 (fols. 9r–21r, 1052H); ZK 714,1 (fols. 1v–15r, *ca.* 1250H). Another copy is MS Istanbul Şehid Ali Paşa 2774,3 (fols. 10–13, n.d.).

6.3.19 ***K. fi l-Jabr wa-l-muqābala***, a treatise on algebra in 2 *juzʾ* of 3 + 2 *qism*s: MR 1 (40 fols., 746H), unique? Merits study.

6.3.19 *K. al-Maqālāt fi-l-ḥisāb*, a treatise on arithmetic in 4 *maqālas* of 10 + 10 + 7 + 6 *bābs*: MR 33 (50 fols., 847H). An edition based on MS Istanbul Laleli 2720 was published by A. S. Saidan in Amman, 1984.

6.3.19 *K. Talkhīṣ aᶜmāl al-ḥisāb*, a treatise on arithmetic in 2 *juz*ᵓ of 3 + 2 *qisms*: K 8522,1 (fols. 2r–43v, 836H); DR 16,1 (fols. 1r–14r, *ca.* 800H); *etc.* (2 copies); plus a commentary by Ibn Haydūr (no. F37), entitled *al-Tamḥīṣ fī sharh al-Talkhīṣ*: ṬR 103 (187 fols., 875H); *etc.* (1 copy); and a poem based on the *Talkhīṣ* by Muḥammad b. Aḥmad al-ᶜUthmānī (no. F40), entitled *Munyat al-ḥisāb*: DR 355 (12 fols., *ca.* 1150H), and commentary thereon by al-ᶜUthmānī entitled *Bughyat al-ṭullāb* . . . : MR 4 (140 fols., *ca.* 1100H); ṬR 122 (165 fols., *ca.* 1150H); *etc.* (2 copies); and a commentary by Ibn al-Hāᵓim (no. C58), entitled *al-Ḥāwī*: J 3964,1 (fols. 1r–7v, 862H); *etc.* (1 copy); and several commentaries thereon (see no. C58); and by ᶜAbd al-ᶜAzīz al-Hawarī (no. F24X) entitled *al-Lubab*: K 6829,1 (fols. 1r–53r, *ca.* 1050H); and by Ibn al-Majdī (no. C62), entitled *Ḥāwi l-lubab*: K 6829,1 (fols. 1r–53r, 1093H); and by al-Ghazzī (no. C115), entitled *Takhṣīṣ ūli l-albāb*: DR 819 (17 fols., *ca.* 1250H).

A new edition of the *Talkhīṣ* has been published by M. Souissi (Tunis, 1969).

6.9.3 *al-Qānūn fi l-farāᵓiḍ*, a short work on the arithmetic of inheritance: J 3964,5 (fols. 25r–29r, 862H).

F24 ANONYMOUS

2.5.16 *(Fuṣūl fī) Tarḥīl al-darārī wa-maᶜrifat burūjihā*, chapters on the determination of planetary longitudes approximately: ṬM 160,5 (fols. 54v–55v, *ca.* 1200H), unique.

F24X ᶜABD AL-ᶜAZĪZ B. ᶜALĪ B. DĀᵓŪD AL-HAWARĪ AL-MIṢRĪ (?)

Renaud, no. 415.

6.3.19 *al-Lubab fī sharḥ Talkhīṣ aᶜmāl al-ḥisāb*, a commentary on the *Talkhīṣ* of Ibn al-Bannāᵓ: see no. F23.

F25 IBN AL-QUNFUDH

Renaud, no. 422.

2.1.17 *Tashīl al-maṭālib fī taᶜdīl al-kawākib*, a small *zīj* for Tlemcen: see no. F23.

5.2.18 A commentary on the poem on astrology by ᶜAlī b. Abi l-Rijāl: see no. B76.

F26 ABŪ ZAYD ᶜABD AL-RAḤMĀN B. ABĪ GHĀLIB MUḤAMMAD AL-JĀDARĪ

Suter, no. 424a; *Renaud*, no. 424a; *GAL*, SII, 217–218.

3.2.7 *Rawḍat al-azhār fī ᶜilm waqt al-layl wa-l-nahār*, a poem on timekeeping compiled in 794H: DM 181,1 (fols. 1r–12r, 1179H); TR 87 (20 pp., *ca.* 1200H); DM 975,1 (fols. 1r–9r, *ca.* 1200H, contains notes about the author); *etc.* (2 copies); plus an anonymous commentary entitled *Natāᵓij al-afkār* . . . : K 4311 (48 fols., defective, 1183H); and a commentary with some tables for Tārūdānt, latitude 30;15°, by Ibn al-Muftī (no. F43), entitled *Qaṭf al-anwār* . . . : K 3854 (49 fols., *ca.* 1050H); DM 1124 (86 fols., 1221H); *etc.* (2 copies); and a commentary without title by an anonymous author: K 7584,2 (fols. 82v–111r, 1257H). The first of these commentaries contains a discussion of Maghribi traditions in trepidation and twilight. The other two contain no historical information of consequence.

F27 AḤMAD B. ABĪ ḤUMAYDA AL-MUṬARRIFĪ

Renaud, no. 534; *GAL*, SII, p. 217.

3.2.7 A note on al-Jādarī (no. F26) by Aḥmad b. Abī Ḥumayda al-Maghribī who was the author of one of the many commentaries on the *Rawdat al-azhār*, is preserved in DM 975,1 (fol. 1r, *ca.* 1200H), and informs us that al-Jādarī died in the vicinity of the old city of Fez (*Fās al-bālī*) and was buried in the cemetery of the gate near al-Ḥamrāᵓ Mosque. There is no copy of this particular commentary in the manuscript now; a copy exists in MS Algiers Fagnan 613,2.

F28 ABŪ ᶜABD ALLĀH MUḤAMMAD B. AḤMAD AL-ḤABBĀK

Suter, no. 435.

4.3.27 ***Bughyat al-ṭullāb fī ᶜilm al-asṭurlāb***, a poem on the use of the astrolabe: DM 169,1 (fols. 1v–6v, *ca.* 1100H); *etc.* (1 copy); plus a commentary by Abū ᶜAbd Allāh Muḥammad b. Yūsuf al-Sanūsī (no. F29): DM 169,2 (fols. 7r–87v, *ca.* 1100H); DM 702,5 (fols. 27v–100v, 1160H).

F29 ABŪ ᶜABD ALLĀH MUḤAMMAD B. YŪSUF AL-MUQRIʾ AL-SANŪSĪ

4.3.27 ***Kifāyat al-ṭullāb fī ᶜilm al-asṭurlāb***, a poem on the use of the astrolabe: DM 780 (8 fols., 828H in author's hand).

4.3.27 A commentary on ***Bughyat al-ṭullāb*** (another poem on the astrolabe) by al-Ḥabbāk: see no. F28.

F30 ANONYMOUS

3.1.19 Auxiliary tables for finding the hour-angle from the solar altitude for all latitudes, preceded by a table displaying the solar longitude over a 4 Syrian year period (incomplete): DM 689 (52 fols., *ca.* 1000H). These tables are based on the minor auxiliary tables of al-Khalīlī (no. C37) extant in the unique copy MS Dublin Chester Beatty 4091. See Pl. XLIa and *SATMI*, I.

F31 ABRAHAM ZACUTO

Born *ca.* 1452 in Salamanca, died *ca.* 1522 in Damascus; see the article "Zacuto" by J. Vernet in *DSB.*

2.2.9 A Maghribi version of the "perpetual almanac" of Zacuto: DM 1081 (326 pp., *ca.* 1200H), unique? See Pl. XXIII. This should now be compared with MS Escorial ar. 966, which is apparently a version for Istanbul prepared by Mūsā Galīnūs (no. H3), on which see further J. Vernet, "Una version arabe resumida del almanach perpetuum de Zacuto," *Seferad*, 10 (1950), pp. 115–133. In the Cairo manuscript the tables (pp. 36–326) are preceded by three different introductions, the first (pp. 1–8) anonymous and arranged in unnumbered sections, the second (pp. 9–21) in 24 *bābs* translated from the Spanish by Aḥmad b. Qāsim al-Jaḥdarī al-Andalusī (no. F33), and the third (pp. 22–30) in 8 *bābs* attributed to ᶜAbd al-Raḥmān al-Fāsī (no. F50).

F32 ABŪ L-ṢALĀḤ JĀBIR B. ᶜABD ALLĀH B. AL-ḤĀJJ AL-GHIYĀTHĪ[1]

Not listed in the modern sources.

2.5.12 ***Muqarrib al-maṭālib fī taᶜdīl al-kawākib***, a poem on planetary astronomy compiled in Fez: DM 1081,4 (pp. 31–35, *ca.* 1200H). Other copies are MS London B.M. Add. 9598,7 (fols. 94v–101v, n.d.) and MS Istanbul Haci Beşir Ağa 669,11 (fols. 191–196, 1080H). al-Ghiyāthī mentions Ibn Isḥāq (no. F74) and Ibn Abi l-Shukr (no. G21) in his introduction. Merits study.

F33 AḤMAD B. QĀSIM AL-JAḤDARĪ AL-ANDALUSĪ

Unidentified.

2.2.9 A translation in 24 *bābs* of the introduction to the tables of Zacuto (no. F31), apparently made from the Spanish (*min lisān al-Rumanṣ*): DM 1081,2 (pp. 9–19, *ca.* 1200H), unique.

F34 ABU L-ḤASAN ᶜALĪ B. MUḤAMMAD AL-QURASHĪ AL-BASAṬĪ AL-QALAṢĀDĪ

Suter, no. 444; *Renaud*, no. 444; *GAL*, II, pp. 343–344, and SII, pp. 378–379; article "al-Qalaṣādi" by A. S. Saidan in *DSB.*

6.3.8 ***Ghunyat al-ṭālibīn ᶜalā judhūr Ibn al-Yāsamīn***, a commentary on the poem on root extraction by Ibn al-Yāsamīn (no. F12): TJ 86,8 (pp. 233–254, 1177H). Perhaps al-Qalaṣādī is also the author of ***Tuḥfat al-nāshiʾīn ᶜalā Arjūzat Ibn al-Yāsamīn***, a commentary on Ibn al-Yāsamīn's poem on algebra: TJ 289,2 (pp. 6–36, *ca.* 900H). Another copy of this anonymous commentary

---

1. In the London manuscript the author is also called al-Tāzī.

exists in MS London India Office Loth 770,2 (*Renaud*).

6.4.4 ***Inkishāf al-julbāb fī funūn al-ḥisāb***, a treatise on arithmetic in 4 *bāb*s in 5 + 6 + 6 + 2 *faṣl*s, compiled in Tunis in 849H: MR 2 (40 fols., *ca.* 1100H).

6.4.4 *Kashf al-julbāb ᶜan ᶜilm al-ḥisāb*, a treatise on arithmetic in 4 *juz*ᵓ of 8 *bāb*s: DR 1039 (39 fols., 1135H); and an abridgement thereof, entitled *Kashf al-asrār ᶜan ᶜilm ḥurūf al-ghubār*: TR 139,2 (pp. 9–68, *ca.* 1100H); ZK 490,1 (pp. 1–72, 1002H); *etc.* (11 copies and 1 fragment). See Pl. XCIIIa.

6.9.11 Part of a commentary on the *farāᵓiḍ* of al-Imām Khalīl (see no. C29): TR 139,3 (pp. 68–88, *ca.* 1100H, lacks end). *GAL*, SII, pp. 98–99 has title.

F35 ABŪ ᶜABD ALLĀH MUḤAMMAD B. MUḤAMMAD B. ᶜABD AL-WAHHĀB ABĪ MUḤAMMAD B. ᶜABD ALLĀH AL-JANNĀD AL-ANṢĀRĪ

Not mentioned in the modern sources.

3.1.16 Prayer-tables for latitudes 34° (Meknes), 31° (Marrakesh), and 30° (Siljilmasa), with an introduction: TR 338,2 (pp. 204–287, *ca.* 1250H), unique? Analysed in *SATMI*, II. See Pl. XXXVII.

F36 ABU L-ḤASAN ᶜALĪ, IBN AL-MAGHRIBĪ[1]

Suter, no. 528 and 528N; *GAL*, SII, 1020, no. 28.

6.3.10 *Lawḥ al-ḍabṭ* (= *Urjūza fī ḥisāb al-yad*), a poem on finger arithmetic: K 3957,4 (fols. 253v–256v, *ca.* 900H); TJ 97,6 (pp. 207–208, 1097H). Commentary by Zayn al-Dīn al-Ṣūfī (no. C93): DR 674 (8 fols., *ca.* 1150H). The poem, without the introduction, was published by A. Saidan, *Arabic Arithmetic: The Arithmetic of Abu al-Wafāᵓ al-Būzajānī* (Amman, 1972), pp. 417–418; and the commentary by J. Ruska in *Der Islam*, 10 (1920), 90 ff.; *cf.* also *Isis*, 64 (1973), p. 125.

1. In both Cairo copies of the poem the author is called Ibn al-ᶜArabī. In the Cairo copy of the commentary the author is called Ibn al-Maghribī.

F37 ᶜALĪ B. ᶜABD ALLĀH B. MUḤAMMAD B. HAYDŪR

6.3.19 *al-Tamḥīṣ fī sharḥ al-Talkhīṣ*: a commentary on the *Talkhīṣ* of Ibn al-Bannāᵓ: see no. F23 (one copy dated 875H).

F38 ABŪ ARAWĪ (?) MUḤAMMAD B. ᶜIMRĀN B. ABĪ MUḤAMMAD AL-QUṬB SAYYIDĪ ᶜABD AL-SALĀM AL-FĪTŪRĪ (?)

Not mentioned in the modern sources.

1.4.13 *al-Qawāᶜid al-mufīda li-l-adhhān al-balīda* = *R. tashtamilu ᶜalā baᶜḍ qawāᶜid ḥisābiya fī maᶜrifat al-shuhūr wa-l-sinīn wa-l-manāzil wa-l-burūj wa-awqāt al-ṣalawāt wa-l-qibla*, a treatise on folk astronomy in 11 *bāb*s: ṬM 160,3 (fols. 39v–55v, 1162H), unique.

F39 ABŪ JAᶜFAR ᶜUMAR B. ᶜABD AL-RAḤMĀN B. ABI L-QĀSIM AL-TŪNISĪ AL-TŪZARĪ

Suter, no. 439; see also MS Manchester Rylands 353D.

4.5.10 *Maḥṣalat al-maṭlūb fi l-ᶜamal bi-rubᶜ al-juyūb*, a treatise on the use of the sine quadrant in 40 *bāb*s compiled about 850H: TR 131,5 (pp. 42–81, 1186H), unique. al-Tūzarī quotes the opinions of Abu l-Wafāᵓ al-Būzajānī (no. B54) and Abū ᶜAlī al-Marrākushī (no. C17) on the latitude of Mecca, and Jamāl al-Dīn al-Māridīnī (no. C47), Shams al-Dīn al-Mizzī (no. C34) and ᶜAbd al-ᶜAzīz on twilight. The last mentioned is ᶜAbd al-ᶜAzīz b. Masᶜūd b. ᶜAbd al-ᶜAzīz (no. C46), the author of the treatise on the sine quadrant preserved in MS Escorial 918,14, rather than ᶜAbd al-ᶜAzīz al-Wafāᵓī (no. C61).

F40 MUḤAMMAD B. AḤMAD . . . B. GHĀZĪ AL-ᶜUTHMĀNĪ

Suter, no. 451; *GAL*, II, p. 311, and SII, pp. 337–338.

6.3.19 *Munyat al-ḥussāb*, a poem on arithmetic based on the *Talkhīṣ* of Ibn al-Bannāᵓ (no. F23): DR 355 (12 fols., *ca.* 1150H); plus a commentary entitled ***Bughyat al-ṭullāb fī sharḥ Munyat al-ḥussāb***, compiled in Fez, 895H: MR 4 (140 fols., *ca.* 1100H), ṬR 122 (165 fols., *ca.* 1150H), *etc.* (2 copies). This commentary has recently been published in Aleppo by Dr. M. Souissi.

F41 ᶜABD AL-RAḤMĀN B. ABĪ MUḤAMMAD AL-ṢAGHĪR AL-AKHḌARĪ

GAL, II, pp. 614–615 and SII, pp. 705–706. See also no. F42.

6.9.14 *al-Durra al-bayḍāᵓ*, a poem on inheritance, printed in Cairo, 1310H, together with an anonymous commentary, Cairo, 1309H (Brockelmann). Anonymous commentary thereon: B 22581 (180 fols., *ca.* 1250H); B 20411 (149 fols., *ca.* 1300H).

F42 ᶜABD AL-ᶜAZĪZ B. AḤMAD B. MUSLIM AL-AKHḌARĪ (??)

See different name in *GAL*, SII, pp. 705–706. See also no. F41.

1.4.12 A commentary on the author's poem *al-Sirāj*, dealing with the lunar mansions: K 18361 (26 fols., 1140H); *etc.* (1 copy); plus another commentary entitled *Mufīd al-muḥtāj fī sharḥ al-Sirāj* by Saḥnūn b. ᶜUthmān (no. F61): K 3853 (48 fols., *ca.* 1150H). Brockelmann (*GAL*, SII, p. 706) lists another (?) commentary in MS Algiers 1451.

F43 ABŪ ZAYD ᶜABD AL-RAḤMĀN B. ᶜUMAR B. AḤMAD AL-SŪSĪ AL-JAZŪLĪ AL-BĀᶜUQAYLĪ, IBN AL-MUFTĪ

GAL, SII, p. 217; *Renaud*, no. 535.

3.2.7 *Qaṭf al-anwār min Rawḍat al-azhār*, a commentary on the treatise of al-Jādarī (see no. F26): K 3854 (49 fols., *ca.* 1150H); DM 1124 (86 fols., 1221H); *etc.* (2 copies). Other copies in Rabat (*Renaud*) and London B.L. Add. 12540. Printed Fez, 1326H?

F44 IBN AL-NAJJĀR

See no. F46.

F45 ABŪ ᶜABD ALLĀH MUḤAMMAD KWYNKH (?)

See no. F46.

F46 MUḤAMMAD B. MUḤAMMAD B. MUḤAMMAD AL-IDRĪSĪ = ABŪ MUḤAMMAD ḤAMŪDA AL-IDRĪSĪ (?)

Not mentioned in the modern sources.

4.7.17 *Rawḍat al-nāẓir fī kayfīyat waḍᶜ khuṭūṭ faḍl al-dāᵓir*, a treatise on sundial theory in 4 *bāb*s: DM 1169,2 (fols. 11v–25v, 1223H), unique? The author mentions two other Maghribi writers who are also new to the modern literature (nos. F44 and F45), named Ibn al-Najjār and Abū ᶜAbd Allāh Muḥammad Kwynkh (?), author of a treatise on sundial theory entitled *Iḥyaᵓ al-mawāt fi l-basāᵓiṭ wa-l-munḥarifāt.* He also mentions the Egyptian astronomers Sibṭ al-Māridīnī (no. C97) and Ibn al-Majdī (no. C62), as well as Ulugh Beg (no. G49). The treatise contains various sundial tables for latitude 36;51°, that is, Tunis.

F47 ᶜABD ALLĀH ḤUSAYN QUṢᶜA AL-TŪNISĪ, IBN QUṢᶜA

Not mentioned in the modern sources.

2.2.5 *Ghunyat al-ṭālib fī taqwīm al-kawākib*, a *zīj* for longitude 41;45° (Tunis) based on the *Zīj* of Ulugh Beg (no. G49), arranged in 5 *maqāla*s and compiled in 1091H: DM 814 (316 pp., 1325H from author's copy). Another late copy of this *zīj* is in the Library of the American University in Cairo.

F48 ᶜALĪ B. MUḤAMMAD AL-DĀDISĪ[1]

GAL, II, p. 616 and SII, p. 708; *Renaud*, no. 537; *Azzawi*, pp. 334–335.

3.2.27 *al-Yawāqīt li-mubtaghī maᶜrifat al-mawāqīt*, a poem on timekeeping: K 4029,1 (fols. 1v–7r, *ca.* 1100H); DM 1169,9 (fols. 75v–86r, 1125H); plus an anonymous

---

1. In K 4029,1 the name is given as ᶜAlī b. Aḥmad b. Muḥammad . . . ; in DM 1169,9 as Muḥammad ᶜAli al-Dādisī; in MM 203,1 as Aḥmad b. ᶜAlī al-Dādisī. The name is given in full, as in *GAL* and in *Renaud*, in MM 236.

commentary, probably by the author, entitled *Fatḥ al-Mūqīt fī sharḥ al-Yawāqīt*: MM 203,1 (fols. 1r–108r, 1107H).

F49 ABŪ ZAYD MUḤAMMAD B. SAᶜĪD B. MUḤAMMAD AL-SŪSĪ AL-MARGHĪTHĪ

Renaud, no. 540.

3.2.25 ***al-Muqniᶜ fī ᶜilm Abī Miqraᶜ***, a poem on the methods of timekeeping used by Abū Miqraᶜ—see no. F17: ṬJ 201,2 (fols. 7r–12v, *ca.* 1000H); DM 178,2 (fols. 17r–19v, *ca.* 1200H); *etc.* (6 copies); plus commentary by author entitled ***al-Mumtiᶜ*** in DM 415 (55 fols., *ca.* 1100H), DM 1053 (53 fols., 1220H), *etc.* (6 copies); plus shorter commentary entitled ***al-Muṭliᶜ*** in ḤM 10 (16 fols., *ca.* 1150H), MJ 10,1 (fols. 1r–17v, 1177H), *etc.* (5 copies); plus commentaries on ***al-Muqniᶜ*** by al-Rasmūkī (no. F59), compiled in 1143H: K 4324 (88 fols., 1280H), and Muḥammad Efendī (no. D171): ṬM 128,2 (fols. 64v–69v, 1275H).

F50 ᶜABD AL-RAḤMĀN B. ᶜABD AL-QĀDIR AL-FĀSĪ

Renaud, no. 541; *GAL*, II, pp. 612 and 615, and SII, pp. 694–695; article "ᶜAbd al-Raḥmān al-Fāsī" by L. Provencal in EI_2.

1.6.7 ***K. al-Uqnūm fī mabādiᵓ al-ᶜulūm***, an encyclopaedia in verse containing sections on various scientific subjects: J 3664 (287 fols., *ca.* 1250H, Maghribi script, fine copy); *etc.* (1 copy from J 3664). Renaud lists MSS Rabat 284 and 286 of this poem; Brockelmann mentions two fragments in Tlemcen. See below.

2.2.9 ***Tuḥfat al-muhtāj fī ᶜilm al-taᶜdīl wa-l-azyāj***, a treatise on the use of tables of Zacuto, arranged in 8 *bābs*: DM 1081,3 (pp. 20–28, *ca.* 1200H), unique. The treatise is anonymous, but the copyist writes that he thought that ᶜAbd al-Raḥmān al-Fāsī was the author.

4.3.37 An extensive commentary on al-Fāsī's section on the astrolabe in the ***Uqnūm*** by Muḥammad Bannānī b. ᶜAbd al-Salām b. Ḥamdūn (no. F57): TR 113 (144 pp., 1327H), unique.

4.8.7 Poem on the magnetic compass (*bayt al-ibra*): ṬJ 367,2 (fols. 11r–12r, *ca.* 1250H). Other copies in Vienna and Rabat (*Renaud*). This poem is not part of the ***K. al-Uqnūm***.

F51 ABŪ ZAYD ᶜABD AL-RAḤMĀN AL-AKHḌARĪ = (??) ABŪ ZAYD ᶜABD AL-RAḤMĀN B. ᶜABD AL-QĀDIR AL-FĀSĪ

Renaud, no. 541.

6.4.36 ***Manẓūma fi l-ḥisāb***, a poem on arithmetic in 8 *bābs*: K 17027,1 (fols. 1r–4v, *ca.* 1300H, attributed to Abū Zayd ᶜAbd al-Raḥmān al-Akhḍarī).

F52 MUḤAMMAD ABU L-AMDĀD (?) AL-MAGHRIBĪ AL-TŪNISĪ, a student of AḤMAD AL-MUQRIᶜ AL-MAGHRIBĪ

5.3.38 ***R. al-Anwār al-saṭiᶜāt fī aḥkām al-dalāᵓil al-qāṭiᶜa***, a treatise on astrology: DM 759 (11 fols., *ca.* 1150H).

F53 ABŪ ᶜABD ALLĀH MUḤAMMAD AL-SHARĪF AL-TŪNISĪ, SANJAQ DĀR

Not listed in the modern sources.

2.2.5 The mean motion tables only, from his *zīj* for longitude 41;45° (Tunis) based on the parameters of Ulugh Beg (no. G49): TR 319,1 (pp. 1–16, *ca.* 1300H). A complete copy of this *zīj* (not listed in *Kennedy*) is in MS Paris B.N. ar. 2536 (48 fols., *ca.* 1150H). Commentaries on the introduction by ᶜAlī b. Māmī (no. F54): DM 1046,4 (fols. 77v–92r, 1163H); and by Bū Daydaḥ (?) (no. F55): DM 1046,1 (fols. 1r–65v, 1169H); DM 1121 (40 fols., *ca.* 1150H, incomplete).

F54 ᶜALĪ B. MĀMĪ (??) AL-ḤANAFĪ

GAL, SII, p. 218. Not listed in *Renaud*.

2.2.5 A commentary on the introduction of the Tunisian recension of the ***Zīj*** of Ulugh Beg by Sanjaq Dār (see no. F53): DM 1046,4 (fols. 77v–92r, 1163H).

Brockelmann notes the existence of a commentary by him on the treatise ***Mujmalat al-maṭlūb*** by al-Māridīnī (see no. C47).

F55 AḤMAD B. MUḤAMMAD BŪ DAYDAḤ (?) AL-QĀDIRĪ (?) AL-QAYRAWĀNĪ

Not listed in the modern sources.

2.2.5 A commentary on the introduction of the Tunisian recension of the *Zīj* of Ulugh Beg by Sanjaq Dār (no. F53): DM 1046,1 (fols. 1r–65v, 1169H); DM 1121 (40 fols., *ca.* 1150H, incomplete).

F56 ᶜUMAR AL-MAGHRIBĪ

Not identified.

4.5.11 *R. fi l-ᶜAmal bi-l-rubᶜ al-mujayyab*, a treatise on the use of the sine quadrant in 45 *bābs*: ṬM 209 (13 fols., 1158H); TR 160 (23 fols., *ca.* 1150H, last *bāb* is numbered 44). Both of these manuscripts are anonymous. The first was copied in Egypt by al-Judarī: in ch. 27 the parameters for twilight are 19°/17°, standard in Egypt. The second of these manuscripts was copied in the Maghrib; in ch. 27 the parameters for twilight are 20°/18°. The same treatise in 45 *bābs*, with twilight parameters 20°/18°, is in MS Alexandria 2293J,6, copied in 1165H, and attributed to ᶜUmar al-Maghribī.

4.5.11 A related anonymous treatise *al-Durr al-masbūb fi l-ᶜamal bi-rubᶜ al-juyūb* probably by the same author, arranged in 23 *bābs*: ṬM 242,7 (fols. 60r–77r, 1071H).

F57 MUḤAMMAD B. ᶜABD AL-SALĀM B. ḤAMDŪN AL-BANNĀNĪ

GAL, II, 615, & SII, p. 686 (*etc*).

4.3.37 A commentary on a poem on the astrolabe by ᶜAbd al-Raḥmān al-Fāsī (see no. F50): TR 113 (144 pp., 1327H); another copy is MS Alexandria *ḥisāb* 50 (see *GAL*, II, p. 615). The commentary contains an anecdote about the discovery of the astrolabe by Ptolemy (no. A18), also recorded by the celebrated thirteenth century Syrian historian Ibn Khallikān: see D. A. King, "The Origin of the Astrolabe according to the Medieval Arabic Sources," *JHAS*, 5 (1981), pp. 43–83, esp. pp. 60–61.

F58 ABŪ MUḤAMMAD ᶜABD AL-ᶜAZĪZ B. ᶜABD AL-SALĀM B. AḤMAD AL-WAZKĀNĪ AL-WAZZĀNĪ

Not mentioned in the standard modern sources.

3.2.30+ An extensive commentary on his poem on timekeeping entitled *Taḥrīr al-mawāqīt* was written by ᶜAbd al-Salām al-ᶜlmy (?) (no. F62). al-Wazkānī mentions al-Jādarī (no. F26) and al-Dādisī (no. F48) and so must postdate them.

F59 MUḤAMMAD B. ᶜABD AL-ᶜAZĪZ AL-JAZŪLĪ AL-RASMŪKĪ

Renaud, no. 542.

3.2.25 A commentary on *al-Muqniᶜ* by al-Sūsī (no. F49): K 4324 (88 fols., 1280H).

F60 AL-JAZŪLĪ

Not identified. See no. F59.

3.2.33 A poem on timekeeing: ṬJ 201,3 (fols. 12v–14v, *ca.* 1100H, lacks first 18 *bābs*).

F61 SAḤNŪN B. ᶜUTHMĀN B. SULAYMĀN . . . AL-WANSHARĪSHĪ

GAL, SI, p. 854, and SII, pp. 706 and 715.

1.4.12 *Mufīd al-muḥtāj fī sharḥ al-Sirāj*, a commentary on the poem on folk astronomy by ᶜAbd al-Raḥmān (!) al-Akhḍarī (no. F42): K 3853 (48 fols., *ca.* 1150H). This work was printed in Cairo in 1314H (see *Suter*, no. 204N).

Note: The following authors have not been identified.

F62 ᶜABD AL-SALĀM B. MUḤAMMAD B. AḤMAD AL-ḤUSNĪ AL-ᶜLMY (?)

3.2.30+ *Abdaᶜ al-yawāqīt ᶜalā taḥrīr al-mawāqīt*, a treatise on timekeeping, essentially a commentary on the poem of *Taḥrīr al-mawāqīt* of al-Wazzānī (no. F58): TR 132,2 (pp. 2–139, *ca.* 1250H), unique. See also an anonymous *bāb* on the prayer-times in TR 132,1 (p. 1, *ca.* 1250H).

4.6.49 ***Irshād al-khall li-taḥqīq al-sāᶜa bi-rubᶜ al-shuᶜāᶜ wa-l-ẓill***, a treatise on an unusual quadrant in 4 *bābs*: DM 207 (7 fols., *ca.* 1300H), unique.

F63 ᶜALĪ AL-BĀṢO (?) AL-TŪNISĪ

4.5.12 A commentary on the ***R. Fatḥīya*** by Sibṭ al-Māridīnī (no. C97): DM 521,14 (fols. 248r–294r, *ca.* 1175H).

F64 ANONYMOUS

1.4.14 *K. al-Istibṣār fī maᶜrifat al-awqāt wa-l-sinīn wa-l-ruᶜūd wa-l-anwāʾ*, a treatise on folk astronomy in which the anonymous author cites poetry by al-Fazārī (no. B3) and quotes Ibn Qutayba (no. B25): K 8536 (51 fols., 1111H, Maghribi hand). This title is not mentioned by Brockelmann.

F65 ANONYMOUS

1.4.14 *al-Qawl fī maᶜrifat awqāt al-tajahhud al-thalātha al-madhkūra fi l-Qurʾān wa-huwa maᶜrifat awqāt al-suḥūr fī jamīᶜ azmān al-sana*, on the times of night vigil: DM 917 (photos of MS London B.L. Add. 9599, fols. 149v–203r, late, Maghribi hand). Another copy is MS Istanbul Hamidiye 1447,25 (fols. 117–118).

F66 ANONYMOUS

2.1.19 ***Rāᶜi l-kawākib***, an anonymous *zīj* with tables for the longitude of Algiers 33;30° based on the parameters of Ibn al-Shāṭir (no. C30): DM 533 (43 fols., *ca.* 1150H, Maghribi script), unique?

F67 ANONYMOUS

2.3.6+ A disordered fragment from the middle of a poem on the compilation of ephemerides: ZK 814 (2 fols., *ca.* 1200H in a Maghribi hand).

F68 ANONYMOUS

3.1.33 ***Ruznāme-yi-jadīd***, an almanac containing an introduction in Turkish in 4 *bābs* and prayer-tables for latitude 36;40° (Algiers): ṬFT 9,1 (fols. 1v–48v, *ca.* 1300H). No other copies are known. The tables are analyzed in *SATMI*, II. See Pls. XLIb.

F69 ANONYMOUS (Tunisian treatise on arithmetic and timekeeping)

3.2.30 Part of an extensive treatise on sexagesimal arithmetic (6 *maṭlabs*) and timekeeping (26 *bābs* + 17 *faṣls*) without any indication of author or title, containing some timekeeping tables for latitude 34;48°, Sfax: K 7584,1 (fols. 1r–81r, 1257H in an elegant Maghribi hand, lacks part of the introduction on sexagesimal arithmetic), unique? The author uses the value of the obliquity of Ulugh Beg, so that the work was compiled between 850H and 1250H.

F70 MISCELLANEOUS ANONYMI (treatises on the astrolabe)

4.3.30 Treatise in unnumbered *faṣls*: DM 1169,6 (fols. 45r–47v, 1158H).

4.3.30 Treatise in 30 unnumbered *bābs*: TR 175 (24 pp., *ca.* 1000H).

F71 MISCELLANEOUS ANONYMI (works on sundials)

4.7.17 A short treatise on sundial theory with simple tables for latitude 33;40°, Fez: TR 141,6 (pp. 146–156, *ca.* 1100H), unique.

4.7.17 A table for marking horizontal sundials for latitude 31;30°, Marrakesh: TR 131,2 (p. 1, *ca.* 1200H), unique. See Pl. CIc.

On these tables, see *JHAS*, 2 (1978), p. 369.

F72 MISCELLANEOUS ANONYMI (treatises on arithmetic)

6.4.26 *Muqaddima fī Ḥall al-aᶜdād al-murakkaba*: DR 81,2 (fols. 28r–28v, *ca.* 1250H, Maghribi script).

6.4.38 A short treatise on fractions: MJ 195,6 (fols. 62v–63v, *ca.* 1100H, Maghribi script).

6.4.44 An anonymous *urjūza* on arithmetic: TR 125,1 (pp. 2–15, 1217H, Maghribi script).

F73 ANONYMOUS

7.2.8 A note about a reddish glow which appeared above the horizon of Marrakesh shortly after midnight on October 28 / 29 (?), 1772, and changed to a bluish-white: TR 131,1 (p. 1, *ca.* 1250H). See Pl. CIc.

ADDENDA

F74 AḤMAD B. ᶜALĪ, IBN ISḤĀQ

Suter, no. 356.

See no. F32.

A copy of his *Zīj* has been discovered in Tls Hyderabad Andra Pradesh State Library 298 (*ca.* 200 fols., *ca.* 800H). Merits detailed investigation.

F75 ABŪ ᶜABD ALLĀH MUḤAMMAD AL-BĀRĪ AL-TŪNISĪ

See no. H23.

G. Authors from Iraq and Iran

References are given in *Suter*, *Brockelmann*, *Azzawi*, and *Storey*. Many of the Persian manuscripts in the ENL were identified by Dr. Cornell Fleischer, who prepared the Persian section of vol. II of the *Catalogue*. The most important Persian manuscripts have been studied by Dr. E. S. Kennedy.

G1 ABŪ JAᶜFAR MUḤAMMAD B. ᶜABD ALLĀH AL-SHARĪFĪ

Azzawi, p. 119. Not listed in the other modern sources.

5.3.1 *Tāj al-madākhil*, a treatise on astrology in 3 *maqāla*s of 18 *bāb*s each, compiled *ca.* 490/1095 for the Seljuk *amīr* Tāj al-Dīn: ṬM 233,1 (fols. 1v–89r, copied 1681 Alexander = 1369 A.D. (??), fine copy), *etc.* (2 wretched copies and 2 fragments and 2 copies of selected chapters). Another copy is MS Princeton Mach 5066 = Yehuda 367 (fols. 1v–83v, 1134H). For al-Sharīfī's planetary apogees for the year 1405 Alexander (= 1093/94), see ṬM 233, fol. 67v. The work also contains some spherical astronomical tables for the latitudes of Marw and Tirmidh. Merits study.

A copy of a Persian/Turkish (?) translation by Ghiyāth al-Dīn b. Fatḥ Allāh al-Kātib al-Baghdādī is MS Istanbul Hamidiye 829,1. A work entitled *al-Durr al-manẓūm* by ᶜAbd al-Raḥīm Manṣūr (see no. D126) is based on the *Tāj al-madākhil* and Abū Maᶜshar (no. B22).

G2 ABU L-FATḤ ᶜUMAR B. IBRĀHĪM AL-KHAYYĀMĪ

Suter, no. 266; *GAL*, I, pp. 620–621, SI, pp. 855–856; article "al-Khayyāmī" by A. P. Youschkevitch and B. A. Rosenfeld in *DSB* (with detailed bibliography).

6.3.2 *M. fi l-Jabr wa-l-muqābala*: DR 898,3 (photos of Leiden 14, pp. 175–218, *ca.* 1050H). The Leiden manuscript was used by F. Woepcke in his edition of al-Khayyāmī's algebra published in 1851. The text corresponds to pp. 1–48 of the published Arabic text, lacking a passage about Abu l-Jūd (no. B72) which occurs in the other manuscripts used by Woepcke.

G3 ABŪ ḤĀTIM AL-MUẒAFFAR B. ISMĀᶜĪL AL-ISFAZĀRĪ

Suter, no. 268; *Krause*, no. 268; *GAL*, SI, p. 856; *GAS*, V, p. 110.

6.8.1 *Irshād dhawi l-ᶜirfān ilā ṣinaᶜat al-qabbān*, a treatise on the balance prepared for Abū Saᶜīd Muḥammad b. Manṣūr b. Muḥammad al-Isfazārī: DR 1021 (photos of [MS Damascus Ẓāhirīya 4460, *cf. Cat.*, pp. 102–103], 17 plates, *ca.* 1000H).

G4 HIBAT ALLĀH B. AL-ḤUSAYN AL-ASṬURLĀBĪ

Suter, no. 278.

4.6.15 See al-Khujandī (no. B50) and Abū Jaᶜfar al-Khāzin (no. B43).

G5 MAḤMŪD B. QĀSIM B. AL-FAḌL AL-IṢFAHĀNĪ

GAS, V, p. 140.

6.1.10 *Talkhīṣ al-Makhrūṭāt*, an abridgement of Apollonius' *Conica* (see no. A8): ṬR 110 (49 fols., *ca.* 1000H). *Cf. GAS*, V, p. 140.

G6 ABŪ MUḤAMMAD ᶜABD AL-JABBĀR B. ᶜABD AL-JABBĀR B. MUḤAMMAD AI-THĀBITĪ AL-KHARAQĪ

Suter, no. 276 and also no. 401; *Krause*, no. 276; *Wiedemann*, II, pp. 628–643; *GAL*, I, p. 624 and SI, p. 863.

1.2.6 *K. Muntaha l-idrāk fī taqāsīm al-aflāk*, on theoretical astronomy in 3 *maqālas* of 20 + 17 + 11 *bābs*: TR 111 (166 fols., 580H) and photos thereof in K 7196; DH 74 (photos of MS Paris B.N. ar. 2499, 157 pl., 633H).

1.2.6 *al-Tabṣira fī ᶜilm al-hayᵓa*, on theoretical astronomy in 2 *qisms* of 22 + 14 *bābs*: ṬH 35 (78 fols., *ca.* 550H, lacks end); ṬH 38,2 (fols. 63r–121v, 1076H); *etc.* (1 copy and 1 fragment). See Pl. V.

G7 ABU L-FUTŪḤ AḤMAD B. MUḤAMMAD B. AL-SĀRĪ, IBN AL-ṢALĀḤ

Suter, no. 287; *Krause*, no. 287; *GAL*, SI, p. 875; *GAS*, V, *sub index*; and the two studies of P. Kunitzsch reviewed in *JHA*, 8 (1977), pp. 204–210

6.3.4+ *Masāᵓil handasīya*, a discussion of some of the problems in the *Īḍāḥ al-burhān* of Jābir b. Ibrāhīm al-Ṣābiᵓ (no. B17): DR 898,10 (photos. of MS Leiden 14, pp. 241–245, *ca.* 1050H). *Cf. GAS*, V, p. 254.

G8 ᶜALĪ B. MUḤAMMAD SHARĪF AL-BAKRĪ AL-MAWṢILĪ AL-MUNAJJIM

Storey, no. 86.

5.3.3 *Burhān al-kifāya*, a treatise on astrology in 12 *qisms* corresponding to the zodiacal signs, each subdivided into *faṣls*: S 4657 (455 pp., *ca.* 1000H); *etc.* (1 fragment); plus a Turkish translation by Ṣubḥī Zāde ᶜAbd al-ᶜAzīz (no. H31), compiled in 1135H: ṬFT 41 (313 fols., 1206H); *etc.* (1 copy). See also Abu l-Khayr al-Fārisī (no. G62).

G9 SAMAWᵓAL B. YAHŪDA AL-ISRĀᵓĪLĪ = SAMAWᵓAL B. YAḤYĀ AL-MAGHRIBĪ

Suter, no. 302; *Krause*, no. 302; *GAL*, I, p. 643, and SI, p. 892.

6.3.5 *K. al-Bāhir*, a treatise on algebra: DR 702,1 (fols. 1r–26r, selected chapters only, *ca.* 1000H). This manuscript was not used by S. Ahmad and R. Rashed in their edition *Al-Bāhir en Algèbre d'As-Samawᵓal*, Damascus: University Press, 1972, which was based on the two manuscripts listed by Krause.

6.3.5 *al-Maṣḥaf al-mukhtaraᶜ fī muᶜjizāt ṣināᶜat al-ᶜadad*, on number theory: DR 702,2 (fols. 26r–38v, *ca.* 1000H), unique? This is not part of the *K. al-Bāhir*.

6.3.5 *K. al-Tabṣira fī ᶜilm al-ḥisāb*, on arithmetic in 2 *maqālas* of 9 + 7 *bābs*: DJ 713,14 (fols. 136v–161r, in two hands, 1013H, defective, Yemeni copy).

6.2.8 Remarks on the *Burhān ᶜalā ᶜamal ḥisāb al-khaṭaᵓayn* of Qusṭā b. Lūqā (no. B38): DR 702,3 (fols. 38v–39v, *ca.* 1100H).

G10 MUḤĪY L-DĪN ᶜABD AL-QĀDIR B. ᶜABD ALLĀH AL-KĪLĀNĪ, a celebrated founder of the Sufi Qādirīya order, born 471/1078 in Gilan, studied in Baghdad, died 561/1167.

GAL, I, 560–563, and SI, 777–779; article "ᶜAbd al-Ḳādir al-Djīlānī" in EI_2 by W. Braune.

3.5.3 Some passages on reckoning time by shadow lengths taken from his major work *al-Ghunya*: MJ 180,17 (fols. 146 + 148r–149r, *ca.* 1150H), corresponding to pp. 89–90 of vol. II of the published text (Cairo, 1304H). See further *SATMI*, III.

G11 SIRĀJ AL-DĪN ABŪ ṬĀHIR MUḤAMMAD B. MUḤAMMAD AL-SAJĀWANDĪ

GAL, I, pp. 470–471, and SI, pp. 650–651; *Suter*, p. 192.

6.3.6 ***(R. fi) Uṣūl yustaᶜānu bihā fī masāᵓil al-jabr wa-l-muqābala***, on the principles of arithmetic and algebra: B 23447,7 (fols. 60v–65r, *ca.* 1050H, anon.), plus an anonymous commentary thereon in which the original is attributed to al-Sajāwandī: B 23447,8 (fols. 65v–79v, *ca.* 1050H). Another copy of al-Sajāwandī's treatise and the same commentary is in MS Istanbul Nurosmaniye 2547,1+2. Brockelmann (SI, 651) mentions a work by al-Sajāwandī entitled ***al-Tajnīs fi l-ḥisāb***, extant in Leningrad, which may be identical to the above since the first section in the Cairo copy is entitled *faṣl fi l-tajnīs*. Ḥājjī Khalīfa (I, 353) informs us that commentaries on the ***Tajnīs*** were written by Masᶜūd b. al-Muᶜtamar al-Mashhadī in 824H, and by Taqi l-Dīn b. Maᶜrūf (no. H12).

G12 FAKHR AL-DĪN MUḤAMMAD B. ᶜUMAR B. ḤUSAYN AL-RĀZĪ

Suter, no. 328; *Krause*, no. 328; *GAL*, I, pp. 666–669, and SI, pp. 920–924; *Storey*, pp. 49–50.

5.3.4 A fragment of ***K. al-Ikhtiyārāt al-ᶜAlāᵓīya fi l-ikhtiyārāt al-samawīya***, a work on astrology in Persian: ṬM 240,3 (fols. 40v–58v, 1303H).

5.3.4 ***K. al-Sirr al-maktūm fī mukhāṭabat al-nujūm***, an Arabic work on astrology and magic in 5 *maqāla*s subdivided in *faṣl*s, *bāb*s, *qism*s, *baḥth*s, *etc.*: TGh 129 (301 pp., different hands, mainly *ca.* 1000H), and selected chapters: S 4349,5 (fols. 53r–61v, *ca.* 1100H); plus a commentary entitled ***al-Durr al-manẓūm wa-khulāṣat al-Sirr al-maktūm*** by al-Kashnāwī (no. D74): DḤ 14 (241 fols., 1246H); plus a Turkish translation by Muḥammad Efendī (no. H21): DMT 1,1 (fols. 1r–177v, *ca.* 1100H). This work merits detailed investigation.

5.3.4 A fragment on the duration of the planets in the zodiacal signs stated to be from *Sharh Rāzī li-l-maqāmat*: DM 126,1A (fol. 1r, *ca.* 1150H).

5.3.4 Verses of poetry to be recited at the first(?) visibility of Mercury, taken from the handwriting of Ibn ᶜAbd al-Salām: MM 248,1 (fol. 1r, *ca.* 750H).

6.3.2+ A fragment from his Persian work *K. Ḥadāᵓiq al-anwār fī ḥaqāᵓiq al-asrār* dealing with geometry: DR 897,4 (photos of MS Leiden 513, fols. 100–106r, *ca.* 900H).

G13 KAMĀL AL-DĪN MŪSĀ B. YŪNUS

Suter, no. 354; *Krause*, no. 354; *GAL*, SI, p. 859; *GAS*, V, pp. 134, 141, & 324.

6.3.9 ***R. fī Bayān annahu lā yumkinu an yujada ᶜadadān murabbaᶜān fardān majmuᶜuhumā murabbaᶜ***, on the theorem that the sum of the squares of two odd numbers is not a perfect square: DR 703,4 (fols. 130r–130v, 1006H). Other copies are listed by Suter and Krause. See also al-Ṭūsī's treatise on the same subject (no. G23). In DR 703 both treatises are appended to a copy of al-Ṭūsī's recension of Euclid's ***Elements***.

G14 MUḤAMMAD (B?) ᶜABD AL-KARĪM AL-GHAZNAWĪ

Storey, p. 6.

6.3.6+ ***Tuḥfat al-ṣudūr***, a Persian treatise on arithmetic in 5 *maqāla*s each of 4 *bāb*s: DRF 1 (112 fols., 670H, anonymous). Merits investigation. See Pls. XC.

On early Persian treatises on arithmetic, see H. Hermelink, "The Earliest Reckoning Books Existing in the Persian Language," *HM*, 2 (1975), pp. 299–303.

G15 SHARAF AL-DĪN AL-MUẒAFFAR B. MUḤAMMAD B. AL-MUẒAFFAR AL-ṬŪSĪ

Suter, no. 333; *Krause*, no. 333; and *GAL*, I, p. 622 and SI, pp. 858–859.

6.3.7 ***Masᵓala handasīya***, geometrical problems compiled 606H in Hamadan: DR 898,18 (photos. of MS Leiden 14, pp. 322–327, *ca.* 1050H, unique).

G16 ATHĪR AL-DĪN AL-MUFAḌḌAL B. ᶜUMAR AL-ABHARĪ

Suter, no. 364; *Krause* no. 364; *GAL*, I, pp. 608–611, and SI, pp. 839–844.

1.2.9 ***K. fī ᶜIlm al-hayᵓa***, on theoretical astronomy, in 22 *jumla*s with numerous diagrams, compiled in 635H: ṬH 48,2 (fols. 1v–62r, anonymous, *ca.* 700H); DH 57 (photos. of MS Paris B.N. ar. 2515, 68 pls., *ca.* 800H). Merits detailed study.

Krause mentions one copy of this "Kompendium der Astronomie in 22 Teilen", namely, MS Istanbul Carullah 1499,2 (84 fols., 677H in Cairo), and the title reads: *ma khtaṣarahu fī ᶜilm al-hayᵓa min hayᵓat Kūshyār wa-min hayᵓat Ibn Aflaḥ al-Ishbīlī*, implying that the treatise is based on Kūshyār (no. B70) and Jābir b. Aflaḥ (no. F1). Krause also mentions a unique copy of another treatise on astronomy by al-Abharī in 20 *faṣl*s: MS Istanbul Feyzullah 1339,2 (fols. 26v–39r, 788H).

4.3.11 ***R. fī l-ᶜAmal bi-l-asṭurlāb***, on the use of the astrolabe in 14 *bāb*s: TR 165,3 (pp. 61–64, 992H, anonymous).

G17 MAḤMŪD B. MUḤAMMAD B. ᶜUMAR AL-JAGHMĪNĪ

Suter, no. 403; *Krause*, no. 403; *GAL*, I, pp. 624–625, and SI, p. 865; *Storey*, pp. 50–51; misdated in *GAS*, V, p. 115.

1.2.7 ***al-Mulakhkhaṣ fī l-hayᵓa***, a treatise on theoretical astronomy in 2 *maqāla*s of 10 + 3 *bāb*s, compiled in 618H: DH 69 (18 fols., *ca.* 900H); ṬJ 966,2 (fols. 9v–24r, 891H); DH 68,1 (fols. 1v–26r, 853H); MH 6 (23 fols., *ca.* 1050H); *etc.* (8 copies). This work was translated into German by A. Hochheim and G. Rudloff in *ZDMG*, 47 (1893), pp. 213–275. See Pl. VI.

Persian translation by Maḥmūd b. Muḥammad b. Qawwām al-Wālishtānī (no. G44) (not mentioned in *Storey*): TR 347,3 (pp. 68–149, 832H in Herat).

Commentaries by al-Sayyid al-Sharīf ᶜAlī b. Muḥammad al-Jurjānī (no. G42): DH 68,2 (fols. 26v–108r, 853H); ṬH 31 (132 fols., 855H, defective); DH 21 (88 fols., 949H); *etc.* (4 copies); and by Jalāl al-Dīn Faḍl Allāh al-ᶜUbaydī (no. G106): ṬH 33,3 (fols. 37v–83v, *ca.* 1000H); and by Muḥammad b. Muḥammad known as Humām al-Ṭabīb (no. G46): DH 41 (82 fols., 1125H); and by an anonymous: ṬH 33,1 (fols. 1r–36r, *ca.* 900H by author); and by Qāḍī Zāde al-Rūmī with several commentaries thereon: see no. G50.

1.2.7 A treatise on planetary distances and sizes, dedicated to al-Imām Badr al-Dīn al-Falāsitī (?): ṬJ 429,2 (fols. 4r–4v, *ca.* 1100H), unique?

6.3.11 ***R. mukhtaṣara fī l-Ḥisāb***, on simple arithmetic: ṬJ 635,35 (fols. 149r–153r, 1170H). Another copy in Princeton is listed by Brockelmann.

G18 ᶜIZZ AL-DĪN AL-ZANJĀNĪ

GAL, I, pp. 336–337, and SI, pp. 497–498.

6.3.12 A treatise on arithmetic in 12 *faṣl*s: DJ 713,10 (fols. 104v–113v, *ca.* 1000H). The treatise is anonymous and bears the title ***al-Risāla al-Yamanīya***; also, the manuscript is of Yemeni provenance. However, another copy of this treatise, apparently bearing the title ***al-Risāla al-ᶜIzzīya fī l-ḥisāb al-hawāᵓīya*** (!), is preserved in MS Damascus Ẓāhiriya 6000,3 (98v–114r, 799H), where it is attributed to ᶜIzz al-Dīn al-Zanjānī. Two more copies of the same treatise are MSS Damascus Ẓāhiriya 7759 (fols. 123v–140v, 955H, attributed to Sibṭ al-Māridīnī (no. C97)!) and 8358 (fols. 1v–7v, 1150H, anonymous). At the end of his treatise al-Zanjānī mentions larger works of his entitled ***Baḥr al-fawāᵓid fī ᶜilm al-ḥisāb*** and ***al-Qawāᶜid*** in DJ 713 the title reads incorrectly ***Baḥr al-qawāᶜid***). The first of these titles is listed by Ḥājjī Khalīfa (I, p. 225) but no author is mentioned. An anonymous fragment of this work may be in MS Leiden Or. 6810,2 (Voorhoeve's *Cat.*, p. 161): the title is here ***Baḥr al-ḥisāb***. Ḥājjī Khalīfa also lists (I, p. 877) ***al-R. al-ᶜIzzīya fī l-ḥisāb*** by Ibn Ḥajar al-ᶜAsqalānī and dealing with inheritance. al-Zanjānī also wrote a work entitled ***al-Risāla al-kāfīya fī l-ḥisāb*** in 10 *faṣl*s, which has been falsely attributed to Ibn al-Samḥ (no. B115): see *JAOS*, 99 (1979), p. 457.

G19 AL-BAYHAQĪ (?)

Not mentioned in the modern sources. Not identified. See no. G39.

1.2.8 ***Mukhtaṣar fī ᶜilm al-hayᵓa***, on theoretical astronomy, arranged with a *muqaddima* and

3 *maqṣads* in 10 + 9 + 6 *mabḥaths*): DH 75 (95 fols., 1075H), unique. Merits investigation.

G20 ZAKARĪYĀ B. MUḤAMMAD B. MAḤMŪD AL-QAZWĪNĪ

Suter, no. 373aN; *GAL*, I, pp. 633–634 and SI, pp. 882–883; article "al-Ḳazwinī" by P. Lewicki in EI_2.

1.3.3 A chapter on fixed stars from his cosmography *K. ᶜAjāʾib al-makhlūqāt wa-gharāʾib al-mawjūdāt*: DM 734 (10 fols., *ca.* 1050H). There are several copies of the ***K. ᶜAjāʾib al-makhlūqāt*** and the Persian translation thereof in the Egyptian National Library.

G21 MUḤYI L-DĪN YAḤYĀ B. MUḤAMMAD B. ABI L-SHUKR AL-MAGHRIBĪ

Suter, no. 376; *Krause*, no. 376; *GAL*, p. 626, and SI, pp. 868–869.

2.1.11 *ᶜUmdat al-ḥāsib wa-ghunyat al-ṭālib*, a *Zīj* for Maragha (not listed in *Kennedy*), compiled *ca.* 630Y (= 1262 A.D.) with a brief introduction in 25 *bābs* followed by an incomplete set of tables: MM 188,1 (fols. 1r–137r, *ca.* 950H by Aḥmad b. Aḥmad b. Timurbay (no. C94)), unique. Merits detailed study. See Pl. XIIIb.

2.1.11 ***Adwār al-anwār mada l-duhūr wa-l-akwār***, another *zīj* for Maragha (*Kennedy*, no. 108): fragments from the introduction in DM 639,18 (fols. 159r–170r, *ca.* 900H). The complete copies in MSS Dublin Chester Beatty 3665 and Meshed 332 (103) merit detailed study.

5.3.7 ***al-Madkhal al-mufīd wa-ghunyat al-mustafīd fi l-ḥukm ᶜala l-mawālīd***, an astrological treatise: DM 882 (179 fols., 939H, fine copy); plus 1 late copy of ***al-Madkhal*** and 2 late copies of ***Ghunyat al-mustafīd***.

5.3.7 ***R. fī Kayfīyat al-ḥukm ᶜalā taḥāwīl sini l-ᶜālam*** and variations on this title (in 23 *bābs*): S 5991,2 (fols. 14v–82v, 864H); DM 8,1 (fols. 1r–92v, *ca.* 900H); DM 883 (63 fols., 939H, originally together with DM 882—see above); *etc.* (4 copies).

5.3.7 ***Risāla/Khātima fi l-aḥkām al-juzʾīya***: (in 8 *bābs*, sometimes appended to the previous treatise as a *khātima*): DM 8,2 (fols. 92v–95v, *ca.* 900H + fols. 97v–103r, late); DM 7,2 (fols. 72v–107v, 1191H); *etc.* (1 copy).

5.3.7 ***M. fi l-Ḥukm ᶜala l-masāʾil al-mutakhaṣṣaṣa bi-aḥwāl al-sāʾil***: ṬM 85,2 (fols. 103v–179v, *ca.* 1100H).

6.1.10 A commentary on the *Conica* of Apollonius (no. A8): DR 696 (photos of MS London B.L. Add. 14332,4, pp. 73–401, 1832 A.D.). *Cf. GAS*, V, p. 141.

6.3.15 A fragment from ***R. fi l-Nisba al-muʾallafa***: TR 140,6 (pp. 37 and 39, *ca.* 1150H). See also no. B94.

G22 ḤUSĀM AL-DĪN ᶜALĪ B. FAḌL ALLĀH AL-SĀLĀR

Suter, no. 482; *Krause*, no. 482; *GAS*, VII, p. 401.

6.1.3 ***Ikhtiṣār daᶜāwa l-maqālatayn al-ūlā wa-l-thāniya min K. al-Uṣūl li-Uqlīdis***: DR 700 (6 pp., 1342H); and ***Muqaddimāt li-tabyīn al-muṣādara fi l-maqāla al-ūlā li-Uqlīdis***: DR 701 (9 pp., 1342H). Both of these manuscripts were copied from what is now MS Chester Beatty Dublin 3045, mentioned in *GAS*, VII, p. 401.

G23 NAṢĪR AL-DĪN MUḤAMMAD B. MUḤAMMAD AL-ṬŪSĪ

Suter, no. 368; *Krause*, no. 368; *GAL*, I, pp. 670–676, and SI, pp. 924–933; *Storey*, pp. 6–7 and 52–60.

1.1.2–6 For his recensions of astronomical works by Euclid, Aristarchus, Hypsicles, Theodosius, Ptolemy, see the entries for the original authors.

1.2.10 ***al-Tadhkira fī ᶜilm al-hayʾa***, a treatise on theoretical astronomy in 4 *bābs* of 2 + 14 +12 +7 *faṣls*: DH 51 (photos of MS Paris B.N. ar. 2509, 83 pls., 791H); K 3957,3 (fols. 155v–253r, *ca.* 900H); TJ 181,1 (pp. 1–120, 1114H); ṬH 38,1 (fols. 1r–62r, *ca.* 1100H). See Pl. VII. Commentaries by Niẓām al-Dīn al-Nīsābūrī (no. G38): DH 54 (photos of MS Paris B.N. ar. 2510, 253 pls., *ca.* 1000H); DH 66 and 88 (two sets of photos of a manuscript in Damietta

Institute, 341 pls., 1243H); a ***Takmila*** by ᶜUmar b. Daᵓūd b. Sulaymān al-Fārisī (no. C21) compiled *ca.* 710H for the prince of Hama Abu l-Fidāᵓ and not listed in any of the modern sources: TR 128 (99 fols., *ca.* 800H), K7193 (photos of TR 128), unique; a commentary by al-Sayyid al-Jurjānī (no. G42): K8529 (225 fols., 857H, incorrectly attributed to al-Nīsābūrī); *etc.* (8 late copies); a commentary by Muḥammad al-Ḥafrī (no. C89), entitled ***al-Takmila fī sharḥ al-Tadhkira***: TR 230 (388 pp., 1053H); and an anonymous commentary: TR 136 (198 pp., *ca.* 1100H).

On the *Tadhkira*, see now F. J. Ragep, *Cosmography in the Tadhkira of Naṣīr al-Dīn al-Ṭūsī*, Ph.D. dissertation, Harvard University, 1982, to be published by Springer-Verlag.

1.2.10 ***Zubdat al-idrāk fī hayᵓat al-aflāk***, a treatise on theoretical astronomy in 2 *maqāla*s of unnumbered *faṣl*s: DH 50 (photos of MS Paris B.N. ar. 2511,1, 20 pls., 912H, anonymous).

1.2.10 ***Risāla Muᶜīnīya***, a treatise in Persian on theoretical astronomy arranged in 4 *maqāla*s of 2 + 14 + 12 + 6 *bāb*s: DHF 1 (77 fols., 869H). Krause lists 3 copies in Istanbul. On this work, see now E. S. Kennedy, "Two Persian Astronomical Treatises by Naṣīr al-Dīn al-Ṭūsī," *Centaurus*, 27 (1984), pp. 109–120.

2.1.12 ***Zīj-i-Īlkhānī*** (*Kennedy*, no. 6): DMF 1 (136 fols., 692H at Maragha), *etc.* (1 inconsequential fragment). See Pl. XIXa. Merits detailed study. On the Persian recension by al-Simnānī and the Arabic recension by al-Ḥalabī, see nos. G40 and C69.

2.3.1 ***Sī faṣl = Mukhtaṣar dar maᶜrifat-i-taqwīm***, a Persian treatise on the compilation of ephemerides in 30 *faṣl*s: S 4485,2 (fols. 3v–12r, 1134H); an anonymous Arabic translation probably by al-Ṭūsī himself entitled ***Mukhtaṣar fī maᶜrifat al-taqwīm***: ṬJ 966,4 (fols. 82v–91v, 891H); ṬM 108 (33 fols., 930H); MJ 180,1 (fols. 1r–28v, 1064H); *etc.* (1 copy); plus an anonymous commentary: TR 43 (122 pp., 995H); DM 941 (photos of MS Paris B.N. ar. 2512, 64 pls., *ca.* 950H); DM 942 (79 fols., *ca.* 1050H); *etc.* (5 copies); and a *ḥāshiya* by ᶜAbd al-Wājid b. Muḥammad (no. G45): DM 933 (photos of MS Paris B.N. ar. 2511,2, 47 pls., *ca.* 950H); KhM 12 (90 fols., *ca.* 950H). A Persian commentary on ***Sī faṣl*** by Ḥusām al-Ahjānī (no. G86): DMF 7 (100 fols. 1199H). A Turkish translation of ***Sī faṣl*** by Aḥmad al-Dāᶜī (no. H66): S 4382 (12 fols., 1208H); *etc.* (2 copies).

4.3.14 ***Bīst bāb = Mukhtaṣar dar maᶜrifat-i-asṭurlāb***, a treatise on the use of the astrolabe in Persian arranged in 20 *bāb*s: ṬJ 966,5 (fols. 93v–106v, 891H); ṬM 255,3 (fols. 14r–38r, *ca.* 1150H); *etc.* (1 copy). Persian commentary by ᶜAbd al-ᶜAlī al-Birjandī (no. G58): ṬJ 398,1 (fols. 1v–55v, *ca.* 1150H); ṬMF 2,2 (fols. 27v–136v, 1120H), *etc.* (2 copies). For other copies, see *Storey*.

5.1.3 An Arabic commentary on the ***K. al-Thamra*** (***Centiloquium***) attributed to Ptolemy (see no. A18): MM 200,1 (fols. 1r–37v, 1182H); *etc.* (2 copies). Persian commentary in ṬFF 17,2 (fols. 62r–95v, 947H); DR 782,2 (fols. 49v–119v, *ca.* 1050H).

5.3.7+ ***Madkhal-i-manẓūm***, a *mathnawī* poem on astrology: MJF 2,2 (fols. 112v–123r, *ca.* 1200H, anonymous). This poem is elsewhere attributed, perhaps falsely, to al-Ṭūsī. *Cf.* Storey, no. 91(4).

5.5.3 A fragment on horoscopes attributed to al-Ṭūsī: DM 167,4 (fols. 81r–82v, *ca.* 1250H).

6.1.1,3,7 For his recensions of mathematical works by Autolycus, Euclid, Archimedes, and Apollonius, see the entries for the original authors.

6.3.9 ***R. fī Bayān annahu lā yumkinu an yajtamiᶜa min ᶜadadayn murabbaᶜayn fardayn ᶜadad murabbaᶜ***, on the theorem that the sum of the squares of two odd numbers is not a perfect square: DR 703,4 (fol. 130r, 1006H). Other copies listed by Suter and Krause, to which add Paris B.N. ar. 2467,14 (fol. 196r, *ca.* 950H). This treatise is copied with another by Kamāl al-Dīn Ibn Yūnus (no. G13) in DR 703 as an appendix to al-Ṭūsī's recension of Euclid's ***Elements***. This appears also to be the case for the other manuscripts. Both treatises merit investigation.

7.2.1 A recension of Euclid's *Optics*: see no. A5.

7.2.5 ***R. fi Inᶜikās al-shuᶜāᶜāt wa-nᶜiṭāfihā***, on optics: K 3957,2 (fols. 152v–155r, ca. 900H, anon.). See also Krause, no. 378(16) for four copies in Istanbul. An English translation of this treatise by H. J. J. Winter and W. ᶜArafat is in

ISIS, 42 (1951), pp. 138–142.
See also nos. G34 and G109.

G24 ᶜABD ALLĀH B. MUḤAMMAD, IBN AL-KHAWWĀM, a student of AL-ṬŪSĪ

Suter, no. 494; *Krause*, no. 494; *GAL*, SII, p. 215; *GAS*, V, p. 115.

6.1.3 *Sharḥ al-Maqāla al-ᶜāshira min K. (al-Uṣūl l-) Uqlīdis*: DR 300,1 (fols. 1r–3v, 1150H, poor Maghribi hand), not listed in *GAS*, V, p. 115, no. 55.

6.3.17 *al-Fawāʾid al-Bahāʾīya fi l-qawāʾid al-ḥisābiya*, on arithmetic and algebra in 5 *maqāla*s, compiled in 675H: K 3956 (38 fols., 721H). MS London B.L. Or. 5615,1 of this work was copied from a copy by the author in 712H. Commentary by [Kamāl al-Dīn al-Fārisī] (no. G30): TR 135,1 (pp. 2–348, mainly 828H in Samarqand); QR 111,1 (fols. 1r–275r, 891H), both anonymous.

G25 JAMĀL AL-DĪN ABU L-QĀSIM B. MAḤFŪẒ AL-BAGHDĀDĪ

Suter, no. 490; *GAL*, I, p. 252.

2.1.13 *al-Zīj al-Waqbīya* (?), an extensive *zīj* for Baghdad (*Kennedy*, no. 3): DM 905 (photos. of MS Paris B.N. ar. 2486, 256 pls., 684H, unique!). This important work awaits detailed study. See already C. Jensen, "The Lunar Theories of al-Baghdādī," *AHES*, 8 (1972), pp. 321–328. al-Baghdādī presents an astrological calculation for an individual, perhaps himself, born in Baghdad in 643H (fol. 197r).

4.3.13 *K. fī ᶜIlm al-asṭurlāb = al-R. al-Kāmila fi ᶜamal al-asṭurlāb*, on the use of the astrolabe in 66 *bāb*s: TR 165,1 (pp. 1–56, *ca.* 1200H). Another copy is in Alexandria Municipal Library. The only other known copy of this work, namely, MS London B.L. 5734,2, is incomplete. Brockelmann also lists B.L. 1002,4, perhaps the same?

G26 SHAMS AL-DĪN MUḤAMMAD B. ASHRAF AL-SAMARQANDĪ

Suter, no. 382; *GAL*, I, pp. 615–617, and SI, pp. 849–850; *Storey*, no. 12; *GAS*, V, pp. 114–115.

1.1.6 See Pl. Ia.

6.1.4 *Ashkāl al-taʾsīs*, selected theorems from Euclid's *Elements*: DR 824 (11 fols., *ca.* 1050H); ZK 127,2 (pp. 4–19, *ca.* 1150H); ṬJ 425,3 (fols. 30v–40v, *ca.* 1150H), *etc.* (6 copies). Anonymous extended version: ṬR 143 (34 fols., 1312H). Commentaries by Qāḍī Zāde al-Rūmī (no. G50): TR 143 (60 pp., 884H); ṬJ 502,2 (fols. 106r–152r, 966H); *etc.* (21 copies); and *ḥāshiya*s thereon by Tāj al-Saᶜīdī (no. G52): DR 26 (23 fols., *ca.* 1100H); ṬR 118,2 (fols. 37r–68v, 1095H); *etc.* (2 copies); and by Ḥaydar al-Ḥusaynābādī (no. G53): DM 1082,9 (fols. 104v–133v, *ca.* 1150H); and by Ḥasan al-ᶜAṭṭār (no. D144): DJ 219,2 (fols. 40v–100v, 1237H); and by Faṣīḥ al-Dīn al-Hindī (no. X14): DR 58 (39 fols., *ca.* 1100H). Also a poem based on *Ashkāl al-taʾsīs* by Muḥammad b. al-Ḥurr al-ᶜĀmulī (no. G83): DJ 846,4 (fols. 61v–67v, 1100H).

G27 QUṬB AL-DĪN MAḤMŪD B. MASᶜŪD AL-SHĪRĀZĪ

Suter, no. 387; *Krause*, no. 387; *GAL*, II, pp. 274–275, and SII, pp. 296–297.

1.2.11 *Nihāyat al-idrāk fī dirāyat al-aflāk*: ṬH 45 (168 fols., *ca.* 850H); MH 7,1 (fols. 1r–271r, 741H in Mosul); DH 56 (photos. of MS Paris B.N. 2517, lacks end, 126 pls., *ca.* 800H). See Pl. XCIIIc.

1.2.11 *K. al-Tuḥfa al-Shāhīya*, compiled in 684H in Sivas: DH 55 (photos. of MS Paris B.N. ar. 2516, 118 pls., 684H in Sivas!); K 3758 (158 fols., *ca.* 900H); DH 80 (265 fols., *ca.* 1000H, lacks last page); plus a commentary attributed to ᶜAlī Qwry (?) Zāde (no. G88): DH 7 (37 fols., *ca.* 1200H, incomplete). See also no. B78 (*sub* 7.2.3) and Pl. VIII.

1.2.11 An anonymous treatise on planetary astronomy containing diagrams very similar to those of al-Shīrāzī: K 3823,4 (fols. 19r–38v, *ca.* 1150H). Merits investigation.

6.1.11 See the entry for Theodosius (no. A12). See also the next entry.

G28 ANONYMOUS

6.3.1 *Ghāyat al-ʿudad fī ʿilm al-ʿadad*, an extensive treatise on number theory in 4 *maqāla*s of 6 + 2 + 3 + 10 *bāb*s: ṬR 130 (60 fols., 1209H from a copy dated 1112H, anonymous), unique. Not listed in the modern sources. The thirteenth century scholar (Quṭb al-Dīn) al-Shīrāzī (no. G27) is mentioned in the introduction.

G29 JAMĀL AL-DĪN MUḤAMMAD B. ʿUMAR B. AḤMAD [B.] HIBAT ALLĀH, IBN ABĪ JARĀDA

Krause, p. 440; *GAS*, V, p. 129.

6.2.6 On his *taḥrīr* of Thābit b. Qurra's *K. Quṭūʿ al-usṭuwāna wa-basīṭuhā*, compiled in 691H, see no. B30.

Note: Ibn Abī Jarāda copied MS Istanbul Fatih 3414 between 676 and 684H (*Krause*, p. 440).

7.2.1 *Tajrīd Uqlīdis fi l-manāẓir*, apparently an abridgement of Euclid's *Optics* (see no. A5): DR 638 (12 fols., *ca.* 1100H), unique. Neither the title nor this Cairo manuscript is mentioned in *GAS*, V, p. 117.

G30 KAMĀL AL-DĪN MUḤAMMAD B. AL-ḤASAN AL-FĀRISĪ

Suter, no. 389; *Krause*, no. 389; *GAL*, SII, p. 295; article "Kamāl al-Dīn al-Fārisī" by R. Rashed in *DSB*.

6.3.17 *Asās al-qawāʿid fī uṣūl al-Fawāʾid*, a commentary on *al-Fawāʾid al-Bahāʾīya* of al-Khawwām (no. G24): TR 135,1 (pp. 2–348, mainly 828H in Samarqand); QR 111,1 (fols. 1r–275r, 891H); MR 38 (213 fols., *ca.* 1100H, lacks first *maqāla*). Each of these three manuscripts is anonymous. Krause lists 5 copies in Istanbul. See Pl. LXXXIX.

6.3.18 *R. fi stikhrāj al-aʿdād al-mutaḥābba = Tadhkirat al-aḥbāb fī bayān al-taḥābb*: TR 135,2 (pp. 349–371, 828H in Samarqand); QR 111,2 (fols. 275v–291r, 891H); both anonymous. The title *Tadhkirat al-aḥbāb* . . . is attributed to al-Fārisī by Ḥājjī Khalīfa and the treatise in TR 135,1 is also by al-Fārisī. Krause lists another copy in Istanbul. See Pl. LXXXIX.

See also no. G92.

7.2.4 *Tanqīḥ al-manāẓir li-dhawi l-abṣār wa-l-baṣāʾir*, on optics: DṬ 368 (photos of MS Istanbul Aya Sofia 2598, 634 pls. in 3 vols., 1079H—see *Krause*, no. 389). Other copies in Leiden, Tehran, Patna, and the U.S.S.R. (see Rosenfeld, *HM*, 2 (1975), p. 68), and Jaipur (see *JHAS*, 4 (1980), p. 82). This treatise was published in Hyderabad. See also Pl. Cb.

Note: The Cairo copy of al-Fārisi's treatise on the rainbow has disappeared from the Egyptian National Library. For a reference to it, see R. Rashed's article in *DSB*.

See also no. G92.

G31 NĀṢIR AL-DIN AḤMAD B. ḤAYDAR B. MUḤAMMAD AL-SHĪRĀZĪ

Storey, no. 95.

4.3.17 *Irshād*, a treatise in Persian on the astrolabe in 50 chapters: ṬJ 912,3 (fols. 26v–45v, lacks chapter 18 to end, *ca.* 1250H). Other copies listed by Storey, to which add MS Instanbul Fatih 5400,3 (fols. 107–151, n.d.).

G32 ʿALĪ SHĀH B. MUḤAMMAD B. QĀSIM AL-KHWĀRIZMĪ, ʿALĀʾ AL-DĪN AL-BUKHĀRĪ

Suter, no. 396; *Storey*, no. 94.

5.3.10 *Ashjār wa-athmār*, a Persian treatise on astrology: ṬFF 13,1 (fols. 1v–117v, 910H); ṬHF 8,2 (fols. 18v–170v, 1047H); plus a Turkish translation by ʿAbd al-ʿAzīz (no. H31): DMT 26 (434 pp., *ca.* 1200H); *etc.* (2 copies).

5.3.10 *Aḥkām al-aʿwām*, a Persian treatise on astrology: ṬFF 4,1 (fols. 1r–104v, *ca.* 1200H).

G33 ABU L-MAḤĀMID MUḤAMMAD B. MASʿŪD B. MUḤAMMAD ZAKĪ AL-GHAZNAWĪ

Suter, no. 496; *Krause*, no. 496; *GAL*, SI, pp. 863–864; *Storey*, no. 84.

5.3.2 *Kifāyat-i-taʿlīm dar ṣināʿat-i-tanjīm*, an extensive Persian treatise on astrology: ṬFF 8 (253 fols., 1199H); ṬFF 12 (218 fols., 1185H); plus an anonymous Arabic commentary (see MS

Berlin Ahlwardt 5891 for details): ṬM 244 (88 fols., 998H); ṬM 153 (129 fols., 1134H); *etc.* (2 copies and 1 fragment).

G34 MUWAFFAQ AL-QAYṢARĪ

Not listed in the modern sources.

1.2.12 ***Zubdat al-hay*ᵓ*a***, a Persian treatise on theoretical astronomy in 30 *bābs*: MHF 1 (77 fols., 741H in Kayseri). This work is attributed in *Krause*, p. 497, and *Storey*, p. 60, to Naṣīr al-Dīn al-Ṭūsī but Ḥājjī Khalīfa (II, p. 593) mentions a work ***al-Zubda fi l-hay*ᵓ*a*** by Muwaffaq al-Qayṣarī. An Arabic translation, apparently prepared by ᶜImād al-Dīn ᶜAlī al-Qāshī is in MS Princeton Mach 4884 = Yehuda 4066 (71 fols., *ca.* 750H). An Arabic version of the same work (?) apparently attributed to al-Ṭūsī is in MS Chester Beatty 4933 (69 fols., 677H).

G35 ANONYMOUS

Storey, p. 108, no. 45.

6.6.1 ***K. al-Mi*ᵓ*a wa-l-*ᶜ*ishrīn* ᶜ*alā ṭarīq jadwal al-sittīn***, extensive sexagesimal quotient tables with an introduction: MMF 8,1 (fols. 1r–67r, *ca.* 700H)—see Pl. XCIb. Other copies in MSS Istanbul Aya Sofia 2698 and Bayezit Umumi 4645 not listed by Storey. See further D. A. King, "Supplementary Notes on Medieval Islamic Multiplication Tables," *HM*, 6 (1979), pp. 405–417, esp. p. 411.

G36 SHAMS AL-AᵓIMMA MUḤAMMAD B. ᶜALĪ AL-WĀBKANAWĪ

Krause, no. N18; *Storey*, no. 99; *Kennedy*, no. 35.

2.1.12 A chapter in Arabic on the computation of the difference in setting times of the sun and moon (*makth al-qamar*): DM 1199,2 (fols. 59v–60v, *ca.* 1200H). al-Wābkanawī's ***al-Zīj al-muḥaqqaq al-Sulṭānī*** is in Persian; for copies, see *Storey*.

G37 SHAMS AL-DĪN MUḤAMMAD B. SHARAF AL-KALLĀᵓĪ

GAL, II, p. 207, and SII, p. 200.

6.9.6 ***al-Majmū*ᶜ *fi l-farā*ᵓ*iḍ*** (copies in Cairo). For an abridgement and commentary by Sibṭ al-Māridīnī, see no. C97.

G38 NIẒĀM AL-DĪN AL-ḤASAN B. MUḤAMMAD AL-NĪSĀPŪRĪ

Suter, no. 395; *GAL*, II, p. 256, and SII, p. 273.

1.1.6 ***Tafsīr al-Taḥrīr***, a commentary on the edition of Ptolemy's ***Almagest*** by al-Ṭūsī (no. G23): QH 1 (237 fols., *ca.* 1000H); ṬH 46 (225 fols., *ca.* 1100H), both anonymous.

1.2.10 A commentary on ***al-Tadhkira*** by al-Ṭūsī: see no. G23.

3.3.3 ***R. fī Ma*ᶜ*rifat samt al-qibla***, a treatise on the determination of the qibla in 5 *faṣls*: TR 278,4 (pp. 96–107, *ca.* 1100H, anonymous), unique. The author refers to his work ***Hay*ᵓ*at al-Niẓāmī*** and presents a qibla computation for Meshed. One could take this title to refer to al-Nīsābūrī's commentary on the ***Tadhkira*** of al-Ṭūsī (no. G23). On the other hand, this copy was made from one by Quṭb al-Dīn ᶜAlī al-Tabrīzī, who may be the author. *Storey*, p. 132, no. 1 mentions a (Persian?) treatise on the qibla in 5 *faṣls* entitled ***Ḥātimīya*** or ***Qiblat al-Āfāq*** attributed to al-Junābadī (no. G67).

6.3.20 ***al-R. al-Shamsīya fi l-uṣūl al-ḥisābīya***, a treatise on arithmetic in 2 *fanns* of 2 + 4 *bābs* subdivided in *faṣls*: K 3957,5 (fols. 257v–304v, *ca.* 1050H); K 8531,2 (fols. 34r–57r, *ca.* 900H); DR 823,1 (fols. 1r–76r, 1314H); TR 278,3 (pp. 34–95, *ca.* 1100H, purportedly 757H, lacks beginning and end). A related treatise by Ḥamza b. ᶜAlī al-Bayhaqī (no. G39): DR 823,2 (fols. 77r–81r, 1314H). A commentary by ᶜAlī al-Birjandī (no. G58) on al-Nīsābūrī's treatise is in MS Manchester Rylands 352C.

G39 ḤAMZA B. ᶜALĪ AL-BAYHAQĪ

GAL, SII, p. 1020, no. 34. See also no. G19.

6.3.20 ***R. fi l-Ḥisāb mulḥaqa bi-l-Shamsīya***, a supplement to al-Nīsāpūrī's treatise on arithmetic (see no. G38): DR 823,2 (fols. 77r–81r, 1314H). Another copy may be the ***R. bi-ta*ᶜ*rīf su*ᵓ*ūl wa-fawā*ᵓ*id fi l-ḥisāb*** in MS Rampur I. 414_{33} listed by Brockelmann.

al-Bayhaqī copied MS London India Office 1148 of al-Ṭūsī's recension of the *Almagest* at Sulṭānīya in 722H (1322/23).

G40 ḤASAN B. ḤUSAYN B. ḤASAN SHĀHINSHĀH AL-SIMNĀNĪ

Suter, p. 149; *Storey*, p. 59.

2.1.12 A fragment from the tables in his *K. Tawḍīḥ Zīj Īlkhānī*, a recension of the *Īlkhānī Zīj* of al-Ṭūsī (no. G23), compiled in 795H: ṬFF 13,5 (fols. 123r–129r, 845H). Another fragment in the author's hand is in the British Library (*Rieu*, ii, 455a—*Storey*). The Cairo manuscript contains some tables of the astrological houses for latitude 34;30°.

G41 RUKN AL-DĪN B. SHARAF AL-DĪN AL-ḤUSAYNĪ AL-ᶜĀMULĪ

Storey, no. 108.

4.3.28 [*Panjah bāb-i-Sulṭānī*], a Persian treatise on the astrolabe in 2 *maqāla*s of 38 + 12 *bāb*s, compiled in Herat *ca.* 850H: S 4792 (136 pp., *ca.* 1100H, lacks first folio, no author or title mentioned). Other copies listed by Storey, to which add MS Istanbul Lale Ismail 289 (105 pp., 915H). See Pl. LIVb.

G42 ᶜALĪ B. MUḤAMMAD, AL-SAYYID AL-SHARĪF AL-JURJĀNĪ

Suter, no. 424; *GAL*, II, pp. 280–281, and SII, pp. 305–306; article "al-Djurdjānī" by A. S. Tritton in EI_2.

1.2.7 A commentary on the *Mulakhkhaṣ* of al-Jaghmīnī: see no. G17.

1.2.10 A commentary on *al-Tadhkira* by al-Ṭūsī (no. G23): K 8529 (225 fols., incorrectly attributed to al-Nīsābūrī (no. G38), 857H); *etc.* (8 copies).

6.1.3 *Ḥāshiya ᶜalā Taḥrīr . . . al-Ṭūsī li-K. al-Uṣūl li-Uqlīdis* (see no. A5): DR 530 (66 fols., 1308H in Rangoon!), not mentioned by Sezgin.

See also no. B108.

G43 MUḤAMMAD AL-ḤUSAYNĪ, SAYYID AL-MUNAJJIM

Storey, no. 100.

5.3.12 *Laṭāʾif al-kalām fī aḥkām al-aᶜwām*, a Persian treatise on astrology: ṬFF 3,2 (fols. 2v–63r, 909H); ṬFF 17,1 (fols. 1r–61r, 947H); *etc.* (1 copy).

G44 MAḤMŪD B. MUḤAMMAD B. QIWĀM AL-WĀLISHTĀNĪ

Storey, no. 18.

1.2.7 *al-Risāla al-Ghiyāthīya*, a Persian translation of al-Jaghmīnī's *Mulakhkhaṣ* (see no. G17) compiled for Ghiyāth al-Dīn Aḥmad: TR 347,3 (pp. 68–149, 832H in Herat), unique? This title is mentioned by Ḥājjī Khalīfa and cited by Storey who lists no copies. Suter (no. 403N) mentions another Persian translation by Muḥammad b. ᶜUmar Asdafānī(?) (*GAL*: Asadabādī) (no. G127) in MS Aya Sofia 2592: this individual is not listed by Storey.

G45 ᶜABD AL-WĀJID B. MUḤAMMAD

Not identified.

2.3.1 A *ḥāshiya* on the Arabic version of al-Ṭūsī's *Sī faṣl* compiled in 797H: see no. G23.

G46 MUḤAMMAD B. MUḤAMMAD, HUMĀM AL-ṬABĪB

GAL, SI, p. 865.

1.2.7 A commentary on the *Mulakhkhaṣ* of al-Jaghmīnī (no. G17): DH 41 (82 fols., 1125H). Brockelmann (*GAL*, SI, p. 865) states that this was compiled in 808/1405.

6.4.13 A treatise in Persian on sexagesimal arithmetic, arranged in 6 *bāb*s: MMT 7,3 (fols. 93v–103v, 1002H), unique? (This should have been listed in Section 6.5 of the Arabic-Persian-Turkish catalogue.)

G47 SAYYID ḤASAN B. SAYYID ᶜALĪ AL-QUMNĀTĪ

GAL, SI, p. 400; not listed in *Suter*.

2.1.10 ***al-Kāmil fī sharḥ al-Zīj al-Shāmil***, a commentary on the ***Shāmil Zīj*** (see no. B100) compiled in 822H: DM 951 (photos of MS Paris B.N. ar. 2530, 118 pl., 1121H). al-Qumnātī does not mention the author of the ***Shāmil Zīj*** and his commentary contains no tables. Another copy of this commentary is MS Istanbul Laleli 2137 (not examined).

G48 GHIYĀTH AL-DĪN JAMSHĪD B. MAS^{c}ŪD B. MAḤMŪD AL-KĀSHĪ

Suter, no. 429; *Krause*, no. 429; *GAL*, II, p. 273 and SII, p. 295; *Kennedy*, no. 20; article "al-Kāshī" by A. P. Youschkevitch and B. A. Rosenfeld in *DSB*.

2.1.20 ***Zīj-i-Khāqānī***, a Persian *zīj* for Samarqand (*Kennedy*, no. 20): TR 149 (370 pp., mainly *ca.* 950H, fine copy); plus a fragment in ṬJ 515,2 (fols. 28r–47v, *ca.* 1200H). The section on correcting the lunar mean motion by eclipse observations in al-Kāshī's introduction is copied separately in DR 898,23 (photos of MS Leiden 14, pp. 389–396, *ca.* 1050H). A study of this work by Dr. E. S. Kennedy is in progress.

6.3.22 ***Miftāḥ al-ḥisāb***, a treatise on arithmetic in 5 *maqālas* of 6 + 12 + 6 + 9 + 4 *bābs*: ṬR 134 (107 fols., 1205H); plus an abridgement *Talkhīṣ al-Miftāḥ* in 30 *faṣls*: DR 306 (13 fols., *ca.* 1250H, incomplete). Published: see the secondary literature for details. See also Pl. XCIIb.

6.4.2 Notes on linear interpolation attributed to al-Kāshī are in ZK 917,4 (pp. 50–54, 1163H).

See also Qāḍī Zāde (no. G50).

G49 ULUGH BEG

GAL, II, pp. 275–276, and SII, p. 298; *Kennedy*, no. 12; *Storey*, pp. 66–72; article "Ulugh Beg" by T. N. Kari-Niazov in *DSB*.

2.2.1 ***al-Zīj al-Sulṭānī = Zīj Ulugh Beg = Zīj al-Raṣad al-jadīd***, a *zīj* in Persian in 4 *maqālas* of 7 + 22 + 10 + 2 *bābs* with many tables, some specifically for Samarqand: ṬFF 18,2 (fols. 2r–229v + 247r, *ca.* 850H); ṬMF 4,1 (fols. 1v–194v, *ca.* 950H); ṬFF 10 (251 fols., *ca.* 1000H); S 5997 (171 fols., 1094H); ṬMF 8,1 (fols. 1v–192r, *ca.* 1100H); *etc.* (2 copies and numerous fragments and extracts).

This important work awaits detailed study. The Persian text of the introduction and a French translation were published in L. A. Sédillot in 1847—see *Storey*, p. 69, for details. A brief account of the tables is contained in *Kennedy*, pp. 166–167.

A copy of the *zīj* containing some tables for Tabriz (latitude 38°) is DMF 5 (52 fols., *ca.* 1150H). A copy containing tables for Istanbul (latitude 41;15°) and Bosnia (latitude 43;48°) is S 4350 (244 fols., *ca.* 1300H).

An anonymous Persian commentary entitled ***Khulāṣat al-khulāṣa*** arranged in 3 *maqālas*: ṬMF 7 (78 fols., *ca.* 1100H). This title is not listed by Storey.

A Persian commentary by Mīram Chelebī (no. H4) entitled ***Dastūr al-^{c}amal wa-taṣḥīḥ al-jadwal***: TR 150 (218 fols., *ca.* 1100H, purportedly by author); MMF 1 (168 fols., 1087H); S 4346 (154 fols., 1093H).

2.2.1 An Arabic translation, of unknown authorship, with the same tables: KhM 8 (186 fols., 1131H, lacks some tables); MM 205,1 (fols. 1r–132r, *ca.* 1150H); DM 95 (143 fols., *ca.* 1250H); *etc.* (numerous extracts). See also the translation of al-Rifā^{c}ī (no. C99) and the recension of Ibn Abi l-Fatḥ al-Ṣūfī for Cairo (no. C98), the translation of Qāḍī Ḥasan (no. G99), and the recensions for Tunis by Sanjaq Dār (no. F53) and Husayn Quṣ^{c}a (no. F47).

2.2.2 An anonymous Turkish commentary: ṬMT 4 (239 fols., *ca.* 1250H), and another by Darwīsh ^{c}Abbās Wasīm entitled ***al-Nahj al-bulūgh*** (*sic*): see no. H32.

A Turkish translation with tables prepared for Ḥasan Efendī in Cairo by ^{c}Abd al-Raḥmān ^{c}Uthmān (no. D139): ṬFT 33 (183 fols., *ca.* 1250H).

2.2.14 See also DM 678 (71 fols., *ca.* 1150H), a collection of tables including some by Ulugh Beg.

G50 MŪSĀ B. MUḤAMMAD B. MAḤMŪD, QĀḌĪ ZĀDE AL-RŪMĪ

Suter, no. 430; *GAL*, II, p. 275, *etc.*; *GAS*, V, pp. 114–115; article "Qāḍī Zāda al-Rūmī" by H. Dilgan in *DSB*.

1.2.7 A commentary on the ***Mulakhkhaṣ*** of al-Jaghmīnī (no. G17), compiled in Samarqand in 841H: K 8534 (141 fols., 857H); K 3957,1 (fols. 1v–151r, *ca.* 900H); ṬJ 966,3 (fols. 25v–81v, 891H); *etc.* (20 copies). Note especially TR 338,1 (p. 1, copied in 1246H) where Ḥājjī Khalīfa's entry on the *Mulakhkhaṣ* is recorded, including the fact that al-Jaghmīnī wrote it in 618H (not recorded in the printed editions).

Ḥāshiya by Faṣīḥ al-Dīn b. ᶜAbd al-Karīm al-Niẓāmī (no. X8): QH 3,2 (fols. 99v–154r, *ca.* 1200H); DH 89 (43 fols., *ca.* 1100H); and by Yūsuf b. Khiḍrbāy known as Sinān Pāshā (no. H5): DH 65 (158 fols., *ca.* 1100H); QH 3,3 (fols. 155v–247r, *ca.* 1200H); and by ᶜAlī al-Birjandī (no. G58); DH 19 (148 fols., 1078H); MH 2,1 (fols. 1r–55r, 1019H); ṬJF 26,1 (fols. 1r–73r, 1043H); ṬJ 162,1 (fols. 1r–114v, 1073H); *etc.* (17 copies).

2.1.20 ***R. fī Istikhrāj jayb daraja wāḥida***, on the derivation of Sine 1° by al-Kāshi (no. C48): MR 37 (6 fols., *ca.* 1150H, anonymous).

The attribution to Qāḍī Zāde is suggested by a reference in Sédillot-fils, *Prolegomènes des Tables Astronomiques d'Oloug-Beg*, (Paris, 1853), p. 77. The Cairo manuscript is listed by Suter (p. 174) as a work of al-Kāshī. The mathematical method is described in A. Aaboe, "al-Kāshī's Iteration Method for the Determination of Sin 1°," *SM*, 20 (1954), pp. 24–29.

3.3.5 A treatise on the determination of the meridian and the azimuth of the qibla, in Persian: MJ 116,10 (fols 76v–80v, 1071H).

6.1.4 A commentary on the ***Ashkāl al-taᵓsīs*** of al-Samarqandī with various later *ḥāshiyas*: see no. G26.

6.7.4 ***R. fi l-Misāḥa***, a treatise in Persian on surveying: MJ 116,9 (fols. 73r–76r, 1071H), unique. The title is not listed by Storey.

G51 ᶜALĀᵓ AL-DĪN ᶜALĪ B. MUḤAMMAD AL-QŪSHJĪ

Suter no. 438; *GAL*, II, p. 305, and SII, pp. 329–330; *Storey*, nos. 20 and 112; article "ᶜAli al-Ḳushdji" in EI_2 by A. Adnan. See also M. Cunbur, *Ali Kuşçu Bibliyografyası*, Ankara: Başbakanlık Basımevi, 1974.

1.2.17 ***al-Risāla al-Fatḥīya***, an Arabic treatise on the theoretical astronomy in 3 *maqālas* in 6 + 10 + 6 *bābs*: fragment in ṬJ 366,8 (fols. 92v–102v, *ca.* 1250H); *etc.* (1 fragment). A commentary by Mīram Chelebī (no. H4): DH 70 (122 fols., *ca.* 1250H, purports to be 925H), ZK 477 (205 fols., 1249H), *etc.* (2 fragments). A Turkish translation by Sayyid ᶜAlī b. Ḥusayn (no. H11): TR 137 (148 pp., *ca.* 1000H); MMT 5,3 (fols. 65v–154v, *ca.* 1100H).

1.2.17 ***R. fi l-Hayᵓa***, a Persian treatise on theoretical astronomy in 2 *maqālas*: S 4348 (75 fols., *ca.* 950H); TRF (31 fols., 1098H). Commentary by al-Lārī (no. G63): ṬJF 26,4 (fols. 122r–171v, 1043H) and ṬFF 15 (134 fols., 1085H).

1.2.17 Also a fragment on the motion of the sphere in ḤF 14,2 (fols. 107r–107v, *ca.* 900H) attributed to ᶜAlāᵓ al-Dīn al-ᶜwsy, perhaps al-Qūshjī.

6.4.18 A supplment to his treatise on algebra by Malik-Muḥammad b. Sulṭān-Ḥusayn Iṣfahānī (no. G64): ZK 91,1 (pp. 1–44, *ca.* 1050H). Qūshjī's treatise is extant in numerous copies (see *Storey*, pp. 9–10).

6.4.18 Also a solution to a geometrical problem (in Arabic): ṬR 111,2 (fol. 31v, *ca.* 1150H).

G52 ABU L-FATḤ MUḤAMMAD B. AL-HĀDĪ ABĪ NAṢR B. ABĪ SAᶜĪD AL-ḤUSAYNĪ AL-ᶜIRĀQĪ, TĀJ AL-SAᶜĪDĪ[1]

GAL, SI, p. 850; *GAS*, V, p. 115.

6.1.4 A *ḥāshiya* on Qāḍī Zāde's commentary on al-Samarqandī's ***Ashkāl al-Taᵓsīs***: see no. G26.

G53 ḤAYDAR B. AḤMAD AL-KURDĪ AL-ḤUSAYNĀBĀDĪ

6.1.4 A commentary on Qāḍī Zāde's commentary on al-Samarqandī's ***Ashkāl al-taᵓsīs***: see no. G26.

---

1. TR 125,2 has Abu l-Fatḥ Muḥammad al-Hādī Abū Naṣr b. Abī Saᶜīd al-Ḥusnī al-ᶜIrāqī, Tāj al-Saᶜdī. *GAL*, SI, p. 850 and *GAS*, V, p. 115 have . . . Muḥammad al-Hādī b. Naṣr . . .

G54 ḤUSĀM AL-DĪN AL-TŪQĀTĪ

GAL, SII, 323.

7.2.6 *R. mukhtaṣara fī Qaws Quzaḥ*, a short treatise on the rainbow: TK 52,2 (pp. 102–105, *ca.* 1000H); ṬJ 429,14 (fols. 83v–85r, *ca.* 1200H). Brockelmann lists a copy in Beirut and states that the treatise has been published in *Mashriq*, 15 (1912), pp. 742–744. Another copy not listed by Brockelmann is MS Berlin Ahlwardt 5691. See also al-Būzajānī (no. B54).

G55 MISCELLANEOUS ANONYMI

5.3.14 A nativity book, including elaborate astrological calculations and predictions for each year of the life of ʿImād al-Dīn Masʿūd b. al-Mawlā Ḍiyāʾ al-Dīn Muḥammad, born in Yazd in 1428: DMF 8 (357 pp., 864H), unique—see Pl. LXXXIIb. A similar work prepared for Amīrzāde ʿAlī b. al-Sulṭān Khalīl Bahādur, born in Ḥisn Kayf in 1468: MMF 4 (189 fols., *ca.* 900H), also unique. A detailed analysis of these manuscripts has been made by Dr. E. S. Kennedy.

G56 SHARAF AL-DĪN ʿALĪ AL-YAZDĪ

Storey, no. 19 and p. 30 (no. 47); *Suter*, no. 528N.

6.4.5 *R. fī Bayān ḥisāb al-yad*, a short Persian treatise on finger arithmetic: DJ 134,11 (fols. 196v–197r, 928H). Other copies in Meshed and Paris are listed by Storey.

G57 ḤUSAYN AL-WĀʿIẒ B. ʿALĪ AL-BAYHAQĪ, AL-KĀSHIFĪ

Storey, no. 116.

5.3.17 *Lubab al-ikhtiyārāt fī taʿyīn al-awqāt*, a Persian treatise on astrology in a *muqaddima*, two *maqālas*, and a *khātima*: ṬFF 7,1 (fols. 1r–129v, 1128H); ṬFF 16 (120 fols., 1285H); ṬFF 9,1 (fols. 1r–92v, 1302H); plus a Turkish translation by Muṣṭafā b. Muṣṭafā Ḥisān (no. H62): ṬFT 40,1 (pp. 1–198, *ca.* 1300H).

5.3.17 A discussion between Makhdūmī Janāb and the astronomers of Herat about the *ikhtiyārāt* associated with a campaign of Shāhrukh stated to be taken from the end of the treatise *K. Lawāʾiḥ al-Qamar* by Ḥusayn al-Wāʿiẓ: ṬFF 7,2 (fols. 129v–136v, 1128H).

The works of this author merit investigation. See already *Storey*.

G58 ʿALĪ/ʿABD AL-ʿALĪ B. MUḤAMMAD B. AL-ḤUSAYN AL-BIRJANDĪ

Suter, no. 456; *Storey*, pp. 80–82; *Azzawi*, pp. 136–137.

1.2.7 A *ḥāshiya* on the commentary of Qāḍī Zāde on al-Jaghmīnī's *Mulakhkhaṣ*: see no. G50.

2.3.4 A commentary on his *Bīst bāb man dar maʿrifat-i-taqwīm-i-tāmm* by Junābadī (no. G67): MMF 3 (155 fols., 1067H). ṬMF 1,2 (fols. 2r–190r, *ca.* 1100H). Other copies listed by Storey.

4.3.14 A commentary on *Bīst bāb* by al-Ṭūsī: see no. G23.

G59 ʿIMĀD (AL-DĪN) B. JAMĀL (AL-DĪN) BUKHĀRĪ

Storey, p. 70 and no. 111.

2.2.2 *Tashīlāt Zīj al-ʿImādī*, a collection of planetary mean motion and double argument equation tables: DMF 10 (222 fols., 1073H). Other copies listed by Storey. Merits investigation and comparison with copies elsewhere. A different set of similar tables is in ṬMF 11 (123 fols., 1080H).

G60 ḤUSAYN B. MUʿĪN AL-DĪN AL-MAYBŪDĪ

GAL, II, p. 272, and SII, p. 294.

1.2.18+ *Jām-i-Gītī-ye-numā*, a Persian treatise on cosmology and astronomy in 30 *maqṣads*: ṬJ 873,2 (fols. 36v–52v, 1036); ṬJF 26,2 (fols. 74v–85r, 1043H); TK 111 (21 pp., *ca.* 1100H).

G61 ANONYMOUS

1.2.7 *Sharḥ al-Mulakhkhaṣ li-l-Jaghmīnī* (see no. G17): ṬH 33,1 (fols. 1r–36r, *ca.* 900H in the author's hand).

G62 ABU L-KHAYR MUḤAMMAD AL-TAQĪ B. MUḤAMMAD AL-FĀRISĪ

Storey, no. 123.

1.2.18 (*R. fī Muṣṭalaḥāt* ᶜ*ilm al-hay*ᵓ*a*), a Persian treatise on astronomical terminology arranged in unnumbered *fā*ᵓ*ida*s: S 4467,2 (fols. 9v–20v, *ca.* 1100H). This is perhaps no. 8 in Storey's list of al-Fārisī's astronomical works.

5.3.18 *Ḥall al-taqwīm fī* ᶜ*ilm al-tanjīm*, a Persian treatise on astrology without subdivisions: S 4349,2 (fols. 20r–30v, *ca.* 1100H). See *Storey*, pp. 83–84, no. (3).

5.3.18 [*Iḥkām al-aḥkām*], a Persian treatise on astrology arranged in a *muqaddima* and 12 unnumbered *bāb*s: S 4467,3 (fols. 20v–34r, *ca.* 1100H). al-Fārisī mentions the ***Burhān al-kifāya*** of al-Bakrī (no. G8). See *Storey*, p. 84, no. (5).

G63 MUṢLIḤ AL-DĪN MUḤAMMAD B. ṢALĀḤ AL-LĀRĪ AL-ANṢĀRĪ

Suter, no. 467; *Storey*, p. 77 and the references there cited.

1.2.17 A commentary on the Persian treatise on theoretical astronomy in 2 *maqāla*s by ᶜAlī Qūshjī: see no. G51.

G64 MALIK-MUḤAMMAD B. SULṬĀN-ḤUSAYN IṢFAHĀNĪ

Storey, no. 23.

6.4.18 A supplement to the treatise on algebra by ᶜAlī Qūshjī (no. G51): ZK 91,1 (pp. 1–44, *ca.* 1050H). Storey lists one other copy in Meshed.

G65 ĪRĀNSHĀH B. ᶜALĪ AL-NĪSĀBŪRĪ

Not mentioned in *Storey*.

5.3.31 *Qirānāt-i-Īrānshāhī*, an astrological world-history based on Jupiter-Saturn conjunctions, stressing the career of Chingiz Khan: ṬMF 3 (168 fols., 1103H), unique? A study of this work by Dr. E. S. Kennedy is to appear in a Memorial Volume for Malcolm Kerr.

G66 QIWĀM AL-DĪN ḤUSAYN B. SHAMS AL-DĪN MUḤAMMAD AL-KHAFRĪ

Storey, p. 27, no. 20.

6.4.14 *Ja*ᶜ*farīyat-i-ḥisāb*, a Persian treatise on arithmetic in 5 *maqāla*s: ZK 91,2 (pp. 45–113, different hands the last dated 1087H, defective). Another copy is in Najaf (*Storey*).

G67 MUẒAFFAR MUNAJJIM B. MUḤAMMAD QĀSIM JUNĀBADĪ

Storey, no. 132.

2.3.4 A commentary on *Bīst bāb* by al-Birjandī: see no. G58.

5.3.29 *Tanbīhāt al-munajjimīn*, a Persian treatise dealing mainly with astrology arranged in a *muqaddima*, 6 *bāb*s, and a *khātima*, compiled *ca.* 1025H: MMF 2 (216 fols., *ca.* 1200H); ṬMF 13 (162 fols., 1168H); ṬMF 10 (115 fols., *ca.* 1150H); plus a Turkish translation by ᶜUmar b. ᶜAbd Allah known as Nuzhat (no. H55): ṬFT 13 (191 fols., 1197H); *etc.* (2 copies). For other copies, see *Storey*. On this work, see now E. S. Kennedy, "Astronomical Events from a Persian Astrological Manuscript," *Centaurus*, 24 (1980), pp. 162–177, and the appendix by O. Gingerich on pp. 178–180.

G68 BAHĀᵓ AL-DĪN MUḤAMMAD B. AL-ḤUSAYN AL-ᶜĀMILĪ

Suter, no. 480; *GAL*, II, pp. 546–547, SII, pp. 595–597, and SIII, p. 1301; *Storey*, nos. 24 and 130; and the anonymous article "al-ᶜĀmilī" in EI_2.

1.2.19 *Tashriḥ al-aflāk*, a treatise on theoretical astronomy in 5 *faṣl*s: TJ 246,7 (pp. 49–64, 1146H); *etc.* (3 copies); plus a *ḥāshiya* by al-ᶜĀmilī (?): DH 83 (14 fols., 1292H); and a commentary by Muḥammad al-Bāqir al-Bānī (?) al-Kurdī (no. G103): TR 130,2 (pp. 38–82, 1274H).

4.3.34 *Tuḥfat-i-Ḥātimī*, a Persian treatise on the use of the astrolabe in 70 *bāb*s based on the treatise of Naṣīr al-Dīn al-Ṭūsī (no. G23): ṬMF 2,1 (fols. 1v–26v, 1119H). Arabic version in 60 *bāb*s by Mullā ᶜAlī Efendī al-Dāghistānī (no. X10): K 3824,24 (fols. 123v–145r, *ca.* 1100H); DM 1083 (19 fols., 1291H).

4.3.34 ***al-Ṣafīḥa***, an Arabic treatise on the astrolabe short enough to be written on the *ṣafīḥa* of an astrolabe: B 19118,7 (fols. 183r–185r, 1102H); *etc.* (2 copies); plus commentaries by ʿAbd Allāh al-Mawṣilī (no. G80), compiled in 1149H: Sh 56,3 (fols. 14r–32v, 1168H) and DM 1188,4 (fols. 109r–117r, 1323H); and by Abu l-Yumn al-Ālūsī (no. G84), compiled in 1275H: MJ 172,2 (fols. 77v–122r, 1278H).

6.4.23 ***Khulāṣat al-ḥisāb = al-Risāla al-Bahāʾīya***, a treatise on arithmetic in 9 *bāb*s subdivided in *faṣl*s: K3789,1 (fols. 1v–52r, *ca.* 1000H, may be author's copy?); TR 84 (99 pp., *ca.* 1050H); ṬJ 635,34 (fols. 136r–148v, 1170H); *etc.* (26 copies). Persian translation by Maʿbūdī al-Shāhrūdī (no. G96), entitled ***Tuḥfat-i-quṭb-i-shāhi***: ṬRF 1,1 (fols. 1r–76v, 1070H). Commentaries by al-Bursawī (no. H57): KA 23 (79 fols., 1289H), DR 1034 (97 fols., 1021H, anonymous); by Muḥamad Shaykh Zāde (no. X18), entitled ***Fatḥ al-Lubab*** . . . : TR 127,1 (pp. 1–412, *ca.* 1200H); by Muḥammad Bāqir (no. G104): DR 1044 (99 fols., 1091H); by ʿUmar al-Jalī (no. D54): ṬR 120 (68 fols., 1092H); ḤR 4 (59 fols., 1131H); *etc.* (8 copies); by Jawād b. Saʿd (no. G73): DR 289 (214 fols., 1073H); DR 1035,1 (fols. 1r–164r, 1111H); *etc.* (3 copies); by Ramaḍān al-Jazarī (no. G75), entitled ***Ḥall al-Khulāṣa*** or ***Aʿla l-Khulāṣa li-ahl al-riʾāsa***: ṬR 129 (174 fols., 1076H); MR 9 (107 fols., 1116H); *etc.* (5 copies); by ʿAbd al-Raḥīm al-Marʿīshī (no. G101): ZK 574 (239pp., 1146H); J 3488,3 (fols. 54r–154v, 1163H); *etc.* (4 copies); by Muḥammad Ḥakīm al-Ḥusaynī (no. G100): DR 818 (64 fols., *ca.* 1150H); *etc.* (1 copy); and a commentary on the third *bāb* by Nūr al-Din al-Wāʾiẓ (no. D184): ṬJ 635,4 (fols. 21r–26r, 1170H).

G69 ḤUSAYN AL-ḤUSAYNĪ AL-KHALKHĀLĪ

GAL, II, pp. 544–545, and SII, p. 591.

3.5.11 ***R. al-Dāʾira al-hindīya***, a treatise on the use of the Indian circle (a geometrical construction for determining the meridian), compiled in 1006H: ṬJF 26,5 (fols. 172r–176r, 1043H); DM 490,1 (fols. 1v–7r, *ca.* 1100H); *etc.* (5 copies). al-Khalkhālī refers in his introduction to a work entitled ***Sharḥ al-Wiqāya*** by Burhān al-Dīn (on which see *GAL*, SII, p. 646). See also nos. H17, and H58 and Z26 below, the anonymous verses of poetry on the Indian circle in ZK 706,9 (fols. 48v–57r, *ca.* 1100H), and the anonymous *faṣl* on the Indian circle (lost). See MS Princeton Mach 4941 for another treatise on this subject by Shāh Efendī al-Fanārī (no. H103). For al-Khalkhālī's other scientific works, see *GAL*.

G70 JALĀL AL-DĪN MUḤAMMAD B. ʿABD ALLĀH YAZDĪ

Storey, no. 129.

5.3.28 ***Tuḥfat al-munajjimīn***, a Persian treatise on astrology in 1 *maqāla* of 3 *bāb*s and a *khātima*: ṬFF 1 (183 fols., 1021H). Storey mentions only one copy in Meshed copied in 1056H. On the tables for Meshed in this treatise, see *SATMI*.

G71 MUḤAMMAD BĀQIR B. ZAYN AL-ʿĀBIDIN YAZDĪ

GAL, SII, p. 591 and p. 1024, no. 72; *Azzawi*, pp. 303–304.

6.4.30 ***ʿUyūn al-ḥisāb***, an extensive treatise on arithmetic in 3 *bāb*s and 2 *maqāla*s with numerous tables: DR 822 (151 fols., *ca.* 1050H); ṬJ 882,1 (fols. 1r–184r, 1060H, incomplete); *etc.* (one defective copy).

G72 IBRĀHĪM B. ZAYN AL-DĪN AL-SHĪRĀZĪ

Not identified.

5.4 A computation relating to the heliacal appearance of Sirius at Mecca in 1041H: DM 59,1 (fol. 1r, 1042H), unique.

G73 JAWĀD B. SAʿD B. JAWĀD

GAL, SII, p. 596.

6.4.23 A commentary on al-ʿĀmilī's ***Khulāṣat al-ḥisāb***: see no. G68 (earliest manuscript dated 1073H).

G74 MALIK MAḤMŪD KARRĀMĪ B. MALIK AḤMAD KHWĀNSĀRĪ

Storey, p. 106, no. 27.

2.2.11 ***Kanz al-fawāʾid fī dhikr al-qawāʾid***, a Persian *zīj* arranged in a *muqaddima* and 14 *maqṣad*s: ṬFF 6 (375 fols., *ca.* 1100H, lacking at end), *etc.* (1 incomplete copy and 1 fragment). Storey lists another copy, MS Istanbul Hamidiye 851. A description of this work has been prepared by Dr. E. S. Kennedy.

G75 RAMAḌĀN B. ABĪ HURAYRA AL-JAZARĪ AL-QĀDIRĪ

GAL, SII, p. 596.

6.4.23 ***Ḥall al-Khūlāṣa li-ahl al-riʾāsa*** or ***Aʿla l-Khulāṣa*** . . . , an extensive commentary on the ***Khulāṣat al-ḥisāb*** of al-ʿĀmilī (no. G68) compiled in 1076H: TR 129 (174 fols., 1076H); MR 9 (107 fols., 1116H); *etc.* (5 copies).

G76 MUḤAMMAD ZAMĀN B. SHARAF AL-DĪN ḤUSAYN AL-MASHHADĪ

Not listed in *Storey* or *Kennedy*.

2.2.12 ***Tuḥfat-i-Sulaymānī***, a Persian *zīj* arranged in 24 *tuḥfa*s, compiled in Meshed in 1078H: ṬFF 14 (111 fols., *ca.* 1100H), unique? Merits detailed investigation. On the spherical astronomical tables for Meshed in this work, see *SATMI*, II.

G77 ḤAYDARA B. ʿABD AL-RAḤMĀN AL-ḤUSAYNĪ AL-JAZARĪ

GAL, SII, p. 1020.

4.3.38 (***R. fi l-ʿAmal bi-l-asṭurlāb***), an Arabic treatise on the use of the astrolabe in 10 *bāb*s: ZK 782,13 (fols. 70v–90v, *ca.* 1120H); TR 134 (19 pp., *ca.* 1100H); *etc.* (3 copies). The first 6 *bāb*s of this treatise are in MS Berlin Ahlwardt 5802 (fols. 49–56, *ca.* 1100H) where the title is given as ***Nuzhat al-ṭullāb fī ʿilm al-asṭurlāb*** and the work is dedicated to Muṣṭafā Pāshā and stated to be translated from a Persian treatise.

G78 MUḤAMMAD BĀQIR B. MUḤAMMAD MUʾMIN KHURĀSĀNĪ SABZAWĀRĪ

Storey, no. 139.

1.2.20 ***R. fī Ithbāt ḥarakat al-shams wa-sukūn al-arḍ***, a treatise in Persian on a proof that the sun moves around the earth: DMF 9 (18 fols., *ca.* 1300H), unique? This treatise is not listed in *Storey*, and might repay investigation.

G79 AḤMAD B. MUḤAMMAD AL-MAHDĪ AL-SHARĪF AL-IṢFAHĀNĪ AL-KHĀTŪNABĀDĪ

Azzawi, p. 307.

2.3.10 ***R. fī Maʿrifat al-taqwīm***, on the compilation of ephemerides in 12 *faṣl*s, written in Karbalāʾ in 1126H: ṬM 227,3 (fols. 72v–86v, 1145H); *etc.* (1 fragment). Other copies are MSS Istanbul Esad Ef. 1975 (21 fols.) and Nurosmaniye 2517,3 (fols. 62v–72r).

G80 ABŪ MUḤAMMAD ʿABD ALLĀH B. FAKHR AL-DĪN AL-ḤUSAYNĪ AL-MAWṢILĪ

Azzawi, pp. 261–262.

4.3.34 ***Sawāniḥ al-qarīḥa fī sharḥ al-Ṣafīḥa*** a commentary on the treatise of al-ʿĀmilī compiled in 1149H: see no. G68.

G81 MUḤAMMAD B. ĀDAM AL-RUSTĀʾĪ

Azzawi, pp. 266–276.

1.2.22 ***Tashrīḥ al-kawākib wa-l-sayyarāt fi l-taqwīm wa-l-zījāt***, a treatise on theoretical astronomy in 10 *bāb*s: DM 947,4 (fols. 108r–147r, 1199H), unique.

2.3.11 / 5.3.37 ***Tafṣīl al-taqwīm fī sharḥ Hidāyat al-tanjīm***, a commentary in 30 *bāb*s on his treatise ***Hidāyat al-tanjīm***, on the compilation of annual ephemerides with daily astrological information: DM 947,1 (fols. 1r–48v, 1200H), unique.

2.5.13+ ***Tashrīḥ al-kawākib wa-l-sayyārāt = Sharḥ al-saḥāʾif***, a Persian introduction in 3 *bābs* followed by auxiliary tables for computing planetary positions: DM 947,5 (fols. 148r–153r, 1200H). Merits investigation.

4.3.40 ***Tadhkirat al-aḥbāb fi l-ᶜamal bi-l-asṭurlāb***, a treatise on the use of the astrolabe in 20 *bābs*: DM 947,2 (fols. 50r–80v, 1199H), unique.

4.5.26 ***Miftāḥ al-mughayyab fi l-ᶜamal bi-l-rubᶜ al-mujayyab***, a treatise on the use of the sine quadrant in 20 *bābs*: DM 947,3 (fols. 82r–107r, 1200H), unique.

G82 ḤASAN B. SHUJĀᶜ B. MUḤAMMAD AL-ḤASAN AL-ḤĀFIẒ AL-TŪNĪ(?)

Storey, p. 105, no. 11.

5.3.30 ***Dalīl al-munajjimīn***, a Persian treatise on astrology, without subdivisions: ṬFF 9 (32 fols., 1222H). Storey lists another copy in Bombay.

G83 MUḤAMMAD B. AL-ḤURR AL-ᶜĀMULĪ[1]

6.1.4 A poem based on the ***Ashkāl al-taᵓsīs*** of al-Samarqandī (see no. G26): DJ 846,4 (fols. 61v–67v, 1100H).

G84 ABU L-YUMN ᶜABD AL-BĀQĪ B. ABI L-THANĀᵓ MAḤMŪD AL-ALŪSĪ, *muftī* of Baghdad

Not mentioned in *Azzawi*! On his father, see *Azzawi*, pp. 269–272.

4.3.34 ***Fayūḍāt al-qarīḥa fī sharḥ al-Ṣafīḥa*** a commentary on the treatise by al-ᶜĀmilī compiled in 1275H: see no. G68.

1. Brockelmann (*GAL*, II, p. 542, and SII, p. 578) lists a Shiᶜite legal scholar Muḥammad b. al-Ḥasan al-Ḥasanī al-Ḥurr al-ᶜĀmulī (1033/1624–1099/1688). *Azzawi*, 304 lists Zayn al-Dīn b. al-Ḥasan b. ᶜAlī known as al-Ḥurr al-ᶜĀmulī, (d. 1078/1667), author of a work on astronomy entitled *Mutawassiṭ al-futūḥ bayn al-mutūn wa-l-shurūḥ*, not extant.

Note: The following authors and miscellaneous anonymi are not listed in *Storey* and have not been identified.

G85 MAḤMŪD MUNAJJIM KĪLĀNĪ

2.4.5 A table for finding the horoscopus for latitude 37°, with a short Persian introduction: ṬFF 20,2 (fols. 156r–158r, *ca.* 1200H).

G86 ḤUSĀM B. SHAMS AL-DĪN AL-KHAṬĪB AL-KHIṬĀBĪ AL-AHJĀNĪ

2.3.1 ***Muḍīḥ al-rusūm fī ᶜilm al-nujūm***, a Persian commentary on ***Sī faṣl*** by al-Ṭūsī: see no. G23. Storey (pp. 58–59) lists various anonymous commentaries on the ***Sī faṣl***.

G87 KARĪM B. IBRĀHĪM

6.4.42 ***al-Wajīza***, a treatise on arithmetic in Arabic, arranged in 3 *bābs* and compiled in 1243H: DJ 726,1 (fols. 1r–20r, 1272H).

6.4.42 ***Kashf al-majhūl***, a treatise on arithmetic in Arabic, arranged in 10 *bābs* and compiled in 1243H: DJ 726,2 (fols. 20v–51v, 1272H).

G88 ᶜALĪ TŪRĪ (?) ZĀDE

1.2.11 The commentary on ***al-Tuḥfa al-shāhīya*** by Quṭb al-Dīn al-Shīrāzī (no. G27) in DH 7 (37 fols., incomplete, *ca.* 1200H) is attributed to Mawlānā Qwry (?) Zāde. The same work is in MS Hyderabad Asafiya 48, attributed to ᶜAlī Tūrī Zāde. Brockelmann (*GAL*, II, p. 274) lists MS Istanbul Aya Sofia 2643 as a copy of a commentary on the ***Tuḥfa*** by ᶜAlī Qūshjī (no. G51), which is probably the same work incorrectly attributed.

G89 QUṬB AL-DĪN ᶜIZZ AL-DĪN (B.?) ᶜABD AL-ḤAYY AL-ZĀHIDĪ AL-BAKRĪ AL-ḤUSAYNĪ AL-LĀRĪ

5.5.3 The sixth *faṣl* of an astrological treatise in Persian entitled ***al-Ḥall wa-l-ᶜaqd***: S 4349,4 (fols. 48r–51v, *ca.* 1100H).

G90 SA^c^D AL-DĪN AS^c^AD B. SA^c^ĪD AL-HAMADĀNĪ

6.2.23 Some notes attributed to al-Hamadānī occur at the end of DR 898,8 of a work by Ibn al-Haytham (no. B77).

G91 ^c^ABD ALLĀH B. NAṢR ALLĀH QARAGOZLU

5.3.39 *Tuḥfat al-mulūk*, a Persian treatise on astrology dedicated to Shāhzāde Ḥasan^c^alī Mirza Qajar, and arranged in a *muqaddima*, 4 *bāb*s, and a *khātima*, subdivided in *faṣl*s: ṬMF 1 (*bis*) (128 fols., 117H). No other copies known to me.

G92 KAMĀL AL-DĪN AL-ḤUSAYN AL-FĀRISĪ

(= no. G30?)

6.1.3 *Ta^c^līq ^c^alā l-maqāla al-thālitha ^c^ashar min Taḥrīr al-Ṭūsī li-Uṣūl Uqlīdis* (see no. A5), not listed by Sezgin: DR 898,15 (photos of MS Leiden 14, pp. 298–300, *ca.* 1050H).

G93 MUḤAMMAD BĀQIR B. MUḤAMMAD TAQĪ

See also nos. G103 and G104.

6.8.12 *R. fi l-Mawāzīn wa-l-makāyīl*, on weights and measures: B 20302,1 (fols. 1v–12r, 1221H); *etc.* (1 copy).

G94 LA^c^LĪ ZĀDE ^c^ABD AL-BĀQĪ AL-ḤUSAYNĪ

5.3.39 *R. fī Ḥisāb ṭāli^c^ al-wilāda wa-mā yata^c^allaqu bihi min al-aḥkām*, a Persian treatise on nativities: ṬFF 19 (43 fols., *ca.* 1250H).

G95 RUSTAM BEY AL-ḤĀRITHĪ AL-BADAKHSHĪ

6.4.46 An Arabic translation of the 8th construction of the first *maqāla* of Clavius' *Elements*: DR 689 (photos of MS London B.L. Add. 14332,1, 20 plates, 1834 A.D.). On Clavius, see no. J1.

G96 MA^c^BŪDĪ B. IBRĀHĪM AL-SHĀHRŪDĪ AL-BISṬĀMĪ

6.4.23 *Tuḥfat-i-quṭb-i-shāhī*, a Persian translation of *Khulāṣat al-ḥisāb* by al-^c^Āmilī (no. G68): ṬRF 1,1 (fols. 1r–76v, 1070H). *Storey*, no. 24 lists numerous Persian translations and commentaries, but not this one.

G97 KHŌJA ^c^AṬĀ^ᵓ^ ALLĀH

4.2.5+ *R. dar Ma^c^rifat-iye-asturlab-i-kurī*, a Persian treatise on the use of the spherical astrolabe in 20 *bāb*s: TR 165,4 (pp. 64–69, 992H).

G98 MENLĀ CHELEBĪ AL-KURDĪ

2.3.10 *R. fi Istikhrāj al-taqwīm*, on the compilation of ephemerides, in 28 *bāb*s (although the introduction mentions 18 *bāb*s): TR 324 (86 pp., *ca.* 1250H); K 4297,2 (fols. 18v–45v, *ca.* 1100H, anonymous); ṬM 89 (43 fols., 1084H, lacks chs. 26–28, anonymous).

G99 ḤASAN B. MUḤAMMAD AL-FAṢĪḤĪ AL-NIẒĀMĪ, QĀḌĪ ḤASAN

2.2.4+ *Durr al-tatwīj bi-ta^c^rīb mu^c^āmarāt al-zīj*, a translation of the introduction of the *zīj* of Ulugh Beg (no. G49): DM 1199,1 (fols. 1v–59v, *ca.* 1200H); DM 52 (40 fols., 1170H), *etc.* (3 copies).

G100 MUḤAMMAD HAKĪM B. ^c^ALĀ^ᵓ^ AL-DAWLA AL-ḤUSAYNĪ

6.4.23 A commentary on al-^c^Amilī's *Khulāṣat al-ḥisāb*: see no. G68.

G101 ^c^ABD AL-RAḤĪM B. ABĪ BAKR AL-MAR^c^ĪSHĪ

GAL, SII, p. 596.

6.4.23 A commentary on al-^c^Amilī's *Khulāṣat al-ḥisāb*: see no. G68.

G102 ᶜABD AL-QĀDIR B. ᶜUTHMĀN AL-ĀDANĪ

4.3.39 A commentary on ***Bahjat al-albāb*** by by al-Qayṣarī: see no. H25. Manuscript copied in 1135H.

G103 MUḤAMMAD AL-BĀQIR AL-BĀNĪ AL-KURDĪ

See also nos. G93 and G104.

1.2.19 A commentary on ***Tashrīḥ al-aflāk*** by al-ᶜĀmilī: see no. G68.

G104 IBN ABI L-QĀSIM AL-ḤUSNĪ MUḤAMMAD BĀQIR AL-TABRĪZĪ

See also nos. G93 and G103.

6.4.23 A commentary on al-ᶜĀmilī's ***Khulāṣat al-ḥisāb***: see no. G68 (manuscript dated 1091H).

G105 ᶜUMAR ᶜABD AL-ᶜAZĪZ

6.4.21 ***Dastūr al-ḥisāb jalālī***, a treatise in Persian on arithmetic in 4 *maqāla*s: ṬRF 1,2 (fols. 77v–101r, 1070H).

G106 JALĀL AL-DĪN FAḌL ALLĀH AL-ᶜUBAYDĪ

1.2.7 A *ḥāshiya* on the ***Mulakhkhaṣ*** of al-Jaghmīnī (no. G17): ṬH 33,3 (fols. 37v–83v, *ca.* 1000H). Another copy is in Istanbul Fatih 3410,2.

G107 ANONYMOUS

1.2.7 A commentary on the commentary on al-Jaghmīnī's ***Mulakhkhaṣ*** by Qāḍī Zāde (no. G50): DH 19 (148 fols., 1078H); MH 2 (56 fols., 1019H in Amud, incomplete); *etc.* (1 copy).

G108 ANONYMOUS

1.3.1 A Persian version of the ***K. Ṣuwar al-kawākib*** by al-Ṣūfī (no. B52): MMF 9 (136 fols., 1043H), not listed in *Storey*. This is a magnificent copy. See Pl. III.

G109 ANONYMOUS (chapter on the fixed stars)

1.3.4 A *faṣl* dealing with the fixed stars from an unidentified work: DM 681,4 (fols. 8r–13v, *ca.* 900H). The author mentions (fol. 13v) the ***K. al-Zubda*** of Naṣīr al-Dīn al-Ṭūsī (no. G23).

G110 MISCELLANEOUS (notes on the fixed stars)

1.3.6 A *faṣl* on the stars and their names: ṬM 103,2 (fols. 4r–6v, *ca.* 1100H).

1.3.6 Notes including star-names in Arabic, Persian, and Turkish, and some symbols for certain fixed stars: ṬM 250,4 (fols. 30v–31r, 1277H).

G111 MISCELLANEOUS (*zīj*es)

2.2.15 Miscellaneous fragments from various Persian *zīj*es: S4493,2 (fols. 11r–104v, late). Tables from unidentified Persian *zīj*es: MM 159,2 + 3 (fols. 19v–28r, *ca.* 1250H). Anonymous solar equation tables: DM 679 (7 fols., *ca.* 1150H).

G112 ANONYMOUS (ephemerides)

2.3.12 Persian ephemerides for 1244H: S 4493,3 (fols. 105r–120v, *ca.* 1244H), and for 1271H: S 4493,1 (fol. 1r–10r, *ca.* 1270H). See Pl. XXVb.

G113 MISCELLANEOUS ANONYMI (Persian treatises on the compilation of ephemerides)

2.3.13 ***Mukhtaṣar dar maᶜrifat-i-shinākhtan-i-taqwīm***, in 5 *bāb*s: S 4349,1 (fols. 1–20r, *ca.* 1100H). *Cf. Storey*, p. 105 no. 15.

2.3.13 ***R. mukhtaṣara fi l-Taqwīm***, in 32 *bāb*s: MJF 2,3 (fols. 123v–131r, *ca.* 1200H).

2.3.13 A fragment from the end of a treatise in 33 *bāb*s: DMF 2,2 (fols. 38r–38v, *ca.* 900H).

2.3.13 A commentary on an anonymous treatise: S 4349,3 (fols. 30v–48r, *ca.* 1100H).

G114 ANONYMOUS (star catalog)

2.4.2 An unidentified table of stellar coordinates computed for the beginning of 1771 Alexander corresponding to the end of 864H: DM 473 (8 fols., *ca.* 1100H!); DM 86 (8 fols., 1101H!). The anonymous compiler states that the longitudes are those of al-Ṣūfī (those of the *Almagest* plus 12;42°), plus 7;30°.

G115 ANONYMOUS

2.5.14 A treatise in Persian on the seven climates: S4485,1 (fols. 1v–2v, 1134H).

G116 ANONYMOUS

2.5.16 *R. fī Maᶜrifat taqwīm al-sabᶜa al-sayyāra bi-l-ṣafāʾiḥ*, a treatise in Persian on the determination of planetary longitudes approximately: K 3824,16 (fols. 148r–155r, 1170H).

G117 ANONYMOUS

3.4.13 Anonymous *ghurra-nāme*s, giving information on the new moons for 1086–1088H: ṬJ 298,7 (fol. 55r, *ca.* 1085H). See Pl. XLVc.

G118 ANONYMOUS (instruments)

4.1.4+ *al-R. al-Ghazānīya fi l-ālāt al-raṣadīya*, a treatise in Persian on observational instruments in 2 *qisms*: DR 702,4 (fols. 40r–63r, *ca.* 1000H). Another anonymous copy is MS Hyderabad Asafiya 153. This treatise differs from the treatise of ᶜAbd al-Munᶜim al-ᶜĀmilī (no. G128).

G119 MISCELLANEOUS (treatises on the astrolabe)

4.3.31 *K. fi l-asṭurlāb*, a Persian treatise in 40 *bāb*s, compiled in 819 Yazdigird (= 1451): DMF 2 (fols. 1r–37v, *ca.* 900H), unique? *Cf. Storey* p. 107 no. 35.

4.3.31 Fragment of an anonymous Persian treatise: TR 165,2 (pp. 57–61, 992H).

G120 MISCELLANEOUS (treatises on the use of the quadrant)

4.4.15 A Persian treatise in 16 *faṣl*s: MMT 7,2 (fols. 76v–92v, 1002H).

4.4.15 *Mukhtaṣar dar maᶜrifat-i-rubᶜ-i-asṭurlāb*, a Persian treatise in 6 unnumbered sections: ṬJ 966,6 (fols. 107r–107v, 891H).

G121 MISCELLANEOUS (treatises on sundials)

4.7.18 A fragment from the end of an unidentified Arabic treatise on sundial theory, probably of Eastern origin: S 4467,1 (fols. 2r–7v, *ca.* 1100H).

4.7.18 *Qiṭᶜa fī maᶜrifat ᶜamal al-rukhāma hisābᵃⁿ wa-takhṭīṭᵃⁿ*, a treatise on sundial theory in Persian, without tables: ṬMF 4,2 (fols. 195v–196v, *ca.* 1000H).

G122 ANONYMOUS

5.1.1 A *malḥama* in Persian arranged in 24 *bāb*s and attributed to Hermes, Daniel, Alexander, and Balᶜam is in ṬJF 12,3 (fols. 29r–56v, *ca.* 1250H).

G123 MISCELLANEOUS ANONYMI

5.4.6 Miscellaneous treatises on astrology: see Section 5.4.6 of the Persian catalogue.

G124 ANONYMOUS

6.4.26 Anonymous Persian *R. fi l-Ḥisāb*: K 3789,2 (fols. 53v–92r, 1008H).

G125 ANONYMOUS

6.8.13 A treatise on weights, in Persian, arranged in 3 *faṣl*s: B 20302,4 (fols. 78v–82r, *ca.* 1250H).

ADDENDA

G126 QUṬB AL-DĪN ᶜALĪ AL-TABRĪZĪ

See no. G38.

G127 MUḤAMMAD B. ᶜUMAR ASDAFĀNĪ (?)/ASADABĀDĪ (?)

See no. G44.

G128 ᶜABD AL-MUNᶜIM AL-ᶜĀMULĪ

Storey, p. 85.

See no. G118.

H. Turkish Authors

Occasional references to Turkish astronomers and their works are given in *Suter*, *Brockelmann*, and *Azzawi*. On Ottoman astronomy, see *Adnan* (much outdated). On astronomical timekeeping, see D. A. King, "Astronomical Timekeeping in Ottoman Turkey," in *Istanbul Symposium Proceedings*, pp. 254–269. Many of the Turkish manuscripts in the ENL were identified by Dr. Cornell Fleischer, who prepared the Turkish section of vol. II of the *Catalogue*.

H1 DARENDELĪ

No biographical information available.

3.1.15 [*Rūznāme Darendelī*], prayer-tables for Istanbul (latitude 41°) with an introduction in Turkish: ṬFT 29,2 (fols. 37v–45r, *ca.* 1200H, anon.). A facsimile of these tables was published in M. D'Ohsson, *Tableau Général de l'Empire Othoman*, (Paris, 1787), vol. 1, facing p. 192. The tables exist in several manuscript copies, and a brief analysis is contained in *SATMI*, II. See Pl. XLa.

H2 MUṢṬAFĀ B. AḤMAD AL-ṢAYRAWĪ AL-ᶜĪSAWĪ, SHAYKH VEFĀ

According to the author of the work entitled *Ṣaḥīfat al-ayyām*, Shaykh Vefā compiled his almanac whilst a prisoner of the Christians in Rhodes, and died in Constantinople in 897H (= 1491).

3.1.15 ***Rūznāme Shaykh Vefā***, an almanac consisting mainly of tables for calendar conversion and computing new moons, with instructions in Turkish in the margins: DMT 14 (10 fols., *ca.* 1050H); QJ 22,1 (fols. 1v–10r, *ca.* 1100H); DMT 2,2 (fols. 26v–36r, *ca.* 1200H). See Pls. XXXVIII and XXXIX. Numerous other copies exist. A fascimile of this almanac was published by G. H. Velschius, *Commentarius in Ruzname Naurus*, 1676, with notes in Latin, but there is no adequate description of the almanac in the modern literature.

An anonymous commentary on the ***Rūznāme*** entitled ***Ṣaḥīfat al-ayyām*** and containing biographical details on Shaykh Vefā: K 4037 (12 fols., *ca.* 1100H, lacks first page); DM 438 (6 fols., *ca.* 1100H). A commentary by Tāj al-Dīn, *muwaqqit* at the Prophet's Mosque in Medina (no. D239), entitled ***Ḥusn al-iqtifāᵓ li-ḥall Rūznāme Shaykh Vefā***: MM 77 (17 fols., *ca.* 1100H). An anonymous Turkish commentary in 5 *bābs*, prepared in 1017H for Meḥmet Pāshā, Governor of Egypt: MMT 1,1 (fols. 1r–3r, *ca.* 1100H); DM 177,4 (fols. 70v–79v, 1103H, with some tables).

Some copies of the ***Rūznāme*** have prayer-tables for Istanbul (latitude, 41;30°!) inside the "main" tables: KhMT 3 (8 fols., *ca.* 1100H); TJ 358,17 (fols. 237r–244v, *ca.* 1150H). See also no. D138.

See *SATMI*, II for a description of the prayer-tables.

H3 MŪSĀ GĀLĪNŪS

Krause, no. N22.

2.2.9 On his recension of the "perpetual almanac" of Zacuto for the longitude of Istanbul, see no. F31. Krause lists another work by Mūsā Gālīnūs on planetary models which deserves investigation.

H4 MAḤMŪD B. MUḤAMMAD B. QĀDĪ ZĀDE, MĪRAM CHELEBĪ

Suter, no. 457; *GAL*, II, p. 593 and SII, p. 665; *Storey*, no. 118; *Azzawi*, pp. 147–149.

1.2.17 A commentary on the *Risāla Fatḥīya* of ᶜAlī al-Qūshjī: see no. G51.

2.2.2 *Dastūr al-ᶜamal wa-taṣḥīḥ al-jadwal*, a commentary in Persian on the *Zīj* of Ulugh Beg (no. G49) compiled in 904H: TR 150 (218 fols., *ca.* 1100H, purports to be copied by the author); MMF 1 (168 fols., 1087H); S 4346 (154 fols., 1093H). For other copies, see *Storey*, pp. 70–71.

4.5.15 *Muhadhdhab dar ᶜamal-i-rubᶜ-i-mujayyab*, a Persian treatise on the use of the sine quadrant in 31 *bābs*: MJ 180,8 (fols. 64v–85r, *ca.* 1100H).

4.4.19 A Persian treatise on the use of the almucantar quadrant in 20 *bābs*: MJ 180,6 (fols. 53v–60v, *ca.* 1100H).

H5 YŪSUF B. KHIḌR BEG, SINĀN PĀSHĀ

Suter, no. 443.

1.2.7 A *ḥāshiya* on the commentary of Qāḍī Zāde on the *Mulakhkhaṣ* of al-Jaghmīnī: see no. G50.

H6 ḤAMZA B. ARSLĀN

Not listed in the modern sources.

6.4.10 *Miṣbāḥ-i-kunūz*, a treatise on arithmetic in Turkish, arranged in 3 *bābs* of 7 + 5 + 4 *faṣls*, dedicated to Maḥmūd, a son of Sultan Bāyazīd: ṬRT 10 (66 fols., 909H). No other copies known to me.

H7 MUḤYI L-DĪN MUḤAMMAD B. AL-ḤĀJJĪ ĀTEMJEH AL-KĀTIB

Not listed in the modern sources.

6.4.11 *K. Majmaᶜ al-qawāᶜid fī bayān muntakhab al-fawāʾid*, a Turkish treatise on arithmetic in 2 *shaqqs* of 16 *faṣls* each and a third *shaqq* of 40 *masʾalas*, compiled in 899H: ṬRT 7 (109 fols., *ca.* 950H). No other copies known to me.

H8 MUḤAMMAD B. KĀTIB SINĀN

Suter, no. 455.

4.4.18 *Hidyat al-mulūk*, a Turkish treatise on the construction and use of the almucantar quadrant in 20 *bābs* with tables, dedicated to Bāyazīd II: DMT 13,1 (fols. 1v–15v, 1183H); TR 109,2 (pp. 258–302, *ca.* 1200H); MR 40,9 (fols. 123v–130r, *ca.* 1250H); *etc.* (2 copies).

4.4.18 *Mūḍiḥ al-awqāt*, an Arabic treatise on the almucantar quadrant in 25 *bābs*, dedicated to Bayazīd II: DM 573,1 (fols. 2r–18r, *ca.* 1200H); K 3824,3 (fols. 12v–22r, *ca.* 1150H).

4.7.18+ A treatise in Arabic on horizontal sundials in 16 *bābs* with tables for latitude 41°: K 4059 (20 fols., 1059H).

H9 AḤMAD B. SULAYMĀN, KAMĀL PĀSHĀ ZĀDE

GAL, II, 597–602, and SII, pp. 668–673, and SIII, p. 1306. Brockelmann lists 179 titles by Kamāl Pāshā Zāde.

2.5.15 A short anonymous treatise explaining the method used by Kamāl Pāshā Zāde to express the date of some of his compilations using fractions: ṬJ 635,11 (fols. 57v–60r, 1170H).

H10 MUṢṬAFĀ B. ᶜALĪ

Azzawi, pp. 286–287.

2.5.14 A fragment from the Arabic version of his treatise on the seven climates: DM 8,3 (fols. 96r–96v, *ca.* 1200H). Complete copies are, *i.a.*, MSS Istanbul Mihrişah S. 304,3 (fols. 209–235, entitled *Iᶜlām al-ᶜibād fī aᶜlām al-bilād*), and Kiliç Ali Paşa 682,9 (fols. 53–71, 1202H).

2.5.14 A treatise in Turkish on the seven climates: QM 6,6 (fols. 46v–53r, *ca.* 1200H); MMT 7,5 (fols. 118v–124r + 125r–125v, 1002H, with some tables).

2.5.14 Some anonymous notes in Turkish on the direction of Damascus from Istanbul: TR 108,2 (pp. 11–21, *ca.* 1275H).

3.2.19 *Tashīl al-mīqāt*, a treatise in Turkish on timekeeping arranged in 25 *bābs*: ṬJ 366,2 (fols. 7v–17v, 1257H).

4.2.8 *R. fi l-ᶜAmal bi-dhāt al-kursī*, a treatise in Turkish on the armillary sphere in 30 *bābs*: KhM 10,11 (fols. 140v–156r, 1095H); MJ 180,20 (fols. 153v–175v, *ca.* 1100H).

4.3.33 *R. al-Asturlāb-i-Salīmī*, a Turkish treatise on the astrolabe in 36 *bābs* (the author promises 45 *bābs* in the introduction): ṬFT 32,1 (fols. 1v–31r, 1117H); *etc.* (2 fragments (Ch. 35 on checking the accuracy of the markings on an astrolabe)).

4.4.20 *Kifāyat al-waqt li-ma*ᶜ*rifat al-dā*ᵓ*ir wa-faḍlihi wa-l-samt*, a Turkish treatise on the use of the almucantar quadrant in two versions, one in 12 *bābs*: MMT 7,4 (fols. 105v–117v, 1002H); MJ 180,15 (fols. 135v–143r, *ca.* 1100H); *etc.* (2 copies); and the other in 24 *bābs*: KhM 10,12 (fols. 157v–171v, 1095H); ṬMT 5,2 (fols. 9r–13v, 1116H); *etc.* (1 fragment).

4.5.19 *R. Rub*ᶜ*-i-mujayyab*, a Turkish treatise on the use of the sine quadrant in 22 *bābs*: ṬMT 21,2 (fols. 10r–28r, *ca.* 1150H); TR 301,2 (pp. 28–82, *ca.* 1250H).

4.6.45 *R. ki ḥall Dā*ᵓ*irat-i-mu*ᶜ*addal*, a treatise in Turkish on the "equatorial (semi-)circle" arranged in 9 *bābs*, compiled in 938H and dedicated to Īyās Pāshā: KhM 10,9 (fols. 116v–133r, 1097H); ṬM 154,5 (fols. 29r–43v, 1151H).

H11 SAYYID ᶜALĪ B. ḤUSAYN

Suter, no. 463; *Adnan*, pp. 67ff.; *Azzawi*, pp. 254–257.

1.2.17 *Khulāṣat al-hay*ᵓ*a*, a Turkish translation of the ***Risāla Fatḥīya*** of ᶜAlī al-Qūshjī dedicated to Sulaymān I: see no. G51.

4.1.7 *Mirāt kā*ᵓ*ināt*, a treatise on instruments in Turkish in 5 *maqālas* dealing with the astrolabe, sine quadrant, almucantar quadrant, armillary sphere, and equatorial semicircle: MMT 7,1 (fols. 1r–75v, 1002H); DJF 9,10 (fols. 177v–205r, *ca.* 1250H); plus extracts as follows.
4th *maqāla* only: MMT 6,4 (fols. 39v–46v, 1139H); *etc.* (2 copies).
4th and 5th *maqālas*: K 3824,10 (fols. 67v–84v, *ca.* 1150H).
Suter lists two copies in Istanbul, MS Aya Sofia 2674 and 2675.

H12 TAQI L-DĪN MUḤAMMAD B. MAᶜRŪF

Suter, no. 471; *GAL*, SII, p. 484; *Azzawi*, pp. 315–317; S. Ünver, *Istanbul Rasathanesi*, (*Türk Tarih Kurumu Yayinlarindan*, VII, Seri—Sa. 54), (Ankara, 1969). See also no. C56.

2.2.8 ***Kharīdat al-durar wa-jarīdat al-fikar*** or ***Jarīdat al-durar wa-kharīdat al-fikar***, a small *zīj* for Istanbul completed 1893A (= 1581/82) for Saᶜd al-Dīn Efendī: ṬM 76 (51 fols., *ca.* 1150H); DM 900,2 (fols. 5v–78r, *ca.* 1250H). This *zīj* is not listed in *Kennedy*. See Pl. XXIVb.

2.5.15 A commentary on his own poem dealing with conversion of dates in different calendars: MJ 180,7 (fols. 61v–63v, *ca.* 1100H, attributed to al-Maᶜrūf al-Miṣrī al-Rāṣid!). This passage is taken from the end of the *zīj* ***Kharīdat al-durar***.

3.5.10 The first page only of a treatise on the effect of refraction at the horizon and the differences of opinion of the Cairo *muwaqqits* thereon: ṬM 211,1 (fol. 1v, *ca.* 1200H, anonymous). The complete treatise exists in MS Istanbul Kandilli 415, which is in Taqi l-Dīn's hand. See further *SATMI*, II.

4.3.32 *K. Tasṭīḥ al-ukar*, a treatise on stereographic projection, arranged in 2 *maqālas* of 3 + 6 *bābs*, and dedicated to Saᶜdī Efendī: ṬM 135 (11 fols., 1303H, lacks tables, anonymous). A complete copy of this work in the author's hand is MS Istanbul Kandilli 415,5.

4.5.22 A poem on the use of the sine quadrant, extant in a commentary by Muḥammad b. Muḥammad al-Ḥanafī (no. D135): DM 626 (10 fols., 1069H); MM 138 (9 fols., *ca.* 1100H). A copy of the poem is MS Istanbul Hüsnü Paşa 1135,2 (fols. 38–39, 1184H).

4.6.5 *al-Thimār al-yāni*ᶜ*a min quṭūf al-āla al-jāmi*ᶜ*a*, a commentary on ***al-Ashi***ᶜᶜ***a al-lāmi***ᶜ***a*** of Ibn al-Shāṭir (no. C30): DM 557,2 (fols. 35v–43v, 1058H, anonymous).

4.7.20 *Rayḥānat al-rūḥ fī rasm al-sā*ᶜ*āt* ᶜ*alā mustawi l-suṭūḥ*, a treatise on sundial theory in 3 *bābs* with tables and diagrams: DM 1140 (58 fols., *ca.* 1100H, purportedly 995H in author's hand—see Pl. CXa); *etc.* (3 poor copies); plus a commentary by al-Fāriskūrī (no. D2) entitled ***Nafḥ al-fuyūḥ*** . . . : MM 239 (148 fols., 1250H, contains a copy of a note about the commentary by Taqi l-Dīn); plus an anonymous Turkish translation: ṬFT 75 (66 fols. *ca.* 1200H, incomplete).

6.4.17 ***Bughyat al-ṭullāb fī ᶜilm al-ḥisāb***, a treatise on arithmetic and algebra in 3 *maqālas*: DR 1023 (15 fols., *ca.* 1300H, incomplete and defective, anonymous), unique? This title is attributed to Taqi l-Dīn by Ḥājjī Khalīfa.

6.4.17 ***al-Nisab al-mushākala (?) fi l-jabr wa-l-muqābala***, a treatise on algebra in 3 *bābs*: DM 557,3 (fols. 44v–48r, 1058H, no title, anonymous); TR 140,10 (pp. 52–61, *ca.* 1150H). Another copy is in Oxford I.881,3 (Suter).

On his commentary on the *Tajnīs* of al-Sajāwandī, see no. G11.

7.1.6 [***al-Binkāmāt al-dawrīya***], a treatise on mechanical devices in 2 *maqālas*: DM 557,1 (fols. 1v–35r, 1058H); DṢ 166,1 (fols. 1r–39v, *ca.* 1125H, fine *naskhī*, lacks end of 2nd *maqāla* and *khātima*). This treatise was published with translation in S. Tekeli, "The Clocks in the Ottoman Empire in the 16th Century" and Taqi al-Dīn's "The Brightest Stars . . . ," Ankara, 1966 (reviewed by O. Pedersen in *Centaurus*, 16 (1972), p. 328). See Pl. XCVa.

7.1.6 ***al-Ṭuruq al-sanīya fi l-ālāt al-ruḥānīya***, in 7 *bābs*: K 3845 (photos of the manuscript now in Dublin Chester Beatty 5232, 87 pls., 959H by author, lacks numerous diagrams); DM 557,4 (fols. 50r–91v, *ca.* 1000H, purports to be in author's hand, lacks first page and all diagrams). The Dublin manuscript is reproduced photo-mechanically with extensive commentary in A. Y. al-Hasan, *Taqi l-Din and Arab Mechanical Engineering* (in Arabic), Aleppo, 1976. Another copy of this treatise is in Kandilli Observatory near Istanbul.

7.2.7 ***Nūr ḥadaqat al-abṣār wa-nawr ḥadīqat al-anẓār***, an extensive treatise on optics in 3 *marṣads* subdivided in 6 + 6 + 5 main *faṣls*: DR 893 (52 folls., *ca.* 1000H, fine *naskhī*, carefully drawn diagrams). See Pl. XCVI. Other copies are MSS Oxford I.930 = Marsh 119 (fols. 3v–83r, 982H) (Suter) and Istanbul Laleli 2558. This work merits detailed investigation. The Oxford copy may be in the author's hand (see H. J. J. Winter in *Centaurus*, 5 (1956), p. 87).

In this treatise (fol. 52r) Taqi l-Dīn mentions another work of his on the visible and true horizons. Ünver (pp. 99 and 100) lists MSS Istanbul Kandilli Observatory 122 of a treatise on the *daqāʾiq al-ikhtilāf* (see no. C61 under 3.5.4) and 45 of a treatise on the reason for the delay of sunset due to refraction at the horizon.

See also the next two entries.

H13 NUQṬA (?) B. MAᶜRŪF, a brother of TAQI L-DĪN (no. H12)

3.4.7 Tables for computing crescent visibility, preceded by a short introduction: K 4038,2 (fols. 3r–6v, 13ca. 1050H, anonymous). In MS Istanbul Esat Ef. 2979 (fols. 273v–276r, 1067H) the same tables occur with a note by the copyist ᶜUmar b. Ismāᶜīl which translates: "I edited this table from the handwriting of . . . Taqi l-Dīn b. Maᶜrūf b. Muḥammad b. Aḥmad al-Ṣahyūnī (!) al-Dimashqī who had copied them from the handwriting of his brother who was their noble compiler, the scholar Nqṭh (?) b. Maᶜrūf."

H14 ANONYMOUS

4.1.6 *Ālāt-i-raṣadīya li-Zīj-i-Shāhinshāhīya*, a treatise in Turkish on the observational instruments used by Taqi l-Dīn (no. H12): MR 40,3 (fols. 65v–70r, anon., 1144H by Muṣṭafā Ṣidqī (no. D81)). This treatise has been published by S. Tekeli in *Araştirma*, 1 (1963). See Pl. LXIX.

H15 ḤUSAYN B. KHALĪL, QRLFJ (?) ZĀDE

4.5.20 ***R. fi l-Rubᶜ al-mujayyab***, a Turkish treatise on the use of the sine quadrant in 3 *bābs*, compiled in 974H: KhM 10,7 (fols. 108r–112r, 1097H).

H16 ᶜALĪ B. WALĪ B. ḤAMZA AL-JAZĀʾIRĪ AL-MAGHRIBĪ

Suter, no. N479a; GAL, SII, p. 536.

6.4.22 ***Tuḥfat al-aᶜdād li-dhawi l-rushd wa-sadād***, a Turkish treatise on arithmetic in 4 *aṣls* of 4 + 8 + 3 + 4 *bābs*: ṬRT 1 (271 fols., 999H). On this work, see Salih Zeki Ef. "Notation algébrique chez les Orientaux," *JA*, IX ser., vol. 11 (1898), pp. 35–52.

H17 AL-KNJDY (?), *muftī* in Amasia

3.5.11 A treatise on the Indian circle: TJ 391,4 (fols. 99–106, *ca.* 1150H); MJ 143,33 (fols. 255v–258r, 1010H, anon.). The author refers to the ***Sharḥ al-Wiqāya*** of Burhān al-Dīn in his introduction: see also al-Khalkhālī (no. G69 above).

H18 ṢALĀḤ AL-DĪN YĀZĪJĪ

5.3.41 ***Shamsīya***, a poem in Turkish on astrology: ṬFT 48 (193 fols., *ca.* 1000H); ṬFT 17 (134 fols., *ca.* 1000H). *Cf.* MS Istanbul Topkapi T 1623, *etc.* See also no. H19.

H19 IBRĀHĪM JŪRĪ (?)

5.3.41 ***Malḥama-ye-Jūrī***, a poem in Turkish on astrology, written in 1045H: ṬFT 37 (118 fols., 1047H); ṬFT 52 (115 fols., 1045H, lacks end). The author complains of the poor style of the poetry of the ***Shamsīya*** of Ṣalāḥ al-Dīn Yāzījī (no. H18).

H20 MUḤAMMAD B. ᶜALĪ, *raᵓīs al-munajjimīn* = MUḤAMMAD CHELEBĪ (d. 1040H)

5.3.40 ***Uṣūl al-iḥkām fī aḥkām al-nujūm***, a Turkish treatise on astrology commissioned by Ismāᶜīl b. Muḥammad al-Malaṭī: ṬFT 21 (63 fols., 1180H); *etc.* (2 copies).

5.3.40 ***Uṣūl-i-aḥkām-i-sāl***, on the same subject: ṬFT 22,4 (fols. 137v–178r, *ca.* 1300H).

5.3.40 A Turkish treatise on astrology: ṬFT 46 (60 fols., *ca.* 1200H).

H21 MUḤAMMAD EFENDĪ, a medical scholar of the time of Sultan Murād IV

5.3.4 A Turkish translation of ***K. al-Sirr al-maktūm*** by al-Rāzī: see no. G12.

H22 IBRĀHĪM AL-QARAMĀNĪ thumma AL-ĀMUDĪ

GAL, SII, pp. 185 and 939.

1.5.1 A commentary (?) with various titles[1] on ***al-Hīᵓa al-sanīya*** of al-Suyūṭī (see no. C103), compiled 1064H: ḤH 6 (31 fols., *ca.* 1100H); ḤH 5 (43 fols., 1104H); *etc.* (4 copies). This treatise is currently being investigated by my student Eliezer Bryski.

H23 ABŪ ᶜABD ALLĀH MUḤAMMAD B. AḤMAD B. MUḤAMMAD B. ḤASAN B. ḤUSAYN, MUNAJJIMAK = MUḤAMMAD FAZĀᵓĪ (?), RAᵓĪS AL-MUNAJJIMIN MUNAJJIMAK

Not mentioned in the modern sources.

4.1.9 Fragments of a treatise comprising at least 5 *maqāla*s as follows. The 3rd *maqāla* on sundials (*fī rasm al-ālāt al-ẓillīya*): DM 692 (7 fols., *ca.* 1200H, attributed to al-Munajjimak): DM 85 (10 fols., *ca.* 1200H, attributed to al-Munajjimak); DM 639,30 (fols. 254r–258r, *ca.* 1200H, anonymous). A note on the construction of sundials for latitude 41° taken from this *maqāla* and stated to be *min ᶜamal Munajjimak nuqila min khaṭṭihi*: TR 140,4 (pp. 32–33 + 43(?), *ca.* 1150H). *Cf.* MS Berlin Ahlwardt 5870 for an anonymous and incomplete copy of this *maqāla*. The fifth *maqāla* on instruments based on stereographic projections (*fī rasm al-ālāt al-ḥāditha ᶜalā tasṭīḥ al-kura*): DM 735 (6 fols., *ca.* 1200, anonymous), DM 70,1 (fols. 1v–3v, *ca.* 1200H, anonymous). All but TR 140,4 of the above Cairo manuscripts are copied in the same hand, and each of them is deficient in diagrams and incomplete.

The author's full name is given in a notice of possession on the title folio of DM 91 of the ***Zād al-musāfir*** of Ibn al-Majdī (see no. C62). In K 3988 of Taqi l-Dīn's ***Rayḥānat al-rūḥ*** (see no. H12), the copyist states that he made this copy from a manuscript in the hand of Muḥammad Fazāᵓī (?) *Raᵓīs al-munajjimīn* (= chief astronomer) Munajjimak (= little astronomer) dated 1082H in the *maḥkama* of Menla Akhrīn (?) in Constantinople.

---

1. ḤH 6: *K. al-Hayᵓa al-islāmīya ᶜalā madhhab ahl al-sunna wa-l-jamāᶜa*; ḤH 5: *Sharḥ mukhtaṣar al-Hayᵓa al-sanīya wa-yuqālu l-Hayᵓa al-islāmīya*; K 3808: *K. Hayᵓat al-islām wa-kalimat ahl al-īmān*; *etc.*

MMF 1 of Mīram Chelebī's commentary on the introduction of the ***Zīj*** of Ulugh Beg (see no. H4) was copied by Muḥammad Fazāʾī al-Kātib (= the scribe) in 1087H.

MS Istanbul Kandilli 63 (63 pp., 1306H) is a copy of an illustrated treatise on sundial theory entitled ***Tawḍīḥ al-mushkilāt fī rasm al-māʾilāt*** written by Abū ᶜAbd Allāh Muḥammad al-Bārī al-Tūnisī (no. F75), which is based on the treatise by Munajjimak, here refered to as the *muwaqqit* in the Mosque of Sultan al-Ghāzi Meḥmet Khān (in Istanbul).

H24 IBRĀHĪM AL-ḤALABĪ, a *muḥaddith* in the Aya Sofia Mosque in Istanbul

6.5.1 *Ḥawāshī* on Sibṭ al-Māridīnī's ***Raqāʾiq al-ḥaqāʾiq*** (see no. C97): DM 877 (22 fols., *ca.* 1100H by the author).

H25 ᶜABD AL-ḤALĪM SUWAYLAM ZĀDE AL-QAYṢARĪ

GAL, SII, pp. 487 and 1017.

4.3.39 ***Bahjat al-albāb fī ᶜilm al-asṭurlāb***, a treatise on the use of the astrolabe in 18 *bābs*: TR 106,9 (pp. 174–217, 1141H); ṬH 40,2 (fols. 29v–52v, 1143H); *etc.* (1 copy); plus a shorter version in 15 *bābs* by al-Qayṣarī: ṬM 154,2 (fols. 8r–15v, 1138H); and commentaries by al-Ḥamīdī (no. D57): ṬM 116 (70 fols., *ca.* 1250H); ṬM 202,1 (fols. 1r–65r, *ca.* 1150H); *etc.* (1 copy); and by al-Ādanī (no. G102): TJ 811,17 (fols. 157v–196r, 1135H). Brockelmann (*GAL*, SII, 1021) lists a work ***Bahjat al-albāb*** . . . in Rabat attributed to Ibrāhīm Efendī.

H26 QŌJA SĀᶜATJĪ, *muwaqqit* at the Salīmī Mosque in Edirne

4.5.31 ***R. Jayb āfāqī***, a treatise in Turkish on the use of the sine quadrant in 25 *bābs*: DM 573,2 (fols. 18v–43r, 1120H, anonymous); MJ 180,10 (fols. 89v–112v, *ca.* 1100H); *etc.* (1 copy).

H27 MUṢṬAFĀ ZAKĪ, *raʾīs al-munajjimīn*

5.3.40 An astrological treatise, with many horoscopes, and concluding with anecdotes about the author's life as court astrologer, dedicated to the Grand Vezir ᶜAlī Pāshā: ṬFT 25 (85 fols., 1134H); ṬFT 3 (50 fols., *ca.* 1200H).

H28 AḤMAD EFENDĪ, *khōja* of ISḤĀQ EFENDĪ BRŪSĪ

4.4.28 ***R. Rubᶜ-i-dāʾira***, a Turkish treatise on the use of the almucantar quadrant: MMT 5,1 (fols. 1v–49r, 1102H by author). See also no. H87.

H29 BADR AL-DĪN MUḤAMMAD B. ASᶜAD AL-ISLĀMBŪLĪ

Not mentioned in the modern sources. *Cf. GAL*, SII, p. 632.

6.4.39 ***K. Tathlīth al-zāwiya wa-tasbīᶜ al-dāʾira***, on the trisection of an angle and the construction of a regular heptagon: MR 41,7 (fols. 65v–66v, 1153H), unique.

6.4.39 ***K. ᶜAmal al-musabbaᶜ wa-ghayrihi min dhawāt al-aḍlāᶜ al-kathīra fi l-dāʾira***, on the construction of regular heptagons and other polygons in a circle: MR 41,8 (fols. 67v–68v, 1153H), unique.

Both of these treatises are summarized in *AHES* 30 (1984), pp. 280–281.

H30 ASᶜAD EFENDĪ B. ᶜALĪ B. ᶜUTHMĀN AL-YĀNIYAWĪ

GAL, SII, pp. 665–666. See also *Adnan*, pp. 126–127.

6.4.40 ***K. ᶜAmal al-murabbaᶜ al-musāwi li-l-dāʾira***: MR 41,22 (fols. 144v–147v, 1153H); DM 172,2 (fols. 10v–16r, *ca.* 1150H, anonymous, lacks diagrams and the last passage found in MR 41,22); no other known copies. This work is not listed by Brockelmann, and merits investigation. It begins: " . . . Now I shall begin with what was intended by saying that for every circle there is a square which equals it (in area), but when I(?) looked at some of the premises (*muqaddimāt*) mentioned in *K. al-Khuṭūṭ al-ḥawlānīya* and *K. Qaṭṭāᶜ al-kura wa-l-makhrūṭ* and ***K. al-Kura wa-l-usṭuwāna*** by the geometer Archimedes (no. A7) (I thought that) it was apropriate that we should state the premises mentioned in the first two books of Archimedes which are not translated into Arabic . . . " al-Yāniyawī then discusses ten such premises.

6.4.40 Fragments from "the book on philosophy which Asᶜad al-Yāniyawī translated from Latin to Arabic dealing with squaring the circle": TR 140,16 (p. 84, *ca.* 1150H). The book referred to is the ***Physics*** of Aristotle (no. A3) (see *GAL*, SII, pp. 665–666).

H31 ṢUBḤĪ ZĀDE ᶜABD AL-ᶜAZĪZ, *raᵓīs al-aṭibbāᵓ*

5.3.3 A Turkish translation of al-Bakrī's ***Burhān al-kifāya***, compiled in 1135H for the Sultan Muṣṭafā b. Aḥmad: see no. G8.

5.3.10 A Turkish translation of ***Ashjār wa-athmār*** by Alāᵓ al-Dīn al-Bukhārī, prepared for the Sultan Muṣṭafā b. Aḥmad: see no. G32.

H32 DARWĪSH ᶜABBĀS WASĪM

2.2.6 ***al-Nahj*** (*sic*, read *Nahj*?) ***al-bulūgh fī Zīj Ulugh*** (!), a Turkish commentary on the *Zīj* of Ulugh Beg (no. G49), without tables, compiled in 1158H for Rāghib Mehmed Pāshā, Grand Vezir of Muṣṭafā III: ṬFT 42 (275 fols., 1192H).

H33 AḤMAD, *muftī* of Filibe (= Plovdiv)

See al-Ḥamīdī (no. H34).

H34 MUḤAMMAD B. ᶜALĪ AL-ḤAMĪDĪ

6.4.42 ***Bawādir fawāᵓid al-wasāᵓil . . . fī nawādir farāᵓid al-masāᵓil***, a commentary on a treatise by Aḥmad, *muftī* of Filibe (= Plovdiv) (no. H33) on dates expressed in terms of a problem of algebra: ṬJ 635,12 (fols. 61r–64v, 1170H).

H35 ᶜALĪ AL-RĀBIṬ AL-MŪYTĀBĪ (?)

4.6.24 ***Takwīr al-amṣār fi l-awqāt wa-l-aᶜṣār***, a treatise in Turkish arranged in 40 *bāb*s on the "complete" quadrant (of Ibn al-Shāṭir (no. C30)), compiled in 1163H in Istanbul: DMT 25 (34 fols., 1163H by author).

H36 ṢALIḤ EFENDĪ MIᶜMĀRĪ

Not identified.

3.1.33 ***al-Jadwal al-kabīr***, extensive tables for timekeeping and prayer-tables for latitude 41° (Istanbul): K 18199 (136 fols., 1165H, by author? lacks prayer-tables); ṬM 151 (144 fols., 1164H, complete); ṬM 215 (144 fols., 1169H, complete); plus a shorter version prepared by Muḥammad Ṣādiq Jihāngīrī (no. H37): ṬM 120 (31 fols., *ca.* 1200H, anonymous). These tables are analyzed in *SATMI*.

H37 MUḤAMMAD ṢĀDIQ JIHĀNGĪRĪ

3.1.33 A shorter version of the tables of Ṣāliḥ Efendī (no. H36); ṬM 120 (31 fols., *ca.* 1200H, anon.). Another copy of the same tables attributed to Jihāngīrī is MS Istanbul Husrev Paşa 232.

H38 IBRĀHĪM ḤAQQĪ

1.6.8 ***K. Maᶜrifatnāme***, an encyclopaedia in Turkish arranged in 3 *fann*s compiled in 1170H: ṬOT 28 (373 pp., 1250H). Printed? See Pl. IX. On the prayer-tables in this work, see *SATMI*. See also Ismāᶜīl Fahīm (no. H39), the author's son.

H39 ISMĀᶜĪL FAHĪM B. IBRĀHĪM ḤAQQĪ

3.1.33 ***Miᶜyār al-awqāt***, prayer-tables for latitude 38;30° (locality unspecified) with an introduction in Turkish arranged in 15 *faṣl*s, compiled in 1193H and dedicated to the author's father Ibrāhīm Ḥaqqī (no. H39): DMT 2,1 (fols. 1r–26v, 1202H), no other known copies. The tables are analyzed in *SATMI*.

H40 ISMĀᶜĪL B. MUṢṬAFĀ KALAMBAWĪ

Azzawi, pp. 290–291.

4.5.27 (*R. fi l-Rubᶜ al-mujayyab*), a treatise on the use of the sine quadrant in 17 *marṣads*: MM 165,1 (fols. 1v–56r, *ca.* 1200H); ṬJ 410,6 (fols. 28v–49r, *ca.* 1200H); *etc.* (4 copies).

4.7.31 A treatise on a kind of sundial called *al-muthallath*, with diagrams and tables: ṬM 211,4 (fols. 9r–14v, *ca.* 1200H). Merits investigation.

4.7.31 A treatise on sundial construction with tables for latitude 41°, Istanbul: ṬM 211,3 (fols. 5r–8v, *ca.* 1200H).

6.4.45 *Sharḥ-i-lūghūrītmah*, a treatise in Turkish on logarithms: ṬRT 11 (97 fols., 1201H); ṬRT 5 (85 fols., *ca.* 1250H).

H41 ISMĀᶜĪL CHINĀRĪ, KHALĪFA ZĀDE

Azzawi, p. 290.

2.2.17 *Tuḥfat bahīj raṣīnī tarjamat Zīj-i-Qassīnī*, a translation of the *Zīj* of Cassini (no. J2): ṬFT 28 (159 fols., *ca.* 1150H); *etc.* (2 copies).

H42 SULAYMĀN MAQĀMĪ B. AḤMAD

Azzawi, p. 291.

2.4.7 Astrological tables: ṬFT 15 (18 fols., *ca.* 1250H).

5.5.1 *R. fī Ṭāliᶜ mawlūd al-Sulṭān Salīm al-thālith*, a treatise in Turkish on the horoscope of Sultan Salīm III: ṬFT 6,1 (fols. 1v–47v, *ca.* 1200H); *etc.* (1 copy).

5.5.1 A treatise in Turkish on the horoscopes of the Ottoman Sultans, perhaps also by Mūsā Zāde: ṬFT 6,2 (fols. 48v–59v, *ca.* 1200H, anonymous); *etc.* (1 copy).

H43 ᶜUTHMĀN B. ᶜABD AL-MANĀN, *turjumān* in Belgrade

GAL, SN II, p. 665.

6.4.48 *Hidyat al-muhtadī li-īqād al-sirāj al-munṭafī*, a treatise on mathematics and surveying translated from European languages, and arranged in a *muqaddima*, 2 *qisms*, and a *khātima*: DR 695 (127 fols., 1206H); ṬR 137 (209 fols., 1211H); plus an abridgement by ᶜAbd al-Fattāḥ al-Bannāʾ (no. D202): DR 628 (110 pp., 1311H by author).

H44 ᶜUTHMĀN B. MUḤAMMAD, *muwaqqit* at the Sultan Meḥmet Khān Qāsim Mosque (in Istanbul)

5.3.40 *Jāmiᶜ al-aḥkām fī masāʾil al-khāṣṣ wa-l-ᶜāmm*, a Turkish treatise on astrology in 2 *maqālas*: DHT 4 (151 pp., *ca.* 1250H).

H45 MŪSĀ ZĀDE MUḤAMMAD ᶜUBAYD ALLĀH

2.2.16 Planetary equation tables with entries to 4 and 5 sexagesimal digits: ṬM 134 (272 fols., *ca.* 1200H).

H46 IBRĀHĪM KHŌJA AL-MUHANDISĪ

Azzawi, pp. 292–293 (???)

4.7.2 See no. F5.

H47 ḤUSAYN ḤUSNĪ, *al-munajjim al-thānī*

Azzawi, pp. 284–285.

2.2.17 *Tahdhīb al-anām fī taᶜrīb Zīj Lalande nām*, an Arabic version of the *Zīj* of Lalande (no. J3), prepared in 1229H: TR 182 (148 pp., *ca.* 1300H). Also, a Turkish version of the *Zīj* of Lalande: ṬFT 1 (81 fols., *ca.* 1285H); *etc.* (2 copies).

3.1.33 Tables for timekeeping for latitude 21;45° (Mecca): K 4002 (40 fols., 1249H), unique?

H48 AḤMAD MUḤSIN, *munajjim thānī*

5.5.2 An astrological almanac for 1266H: ṬFT 45 (13 fols., *ca.* 1266).

H49 ḤUSAYN ZĀʾID

Azzawi, p. 329.

2.2.17 ***al-Maṭlaᶜ al-saᶜīd fī ḥisābāt al-kawākib ᶜala l-raṣad al-jadīd***, an introduction in 9 *bāb*s and tables based on European "*zīj*es": ṬM 193 (90 pp., *ca.* 1325H).

H50 MUḤAMMAD ᶜĀRIF, *munajjim bāshī*

5.3.40 A Turkish treatise on astrology: ṬFT 19 (32 fols., *ca.* 1300H).

5.3.40 A Turkish treatise on astrological predictions for the year 1324H: ṬFT 26 (57 pp., *ca.* 1325H).

H51 MUḤAMMAD RAḤMAT AL-MAKKĪ, "astronomer" in the province of Diyar Bakr

5.3.36 ***al-Nūr al-sāṭiᶜ fī aḥkām al-ṭāliᶜ***, a treatise on astrology with tables: DM 1203 (70 fols., 1318H by author).

H52 ḤUSAYN B. JAMĀL (?)

GAL, SII, p. 929 (where the name is Ḥusayn b. Kamāl al-Qādirī).

5.3.32 ***Durr al-munṣān fīmā yaḥduthu fī dawlat Āl ᶜUthmān***, an astrological treatise: ṬM 179 (104 fols., *ca.* 1300H). Brockelmann mentions the title but lists no manuscripts.

Note: The following authors have not been identified.

H53 MUʾMIN ZĀDE

6.4.47 ***Mirʾāt al-qulūb***, a treatise in Turkish on arithmetic arranged in 21 unnumbered *faṣl*s: TJ 358,10 (fols. 203v–212v, *ca.* 1150H).

H54 IBRĀHĪM QAṢṢĀB BĀSHĪ ZĀDE

1.4.19 ***R. fī Bayān manāzil al-qamar wa-makthihi fīhā***, on the lunar mansions: MJ 59,4 (fol. 5v, *ca.* 1100H).

H55 ᶜUMAR B. ᶜABD ALLĀH, NUZHAT

5.3.29 A Turkish translation of al-Junābadī's *Tanbīhāt al-munajjimīn*: see no. G67.

H56 ĀKOVALĪ ZĀDE KHĀTIM (?)

6.4.1 A commentary on *al-Lumaᶜ* by Ibn al-Hāʾim (no. C58): DR 1067 (29 fols., 993H in the hand of the author).

H57 AL-BURSAWĪ / AL-BRUSAWĪ

6.4.23 A commentary on the *Khulāṣat al-ḥīsab* of al-ᶜĀmilī (no. G68) compiled in 1021H: KA 23 (79 fols., 1289H); DR 1034 97 fols., 1021H, anon.).

H58 AḤMAD, *khōja* of ISḤĀQ EFENDĪ

3.5.11 ***R. maᶜmūla fī Bayān al-ẓill wa-taḥdīd al-jihāt wa-taᶜyīn al-qibla bi-l-dāʾira al-handasīya*** (!), in Turkish: MMT 5,2 (fols. 49r–63r, *ca.* 1200H). The construction was usually called *al-dāʾira al-hindīya* (see no. G69). The treatise contains material relating to the nature of light and shadow, and a computation of the qibla at Bursa.

H59 MUḤAMMAD B. IBRĀHĪM AL-ISTĀNBŪLĪ AL-ḤANAFĪ

5.3.40 ***Fatḥ al-saᶜādāt wa-l-mawāhib fī maʾrifat maṭāriḥ shuᶜāᶜāt al-kawākib***, a treatise in 3 *bāb*s on the astrological doctrine of the projection of the rays: K 3999 (5 fols., *ca.* 1150H).

H60 ᶜALĪ CHELEBĪ AL-RŪMĪ

5.3.38 ***Kashf al-asrār fī hatk al-astār***, a treatise on astrology and magic: ZK 653 (22 fols., *ca.* 1325H, lacks end).

H61 PĪR ḤASAN

5.3.22 A treatise on the astrological significance of conjunctions, mentioning Dorotheus (no. A15): K 4297,4 (fols. 48v–105r, *ca.* 1150H, purportedly 956H).

H62 MUṢṬAFĀ B. MUṢṬAFĀ ḤISĀN

5.3.17 A Turkish translation of al-Kāshifī's *Lubab al-ikhtiyārāt*: see no. G57.

H63 MAḤMŪD EFENDĪ QARSHAHRĪ

4.4.25 *al-R. al-Ṣāliḥīya fi l-ᶜamal bi-l-rubᶜ al-sharqī al-shamālī al-maqṭūᶜ al-muqanṭar*, a treatise on the use of the almucantar quadrant in 13 *bābs*: TJ 177,6 (pp. 99–109, *ca.* 1150H).

H64 NAJM AL-DĪN B. SAYYID MUḤAMMAD AL-MUNAJJIM

3.4.2 A note in Turkish on lunar crescent visibility: ṬMF 8,2 (fol. 192r, *ca.* 1100H).

H65 SULAYMĀN RIYĀḌĪ

5.1.3 A Turkish translation of the commentary of ᶜAlī b. Riḍwān (no. B85) on the *Tetrabiblos* of Ptolemy: ṬFT 34 (298 fols., 1195H); ṬFT 16 (175 fols., 1278H).

H66 AḤMAD AL-DĀᶜĪ

2.3.1 A Turkish translation of al-Ṭūsī's *Sī faṣl*: see no. G23.

H67 YŪNUS EFENDĪ

6.3.1 A Turkish translation of *Raqāᵓiq al-ḥaqāᵓiq* by Sibṭ al-Māridīnī: see no. C97.

H68 MUḤAMMAD AMĪN B. ᶜABD AL-RAḤĪM

4.5.31 *Jawhar al-ḥayāt (!) fī ḥall al-mushkilāt*, a Turkish treatise on the sine quadrant in 25 *maqṣads*: DJ 844,4 (fols. 124r–135v, 1176H).

H69 MAWLŪD B. AL-ṢIDDĪQ AL-ḤĀFIẒĪ AL-YAᶜLĀWĪ

4.7.32 *Dalīl al-awqāt fi l-basāᵓiṭ wa-l-munḥarifāt*, a treatise on sundial construction in 2 *bābs*: K3998 (27 fols., 1335H).

H70 MISCELLANEOUS ANONYMI (Turkish treatises on *ᶜilm al-hayᵓa*)

1.2.25 Fragment of an unidentified treatise mentioning various earlier scholars and arranged as a series of questions and answers on astronomical topics: ṬJ 410,1 (fols. 1r–11r, 1092H); DJ 103,1 (fols. 1v–6r, *ca.* 1150H, lacks introduction).

1.2.25 Fragment on theoretical astronomy: TR 106,5 (pp. 122–135, *ca.* 1150H).

H71 ANONYMOUS

1.3.5 An anonymous illustrated Turkish treatise on the constellation figures: ṬJ 811,11 (fols. 82v–132v, 1198H); DMT 1,4 (fols. 208v–295r, *ca.* 1100H). See Pl. X and also al-Ṣūfī (no. B52). A Turkish version of the *Ṣuwar al-kawākib* (55 pages) by Aḥmad Ḥamdī was printed in Istanbul (Tamir Press, 1284H). I have not investigated the relationship between the two Cairo manuscripts and this printed work.

H72 ANONYMOUS

1.4.21 A Turkish treatise on the lunar mansions in 7 *faṣls*: ṬFT 2 (63 fols., 913H, defective).

H73 MISCELLANEOUS ANONYMI

2.2.6 A Turkish commentary on the introduction to the *Zīj* of Ulugh Beg (no. G49): ṬMT 4 (239 fols., *ca.* 1250H).

2.2.6 Solar and lunar tables for Istanbul based on the parameters of Ulugh Beg: DM 92 (21 fols., *ca.* 900H) ≠ TR 109,1 (pp. 1–246, *ca.* 1050H) ≠ DM 847,1 (fols. 1r–173r, *ca.* 1200H).

2.2.6 A Turkish copy of the *Zīj* of Ulugh Beg containing tables for Istanbul: S6010,2 (fols. 11v–172v, *ca.* 1100H).

H74 ANONYMOUS

2.2.6 A copy of the *Zīj* of Ulugh Beg (no. G49) containing tables for Istanbul (latitude 41;15°) and Bosnia (43;48°): S 4350 (244 fols., *ca.* 1300H).

H75 ANONYMOUS

2.2.16 A table of positions of the fixed stars and their meridian altitudes for latitude 40°: MMT 6,2 (fol. 23v, *ca.* 1200H).

H76 ANONYMOUS

2.2.16 Fragments from a Turkish *zīj*: DM 847,2 (fols. 174r–183r, *ca.* 1200H).

H77 ANONYMOUS

2.2.17 An Arabic version of the *Zīj* of Cassini (no. J2) for the longitude of Istanbul: ṬM 253 (164 pp., *ca.* 1300H); ṬM 167 (55 fols., 1343H, with tables of the astrological houses for Aleppo). See also ṬM 211,2 (fols. 2v–4r, *ca.* 1200H).

H78 MISCELLANEOUS ANONYMI (ephemerides)

2.3.12 136 Turkish ephemerides for specific years after 1050H: DMT 22.

2.3.12 Turkish ephemerides for 1217H showing steps in calculation: DMT 23 (30 fols.).

2.3.12 Turkish ephemerides for each year from 1175H to 1195H: ṬFT 31 (283 fols., *ca.* 1175–1195H).

2.3.12 Turkish ephemerides for several years *ca.* 1250H computed for the longitude of Istanbul using the *Zīj* of Lalande (no. J3): ṬFT 44 (144 fols., *ca.* 1250H).

H79 MISCELLANEOUS ANONYMI (Turkish treatises on the compilation of ephemerides)

2.3.14 *R. fī Maᶜrifat istikhrāj al-taqwīm*, in 26 *bābs*: ṬFT 20 (100 fols., *ca.* 1100H, lacks beginning).

2.3.14 *Miftāḥ al-taqwīm*, in many unnumbered sections, dedicated to ᶜAbd al-Ḥamīd II: ṬMT 1,1 (fols. 1v–65v, *ca.* 1310H).

H80 ANONYMOUS

2.4.2 A star catalog for *ca.* 1200H: KhM 7,4 (fols. 29v–34v, *ca.* 1200H, with tables of ascensions for latitudes 40;20° and 41°).

H81 ANONYMOUS

2.4.5 Tables of the astrological houses for latitude 43;48°: ṬFT 49,2 (fols. 94v–115r, *ca.* 1150H), and for latitude 41°: TR 280 (366 pp., *ca.* 1200H).

H82 ANONYMOUS

2.5.9 Notes in Turkish on eclipses and on the light of the moon: ṬM 102,5A (fols. 64r–68r, *ca.* 1150H).

H83 MISCELLANEOUS (Turkish tables for timekeeping)

3.1.20 Prayer-tables for latitude 41;15° (Istanbul): DM 36 (4 fols., *ca.* 1200H).

Prayer-tables for Istanbul arranged according to the Syrian months: MM 100 (13 fols., *ca.* 1150H).

A table of the time from sunrise to the culmination of the star Kaff al-Khaḍīb, compiled in 950H: ZK 782,5 (fols. 25v–28r, 1115H).

A fragment of a *ṭaylasān* table for latitude 42°: KhMT 2,1 (fol. 1r, *ca.* 1150H).

Prayer-tables for latitude 42°: ṬRT 5,1 (fols. 1v–8v, *ca.* 1250H).

Prayer-tables for latitude 38°: KhMT 2,3 (fols. 55v–58r, *ca.* 1150H).

A table of the solar altitude as a function of the hour-angle for each of the zodiacal signs, for latitude 41° (Istanbul): ṬM 255,10 (fol. 91r, *ca.* 1100H).

Each of these tables is analyzed in *SATMI*.

H84 MISCELLANEOUS (late Turkish tables for timekeeping)

3.1.33 Table for finding the time when the star *al-Kaff al-khaḍīb* culminates: DM 1206,2 (fols. 54v–55r, *ca.* 1250H).

3.1.33 Prayer-tables for latitude 44°: ṬFT 49,2 (fols. 94v–115r, *ca.* 1150H).

3.1.33 A Turkish almanac containing prayer-tables for an unspecified latitude: ṬFT 11, (57 fols., *ca.* 1300H).

H85 ANONYMOUS (Turkish treatise on the astrolabe)

4.3.42 A short treatise in Turkish on the construction of the markings for the prayer-times, arranged in 5 *bāb*s: K 3824,15 (fols. 146r–147r, *ca.* 1100H).

H86 ANONYMOUS

4.4.12 Turkish translation of *Kifāyat al-qunūᶜ* by Sibṭ al-Māridīnī: see no. C98 (manuscript datable *ca.* 1100H).

H87 MISCELLANEOUS ANONYMI (Turkish treatises on the almucantar quadrant)

4.4.28 *Hidyat al-ikhwān*, in 30 *bāb*s: ṬJ 366,4 (fols. 26v–47r, 1267H). The beginning of this treatise is identical to that of the treatise of Aḥmad Efendī (no. H28).

4.4.28 Treatises without titles in 15 + 2 *bāb*s: MMT 6,7 (fols. 73v–89v, *ca.* 1150H); and in 14 *faṣl*s: KhM 10,8 (fols. 112v–116r, 1097H). Fragments in ṬJ 410,4 (fols. 17v–21r, *ca.* 1200H).

H88 ANONYMOUS

4.6.14 *R. fi l-ᶜAmal bi-l-rubᶜ al-jāmiᶜ*, a treatise on the use of the universal (*jāmiᶜ*) quadrant, arranged in 10 *bāb*s: KhM 10,10 (fols. 134v–139r, 1095H). The anonymous author mentions latitude 41° (Istanbul).

H89 ANONYMOUS (Turkish treatises on the sine quadrant)

4.5.31 Treatise in 18 unnumbered *bāb*s: MMT 6,5 (fols. 49r–69r, *ca.* 1150H).

4.5.31 *al-R. al-Turkīya fi l-aᶜmāl al-jaybīya*, in 24 unnumbered *bāb*s: DMT 19 (9 fols., 1042H).

4.5.31 A late treatise without divisions: TM 141,2 (fols. 36v–43v, *ca.* 1325H).

H90 ANONYMOUS (Turkish treatises on the instrument called *dāʾirat al-muᶜaddil*, the equatorial (semi-)circle)

4.6.46 Treatise in several sections: ṬJ 410,7 (fols. 49v–50r, *ca.* 1200H).

H91 ANONYMOUS (treatise on an instrument called *mīqāt*)

4.6.48 *Munyat al-muwaqqitīn wa-tuḥfat al-mutafakkirīn*, a Turkish treatise in 25 *bāb*s with an Arabic introduction: S 4368 (50 fols., 1227H), unique.

H92 ANONYMOUS

4.8.9 *R. (fi) Istiᶜmāl barkar al-nisba*, a Turkish treatise on the geometric compass (?): ṬR 141,1 (fols. 1v–35v, *ca.* 1325H).

H93 ANONYMOUS

5.1.1 Some Hermetic material in Turkish is in DMT 4,2 (fols. 78v–87r, *ca.* 1200H).

H94 MISCELLANEOUS ANONYMI (Turkish treatises on astrology)

5.3.40 A horoscope of the Sultan ᶜAbd al-Majīd: ṬFT 45 (13 fols., *ca.* 1275H).

5.3.40 *Muẓaffir al-mushkilāt*: ṬFT 22 (189 fols., *ca.* 1300H).

H95 MISCELLANEOUS ANONYMI

5.4.7 Miscellaneous Turkish treatises on astrology: see Section 5.4.7 of the Turkish catalogue.

H96 MISCELLANEOUS ANONYMI

5.5.2 Late Ottoman astrological almanacs: ṬFT 19 (33 fols., *ca.* 1323H); ṬFT 26 (57 pp., *ca.* 1325H).

H97 MISCELLANEOUS ANONYMI

5.5.2 Astrological almanacs (without ephemerides) in Turkish for specific solar years dedicated to the ruling Sultan: ṬFT 74 (12 fols., *ca.* 1200H); TR 306 (16 pp., *ca.* 1122H); TR 192 (16 pp., *ca.* 1165H); TR 307 (28 pp., *ca.* 1240H), *etc.*

H98 ANONYMOUS

6.4.26 A treatise on geometry in 4 *faṣl*s: TR 106,3 (pp. 109–112, *ca.* 1100H).

H99 ANONYMOUS (arithmetic)

6.4.38 ***R. fī wujūh al-aᶜdād al-mutanāsaba wa-wujūh taqsīm al-ghuramāʾ***, a Turkish treatise in 2 *qism*s of 8 + 1 *faṣl*s: ṬJ 635,2 (fols. 3r–10r, 1170H).

H100 ANONYMOUS

6.4.47 A treatise in Turkish on units in arithmetic: DJ 607,3 (fols. 28v–32v, *ca.* 1200H).

H101 ANONYMOUS

6.7.9 A fragment on surveying, in Turkish: ṬR 141,2 (fols. 34v–35v, *ca.* 1325H).

H102 ANONYMOUS (inheritance)

6.9.18 ***R. fi l-Tarikāt***, in Turkish: MR 15,2 (fols. 3v–8r, *ca.* 1200H).

(***R. fi l-Mawārīth***), in Turkish: DR 660,2 (fols. 4r–21v, *ca.* 1200H).

H103 SHĀH EFENDĪ AL-FANĀRĪ

See no. G69.

J. European Authors

J1 CLAVIUS
DSB, *s.v.*
See no. G95.

J2 CASSINI
DSB, *s.v.*
See nos. H41 and H77.

J3 LALANDE
DSB, *s.v.*
See nos. H47 and H78.

J4 EDWARD ABBOTT
Not identified.
See Pl. CVII.

J5 MISCELLANEOUS
See no. H49.

J6 RUDOLF WALP
See no. D117.

X. Miscellaneous Authors

Most of the following authors are not properly identified as far as their locations are concerned. They are arranged in the subject order of Part II of the Arabic catalog.

X1 ABŪ MUḤAMMAD ᶜUMAR B. HĀRŪN AL-SAQALĪ

Not mentioned in the modern sources.

1.2.14 *K. al-ᶜAlam al-munīr fi l-falak al-athīr*, a treatise on theoretical astronomy in 46 *bābs*: K 4046 (47 fols., lacks part of introduction and Chs. 38–46, *ca.* 1050H), unique.

X2 AḤMAD B. ᶜABD AL-GHANĪ AL-SARŪJĪ

GAL, SI, p. 646.

1.4.13 A fragment from his work ***al-Ghāya bi-sharḥ al-Hidāya*** dealing with the determination of the qibla by the sun, moon, and stars: MJ 183,5 (fols. 107r–114v, *ca.* 1150H, anonymous).

6.8.4 A chapter from ***al-Ghāya bi-sharḥ al-Hidāya*** dealing with dirhams: ṬR 144,2 (fols. 13r–17v, 1196H).

X3 ḤĀTIM

1.4.22 A *qaṣīda* on the four seasons entitled *Shamᶜat al-ẓulam fī ṣanᶜat al-fuṣūl wa-l-anwāᵓ*: DM 651,7 (fols. 64v–67r, *ca.* 900H). A poem on the calendar entitled ***Mawāqīt al-mawāqīt***, also attributed simply to Ḥātim, is MS Yale Nemoy 1473 (L-507). Brockelmann (*GAL*, SII, 565) mentions ***al-Qaṣīda al-Ḥātimīya*** by the Yemeni scholar Ḥatim b. Aḥmad al-Ahdal al-Ḥusaynī (d. 1013H).

X4 MUṢṬAFĀ B. NŪR AL-DĪN

1.4.22 *Manẓūma fī Manāzil al-qamar*, a poem on the lunar mansions: K 4006,2 (fols. 15r–15v, 1117H).

X5 ᶜABD AL-WAHHĀB

1.4.22 *Urjūza fī Manāzil al-qamar*, a poem on the lunar mansions: ṬJ 811,2 (fols. 5r–5v, 1198H).

X6 ᶜABD AL-ᶜAZĪZ AḤMAD

1.4.22 A poem on the lunar mansions: MM 149,2 (fol. 10r, *ca.* 1100H).

X7 SHAMS AL-DĪN MUḤAMMAD B. MUḤAMMAD AL-BAGHDĀDĪ

Suter, no. 517 and 517N; *GAL*, SII, p. 1023. Not properly identified.

2.4.1 Sine tables for each minute of arc to 4 digits are attributed to al-Baghdādī (full name) in DM 136,1 (fols. 1r–24r, 1147H). The same tables are contained in ṬM 235,2 (fols. 27v–42r, 866H).

2.5.4 A fragment from a treatise by Shams al-Dīn al-Baghdādī (the 3rd *faṣl* dealing with the Sine) is contained in K 4031,2 (fols. 5v–9v, *ca.* 1150H). Merits investigation. See also C139.

X8 FAṢĪḤ AL-DĪN MUḤAMMAD B. (?) ᶜABD AL-KARĪM AL-NIẒĀMĪ

See no. G50.

X9 AḤMAD B. MĀJID

GAL, SII, pp. 230–231; article "Ibn Mādjid" by S. Maqbul Ahmed in EI_2.

3.3.4 A poem on the qibla entitled *Tuḥfat al-quḍāt* written in 893H: DM 409,1 (fols. 1r–7v, *ca.* 1150H, anonymous). Another copy, which is attributed, is MS Paris B.N. ar. 2292 fols.

128r–137r, on which see G. Ferrand, *Instructions Nautiques et Routiers Arabes et Portugais des XVe et XVIe Siècles*, (Paris: 1928), III, p. 209–211.

See also nos. X27 and Z41. An anonymous medieval treatise on navigation entitled *Furjat al-humūm fi l-ᶜalāmāt wa-l-majārī wa-l-nujūm* was published by H. S. Shihab in Kuwait, 1984.

X10 MULLĀ ᶜALĪ EFENDĪ AL-DĀGHISTĀNĪ

4.3.34 *R. fi l-asṭurlāb*, an Arabic version in 60 *bābs* of *Tuḥfat-i-Ḥātimī* by al-ᶜĀmilī: see no. G68.

X11 IBRĀHĪM B. MUḤAMMAD B. SHUKR

4.6.17 ***R. laṭīfa ᶜala l-Āla al-maᶜrūfa bi-l-shāmila***, on a universal instrument in 7 *faṣls*: DM 879,3 (fols. 7v–9r, 1090H), unique. Merits study.

X12 IMĀM BADR AL-DĪN

5.5.3 Notes on astrology: ṬM 85,3 (fols. 179v–181v, *ca.* 1100H).

X13 ABŪ ISḤĀQ

6.1.3 ***Ilḥāq Abī Isḥāq ᶜalā quṣur al-biḍāᶜa wa-ᶜadam al-istiḥqāq***, a commentary on the first 4 *maqālas* of al-Ṭūsī's *taḥrīr* of Euclid's ***Elements*** (see no. A5): QR 114 (72 fols., 1182H).

X14 FAṢĪḤ AL-DĪN AL-HINDĪ (?)

6.1.4 A *ḥāshiya* on Qāḍi Zāde's commentary on al-Samarqandī's *Ashkāl al-taᵓsīs*: see no. G26.

X15 SHAMS AL-DĪN MUḤAMMAD

6.3.23 A poem on finger arithmetic: J 3964,7 (fols. 31v–32v, 862H), unique.

X16 AL-IMĀM AL-AMĪR AL-MĀLIKĪ

6.3.23 ***Shifāᵓ al-ghalīl bi-zawāl mā ashkala fī masᵓalat ḥawḍ al-salsabīl***, a short treatise on arithmetical problems relating to the filling of watertanks: K 4279 (4 fols., *ca.* 1250H). The treatise is attributed to al-Imām al-Amīr al-Mālikī with no further information on the author's name. The title is not listed by Brockelmann and I know of no other copies.

X17 ᶜABD ALLĀH B. ḤASAN AL-ᶜAFĪF AL-KĀZŪNĪ

6.4.1 ***Rafᶜ al-niqāb ᶜan Nuzhat al-ṭullāb***, a commentary on ***al-Nuzha*** by Ibn al-Hāᵓim: see no. C58 (manuscript dated *ca.* 1100H).

X18 MUḤAMMAD SHAYKH ZĀDE

6.4.23 ***Fatḥ al-lubāb fī sharḥ Khulāṣat al-ḥisāb***, a commentary on the treatise of al-ᶜAmilī: see no. C68 (manuscript dated *ca.* 1200H).

X19 ᶜALĪ B. ᶜABD ALLĀH AL-ᶜASHSHĀQĪ

6.4.25 ***Maṭālib al-adhkiyāᵓ wa-ūli l-albāb***, a treatise on arithmetic in 6 *maṭlabs*: ṬJ 825,5 (fols. 128v–137v, *ca.* 1100H); DJ 607,2 (fols. 22v–27r, 1202H). No other copies known to me.

X20 IBRĀHĪM B. ᶜABD AL-QĀDIR B. IBRĀHĪM AL-ᶜALĀᵓĪ

6.4.32 *R. fi l-ḥisāb*, a treatise on arithmetic in 8 *bābs*: B 21606,6 (fols. 16v–18v, *ca.* 1200H, *naskhī* script, provenance unknown). Another copy of this work is MS Princeton Mach 4838 = Yahuda 3225, fols. 139r–143v, *ca.* 1150H.

X21 AḤMAD AL-ḤUSAYNĪ

6.4.37 ***R. fī Bayān ḍābiṭat ᶜuqūd al-aᶜdād***, a treatise on units in arithmetic arranged in 5 *faṣls*: ṬJ 635,10 (fols. 54v–56v, 1170H). In MS Istanbul Esad Efendi 3673,5 (fols. 27–28) of this work the author is also called al-Yamanī: *cf. MAY*, p. 57.

X22 AḤMAD B. ḤASAN B. ᶜABD AL-KARĪM ḤAMZA[1]

6.4.42 *R. al-Ghurbāl*, a treatise on factors, with tables: TR 320 (65 pp., 1308H).

X23 MUḤAMMAD B. ARNŪṬ AL-QĀḌĪ

6.4.44 A poem on cubing numbers: DR 1103 (2 fols., *ca.* 1200H).

X24 ISḤĀQ AL-ZARKĀNĪ

4.3.41 Marginalia to an anonymous treatise on the astrolabe: see no. Z29.

X25 KAMĀL AL-DĪN MUḤAMMAD B. AL-NĀSIKH AL-MĀLIKĪ

6.9.4 *al-Jawāhir al-thamīnāt fī ᶜilm al-farāʾiḍ wa-qismat al-tarikāt*, on inheritance: B 20490,1 (fols. 1r–54r, 903H). This title is listed by Ḥājjī Khalīfa (I, p. 613). In the middle of the text there is a treatise on arithmetic entitled *Madkhal al-ṭullāb fī ᶜilm al-ḥisāb*. Merits investigation.

1. Not identified, unless he be identical with the Egyptian *muftī* al-Jawharī (d. 1182/1768) listed in *GAL*, II, p. 435 and SII, pp. 459–460.

X26 SHARAF AL-DĪN ABU L-JŪD DAʾŪD B. SULAYMĀN

GAL, SN II, p. 201 (?).

6.9.5 An abridged version of a work of his on inheritance: B 20490,2 (fols. 54v–56r, 903H).

X27 SULAYMĀN B. AḤMAD AL-MAHRĪ[2]

GAL, SII, p. 231, article "al-Mahrī" by G. Ferrand in EI_1; see also I. Khoury's publications mentioned below.

8.8 His two works on navigation *K. al-ᶜUmda al-Mahrīya fī ḍabṭ al-ᶜulūm al-baḥrīya* and *K. al-Minhāj al-fākhir fī ᶜilm al-baḥr al-zākhir* are in TR 309 (88 pp., *ca.* 1325H from a copy dated 1069H), *etc.* (1 copy thereof), and TR 308 (23 pp., 1325H from a copy dated 1069H). Both of these, along with the other major Arabic works on navigation, have been published in I. Khoury, *al-ᶜUlūm al-baḥrīya ᶜinda l-ᶜArab*, in several volumes, Damascus: Arabic Language Academy, 1970 to present.

See also nos. X9 and Z41.

2. Both TR 308 and 309 have Sulaymān b. Muḥammad al-Mahrī.

Z. Anonymous Authors

Z1 ANONYMOUS (treatise on planetary distances)

1.2.16 ***R. fī Ma*ᶜ*rifat al-ab*ᶜ*ād wa-l-ajrām***, a treatise on planetary sizes and distances which the anonymous author says he translated from Persian: DM 573,4 (fols. 57v–59r, 1120H); and a fragment in ZK 782,1 (fol. 3v, *ca.* 1100H).

Z2 MISCELLANEOUS (treatises on theoretical astronomy)

1.2.16 A treatise on theoretical astronomy in 4 *bābs* of 4 *faṣls* each: K 9740 (13 plates from a manuscript in the private collection of Rashīd al-Ṣaffār, 966H).

1.2.16 A treatise on theoretical astronomy in 7 *bābs*: J 3463,5 (fols. 28v–31r, 1000H).

1.2.16 A fragment from a treatise on theoretical astronomy, including a section entitled *al-naẓar al-thālith fī falak* ᶜ*Uṭārid*: DM 681,5 (fols. 14r–16v, *ca.* 800H).

1.2.16 A fragment from a treatise on theoretical astronomy, containing *mabḥaths* 3–5, of which the first is entitled *fī jihāt ḥarakāt al-aflāk*: DM 620,2 (fols. 15r–17v, *ca.* 900H).

Z3 MISCELLANEOUS (treatises on theoretical astronomy)

1.2.24 ***R. fī*** ᶜ***Ilm hay***ᵓ***at al-kawākib wa-maqādīr ab***ᶜ***ādihā***, on theoretical astronomy in 3 *bābs*: DH 45 (11 fols., *ca.* 1250).

1.2.24 Treatise in 60 *bābs*: ṬM 219 (36 fols., *ca.* 1300H).

1.2.24 Treatise in 3 *faṣls*: ṬM 154,4 (fols. 20v–24v, 1138H); ZK 782,11 (fols. 60v–68v, 1121H).

1.2.24 ***Tibyān wa-hidāyāt fī*** ᶜ***ilm al-hay***ᵓ***a wa-ṣṭilāḥ ahlihā***, on theoretical astronomy, arranged in *tibyāns* and *hidāyas*: TJ 246,6 (pp. 44–49, 1146H).

An anonymous treatise in 2 *bābs* is in MS Princeton Mach 4900 = Yehuda 669 (fols. 51v–61v, *ca.* 1300H).

Z4 MISCELLANEOUS (treatises on folk astronomy)

1.4.15 ***Bāb ma*ᶜ*rifat ḥulūl al-shams fi l-manāzil***, on the sun's entry in the lunar mansions: DM 948,3 (fols. 91r–98v, 1320). ᶜAlī b. Ibrāhīm is mentioned on fol. 91v and Abū ᶜAlī on fol. 97r.

1.4.15 ***R. fī Manāzil al-qamar***, on the lunar mansions: DJ 705,5 (fols. 40v–41r, *ca.* 1310H).

1.4.15 ***Qiṭ*ᶜ*a fī Ma*ᶜ*rifat ẓill al-zawāl bi-l-aqdām fi l-ashhur al-Rūmīya***, on shadow lengths: DM 948,4 (fols. 99r–101r, 1320H).

1.4.15 A fragment from a commentary on a poem on the midday shadow lengths of the days of the Syrian year: DM 899,2 (fols. 8r–8v, *ca.* 1100H).

Z5 MISCELLANEOUS (treatises on folk astronomy)

1.4.20 ***R. fī Ma*ᶜ*rifat manāzil al-qamar wa-l-istidlāl bihā fī ma*ᶜ*rifat sā*ᶜ*āt al-layl***, on timekeeping by night using the lunar mansions: K 3824,12 (fols. 96v–106v, 1150H).

1.4.20 A treatise on the lunar mansions: MM 168,3 (fols. 6v–7r, *ca.* 850H).

1.4.20 ***Fā*ᵓ*ida fī ma*ᶜ*rifat al-qamar fī ayy al-manāzil***: DM 534,2 (fol. 19r, *ca.* 900H).

1.4.20 ***Faṣl fī*** ᶜ***ilm bayān qismat al-manāzil*** ᶜ***ala l-fuṣūl***: MM 149,1 (fols. 1r–7v, *ca.* 1100H).

1.4.20 ***Faṣl fī ma*ᶜ*rifat manāzil al-qamar***: MM 142,5 (fols. 10v–12r, *ca.* 1000H).

1.4.20 A note on the lunar mansions: MM 2,2 (fols. 55v–56v, 1019H).

1.4.20 A fragment of a treatise on folk astronomy, from the 9th *bāb* to the 16th: DM 41,2 (fols. 2r–7v, *ca.* 1100H).

1.4.20 Treatises on the lunar mansions, with a table of their positions for 1090H "according to the observations of al-Mnwby (?)": DM 1046,3 (fols. 73v–76r, *ca.* 1200H, Maghribi script). The Egyptian astronomer Shams al-Dīn al-Minūfī (no. C120) may be intended.

Z6 MISCELLANEOUS (poems on folk astronomy)

1.4.23 A poem on finding the lunar mansion rising at daybreak: DR 1066,4 (fol. 20v, *ca.* 1100H), *etc.* (1 copy).

1.4.23 A poem on the lunar mansions: MM 179,2 (fols. 20v–22r, *ca.* 1100H), plus an anonymous commentary thereon: DJ 254,1 (fols. 1r–22v, *ca.* 1000H); MM 179,1 (fols. 1r–20v, *ca.* 1100H), *etc.* (1 copy).

1.4.23 A poem on the risings of the lunar mansions: DM 738,3 (fols. 8r–8v, 982H).

1.4.23 A poem on the names of the zodiacal signs: DM 453,1 (fol. 1r, *ca.* 1100H). I do not understand what the poet intended.

1.4.23 ***al-Mukhtār min ᶜilm al-falak***, a poem on folk astronomy: TR 55,2 (pp. 40–57, 1293H).

1.4.23 A poem on the lunar mansions: DM 499,2 (fols. 8r–13v, *ca.* 1150H).

Z7 ANONYMOUS (treatise on *al-hayᵓa al-sunnīya*)

1.5.2 ***Tafsīr al-āyyatayn al-karīmatayn***, a treatise on Qurᵓānic verses dealing with the celestial spheres: ṬJ 873,1 (fols. 1v–34v, *ca.* 1050H, beautiful *naskhī*, lacks end).

Z8 ANONYMOUS

1.5.3 A treatise written from a religious viewpoint on the place where the sun rises and sets and on eclipses and on the rising of the crescent: MM 103 (8 fols., 1194H).

Z9 MISCELLANEOUS (treatises on ephemerides)

2.3.15 Two short treatises on the symbols used in ephemerides: ḤM 16,3 (fols. 39r–40v, *ca.* 1100H) and MM 190,3 (fols. 9v–10r, *ca.* 1200H).

2.3.15 ***Faṣl fī Maᶜrifat ḥall al-taqwīm ᶜalā sabīl al-ijmāl***: DM 602 (15 fols., *ca.* 1100H, lacks end). The anonymous author mentions Ibn al-Shāṭir (no. C30) and Ibn al-Majdī (no. C62) (fols. 1v and 13v).

Z10 MISCELLANEOUS (tables)

2.4.3 Miscellaneous anonymous planetary tables: MM 76 (18 fols., *ca.* 1100H); DM 725 (7 fols., *ca.* 1050H); TR 319,2 (pp. 17–40, *ca.* 1250H); DM 909,6 (photos. of MS Paris B.N. ar. 2496, fols. 118r–123v, *ca.* 1000H).

Z11 MISCELLANEOUS (tables)

2.4.4 Miscellaneous anonymous tables relating to the lunar mansions: ṬM 83,5 (fols. 43r–44v, *ca.* 900H); MMF 7,9 (fol. 64v, 1060H); DM 681,7 (fols. 28v–29r, *ca.* 900H); DM 711,2 (fols. 14v–15r, *ca.* 1100H); MM 167,12 (fols. 204v–205r, 989H).

Z12 MISCELLANEOUS (tables)

2.4.5 Tables of the ascensions of the zodiacal signs for each degree of latitude from 1° to 60°: DM 637,2 (fol. 45v, *ca.* 850H). Other anonymous ascension tables: MM 167,10 (fols. 201v–202v, 989H); MM 42 (5 fols., *ca.* 1100H); K 3860 (22 fols., *ca.* 1325H); K3864 (112 pp., *ca.* 1325H); DM 57,2 (fols. 5r–6v, *ca.* 1000H); ṬM 224 (681fols., *ca.* 1300H); ṬMF 1,1 (fol. 1v, *ca.* 1100H).

Z13 MISCELLANEOUS (tables)

2.4.5 Tables of the astrological houses for unspecified latitudes: DM 896,2 (photos of MS Paris B.N. ar. 2487, fols. 36v–38r, *ca.* 1000H); ṬM 237 (46 fols., *ca.* 1200H).

Z14 MISCELLANEOUS (tables)

2.4.6 Miscellaneous anonymous tables for the planetary sectors: DM 644,4 (fols. 21r + 26v, *ca.* 1100H); DM 620,1A (fol. 13v, *ca.* 1000H).

Miscellaneous tables for computing planetary conjunctions: DM 554, (7 fols., *ca.* 1100H) = DM 711,1 (fols. 1r–14r, *ca.* 1100H); DM 1108,6 (fols. 30r–32v, 1053H).

Tables of the sines of the latitudes of the planets (??): ṬFT 7 (94 fols., *ca.* 1300H).

Miscellaneous tables: ṬFT 5 (158 fols., *ca.* 1300H).

Z15 MISCELLANEOUS (tables)

2.4.7 Miscellaneous anonymous astrological tables: ṬFF 3,1 (fols. 1v–2r, 909H); K8535,2 (fol. 94r, *ca.* 1000H); MM 54,2 (fols. 9v–10r, *ca.* 1200H); ṬM 248,2 (fols. 21v–28v, *ca.* 1250H); MMF 7,10 (fols. 66r–66v, 1060H); ṬM 148,2 (fols. 138v–140v, *ca.* 1200H); ṬM 113,2 (fols. 129v–132r, 1053H); ṬM 121,2 (fols. 47v–48r, *ca.* 1325H); ṬM 133,1 (pp. 1–132, *ca.* 1300H); DM 1046,2 (fols. 66r–70v, *ca.* 1150H); ṬM 75,2 (fols. 59r–61r, *ca.* 1300H); DM 38 (24 fols., *ca.* 1100H); DM 290 (8 fols., *ca.* 1250H). Tables of particular interest are in MM 99,2 (fols. 5r–43v, *ca.* 1175H); DM 562 (6 fols., *ca.* 1250H).

Z16 MISCELLANEOUS (geographical tables)

2.4.8 Miscellaneous anonymous geographical tables: DM 84,2 (fols. 11r–11v, *ca.* 1150H); DM 258 (8 fols., *ca.* 1150H); DM 702,1B (fol. 5r, 1160H); DM 637,5 (fol. 47v, *ca.* 1000H); MM 101,2 (fol. 12v, 1159H); MMT 6,6 (fol. 71v, *ca.* 1150H); ṬM 158,4 (fols. 26v–28v, 1155H). For information on the stations on the pilgrim routes see DM 651,4 (fol. 40r, *ca.* 1100H); J3487,6 (fols. 66r–68r, *ca.* 1150H).

Z17 MISCELLANEOUS (calendrical tables)

2.4.9 Miscellaneous calendrical tables: ṬJ 339,39 (fols. 163v–164r, *ca.* 1085H); ṬM 228,4 (fols. 11r–15r, *ca.* 900H); ṬM 255,9 (fols. 90r–96r, *ca.* 1060H); ṬM 128,3 (fols. 70r–71r, 1275H); MM 113 (6 fols., *ca.* 1100H); DM 109,2 (fols. 4r–32v, *ca.* 1200H).

Tables relating to the Armenian calendar: TR 296,3 (pp. 164–166, 1123H).

Tables relating to the Jewish calendar: K4011,3 (fols. 10v–19r, 1287H).

Z18 MISCELLANEOUS (almanacs)

2.4.10 Almanacs (*natīja*) for 1083H DM 141,2 (fols. 10v–25v, *ca.* 1085H, defective); and for 936 Jalali (= 1023–24H): DM 141,3 (fols. 26r–38v, *ca.* 1025H).

Z19 MISCELLANEOUS (calendars)

2.5.15 Treatise on calendars, beginning with a list of parameters for the "mean lunar crescent month" (*al-shahr al-hilālī al-wasaṭ*) taken from various earlier sources: MM 151,1 (fols. 1v–7v, *ca.* 1200H).

2.5.15 A treatise on the determination of the *notae* of the months: DM 604 (8 fols. 1187H).

2.5.15 A circle with moveable parts for comparing dates: Sh 76,2 (fol. 6v, *ca.* 1300H).

Z20 ANONYMOUS

2.5.15 A short treatise on the conversion of dates and the determination of solar longitude, with tables: DM 909,3 (photos of MS Paris B.N. ar. 2496, fols. 9r–9v, 973H, lacks some tables).

Z21 ANONYMOUS

2.5.15 A treatise on the date of the Hijra, compiled in 1283H: ZK 145 (11 fols., 1283H by the author).

Z22 MISCELLANEOUS

2.5.16 Notes on trepidation: DM 486,1 (fol. 1r, *ca.* 1140H).

Notes on comets: MM 164,1 (fol. 1r, 1185H).

Notes on the computation of solar longitude: Sh76,6 (fol. 10v, *ca.* 1300H); DM 781,5 (fols. 10v–11r, *ca.* 900H).

2.5.16 ***Fāʾida fī Maᶜrifat ṭulūᶜ al-Shiᶜra l-yamanīya***, on the heliacal rising of Sirius: DM 1108,2 (fols. 4v–5r, 1053H) = MM 244,3 (fols. 9r–9v, *ca.* 1200H) ≠ DM 790,2 (fol. 19v, *ca.* 1150H).

2.5.16 ***Laṭāʾif al-ishāra fī taqwīm al-sayyāra***, a treatise in 3 *faṣl*s dealing with the determination of the longitudes of the sun and moon and on interpolation: ZK 441 (7 fols., *ca.* 1300H). *Cf.* no. D151 for a work with a similar title, and also *GAL*, SII, p. 1018, no. 14.

2.5.16 A short treatise on determining solar longitude: DJ 286,1 (fols. 1v–2v, *ca.* 1300H).

Z23 MISCELLANEOUS (timekeeping)

3.1.20 Prayer-tables for an unspecified latitude, compiled in 978H?: MJ 180,13 (fols. 127v–130r, *ca.* 1100H).

A table for converting time in hours and minutes to degrees: ṬFF 13,6 (fols. 129v–130r, 845H).

3.1.36 Prayer-tables for each day of 131H for an unspecified latitude: ṬMT 1,2 (fols. 66r–71v, *ca.* 1310H).

3.1.36 An *imsākīya* for Ramadan, 1245H for an unspecified latitude: ṬMT 3 (2 fols., 1245H).

3.1.36 A fragment of prayer-tables for an unspecified latitude with values for each 10 days of the Syrian year: DM 177,1B (fol. 3v, *ca.* 1100H).

3.2 A poem on timekeeping without subdivisions: TR 53,2 (pp. 9–10, 1273H).

3.5.15 Notes on finding the time of the times of the prayers of the *ᶜīd* and *al-ḍaḥwa al-kubrā* and *al-imsāk* using an almucantar quadrant, and the method of using such a quadrant for a latitude other than the one for which it was made: TR 62,3 (p. 37, 1228H).

Z24 MISCELLANEOUS (treatises on the qibla)

The first two sources are analyzed in D. A. King, "The Sacred Geography of Islam," *Islamic Art*, 2 (1983).

3.3.4 Treatise on the Kaᶜba and qibla, with illustrations: ṬJ 811,7 (fols. 59r–61r, *ca.* 1150H). Merits detailed study. See Pls. XLVII and Ca.

3.3.4 Qibla circle displaying directions of Mecca for various localities: DM 637,4 (fol. 44v, *ca.* 850H). See Pl. XLIX and also ṬJ 811,7 mentioned above.

3.3.4 Computations of the qibla for various localities in Persia: TR 278,2 (p. 33, *ca.* 1100H).

3.3.4 Fragment of a treatise on the religious significance of facing the qibla: DM 65 (38 fols., *ca.* 800H, beautiful *thulth*, lacks beginning and end).

Z25 MISCELLANEOUS (qibla tables)

3.3.6 A table entitled ***al-jadwal al-ᶜishrīnī li-maᶜrifat inḥirāf al-qibla***, "a table with 20 × 20 entries for finding the qibla" is in MM 167,13 (fols. 205v–206r, 989H, anonymous); DM 1030,2 (fols. 3r–4r, *ca.* 1200H, attributed to Yūsuf al-Damīrī (no. C166)); MM 204,9 (fols. 83v–84r, 1052H, anonymous); TR 103,2 (pp. 100–104, *ca.* 1300H, attributed to Abū Jaᶜfar Muḥammad b. Mūsā al-Khwārizmī (no. B14)). Another table of this kind containing 40 × 20 entries is in DM 707 (3 fols., *ca.* 1100H, anonymous). See no. C54 and Pl. LI, and also *JNES*, 34 (1975), p. 122.

Z26 ANONYMOUS

3.5.11 A commentary on the ***Sharḥ al-Wiqāya*** dealing with the Indian circle: MJ 143,32 (fols. 249v–258r, 1010H). See also no. G69.

Z27 ANONYMOUS

4.2.5 ***al-Maṭlab al-sirrī fī ᶜamal al-asṭurlāb al-kurī***, a treatise on the spherical astrolabe in 6 *bāb*s: ṬM 155,2 (fols. 15v–19r, *ca.* 1050H), unique. Merits study.

Z28 MISCELLANEOUS (treatises on the celestial sphere)

4.2.7 ***Risāla sharīfa fi l-ᶜAmal bi-l-kura dhāt al-kursī***, in 26 *bāb*s: ZK 782,8 (fols. 40v–48r,

1116H); and in 25 *bābs*: ṬM 154,3 (fols. 15v–20r, 1138H).

4.2.7 ***R. fi l-Kura al-musammā bi-dhāt al-kursī***, in 33 *bābs*: TR 106,11 (pp. 232–252, *ca.* 1150H); K 3844,8 (fols. 59v–69v, 1144H); MM 101,1 (fols. 1v–12r, 1159H); *etc.* (3 copies). Other copies are MSS Berlin Ahlwardt 5869, also anonymous, and Istanbul Nurosmaniye 2918,5, fols. 61r–74v, attributed to Ptolemy (no. A5).

4.2.7 ***R. laṭīfa fi l-ᶜAmal bi-l-kura***, in 13 *bābs*: DM 173,4 (fols. 88r–90r, *ca.* 1150H); TR 305 (7 pp., 1106H); *etc.* (2 copies).

4.2.7 ***Ghāyat al-idrāk fi l-ᶜamal bi-kurat al-aflāk***, in 22 *bābs*: MM 198,1 (fols. 1r–18v, 1167H), unique.

4.2.7 ***R. fī Waḍᶜ al-kawākib fi l-kura wa-l-asṭurlāb***, using tables of al-Ṣūfī (no. B52) and Ulugh Beg (no. G49), arranged in 13 *faṣls*: TR 325,1 (pp. 1–10, 1108H), no other copies.

Z29 MISCELLANEOUS ANONYMI (treatises on the astrolabe)

4.3.41 ***(Risāla fi asmāᵓ rusūm al-asṭurlāb wa-baᶜḍ aᶜmālihā)***, in 15 *faṣls*: ḤM 19,1 (fols. 1v–28r, 1144H); DM 573,3 (fols. 44r–57r, 1120H); ṬM 154,1 (fols. 1v–7v, 1138H); MJ 180,3 (fols. 31v–38v, *ca.* 1050H); ṬM 255,2 (fols. 3v–13v, *ca.* 1150H); *etc.* (13 copies, including 3 with marginal commentary by Isḥāq al-Zarkānī (no. X28)). Another anonymous copy is MS Berlin Ahlwardt 5810.

4.3.41 ***(R. fi l-ᶜAmal bi-l-asṭurlāb)***, in 15 *bābs*: DM 716 (20 fols., 1157H); DM 1082,10 (fols. 134v–141v, 1154H); *etc.* (2 copies).

Z30 ANONYMOUS

5.2.32 A handbook of astrological predictions (*qurᶜa*) arranged according to the lunar mansions: TGh 137 (80 pp., *ca.* 1150H). Merits investigation.

Z31 ANONYMOUS

5.2.33 A treatise without title on geometrical concepts, the problem of the duplication of the cube, and the incidence of the plague, arranged in 3 *bābs*, the first of which is subdivided into 9 *maṭlabs*: ṬR 151 (11 fols., *ca.* 1050H); ṬJ 635,9 (fols. 47r–53r, 1170H by Muṣṭafā Ṣidqī (no. D81). The copyist Muṣṭafā Ṣidqī implied that this was an absurd treatise written by an idiot. Merits study, especially in the light of the recent publications of S. Tekeli on the problem of the duplication of the cube, on which see *Araştirma*, 6 (1968), pp. 11–17. See also MS Princeton Yahuda 1809 (Mach 4996), fols. 14v–18r, 1032H) of a treatise on the Delian problem by al-Shubrāmallisī (no. D7).

Z32 MISCELLANEOUS (astrology)

5.4.1 A treatise on astrology, in 8 *maqālas* subdivided in *bābs*: ṬM 238 (121 fols., 1314H). Numerous early Muslim astrologers are mentioned.

5.4.1 A treatise in 12 *bābs* with tables: DM 180,6 (fols. 89v–102v, 1238H); *etc.* (1 copy). Others in 10 *bābs* ṬM 161,3 (fols. 32v–37r, 1231H); and in 3 *bābs*: DM 180,8 (fols. 123r–125r, 1238H).

Z33 MISCELLANEOUS (astrology)

5.4.2 Various treatises on astrology, with titles, as follows.

Hidāyat al-sallāk fī maᶜrifat ḥarakāt al-aflāk, without subdivisions and with some non-numerical tables: ṬM 107 (81 fols., *ca.* 1300H).

al-Mabāhij al-zahīya wa-l-manāhij al-murḍīya, in 7 *bābs*: DM 1148 (54 fols., *ca.* 1100H).

al-Risāla al-Jafrīya fī qirānāt al-milla al-Ḥanafīya wa-l-dawla al-tawḥīdīya, without subdivisions: TR 141,5 (pp. 141–144, 1152H, Maghribi script).

Bughyat al-insān fī maᶜrifat al-kawākib wa-l-azmān, in 10 *bābs*: DM 16 (42 fols., 1263H).

al-Nubdha al-munāra (?) fī aḥkām al-sabᶜa al-sayyāra, without subdivisions: K 4010 (232 fols., *ca.* 1300H).

Z34 MISCELLANEOUS (astrology)

5.4.3 Various fragments of treatises on astrology as follows.

Treatise in unnumbered *faṣls*: DM 681,6 (fols. 17r–27r, *ca.* 1000H); another in which the author mentions al-Ṭūsī (no. G23) and Da'ūd al-Anṭākī (no. C125): K8527,3 (fols. 89v–91r, *ca.* 1100H); another in 5 *faṣls*: ṬJF 21,11 (fols. 183r–187v, 1206H); another in 8 *bābs*: DM 880 (7 fols., *ca.* 1200H); another without numbered divisions: DM 681,2 (fols. 2r–4v, *ca.* 850H); another in unnumbered *bābs*: DM 590,1 (fols. 1r–7v, *ca.* 1100H).

5.4.3 The first *bāb* (subdivided in 35 *faṣls*) of an unidentified treatise: MMF 7,4 (fols. 28v–37v, *ca.* 1100H).

5.4.4 / 5.4.5 More miscellaneous anonymous treatises on astrology are listed in the Arabic catalog, Sections 5.4.4 and 5.4.5.

Z35 MISCELLANEOUS (astrology)

5.5.1 Various late individual horoscopes are listed in the Arabic catalog, Section 5.5.1(6), including one of the Prophet Muḥammad in DM 1063,1 (fols. 1v–2r, *ca.* 1150H). See Pl. LXXV.

5.5.3 Anonymous astrological treatises relating to people buried: DM 639,31 (fols. 259v–262v, *ca.* 1200H); to prices: DM 862,7 (fols. 39v–42r, 1293H); to war: DM 122,7 (fols. 111r–113r, 1183H); to conception: DM 163,3 (fols. 18v–19r, *ca.* 1250H).

5.5.3 Anonymous poems on astrology, relating to Saturn: DM 211 (3 fols., 1273H); to sexual intercourse: DM 817,3 (fol. 85v, *ca.* 1100H); to wedding nights: MM 167,9 (fols. 200v–201r, *ca.* 1100H); to marriage: ṬM 242,5 (fol. 43v, *ca.* 1250H); to medicine: MM 168,5 (fol. 16v, *ca.* 1100H).

Z36 ANONYMOUS

5.5.3 Non-numerical astrological tables entitled *jadāwil baḥr al-ḥuruf*, with a short introduction: MM 189,3 (fols. 110r–122v, *ca.* 1100H). See no. F15.

Z37 MISCELLANEOUS (treatises on mathematics)

6.4.38 ***Kalām fi l-nisba al-suddāsīya***, on proportions involving six quantities, arranged in 5 *bābs*: MR 39,2 (fols. 25v–26r, 1150H).

6.4.38 ***Taqyīd fi stikhrāj al-aᶜmida fi l-muthallathāt***, on finding the perpendicular height of a triangle: MR 39,3 (fols. 26v–28r, 1151H).

6.4.43 An anonymous treatise on geometry in 4 *bābs*: TR 106,3 (pp. 109–112, 1059H).

6.4.43 Notes on arithmetic: MMF 5,4 (fols. 6v–7v, *ca.* 1175H).

6.4.43 A fragment from a commentary on an unidentified work on geometry: DR 1089 (2 fols., *ca.* 1150H).

Z38 ANONYMOUS

6.5.3 A short Arabic treatise on the use of a sexagesimal multiplication table: ṬFF 18,1 (fols. 1r–1v, *ca.* 1000H).

Z39 MISCELLANEOUS (surveying)

6.7.8 A treatise on surveying stated to be taken from ***Sharḥ al-Fatḥ*** and to be the 23rd *bāb*: TR 353,3 (pp. 46–50, 1319H).

6.7.8 A fragment on surveying labelled *al-raṣad al-sādis ᶜashar min al-misāḥa*: DM 567,2 (fols. 8r–9v, *ca.* 1150H).

6.7.8 A fragment from a treatise on surveying in 3 *faṣls*: TR 222,3 (pp. 76–82, 1099H).

Z40 ANONYMOUS (inheritance)

6.9 Notes on inheritance tables: B 25293,2 (fols. 59r–60v, *ca.* 1100H).

Z41 ANONYMOUS (navigation)

8.8 An important collection of late anonymous works on navigation, including numerous astronomical tables of a kind not attested in the standard Islamic astronomical sources, is DM 570 (182 fols., 1276H). See Pl. XCVII. This manuscript merits detailed investigation by a specialist. See also X9 and X27.

CAPTIONS FOR PLATES

1. THEORETICAL ASTRONOMY

I (ṬH 43, fols. 79r and 91v): Extracts from a thirteenth-century copy of Naṣīr al-Dīn al-Ṭūsī's recension of Ptolemy's ***Almagest*** (see nos. A18 and G23, *sub* 1.1.6). Plate b shows part of Book X dealing with Ptolemy's mathematical model for Mars.

Plate c shows part of the last page of the recension, with a table of planetary visibility (there are numerous variants from the table in the ***Almagest***), and a statement that the recension was completed in a year that is now illegible. The Egyptian astronomer Ibn Abi l-Fatḥ al-Ṣūfī (no. C98) "checked and completed" this manuscript in 1492/93.

As in most medieval Islamic tables, the entries are expressed sexagesimally (to base 60) and are written in standard Arabic alpha-numerical (*abjad*) notation. On this, see R. A. K. Irani in *Kennedy et al.*, pp. 710–721.

No systematic research has been conducted on Ptolemy's ***Almagest*** in its Arabic translations. P. Kunitzsch is currently investigating Ptolemy's star catalog, but the time is surely ripe for a detailed study of the whole Arabic tradition of the *Almagest*. The first reliable English translation of the Greek original has just been published by G. Toomer, and commentaries by O. Pedersen and O. Neugebauer are available. Furthermore, F. Sezgin (*GAS*, VI) has now identified the available Arabic manuscripts.

II (DM 390, fols. 110r and 151r): Illustrations of the constellations of Sagittarius and Canis Minor, taken from fine copy of the ***Kitāb Ṣuwar al-kawākib***, a treatise on uranography by the tenth-century astronomer al-Ṣūfī of Shiraz (no. B52). This copy is datable to *ca.* 1500. In al-Ṣūfī's work there are two pictures of each constellation: one as it appears in the sky and the other as it appears on a celestial globe. See also Plates III, IV, and X.

III (MMF 9): Illustrations of Orion and Leo in a splendid copy dated 1633/34 of an anonymous Persian translation of al-Ṣūfī's treatise on uranography. (See nos. B52 and G108).

IV (ṬFF 11, from fol. 150v to 160r): Illustrations of the constellation figures taken from the eleventh-century Persian astronomical and astrological compendium of Shāhmardān Rāzī (no. B91). These have recently been investigated by Z. Tanindi (*Istanbul Congress Proceedings*). See also Plate XIIIa.

V (ṬH 35): Illustrations from a fine copy of the ***Tabṣira fī ᶜilm al-hayʾa*** by the twelfth-century astronomer al-Kharaqī (no. G6) of Marw in Central Asia. The illustration on the left shows a much simplified version of Ptolemy's internesting planetary models. Ptolemy in his ***Planetary Hypotheses*** proposed a model for the universe in which the maximum distance of a planet from the earth was equal to the minimum distance of the next planet. Simplified diagrams of this configuration, in which the planetary orbs were represented by a series of concentric circles, were common in both Islamic and European treatises on astronomy. Only with the rediscovery of the relevant portions of the ***Hypotheses*** by B. R. Goldstein (see *Trans. of the American Philosophical Society*, 57 (1967)) was a clear understanding of Ptolemy's system achieved.

VI (ṬJ 966,2, fol. 11r): Illustrations of individual models for the sun, moon and planets in the survey of Ptolemaic astronomy entitled ***al-Mulakhkhaṣ fi l-hayʾa*** by Maḥmūd ibn ᶜUmar al-Jaghmīnī (no. G17), compiled in Khwārizm in Central Asia in 1221/22. This copy is dated 1486.

VIIa (K3957,3, fol. 184v): A diagram of the lunar model in the *Tadhkira* of the thirteenth-century Persian scholar Naṣīr al-Dīn al-Ṭūsī (no. G23). Almost 30 years ago, E. S. Kennedy discovered that al-Ṭūsī advocated some improvements on Ptolemy's planetary models. In particular, al-Ṭūsī's lunar model attempts to eliminate some of the problems of Ptolemy's model. The ***Tadhkira*** has now been studied in detail by F. J. Ragep (Harvard Ph.D. thesis).

VIIb (TR 128, fol. 45v): Sample illustrations from a unique copy of a commmentary on al-Ṭūsī's ***Tadhkira*** compiled in Syria *ca.* 1310 by ᶜUmar ibn Daᵓūd ibn Sulaymān al-Fārisī.

VIII (K3758): Illustrations of the planetary models in a copy of ***al-Tuḥfa al-shāhīya*** by the thirteenth-century scholar Quṭb al-Dīn al-Shīrazī (no. G27). This copy was made *ca.* 1500. Studies of these non-Ptolemaic models, initiated by E. S. Kennedy in the 1950's, have been resumed by G. Saliba.

IX (ṬOT 28, fols. 90v, 97r, and 57r): Selected illustrations from the Turkish encyclopaedia ***Kitāb Maᶜrifatnāme*** by Ibrāhīm Ḥaqqī (no. H38) showing the heliocentric and geocentric theories of the planetary system and some maps of the world. This work was compiled in 1756/57, and the heliocentric planetary theory is taken from the earlier work of Ibrāhīm Mutafarriqa, and hence indirectly from European sources.

Xa (DMT 1,1, fols. 273v–274r): An extract from an anonymous illustrated Turkish treatise on the constellations.

Xb (ṬJ 811, fols. 106v–107r): Illustrations of the constellations from another copy of the same treatise, copied in Egypt in 1783/84.

2. MATHEMATICAL ASTRONOMY

XI (TR 99, pp. 1–2): The title folio and first page of the unique manuscript of a work by the early tenth-century astronomer ᶜAbd Allāh ibn Masrūr (no. B37), dealing with the ***Zīj*** of the celebrated early ninth-century astronomer al-Khwārizmī (no. B14).

The ***Zīj*** of al-Khwārizmī is no longer extant in its original form, but only in a Latin translation of an Andalusian recension. This Latin version has been published and translated into English (H. Suter; O. Neugebauer). Also some medieval Hebrew commentaries on the original ***Zīj*** have survived and have been published with translation and analysis (B. R. Goldstein). Thus this work is of considerable interest: a preliminary translation of the text has been made by E. S. Kennedy.

The title page indicates that the manuscript once contained some corrections by Abu l-Qāsim ibn (Amājūr) (no. B36) to the ***Zīj*** of Ḥabash (no. B35), who was a late contemporary of al-Khwārizmī.

XII (DM 400, fols. 50v–51r): Extracts from the tables in a fine Yemeni copy of the ***Zīj*** of the tenth-century Persian scholar Kūshyār ibn Labbān (no. B70), with messy marginalia by a later Yemeni astronomer. These particular tables display the motion of the solar and planetary apogees and the equation of time, concepts fundamental to planetary astronomy.

A survey of some 125 Islamic *zīj*es was published by E. S. Kennedy in 1956 in a study which marked the beginning of a new era in the history of Islamic science: see *Kennedy* in the list of bibliographical abbreviations. The number of known *zīj*es has now been increased to about 200. Of these only three are published. The Latin version of a recension the ***Zīj*** of al-Khwārizmī (see Plate XI) has received the optimum scholarly treatment. The ***Zīj*** of the tenth-century Syrian astronomer al-Battani is published with a Latin translation and commentary by C. A. Nallino. The ***Zīj*** of the eleventh-century scholar al-Bīrūnī (see Plate XVII) is published without translation. There are dozens of other ***zīj***es preserved in the manuscript sources, and clearly this is a rich area for future research.

XIIIa (ṬFF 11, fol. 78v): Part of a table of an auxiliary function for timekeeping (the product of the cosines of the solar declination and of terrestrial latitude) computed by the tenth-century scholar ᶜAlī al-Nasawī (no. B67) for the latitude of Rayy in N. Iran. al-Nasawī's astronomical writings have not survived in their original form, and this table, as well as other material due to him, is found in the Persian compendium by his student, Shāhmardān Rāzī (no. B91). See also Plate XLIVb.

XIIIb (MM 188,1, fol. 72v): A table for planetary visibility specifically computed for the latitude of Maragha in a unique copy of a *zīj* prepared there in the mid-thirteenth century by Muḥyi l-Dīn al-Maghribī (no. G21). The manuscript is copied in the distinctive hand of the late fifteenth-century Egyptian astronomer Aḥmad ibn Aḥmad ibn Timurbāy (no. C94).

XIVa (MM 116, fols. 1v–2r): Some eighteenth-century Egyptian planetary tables based on those of the tenth-century Cairo astronomer Ibn Yūnus (B59). His major work, the *Ḥākimī Zīj*, was prepared for the Fatimid Caliph al-Ḥākim, but only about two thirds of it survives. The planetary tables in the *Ḥākimī Zīj* were influential in later Egyptian, Persian and Yemeni astronomical circles, and some of the "lost" material of Ibn Yūnus was located in the Egyptian National Library during the cataloging project.

XIVb (MM 188,2, fol. 81v): The star catalog of Ibn Yūnus (no. B59), missing from the extant fragments of the *Ḥākimī Zīj* and here preserved in a fifteenth-century Egyptian compendium of tables copied by Aḥmad ibn Timurbāy (no. C94).

XV (DM 678): Sample tables from a collection (see no. D207) made from the *Zīj*es of the two tenth-century astronomers al-Ṭabarī (no. B92) and Ibn Yūnus (no. B59) and various later sources. Shown here are tables of positions of pseudo-comets attributed to al-Ṭabarī (on this subject, see *Kennedy et al.*, pp. 311–318) and of the positions of the lunar mansions "according to the opinion of Ibn Yūnus."

XVI (DM 29): The title folio and an extract from some tables displaying the lunar equation as a function of the double elongation and the mean anomaly. The lunar equation can thus be found directly from arguments obtained from mean motion tables, and there is no need for any calculation with auxiliary tables of the Ptolemaic variety. The tables are attributed on the title folio to Ibn Yūnus (no. B59), and the same tables are specifically mentioned by Ibn al-Majdī (no. C62) in the fifteenth century. However, Ibn al-Majdī did not know who had computed these tables. Notice the curious name *ḥabṭaq* which is applied to these tables: see further Plate XXa.

XVII (DM 866, pp. 1 and 82): The title folio and illustrations from the sections on spherical astronomy in a magnificent copy of the *Zīj* of the early-eleventh-century scholar al-Bīrūnī (no. B78). This copy, dated 1274/75, was used by M. Krause in his edition of this work. A table of contents of this important work is provided in *Kennedy et al.*, pp. 573–595.

XVIII (ṬM 138, fols. 1r and 41r): Extracts from a fine copy of the anonymous ***Shāmil Zīj*** (see no. B100), incorrectly attributed in a note in Persian written in 1407/08 on the first page to the tenth-century Egyptian astronomer Ibn Yūnus (no. B59). The tables in this *Zīj* are computed for an unspecified locality in Iran. Plate b shows a star catalog for the year 600 (Hijra or Yazdigird ?). The various copies of the ***Shāmil Zīj*** and its derivatives merit detailed study.

XIXa (DMF 1, fol. 120v): An extract from the tables in a fine copy of the Persian ***Zīj-i-Īlkhānī***, compiled at the observatory in Maragha in the mid-thirteenth century by the astronomers of Naṣīr al-Dīn al-Ṭūsī (no. G23). This extract is a section of a table displaying the time elapsed since sunrise as a function of meridian altitude (entered horizontally) and instantaneous altitude (entered vertically). Such tables were called *ṭaylasān*, *i.e.*, "shawl", in medieval Arabic, because of their trapezoidal shape.

XIXb (MM 54,1, fol. 3v): An extract from a star catalog for the year 1061 Hijra (=1650/51), compiled by the Egyptian astronomer Muṣṭafā ibn Suhrāb (no. D36). The curious title *jadwal saḥītīyat al-kawākib al-thābita* . . . seems to refer to a thirteenth-century Egyptian astronomer named al-Saḥītī (no. C173), who is mentioned by the Yemeni Sultan al-Ashraf (no. E8) in connection with the star catalog in his treatise on instrument construction—see Plates LXII–LXIV.

XIXc (TJ 391, fol. 17v): An extract from the tables appended to a treatise on the use of the sine quadrant by the fourteenth-century Syrian astronomer al-Mizzī (no. C34), copied by the author. The tables, called *shabaka*, *i.e.*, "net", in medieval scientific Arabic, display the solar longitude for each day of the first year of a four-year cycle, and there are three more such tables. None of the numerous other copies of al-Mizzī's treatise contains these tables.

XXa–b (DM 910, fols 3r and 81v): The title page and an extract from the tables in an astronomical treatise by the early thirteenth-century Coptic scholar al-Asᶜad ibn al-ᶜAssāl. The tables in this work are for determining Christian feasts and for determining solar and lunar longitudes. The term *ᵓbqṭy* used in the title (elsewhere also *ᵓbṭqy*) appears to be the origin of the Arabic term *ḥabṭaq* which was applied to double-argument planetary equation tables—see Plate XVI. It derives from the Greek *epaktai*, epact, referring to the number of days by which the solar year exceeds twelve lunar months. This work by Ibn al-ᶜAssāl merits detailed investigation. Another earlier copy exists in the Vatican Library.

XXc (DM 291,2): An extract from a list of units of weight and measures in Coptic and Arabic, copied *ca.* 1500.

XXI (TR 274, pp. 32, 150 and 104–105): Extracts from an anonymous Yemeni almanac and ephemeris for the year 808 Hijra (= 1405/06) (see no. E11). The almanac contains a considerable amount of historical information in the introductory parts (see Plate a), as well as daily positions of the sun, moon, and planets (Plate c, r.h.s.), computed for the longitude of Taiz using the ***Ḥākimī Zīj*** of the tenth-century Egyptian astronomer Ibn Yūnus (no. B59). It also contains astrological predictions based on the relative positions of the planets (Plate c, l.h.s.) and general horoscopes for each month of the year and each eclipse forecast for the year (Plate b).

Ephemerides of this kind were compiled by Muslim astronomers from the ninth century until the nineteenth. They had a high rate of attrition since at the end of the year they had no further practical value. Apart from one other Yemeni almanac in the Egyptian National Library, all other surviving Islamic almanacs are considerably later. Recently B. R. Goldstein and D. Pingree have analyzed some fragments of such almanacs from the Cairo Geniza, dating mainly to the twelfth century—see *JNES*, 38 (1979), pp. 153–175 and 231–256, but these Yemeni almanacs have never been studied.

XXII (DM 405, fols. 7v–8r and 10v–11r): Extracts from the lunar mean motion and longitude increment tables called ***al-Durr al-yatīm***, compiled by the fifteenth-century Egyptian astronomer Ibn al-Majdī (no. C62). The tables can be used to generate annual planetary ephemerides with some facility since they are based upon the periodic motions of the planets—see Plate XXVa. These tables and the four-century long tradition based on them have recently been investigated by E. S. Kennedy and the author.

XXIII (DM 1081, pp. 2 and 191): Extracts from the introduction and tables in a Maghribī version of the perpetual almanac of the Jewish astronomer Zacuto (*fl.* Salamanca, *ca.* 1475—see no. F31). The extensive tables are here preceded by three different Maghribi treatises on their use, not known to exist in any other copies. This manuscript appears to date from *ca.* 1800. Notice that the entries in the tables are written in Western Arabic numerals. Zacuto's tables await proper investigation.

XXIVa (ṬM 80, fols. 1r and 23v): An extract from the trigonometric tables copied *ca.* 1450(?) by the Cairo *muwaqqit* (professional mosque time-keeper) Sayf al-Dīn Sāṭilmish (see nos. C72 and C137), also referred to as *al-amīr al-ajall.* The tables display values of the sine, cotangent, and versed sine functions to three sexagesimal digits for each minute of argument. This particular extract shows values of the cotangent function to base 12 for each minute of arc between 42;1° and 48°. The notices of possession indicate that this manuscript was in Istanbul in 1722/23.

XXIVb (ṬM 76, fol. 4r): Extracts from the sine tables in the *Zīj* of the sixteenth-century Istanbul astronomer Taqi l-Dīn ibn Ma^{c}rūf (no. H12). Entries are expressed *decimally* in the *abjad* or Arabic alphanumerical notation. This is the first attested use of 10 as a base in simple trigonometric tables, although Ibn al-Adamī (tenth century) also used trigonometric functions to base 10 in some tables for constructing sundials.

No survey has been made yet of Islamic trigonometric tables to ascertain their accuracy and interdependence.

XXVa (DM 405,3, fols. 10v–11r): An extract from an anonymous set of ephemerides for the year 1194 Hijra (= 1780) (see no. D209). The ephemeris gives positions of the outer planets only for each ten days, and of Mercury for each five days: this suggests that it was completed using the auxiliary tables entitled ***al-Durr al-yatīm***, associated with Ibn al-Majdī (no. C62). See Plate XXII.

XXVb (S 4493,3, fol. 112v): An extract from a Persian ephemeris for the year 1828/29. al-Bīrūnī (no. B78) in his ***Tafhīm*** wrote on the compilation of such ephemerides some 800 years previously, and there is every reason to suppose that they were compiled annually throughout the Middle Ages, even though so few early examples survive.

3. ASTRONOMICAL TIMEKEEPING

XXVIa (DM 108, fol. 1r): The title folio of one copy of the main Cairo corpus of tables for timekeeping. The corpus consists of about 200 pages of tables for reckoning time by the sun and for regulating the times of the five daily prayers. The times of the daylight prayers are defined in terms of shadows, and the nighttime prayers in terms of sunset, morning and evening twilight, and sunrise. In numerous medieval sources, there is also reference to a sixth prayer, the *ḍuḥā*, performed at midmorning.

These tables were used in Cairo in one form or another from the thirteenth century to the nineteenth century. Some of the tables (for the solar azimuth) were computed by the tenth-century Egyptian astronomer Ibn Yūnus (no. B59), others (for the time since sunrise) were added by al-Maqsī (no. C15) in the thirteenth century, and the corpus was edited by al-Bakhāniqī (no. C28) in the fourteenth century—see Plate XXVIII. In this copy of the tables, the entire set is attributed to Ibn Yūnus.

All known Islamic tables for reckoning time by the sun and stars and for regulating the times of prayer have been surveyed by the author (*SATMI*). On the Cairo corpus, see already *AHES*, 10 (1973), pp. 342–394.

XXVIb (DM 444, fols. 11v–12r): Extracts from the tables of time since sunrise as a function of solar altitude and solar longitude, computed for the latitude of Cairo by Shihāb al-Dīn al-Maqsī (no. C15) in the late thirteenth century. The extract shows part of the subtable for solar altitude 17°, and above the title is a note stating that the duration of evening twilight for a given solar longitude x can be found by entering in the table with argument $x + 180°$.

XXVIc (DM 690, fols. 15v–16r): An extract from the main tables of the Cairo corpus in the edition of the fourteenth-century astronomer al-Bakhāniqī (no. C28). For each degree of solar altitude, in this case, 15°, the functions (1) time since rising, (2) hour angle, and (3) solar azimuth, are displayed for each degree of solar longitude.

XXVII (DM 954,2): Extracts from the prayer-tables in the Cairo corpus. All of the tables have entries for each degree of solar longitude, computed to two sexagesimal digits.

The tables in Plate a display (1) the solar altitude and (2) the time after midday when the sun is in the direction perpendicular to the qibla. The underlying direction for the qibla in Cairo is that computed by Ibn Yūnus (no. B59) in the tenth century. See also Plate L.

The tables in Plate b display (1) the time taken by the sun to move between the true and the visible horizons, based upon the assumption that the two are separated by about 2/3°—see Plate XXXIII, and (2) a factor for converting equatorial degrees (as used by the astronomers) to seasonal hours (as used in popular practice).

The tables in Plate c display (1) the time from sunset until the *ṭafy* (*al-qanādīl*), *i.e.* the moment just before daybreak when the muezzin should extinguish the flames in the oillamps illuminating the minaret during the nights of Ramadan; and (2) the time from sunset until the moment just before daybreak when the muezzin should pronounce the *salām*, *i.e.* a blessing on the Prophet Muḥammad. These tables are thus of considerable interest to the history of Mamluk religious institutions (see already *AHES*, 10 (1973), pp. 370–371).

XXVIII (DM 108, fols. 3v–4r): Notes by al-Bakhāniqī (no. C28) describing the way in which he put together the Cairo corpus of tables for timekeeping.

XXIX (MM 132, fols. 82v–83r): An extract from the universal table for timekeeping by day and night compiled by Najm al-Dīn al-Miṣrī (no. C16) in Cairo in the late thirteenth century. This particular extract shows the time since rising of the sun or any star when the meridian altitude is 38°, as a function of the observed altitude and the half arc of visibility. The table in its entirety contains over a quarter of a million entries and is the largest known table from the medieval period. The Cairo manuscript contains only half of the work; the other half of the same original manuscript is preserved in the Bodleian Library in Oxford.

XXX (DM 291,3, fol. 4v): An extract from a list of formulae for spherical astronomy, taken from the compendium of the thirteenth-century Cairo astronomer Abū ᶜAlī al-Marrākushī (no. C17—see also Plates LIX–LXI). The ratio of the first and second quantities is equal to the ratio of the third and fourth, and there are 62 such formulae in the original work. The fourteenth-century Syrian astronomer Ibn al-Shāṭir (no. C28) presented a list of 184 such formulae in his *Zīj*.

No algebraic notation was used by the astronomers of medieval Islam, but these lists of formulae constitute a convenient alternative. With the highly technical vocabulary relating to spherical astronomy which was developed by Muslim scientists, even the most complicated formulae for timekeeping could be rendered in the form of a simple ratio.

XXXIa (ṬM 228, fols. 31v–32r): An extract from the hour-angle tables for Damascus compiled by Shams al-Dīn al-Khalīlī (no. C37) in the fourteenth century. The extract shows the hour-angle when the sun was altitude 7° (r.h.s.) and 8° (l.h.s.) for each degree of solar longitude. These and other tables by al-Khalīlī were used by the *muwaqqit*s of the Umayyad Mosque in Damascus from the fourteenth to the nineteenth century.

XXXIb (ṬM 228, fols. 16v–17r): An extract from some anonymous prayer-tables for the latitude of Tripoli, taken as 34°. Values of twelve functions are displayed for each degree of solar longitude. These tables occur amidst others from al-Khalīlī's corpus of tables for Damascus.

XXXII (ṬM 228, fols. 5v and 86v): Sample extracts from the universal auxiliary tables for solving problems of spherical astronomy for all latitudes compiled by Shams al-Dīn al-Khalīlī (no. C37). Successive applications of the three functions entitled *al-maḥfūẓ al-awwal* and *al-thānī* and *jayb al-tartīb* yield solutions to any problem which in modern terms can be solved by the cosine formula of spherical trigonometry. These highly sophisticated tables were discovered only in 1970. They represent the culmination of a tradition of compiling such auxiliary tables which started with Ḥabash al-Ḥāsib (no. B35) in the ninth century.

XXXIII (TR 303, fol. 33v): Notes on the difference in setting times of the sun over the true and visible horizons for the latitude of Cairo. The values here recorded by the fifteenth-century Cairo astronomer ᶜIzz al-Dīn al-Wafāʾī (no. C61) are attributed to the fourteenth-century Damascus astronomer Ibn al-Shāṭir (no. C30). In other Egyptian sources they are attributed to the tenth-century Cairo astronomer Ibn Yūnus (no. B59)—see Plate XXVIIb. Before the discovery of this material it was not known that the Muslim astronomers had concerned themselves with quantitative estimates of the effect of refraction at the horizon. See also Plate XXXIV.

XXXIV (MM 123): Diagrams in a treatise by the sixteenth-century Egyptian astronomer ᶜAbd al-Qādir ibn Muḥammad al-Minūfī (no. C121) dealing with the effects of refraction at the horizon. The author, who was not fully in control of his subject, mentions Ibn Yūnus (no. B59), Ibn al-Haytham (no. B77), al-Marrākushī (no. C17) and Ibn al-Shāṭir (no. C30).

XXXV (TR 354, pp. 1 and 40): Extracts from a corpus of tables for timekeeping based on latitude 31°, and intended for use in Alexandria and Maḥalla. The tables display the time since sunrise, the hour-angle, and the solar azimuth for each degree of solar longitude and each degree of solar altitude. These extracts show the entries for solar altitudes 20° (northern declinations) and 79°–82°. On the title folio the tables are incorrectly attributed to (the tenth-century Egyptian astronomer) Ibn Yūnus (no. B59), wrongly called al-Mawṣilī (*i.e.* from Mosul). Their true compiler is unknown (see no. C143).

XXXVIa (DM 1108, fols. 8v–9r): Some tables by the late-sixteenth-century Cairo astronomer Ibn Abi l-Khayr al-Ḥusnī (no. C124) displaying the computation of the duration of morning and evening twilight at the equinoxes and solstices for each half climate.

XXXVIb (MM 167, fols. 159v–160r): Part of a table for finding the times of moonrise and moonset, compiled by Jamāl al-Dīn Yūsuf ibn Ṭūghān (no. C92), an astronomer of Cairo in the fourteenth or fifteenth century. No other tables for this purpose are known from the medieval period.

XXXVIc (MM 72, fols. 5v–6r): An extract from an anonymous set of tables displaying the solar altitude as a function of time of day for the latitude of Cairo. Values are given for each degree of semi-diurnal arc and each degree of time elapsed since sunrise: the extract shows the values for semi-diurnal arc 86°. The compiler may have been the copyist, Ibn al-Kattānī (no. C32).

XXXVII (TR 338,2, pp. 258–259): Extracts from a unique copy of the prayer-tables of al-Jannād (no. F35) for various latitudes serving Morocco. Values of some fourteen functions are displayed for each degree of solar longitude, and the extract shows two pages from the tables for latitude 31°, intended to serve Marrakesh.

XXXVIII–XXXIX (KhMT 3, fols. 1v–2r and 2v, and DMT 2,2, fol. 31v): Extracts from the almanac (*ruznāme*) attributed to Sheikh Vefā (no. H2), a Turkish scholar of the fifteenth century. The almanac contains information on the calendar. Some, but not all, copies contain a set of tables displaying the times of prayer computed for the latitude of Istanbul (*cf.* Plates XXXIXa and b).

The times tabulated are the lengths of day and night and the times from sunrise to the *ẓuhr*, midday to the *ᶜaṣr*, the *ᶜaṣr* to sunset, and from sunset to the *ᶜishāᵓ*. Entries are based on a value of 41° 30′ for the latitude of Istanbul, which is different from the one used by Darendelī (see Plate XLa), namely, the more accurate value 41° 0′. Vefā's almanac remained popular in Istanbul until the nineteenth century. It was copied both in manuscript form and on scrolls.

XLa (ṬFT 29,2, fol. 43v): An extract from the prayer-tables for Istanbul compiled by an individual named Darendelī (no. H1), apparently in the sixteenth century. The tables display for each day of the solar year the length of day and night, and the times (according to the Ottoman convention that sunset is 12 o'clock) of the *ẓuhr*, first *ᶜaṣr*, second *ᶜaṣr*, *ᶜishāᵓ* and *ḍaḥwa* (midmorning) prayers, as well as the time to start fasting in Ramadan (*imsāk*) and the time when the sun is in the direction of Mecca. The tables are based on the accurate value 41° 0′ for the latitude of Istanbul but Darendeli's almanac was far less popular over the centuries than the *Ruznāme* of Shaykh Vefā—see Plates XXXVIII–XXXIX.

XLb (DJ 323,7, fol. 34r): An illustration in a late nineteenth century Egyptian manuscript of a technique for reckoning the time of day in seasonal hours from the shadow cast by a vertical thumb on horizontal fingers. See no. D228.

XLIa (DM 689): Extracts from an anonymous set of Tunisian auxiliary tables for computing the time of day for various latitudes (see no. F30). The tables date from the late fourteenth or fifteenth century and were clearly inspired by the minor auxiliary tables of the Damascus astronomer al-Khalīlī (no. C37), extant in a unique manuscript in Dublin. These particular tables display a function of use in timekeeping calculations (the product of the cosines of the solar declination and the local latitude) for latitudes 36° 30′ and 36° 40′ and the hour angle in terms of this function. The manuscript contains about one hundred pages of tables.

XLIb (ṬFT 9, fols. 34v and 40v): Extracts from an anonymous Ottoman almanac and set of prayer-tables for Algiers (see no. F68). The first table displays for each day of December the lengths of day and night according to the astronomical and religious definitions (that is, measured from sunrise and daybreak, respectively), the duration of morning and evening twilight, and the intervals between midday and the afternoon prayer and sunset. The second displays for each day of January the times of the *ṣanjāq* (?), the afternoon prayer, and the beginning of fasting in Ramaḍān (*imsāk*). The term *ṣanjāq* is puzzling. It appears to be a copyist's error for *ḍuḥā*, the name of the prayer at midmorning, followed by a *q*, the standard medieval abbreviation for *daqāᵓiq*, minutes, in medieval scientific Arabic. But it is hard to imagine such a mistake in a popular almanac.

XLIIa (DM 1207, fol. 57v): An extract from some wretched prayer-tables for the latitude of Rosetta (see no. D217). The tables, which are anonymous, are carelessly computed as well as carelessly copied. The function tabulated on this page is the length of night.

XLIIb (Sh 89, fol. 296r): An extract from the prayer tables for Medina in various treatises by the seventeenth-century religious scholar al-Hattārī (no. D42). This particular table displays the duration of darkness, *i.e.* the interval between nightfall and daybreak.

XLIIc (K 3798): An extract from some prayer-tables for various integral latitudes in Egypt and the Hejaz compiled for the use of Egyptian pilgrims a few decades ago by Muḥammad Ḥifẓī (no. D194), an official in the Egyptian Customs. These particular tables serve latitude 25°. One may wonder whether the author was aware that an unidentified Egyptian astronomer in the fourteenth or fifteenth century had compiled a set of prayer tables for each degree of latitude from Mecca to Istanbul; a unique copy of these is preserved in Princeton University Library.

XLIII (DM 119, fols. 8v–9r): Some notes appended to a copy of a treatise on timekeeping by the fifteenth-century Egyptian astronomer Sibṭ al-Mārīdinī (no. C97). These include a story about the Umayyad Caliph ᶜUmar ibn ᶜAbd al-ᶜAzīz using a (Graeco-Roman) sundial to regulate the prayer-times (see no. B61), a poem on the prayer-times attributed to the Imām al-Shāfiᶜī (no. B2), illustrations of the configurations of the stars in the lunar mansions, and an arithmetical shadow scheme for reckoning the time of day approximately (see no. D228).

There is, in addition to the kind of information about the Caliph ᶜUmar II recorded in these notes, other evidence in the medieval sources that the times of the daylight prayers in Islam were originally regulated in terms of the seasonal hours (twelfth divisions of the length of daylight). The definitions for the midmorning (*ḍuḥā*), midday (*ẓuhr*) and midafternoon (*ᶜaṣr*) prayers in terms of shadows, which have been in standard use for over a millennium, represent practical means of regulating these three times of prayer at the third, sixth and ninth hours of daylight. The *ḍuḥā* was not universally observed, but instances of its use are attested in numerous medieval sources as well as in modern practice in several localities.

XLIVa (TR 321, fol. 1v): A rule for determining the day of the week of the first day of the month in the Hijra calendar, attributed to the fourth Caliph ᶜAlī ibn Abī Ṭalīb (no. B1). The manuscript is datable to *ca.* 1900.

In the Islamic lunar calendar, the months begin with the actual sighting of the lunar crescent. Such rules were applied by the religious authorities when the crescent could not actually be seen because of weather conditions or for other reasons. More sophisticated techniques for predicting the possibility of visibility on a given evening and for computing when and where the crescent would appear were developed by the astronomers—see the next few plates.

XLIVb (ṬFF 11, fol. 61r): A table for determining lunar crescent visibility by the early ninth-century Baghdad scholar al-Khwārizmī (no. B14). Values of the minimum distance between the sun and moon required to ensure visibility are displayed for each zodiacal sign. The table occurs in an eleventh-century Persian compilation by Shāhmardān Rāzī (no. B91) of Rayy. See also Plate XIIIa.

al-Khwārizmī's *Zīj* does not survive in its original form: we have only a Latin translation of an Andalusian recension and some Hebrew commentaries. See already the caption to Plate XI. A quite different table, computed for a latitude different from that of Baghdad, is preserved in the Latin version. These tables have been studied by E. S. Kennedy and M. Janjanian and, more recently by the author (see *Kennedy et al.*, pp. 151–156, and *Kennedy Festschrift*).

XLVa (ṬM 119, fol. 1r): The title folio of an Egyptian copy of the astrological treatise of the Abbasid astrologer ᶜUmar ibn al-Farrukhān al-Ṭabarī (no. B6), on which is copied an early Andalusian (?) table for determining lunar crescenty visibility. For each of the seven climates (*iqlīm*) of classical geography, values of the minimum distance between the sun and moon to ensure visibility are displayed for each zodiacal sign. Several copies of this table are found in various Maghribi, Egyptian, Yemeni and Turkish manuscripts. In each copy the values are extremely corrupt, which argues for an [redacted] of compilation. On this table, see my article in *Kennedy Festschrift*. No information is [redacted] on the original compiler. Note that this manuscript was copied by the owner, Abu l-Itqān [redacted] *ca.* 1745.

[redacted] (MM 183,1, fol. 11v): Part of a set of tables displaying the lunar altitude and azimuth at [redacted] a function of the half arc of visibility of the moon and the difference between sunset and [redacted] compiled by the fifteenth-century Egyptian astronomer Ibn al-Majdī (no. C62). These tables, [redacted] their kind, enable one to predict precisely where in the sky the crescent will be seen.

XLVc (ṬJ 298,7, fol. 55r): Part of an anonymous Persian *ghurratnāme* (see no. G117) giving the day of the week corresponding to the beginnings (*ghurra*) of each lunar month for the years 1086–88 Hijra (= 1675–77).

XLVIa (DṢ 166,2, fol. 40r): Part of an anonymous set of tables displaying calculations for the visibility of the lunar crescent during the years 1125–1130 Hijra (= 1713–18). For each month, the ecliptic coordinates of the moon at sunset, the altitude of the moon over the local horizon (Cairo ?), and the time interval between sunset and moonset are calculated. Then a prediction is made for the first day of the civil calendar: the crescent will be seen, the crescent will be seen faintly, the crescent will probably be seen, or the crescent will not be seen at all. Calculations of this kind underlie the information about the crescents in annual almanacs: see Plates XXI and XXV.

XLVIb (DM 141,3): An extract from an anonymous Egyptian ephemeris for the solar year corresponding to 1614/15, displaying diagrams of the appearance of the lunar crescent at the beginning of each month relative to the horizon of Cairo.

XLVII (ṬJ 811, fols. 59v–60r): Extracts from an anonymous illustrated treatise dealing with the Ka^{c}ba, the qibla, and a mystical circle called *dā^{c}irat rijāl al-ghayb*.

The diagram on the left displays the rectangular base of the Ka^{c}ba (the elongation is exaggerated) with prominent features such as the Black Stone (represented by a black dot) and the waterspout and semi-circular wall on the north-western end. The cardinal winds are associated with the corners of the Ka^{c}ba. There are several statements on the astronomical orientation of the Ka^{c}ba and its relationship with the winds in the medieval Islamic sources. The actual orientation of the Ka^{c}ba, which accords with the information recorded in the medieval sources, is investigated by G. S. Hawkins and the author in *JHA*, 13 (1982), pp. 102–109.

The diagram on the right displays the Ka^{c}ba with associated winds and weather conditions and geographical directions, represented on two diagonally-superimposed squares usually associated in Western writings on physics with the four elements and their qualities. This diagram is unique of its kind in the known Islamic sources.

On the circle for locating the *rijāl al-ghayb*, see Plate Ca.

XLVIII (B 19385, pp. 2–3): The beginning of a treatise on the qibla in early Islamic Transoxania, compiled by the early-eleventh-century *qāḍī* of Samarqand, al-Bazdawī (no. B88). The treatise casts considerable light on the problem of mosque orientation in Samarqand and its environs. For example, al-Bazdawī explains that the Ḥanafīs used due west for the qibla and the Shāfī^{c}īs used due south. Also, the earliest mosque in Samarqand was built towards winter sunset. An analysis of this treatise is to appear in *JHAS*.

XLIX (DM 637,4, fol. 46v): A diagram copied in Egypt *ca.* 1450 displaying the orientation of the Ka^{c}ba and the approximate orientations of the *miḥrāb*s for various localities. This diagram is similar to those found in the treatises on geography by Yāqūt (*ca.* 1200) and al-Qazwīnī (*ca.* 1250). It derives from more complicated schemes formulated *ca.* 1000 by Ibn Surāqa, and is one manifestation of the sacred geography in Islam. Some twenty different schemes in which the world is divided in sectors about the Ka^{c}ba were devised in the medieval period. The qibla for each sector of the world in these schemes is defined in terms of astronomical horizon phenomena, a fact which helps explain the often curious orientation of medieval mosques. See further the article "Makka (centre of the world)" in *EI*$_2$.

La (DM 758, fol. 4r): A table displaying "the altitude (of the sun when it is) in the azimuth of the ventilator (*bādahanj*)," which occurs in many copies of the Cairo corpus—see Plates XXVI–XXVIII. It is here included in an Egyptian copy of the auxiliary tables of al-Khalīlī (no. C37—see also Plate XXXII), and is specifically attributed to the fourteenth-century Egyptian astronomer Ibn al-Rashīdī (no. C39).

The table is based on the direction of winter sunrise, *ca.* 27° S. of E. for Cairo, which means that the open front of the ventilator should face *ca.* 27° E. of N. The discovery of this table and an investigation of other medieval Egyptian texts on the large ventilators which adorned the houses of medieval Cairo (a few Ottoman examples survive to this day) lead to the realization that the roughly orthogonal street plan of the medieval city of al-Qāhira was fortuitously (?) aligned in an astronomical direction. On this, see further *JAOS*, 104 (1984), pp. 97–133.

Lb (MM 114, fol. 5r): A statement by the fifteenth-century astronomer Ghars al-Dīn al-Ḥalabī (no. C88) on the qibla used for the Mosque of ᶜAmr in Fusṭāṭ, the first mosque built in Egypt (639). This value was *ca.* 27° S. of E., that is, toward winter sunrise (see Plate a). The city of al-Qāhira was founded in 969 in the same direction, but Egyptian astronomers, starting with Ibn Yūnus (no. B59) in the late tenth century, computed the value of the qibla at Cairo as 37° S. of E. The Mosque of al-Ḥākim and the Azhar Mosque are oriented in this new direction, skew to the street plan by about 10°. Many Mamluk mosques and madrasas in the old city have their exteriors aligned with the street plan and their interiors aligned with the qibla of the astronomers. The mausolea in the Mamluk "City of the Dead" are aligned internally and externally in the qibla of the astronomers.

LIa (DM 832,1, fols. 36v–37r): Part of a table displaying the qibla and the distance from Mecca as functions of terrestrial latitude and longitude difference from Mecca. This table is found in a unique copy of a treatise on the sine quadrant by ᶜAlī ibn Ṭībughā (no. C54), a *muwaqqit* at the Umayyad Mosque in Aleppo in the fourteenth century. Several different tables for finding the qibla have been located during the past ten years. This one is based on the complicated accurate formula for finding the qibla from terrestrial coordinates.

LIb (MM 204, fol. 84v): Part of an anonymous qibla table in an Egyptian manuscript copied *ca.* 1600. This table is based on a simple approximate formula for finding the qibla.

LIc (TR 103, p. 101): Part of a qibla table attributed to the celebrated ninth-century astronomer Abu Jaᶜfar al-Khwārizmī (no. B14). The table is based on the same simple approximate formula. The attribution of the table is not certain, but other material on the qibla, including a table which may indeed be by al-Khwārizmī, has been located in manuscripts in Istanbul and Tashkent.

LII (MM 167, fols. 203v and 204r): Lists of coordinates of the stations on the two pilgrim routes from Cairo and Damascus to Mecca, together with coordinates and qibla directions of localities mainly in Syria and Palestine. The qiblas were computed by the fourteenth-century Damascus astronomer Shams al-Dīn al-Khalīlī (no. C37).

4. INSTRUMENTS

LIII (ṬM 255, fol. 2v): Two sets of notes on the origin of the astrolabe and the etymology of the Arabic term *asṭurlāb*, one in Arabic and the other in Persian. Various dubious etymologies are [redacted]

LIVb (S 4792, fol. 114v): An illustration of a kind of sine quadrant bearing a grid for performing calculations in trigonometry. This kind of quadrant was first described by al-Khwārizmī (no. B14) in the ninth century. The illustration is found in a copy of a Persian treatise on the astrolabe compiled by Rukn al-Dīn al-ᶜĀmilī (no. G41) in Herat (modern Afghanistan) in 1456.

Illustrations of quadrants are rare in Persian sources. The different varieties of quadrant generally replaced the astrolabe in the central parts of the Islamic world, and only in Iran and India did the construction of astrolabes continue into the nineteenth and, in Isfahan, even into the mid-twentieth century. Trigonometric grids such as this were incorporated on the backs of these late astrolabes, but they were seldom used for any serious scientific purpose.

LV (DM 103, fols. 47v–48r): An extract from a fourteenth-century Egyptian copy of the tables compiled by the early ninth-century Baghdad astronomer al-Farghānī (no. B18). These tables display coordinates for marking astrolabe plates for each degree of latitude: the extract shows such tables for latitudes 19° and 20°. al-Farghānī started a tradition of preparing tables for astrolabe construction which lasted until the nineteenth century. Before the discovery of his tables, it was thought that Muslim astrolabists used geometric construction for marking the curves on astrolabe plates.

LVIa (DM 970, fols. 5v–6r): Diagrams of a "universal" instrument described in a treatise by the tenth-century scholar al-Khujandī (no. B50). This treatise has been investigated by Dr. R. Lorch.

LVIb (ṬM 155, fol. 27r): An illustration of the universal plate designed by the eleventh-century Andalusian scholar al-Zarqāllu (no. B87). The plate serves mainly to convert ecliptic and equatorial celestial coordinates, and was simplified from the universal astrolabe of his contemporary, ᶜAlī ibn Khalaf al-Shakkāz. The illustration occurs in the somewhat mysterious treatise shown in Plate LVII.

LVII (ṬM 155, fols. 1r and 1v): The title folio of a treatise entitled ***Kitāb al-Miqyās al-murajjaḥ fī l-ᶜamal bi-l-asṭurlāb al-musaṭṭaḥ***, falsely attributed to the eleventh-century scholar Abu l-Rayḥān al-Bīrūnī (see nos. B78 and B79). On the first page of the treatise, there occurs a statement that Thābit ibn Qurra (no. B30) had written that the astrolabe was invented by Hipparchus (no. A9). This is the first confirmation in the medieval literature of O. Neugebauer's attribution of the theory of stereographic projection to Hipparchus. See also Plate LIII.

LVIII (QM 2,1, fols. 35v–36r): Sample illustrations from a unique copy of a treatise on the use of the astrolabe by an unidentified author named Abū ᶜAlī al-Fārisī, who worked in Hama in Syria *ca.* 1300. The illustrations show the use of an astrolabe to measure the depth of a well (r.h.s.) and the relative positions of geographical localities with different latitudes and longitudes (l.h.s.). These are, of course, minor uses of the astrolabe, the main function of which is to represent in two dimensions the configuration of the heavens relative to the local horizon and hence to measure time by day or night. For an introduction to the astrolabe, see J. North's article in *Scientific American*, vol. 230 (1974), pp. 96–106, and W. Hartner's article "Asṭurlāb" in EI_2.

LIX–LXI (K 3821): Sample illustrations from the compendium on spherical astronomy and instruments by the thirteenth-century scholar al-Marrākushī (no. C17). Because of his name the author has been thought to have worked in the Maghrib. In fact, he worked in Cairo, and compiled this work there *ca.* 1280. It was highly influential in later astronomical circles in Egypt, Syria, Yemen and Turkey (though not the Maghrib or the Islamic East)—see further *Isis*, 74 (1983), pp. 539–540. The work exists in several copies and a translation of the first half, dealing with spherical astronomy, and a summary of the second half, dealing with other instruments, were published by the Sédillots *père et fils* in the mid-nineteenth century. It has been virtually ignored by historians of science ever since.

This particular manuscript was copied in a variety of different hands *ca.* 1500. Plate LIX displays different kinds of astrolabe retes discussed by al-Marrākushī. This material is taken from al-Bīrūnī (no. B78), who in turn took it from al-Sijzī (no. B56). These non-standard retes have been subjected to a preliminary investigation by J. Frank (*Zur Geschichte des Astrolabs*, Erlangen, 1920), and a new study of them based on all of the material now known to be available would be of considerable interest.

Plate LXa displays a special kind of quadrant for reckoning time of day and Plate LXb displays a geared mechanism for reproducing the apparent relative positions of the sun and moon. Modern accounts of these instruments are quite out of date and there is plenty of new material for a proper investigation.

Plate LXI displays diagrams of plane horizontal, conical and cylindrical sundials and tables of coordinates for constructing them. The writings of C. Schoy on Islamic sundial theory (*Gnomonik der Araber*, Berlin & Leipzig, 1923) are now very much out-of-date, and there is a wealth of new material awaiting any scholar who would take over where Schoy left off.

LXII (TR 105, fols. 137v–138r): Extracts from the tables for constructing horizontal sundials for various latitudes contained in the treatise on instruments by the Yemeni Sultan al-Ashraf (no. E8) in the late thirteenth century. These particular tables serve latitude 13° 37′, that is, Taiz. Tables of this kind were first prepared by al-Khwārizmī (no. B14) in Baghdad in the early ninth century, but al-Ashraf owed his inspiration to the tables for marking sundials for the latitude of Cairo in the compendium of Abū ʿAlī al-Marrākushī (no. C17)—see Plate LXI.

LXIII (TR 105, fols. 74v–75r): Extracts from the tables for constructing astrolabe plates for various latitudes contained in the treatise by the Yemeni Sultan al-Ashraf. Tables of this kind were first devised by the ninth-century scholar al-Farghānī (no. B18)—see Plate LV. al-Ashraf actually used these tables himself to construct astrolabe plates—see Plate CII.

LXIV (TR 105, fols. 143v and 145v): Illustrations of a magnetic compass (called *al-ṭāsa*) appended to al-Ashraf's treatise on instruments. This is the earliest reference to the compass in the medieval astronomical literature, but al-Ashraf makes no claim to have invented it. The earliest known astronomical instrument fitted with a compass was made by Ibn al-Shāṭir (no. C30) in Damascus in the fourteenth century, and the first known measurement of magnetic declination was made by al-Wafāʾī (no. C61) in Cairo in the fifteenth century (see M. Dizer in *JHAS*, 1 (1977), pp. 257–260). The development of the magnetic compass in the Islamic world is currently being investigated by S. Banerjee of the University of Minnesota.

LXV (DM 291,1, fols. 1r and 1v–2r): The beginning of a short treatise by the early-fourteenth-century Aleppo astronomer Ibn al-Sarrāj (no. C26) describing a universal astrolabe of his own invention. A marginal note states that Ibn al-Sarrāj could not have invented this instrument because the writer had seen a similar astrolabe made about 170 years before the time of Ibn al-Sarrāj.

In fact, Ibn al-Sarrāj seems to have "reinvented" the same universal astrolabe as was devised by ʿAlī ibn Khalaf al-Shakkāz in Andalusia in the eleventh century. But, as we learn from Ibn al-Sarrāj's [redacted]

LXVI (DM 504,2, fol. 8r): A list of treatises on quadrants, several of which are no longer extant (see no. C147).

LXVII (MR 40,2, fols. 52v–53r): Examples from a set of illustrations displaying the construction of markings on various astronomical instruments, prepared by Muṣṭafā Ṣidqī (no. D81) in 1747. These and other such diagrams in this manuscript await study.

LXVIII (DR 892): Selected illustrations from the treatise on sundial theory by the fifteenth-century Cairo astronomer al-Karādīsī (no. C90), copied in the elegant hand of ᶜAlī ibn Ḥasan al-Baḥṭīṭī in 1480/81. The illustrations show both horizontal and vertical sundials as well as a universal polar sundial. Most of the sundials which adorned the mosques of Cairo in Mamluk and Ottoman times have disappeared; some exceptions are now housed in the Museum of Islamic Art and the Azhar Mosque Library in Cairo. See also the description of the sundial which once graced the Mosque of Ibn Ṭūlūn in Cairo in *JHAS*, 2 (1978), pp. 331–357.

LXIX (MR 40,3, fols 65v–70r): Illustrations from an anonymous Turkish treatise on the astronomical instruments used by Taqī l-Dīn (no. H12) in the Istanbul Observatory in the late sixteenth century. The copy is in the distinctive hand of Muṣṭafā Ṣidqī (no. D81) and is dated 1731/32. The treatise has been published by S. Tekeli.

LXX (DM 1140): Selected illustrations and tables from the treatise on sundial theory by the sixteenth-century astronomer Taqī l-Dīn (no. H12). The tables in the treatise are for constructing plane sundials, horizontal, vertical and skew to both the horizon and the meridian, for the latitude of Istanbul, although the treatise was written near Nablus in Palestine—see Plate CXa. Several Ottoman sundials are still *in situ* on the walls of various mosques in Istanbul: these have been surveyed by W. Meyer (in *Istanbul Symposium Proceedings).*

5. ASTROLOGY

LXXI (DM 1156, fols. 45v–46r): The first two pages of a unique copy of an astrological treatise in 22 chapters attributed to the sage Hermes (no. A22). The manuscript dates from *ca.* 1400 and is probably of Egyptian provenance. Most other Arabic manuscripts of Hermetic material are considerably later. This fine copy, and much of the extensive Hermetic material preserved in the Egyptian National Library, deserves investigation at the hands of a specialist. See also Plate CVIb.

LXXII (ṬM 243, fols. 117r and 129r): The first and last pages of a treatise on astrology attributed to Aristotle (no. A3), copied by a Christian monk in an unidentified monastery called Dayr Bashīr in 1246–47.

LXXIIIa (DM 122,1, fol. 1r): A short treatise on seating arrangements in royal councils (*majālis al-mulūk*) attributed to Ptolemy (no. A18), copied in Egypt *ca.* 1750.

LXXIIIb (MM 54,2, fols. 9v–10r): Diagrams of the arrangement of the "council of kings" according to opinion of "Jnh the Indian" (no. A30). These occur in an Egyptian manuscript dating from *ca.* 1800.

LXXIVa (Sh 76, fol. 1v): The introduction to a treatise on astrology attributed to "Rymbr the Greek" (no. A25) and stated to have been translated by his student Sarjis (no. A26). The copy was made *ca.* 1900, probably in Egypt.

LXXIVb (MM 105, fol. 1v): The first page of an astrological treatise attributed to Aristus (no. A20).

LXXIVc (DM 1058, fol. 1v): The first page of a copy of the same treatise attributed to Eutochius (see also no. A20).

LXXV (DM 1063,1, fols. 1v–2r): A horoscope of the Prophet Muḥammad, copied *ca.* 1750 by the Imām of the Aqṣā Mosque in Jerusalem (see no. Z34). Astrology, generally frowned upon by the religious authorities, was nevertheless widely practiced in medieval Islamic society. The role of astrology in Islamic society has been the subject of a recent article by G. Saliba (*Arab Studies Quarterly*, 4 (1982), pp. 211–225).

LXXVI (TGh 140): Extracts from a book of astrological lore attributed to the Abbasid Caliph al-Maʾmūn (no. B9). This copy dates from *ca.* 1850. There is no published material on books of this genre. See also the next plate.

LXXVII (Sh 25): Extracts from a book of prognostications attributed to Abū Maʿshar (no. B22), the famous astrologer of ninth-century Baghdad. See also the preceding plate.

LXXVIIIa (MM 84, fol. 1r and 1v): The title page and introduction of an astrological treatise attributed to Teucrus (no. A23).

LXXVIIIb (DḤ 15, fols. 1r and 1v): The title page and introduction of an astrological treatise with the curious title ***Kitāb al-Dwʾrdhmzj*** attributed to the celebrated ninth-century scholar al-Kindī (no. B19).

LXXIX (DM 1108,4B, fol. 20v): A table for computing of life expectancy, stated to have been found written in gold in a book entitled ***al-Kāmil fi l-nujūm*** in the treasury of the ninth-century governor of Khurāsān, ʿAbd Allāh ibn Ṭāhir ibn al-Ḥusayn (see EI_2, *s.v.*). In other copies of this table, the underlying theory is (correctly) attributed to Vettius Valens (no. A21).

LXXX (MM 204, fols. 75v–76r): The text of a treatise on the astrological significance of comets by the celebrated translator and scholar of ninth-century Baghdad, Ḥunayn ibn Isḥāq (no. B20). Only one other copy of this treatise, a later manuscript also preserved in Cairo, is known. See also Plate LXXXIV.

LXXXI (DM 4, fols. 1r and 9v): The title folio and sample tables from a fine copy of a treatise on astrology by the thirteenth-century author, Najm al-Dīn Ayyūb al-Ikhlāṭī (that is, from what is now Ahlat in Eastern Turkey), astrologer to the Ayyubid Sultan al-Ṣāliḥ ibn al-Kāmil. The author (no. C3) refers to numerous earlier sources in his introduction (see *GAS*, VII, pp. 21–22), and his work merits detailed investigation.

[redacted]

LXXXIII (MM 13,3, fol. 49v): A fragment from an astrological history of the campaigns between the Mamluks and Ottomans around Adana during the period 1470–73. The history was written by Yūsuf ibn Qurqmās al-Ḥamzāwī (no. C91), and the manuscript was copied *ca.* 1480 by the Egyptian astronomer Aḥmad ibn Aḥmad Timurbāy (no. C94). The treatise merits detailed investigation.

LXXIV (DM 729, fols. 10v–11r): Illustrations of different types of astrological pseudo-comets in a treatise of ᶜAbd Allāh ibn Aḥmad al-Maqdisī (no. D45), who lived *ca.* 1675, probably in Cairo. See also Plate LXXX.

6. MATHEMATICS

LXXXV (ṬR 107, fols. 134v–135r): The last two pages of a copy of Naṣīr al-Dīn al-Ṭūsī's recension of Euclid's ***Elements***, copied in 1387. The diagram relates to the construction of a regular icosahedron (with 20 sides) inside a sphere.

F. Sezgin (*GAS*, V) has listed all of the available material on the Arabic tradition of the ***Elements***, and specific Books have been investigated recently by J. W. Engroff and G. De Young (Harvard Ph.D theses—see *Berggren* for details), and G. Matvievskaya (in *Kennedy Festschrift*).

LXXXVI (ṬJ 207,4, fols. 52r and 55r): The title folio and an extract from of a commentary by the thirteenth-century Yemeni scholar Abū ᶜAbd Allāh al-Khuzāᶜī (no. E6) on the chapter dealing with inheritance in the ***Algebra*** of the celebrated ninth-century scholar, Abū Bakr (*sic*) al-Khwārizmī (no. B14). The solutions are presented in geometrical terms.

al-Khwārizmī's work was influential in the early development of algebra in the Latin West, but in the Islamic East it was soon superceded by more substantial and innovative works most of which were not known in the West and are known now only through research on manuscripts. Muslim scholars other than specialists in mathematics were in general more interested in the treatment of the algebra of inheritance in al-Khwārizmī's work, and for this reason it was particularly popular in the Yemen and the Hejaz.

Islamic writings on algebra are being published systematically by R. Rashed in Paris (see *Berggren* for details). The numerous publications of A. Anbouba should also be consulted (see, for example, *JHAS*, 2 (1978), pp. 66–100). Islamic writings on the algebra of inheritance have been ignored by historians of science since the time of S. Gandz (see his study "The Algebra of Inheritance" in *Osiris*, 5 (1938), pp. 319–391). There is considerable new material on this topic available in the Egyptian National Library.

LXXXVIIa (MR 42,1): A chapter heading in the treatise on arithmetic by the tenth-century scholar Abu l-Wafāᵓ al-Būzajānī (no. B54). This copy is dated 487 Hijra (= 1094), and is the earliest scientific manuscript in the Egyptian National Library. (A list of dated manuscripts is contained in vol. I of the Arabic catalog.) al-Būzajānī's work has been published by A. S. Saidan.

Several early Arabic treatises on arithmetic have been published by A. Saidan and A. Anbouba (see *Berggren* for details).

LXXXVIIIa (K 8522): The introduction to a poem on algebra by the late twelfth-century Maghribi scholar Ibn al-Yasamīn (no. F12). This poem was extremely popular in later Egypt, Syria, and the Maghrib. This copy, which is dated 1432/33, is of uncertain provenance.

LXXXVIIIb (B 23317, fol. 77v): The first page of one of four treatises on number theory, arithmetic, geometry and algebra by Ibn Fallūs (no. C6), who worked in Cairo and Mecca in the early thirteenth century. None of these works has been studied. The same manuscript contains a unique copy of a treatise on the inheritance laws governing non-Muslim subjects of the Islamic commonwealth, written by Taqi l-Dīn al-Subkī (no. C27), chief judge of Damascus in the fourteenth century.

LXXXIX (TR 135, pp. 224 and 360): Extracts from the tables in two treatises on mathematics by the thirteenth-century scholar Kamāl al-Dīn al-Fārisī (no. G30) of Baghdad. This manuscript was copied in Samarqand in 1424/25. As R. Rashed has shown (*AHES*, 28 (1983), pp. 107–147), al-Fārisī takes the subject of amicable numbers a step further than his predecessor Thābit ibn Qurra (no. B30).

XC (DRF 1, fols. 3r and 3v–4r): The table of contents of a Persian treatise on arithmetic by the thirteenth-century (?) scholar Muḥammad ibn (?) ᶜAbd al-Karīm al-Ghaznawī (no. G14). This manuscript was copied in 1271/72, probably in Iran, and became part of the *waqf* of the Gulshānī *khānqāh* (a Sufi hospice) in Cairo. al-Ghaznawī's treatise, like most later Islamic writings on arithmetic, has not been studied in modern times.

XCIa (TR 119, p. 116): An extract from an enormous sexagesimal multiplication table attributed to Yūsuf ibn Qurqmās al-Ḥamzāwī (no. C91), who was the official responsible for the Aleppo pilgrimage caravan *ca.* 1500. The extract shows sexagesimal products $m \times n$ for $m = 5;42, 5;43, \ldots, 5;47$ and $n = 1, 2, \ldots, 30$. The whole table contains about 216,000 ($60 \times 60 \times 60$) entries. Less extensive multiplication tables with about 3,600 (60×60) entries were used by Muslim astronomers at least from the tenth century onwards. See *HM*, 1 (1974), pp. 317–323 and 6 (1979), pp. 405–417 for details.

XCIb (MMF 8, fol. 57v): An extract from an anonymous set of tables of sexagesimal quotients. The extract shows quotients m/n for $m = 1, 2, \ldots, 120$ and $n = 99$. The table was compiled in Iran, and the manuscript dates from *ca.* 1300.

XCIIa (B 23205): An extract from a much annotated Egyptian copy of a treatise on the algebra of inheritance by Ibn al-Hāʾim (no. C58), a celebrated scholar of Jerusalem in the early fifteenth century. His writings have never been studied, but would repay investigation. Of particular interest might be the tables which he compiled to facilitate the calculation of individual shares of estates. See also Plate LXXXVI.

XCIIb (ṬR 134, fol. 49v): An extract from the ***Miftāḥ al-ḥisāb*** of the early-fifteenth-century scholar Jamshīd al-Kāshī (no. G48) of Samarqand, displaying a table of the circumference of a circle as a function of its radius and of the area of a circle as a function of the square of its radius. al-Kāshī's writings on the determination of *pi* have been studied by P. Luckey (see *Berggren*) and his method for calculating the Sine of 1° has been investigated by A. Aaboe (in *Scripta Mathematica*, 20 (1954), pp. 24–29). al-Kāshī's work represents the culmination of the Islamic achievement in computational mathematics.

7. PHYSICS, *ETC.*

XCIIIc (ṬH 45, fol. 145r): An illustration of a plumb-line in the *Nihāyat al-idrāk* on theoretical astronomy by the thirteenth-century scholar Quṭb al-Dīn al-Shīrāzī (no. G27). Such illustrations may have mislead some modern scholars to believe that Muslim scientists were acquainted with the pendulum: on this, see now *AIHS*, 29 (1979), pp. 35–52.

XCIV (ṬR 123, fols. 15v–16r and 24v–25r): Some illustrations from a copy of the Arabic translation of Heron's treatise on mechanics (see no. A11) in the hand of the Egyptian astronomer Muḥammad ibn Abi l-Khayr al-Ḥusnī (no. C123), dated 1565/66. The Islamic tradition in mechanics as represented by the writings of the Banū Mūsā, al-Jazarī and Taqi l-Dīn (nos. B16, C2, and H12), has been investigated by D. Hill and A. Y. al-Hassan. Newly discovered material of Andalusian provenance which represents a different tradition (see *HS*, 13 (1975), pp. 288–289, and *JHAS*, 1 (1977), pp. 33–46) is currently being investigated by M. V. Villuendas of Madrid.

XCVa (DṢ 166,1, fol. 36v): An illustration from a treatise on mechanical clocks by the late-sixteenth-century scholar Taqi l-Dīn (no. H12). This treatise describes various clocks of European provenance. It has been published and translated by S. Tekeli.

XCVb (DR 260,2, fol. 61r): An extract from the translation of Euclid's *Optics*, probably by Thabīt ibn Qurra (no. B30), copied *ca.* 1200. Islamic contributions in optics have been investigated most recently by A. I. Sabra, whose edition of the *Optics* of Ibn al-Haytham (no. B77) is currently being published in Kuwait.

XCVI (DR 893, fols. 8v and 41v–42r): Illustrations from the treatise on optics by the late-sixteenth-century astronomer Taqi l-Dīn (no. H12). This treatise has yet to be studied.

XCVII (DM 570): Specimens of the illustrations of ships and tables for navigation found in an Ottoman compendium of unknown provenance (no. Z41), apparently compiled in the nineteenth century. This material awaits detailed study at the hands of a specialist. The earlier navigational texts of Ibn Mājid and al-Mahrī (nos. X9 and X27) have been published by I. Khoury in Damascus and the most important treatise of Ibn Mājid has been translated into English with a masterful commentary by G. R. Tibbetts as *Arab Navigation* . . . , London, 1971, repr. 1981.

XCVIII (MNT 3, fols. 61r and 65v–66r): Illustrations from an anonymous Ottoman treatise on warfare and military engineering and architecture.

8. MISCELLANEOUS

XCIXa (ṬFF 20, fol. 0v): A list of the major astronomical observers and their dates, found on the flyleaf of an Ottoman manuscript. The compiler had little idea of the history of astronomy: Hipparchus is put in Alexandria, Ptolemy on Rhodes, Ibn al-Shāṭir in Cairo, ... , and the "dates" associated with each individual are an even worse mess.

XCIXb (DM 145, fol. 56v): A somewhat confused diagram of a model for the sun on the flyleaf of a manuscript dateable to *ca.* 1700. The solar apogee is represented as being at Cancer 0°, which is not unreasonable. However, the center of the eccentric deferent is not on the apogee but on a line perpendicular to the apogee, which is labelled "east-west."

Ca (ṬJ 811,6): An illustration of a circle for locating the mysterious *rijāl al-ghayb*, intermediaries between man and God in popular Ottoman Ṣūfī belief. The material on the *rijāl* in this manuscript has been investigated by my student, Zeina Matar. Such diagrams were popular in Ottoman almanacs—see, for example, Plate XXXVIII.

Cb (TR 140,20, p. 115): Notes on astronomy and mathematics probably made by a student in an astronomy class in Istanbul *ca.* 1750. A cataloger's nightmare, such notes can sometimes repay investigation: see, for example, the third paragraph. Here we find a quotation from the thirteenth-century scholar Quṭb al-Dīn al-Shīrāzī (no. G27) concerning a treatise by the eleventh-century scholar Ibn al-Haytham (no. B77) on the fact that the visible part of the heavens is more than one half. Manuscripts of the original treatise are preserved and are listed by F. Sezgin, but the work has never been studied.

CIa (ṬM 250,4, fol. 30v): A list of symbols for the fixed stars occurring amid some notes on astrology in a manuscript dated 1860/61.

CIb (MM 189, fol. 116v): An extract from an anonymous set of astrological tables copied in Egypt *ca.* 1700. The tables apparently have to do with letter-magic but their precise purpose is not clear.

CIc (TR 131, p. 1): This page, a flyleaf of a Maghribi manuscript copied in 1772/73 (see nos. F71 and F73), contains a record of a reddish glow observed in the sky above Marrakesh one October night in 1772, and a table for constructing a horizontal sundial for the latitude of that city.

CII (TR 105, fols. 147r and 147v): Part of the first of two *ijāza*s by teachers of the Yemeni Sultan al-Ashraf (no. E8) describing six astrolabes which he made, using the tables in his compendium on instrument construction—see Plate LXIII. One of the instruments mentioned is preserved in the Metropolitan Museum of Art in New York. Prior to the identification of these texts, opinions had been expressed doubting the Yemeni provenance of the artefact and even its authenticity! See *MAY*, p. 29 and Plates V and VI, and a detailed study in *ZGAIW*, 2 (1986).

CIIIa (TR 105, fol. 1r): A mark of ownership by the Yemen Sultan al-Muʾayyad (ruled 1297–1321) on a copy of a treatise on instruments by his brother, the Sultan al-Ashraf (no. E8). See also Plates LXII–LXIV and CII. Several of the Rasulid Sultans were interested in astronomy—see *MAY* for details. (Lists of owners and copyists and readers of Cairo scientific manuscripts are presented in vol. I of the Arabic catalog.)

CIIIb (QM 2, fol. 98r): The colophon of a treatise on a special kind of sundial called *mīzān al-Fazārī*, probably compiled by the thirteenth-century Egyptian astronomer Nāṣir al-Dīn Muḥammad ibn Simᶜūn (no. C24), copied in 1306/07 by Ibrāhīm b. Aḥmad b. Khalīl, *muwaqqit* at the Sacred Mosque (*al-ḥaram al-Khalīl*) in Hebron in Palestine.

The earliest known references to the institution of *muwaqqit*, that is, the professional astronomer [redacted]

instruments have been discovered in the Egyptian National Library and the Chester Beatty Library in Dublin—see Plate LXV.

CIVa (TJ 391, p. 73): The colophon of the treatise *Kashf al-ghayb* by the fourteenth-century Damascus astronomer al-Mizzī (no. C34) on the use of the sine quadrant, an instrument bearing an orthogonal grid (like modern graph paper) for performing calculations involving trigonometric ratios. A note in the hand of the author attests that his student al-Karakī (no. C35) studied and mastered the contents under his direction. al-Karakī later compiled a corpus of tables for timekeeping for Jerusalem, extant in a unique copy preserved in Leipzig.

CIVb (DM 832, fol. 1r): The title folio of a collection of works on quadrants by the two fourteenth-century Syrian astronomers ᶜAlī ibn Ṭībughā (no. C54) and Sharaf al-Dīn al-Khalīlī (no. C38).

CV (DM 214, fol. 1r): The title folio of a fine copy of a treatise on the fiscal calendar by Abu l-Baqāᵓ ibn Yaḥyā ibn al-Jīᶜān (no. C83), copied *ca.* 1450 and later part of the *waqf* of the Sultan Qāytbāy.

CVIa (ṬM 80, fol. 1r): The title folio of some trigonometric tables copied *ca.* 1450 (?) by the Cairo *muwaqqit* Sayf al-Dīn Sāṭilmish (no. C72). The notices of possession indicate that the manuscript was in Istanbul in 1722/23.

CVIb (DM 1156, fol. 1r): The title folio of a fine unique copy of a treatise on astrology attributed to Hermes (no. A22). The manuscript dates from *ca.* 1400 and is probably of Egyptian provenance. See also Plate LXXI.

CVIc (ṬR 104, fol. 1r): The title folio of a unique copy of a treatise on arithmetic, geometry, astronomy and astrology compiled by an early Mamluk scholar named Tāj al-Dīn al-Tabrīzī (no. C22), otherwise unknown to the modern literature.

CVII (DM 637, fols. 1r and 0r): The title folio and flyleaf of a fine copy of the *Lumᶜa*, a *zīj* for Cairo by the fourteenth-century astronomer Aḥmad al-Kawm al-Rīshī (no. C41). This copy dates from 1463/64 and bears marks of ownership of several later Egyptian astronomers. (Note even an owner called Fulān ibn Fulān!) The flyleaf bears a note in English by one Edward Abbott, written in Cairo in 1650.

CVIIIa (DM 56, fol. 3r): An example of the handwriting of the fifteenth-century Cairo astronomer al-Wafāᵓī (no. C61). Note also the mark of ownership of the later Cairo astronomer Sulaymān (ibn Bakhshīsh or Ḥashīsh) al-ᶜUthmānī (no. C117). On his curious name, see also Plate CXIa.

One result of the cataloging of the Cairo scientific manuscripts was that individuals who previously had been only names listed by Suter and Brockelmann were recognized for their true scientific worth. However, personal information on most of these astronomers has yet to be extracted from the medieval historical sources and biographical dictionaries.

CVIIIb (DM 121, fol. 1r): The title folio of a treatise on spherical astronomy by the late fifteenth-century Cairo astronomer al-Karādīsī (no. C90), copied by his late contemporary Yūsuf ibn Qurqmās al-Ḥamzāwī (no. C91), who was also an astronomer—see Plate LXXXIII.

CVIIIc (MJ 202,2, fol. 14r): The title folio of a unique copy of an astrological treatise by the thirteenth-century Andalusian scholar Ibn Sabᶜīn (no. F15), copied in 1681, probably in Egypt.

CIXa (DM 751, fol. 5r): The colophon of a copy of a treatise on a special kind of almucantar quadrant by the Egyptian astronomer Ibn al-Mājdi (no. C62), compiled in August, 1409. The copy was made by another Cairo astronomer al-Kawm al-Rīshī (no. C41) about six months later and bears the author's approval of al-Rīshī's understanding of the contents of the treatise.

CIXb (ṬM 82, fol. 37v): The colophon of a copy of the treatise *Ghunyat al-fahīm* dealing with the compilation of annual ephemerides by the early-fifteenth-century Egyptian astronomer Ibn al-Majdī (no. C62), copied by the late-fifteenth-century Egyptian astronomer Ibn Abi l-Fatḥ al-Ṣūfī (no. C98).

CIXc (ṬR 105, fol. 15v): The colophon of a copy of Sibṭ al-Māridīnī's treatise on arithmetic (see no. C97), written in the elegant hand of Aḥmad ibn Yūnus al-Shalabī al-Ḥanafī, and followed by an *ijāza* in the decrepit hand of the ageing author, dated 1498/99.

CXa (DM 1140, fol. 57v): The colophon of a treatise on sundial theory of Taqi l-Dīn (no. H12), purporting to be in his own hand. Notice that the scribe misspelled the name of the town of Nablus in Palestine where Taqi l-Dīn completed his treatise in 1567/68. See also Plate LXX.

CXb (DM 557,4, fol. 91v): The colophon of a collection of treatises on mechanical devices, astronomical instruments, and algebra by Taqi l-Dīn. The colophon suggests that the copyist was the author himself and indicates that the manuscript was acquired by ʿAbd al-Raḥmān al-Ṭulūnī (no. D16), the *muwaqqit* and *imām* of the Mosque of Ibn Ṭūlūn in Cairo *ca.* 1640. A manuscript in Taqi l-Dīn's own hand is in Kandilli Observatory Library near Istanbul and various manuscripts formerly in his possession are preserved in the Universiteitsbibliotheek in Leiden.

CXc (DR 92, fol. 1r): The title folio of a treatise on weights and measures by an unidentified author Īlyā al-Maṭrān (no. D137), copied in Cairo *ca.* 1780 in the distinctive hand of Ḥusayn Bāsh Jāwīsh al-Ishrāfī. Observe the notice of possession of Maḥmūd ibn Sulaymān (no. D96), owner of several astronomical manuscripts in the Egyptian National Library.

CXd (MMF 1, fol. 168v): The colophon of a copy of the commentary on the *Zīj* of Ulugh Beg by Mīram Chelebī (no. H4), copied in 1676/77 by Muḥammad Fazāʾī al-Katīb, apparently identical with the Istanbul astronomer known as Munajjimak (no. H23).

CXIa (DM 633, fol. 1r): The title folio of an astrological treatise by the late-sixteenth-century Egyptian astronomer Sulaymān al-Rūmī (no. C117). Unfortunately the author does not give his full name and it is still a matter of conjecture whether his nickname was Ibn Ḥashīsh or Ibn Bakhshīsh. Both seem rather unlikely, but both are attested in other manuscripts. See also plate CVIIIa.

CXIb (DM 1108, fol. 0r): Notices of possession of Ḥasan ibn ʿAbd Allāh (who signed himself *al-Qiṭṭ (al-)miskīn*, maybe because he looked like a Cairo alley cat) and Maḥmūd ibn Ḥasan al-Nīshī, both collectors of scientific manuscripts in Cairo after 1640.

PLATES

b.

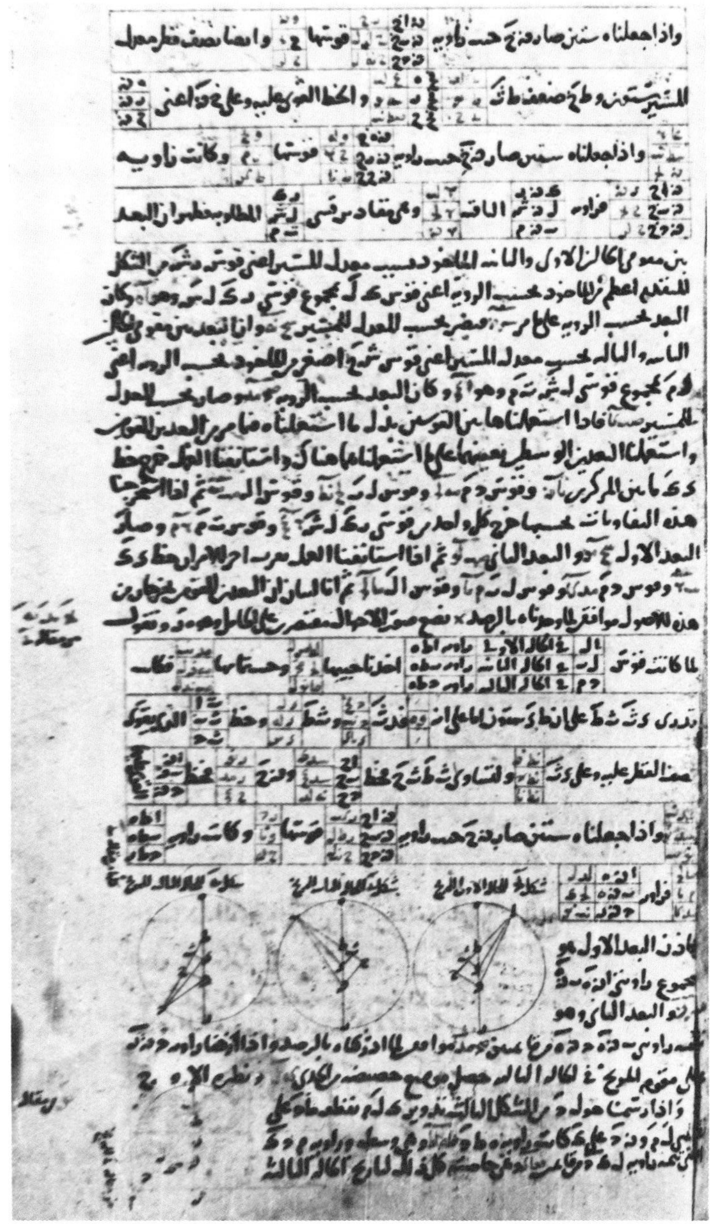

[redacted]tolemy's *Almagest* by Naṣīr al-Dīn al-Ṭūsī

II al-Ṣūfī's treatise on the fixed stars and constellations

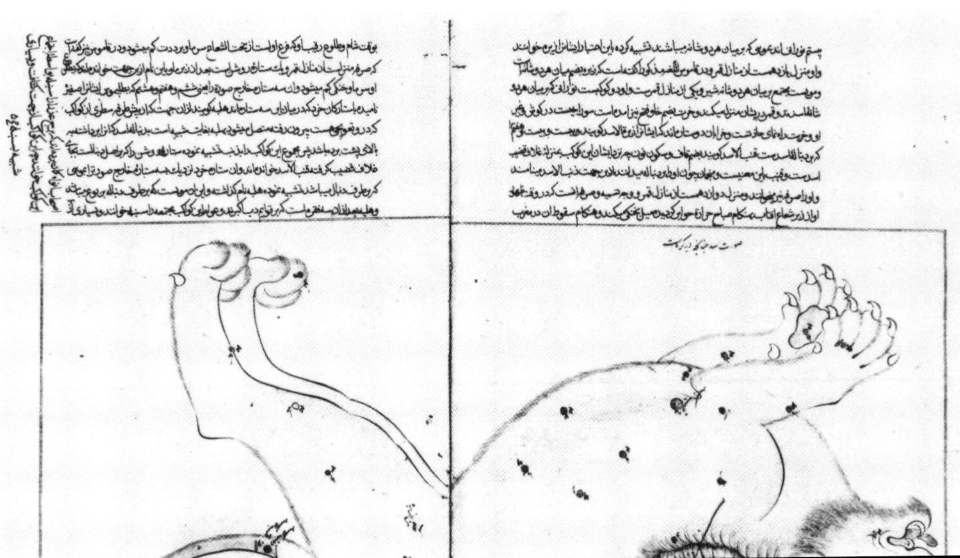

IV Illustrations of the constellation figures in Shāhmardān's compendium of astronomy and astrology

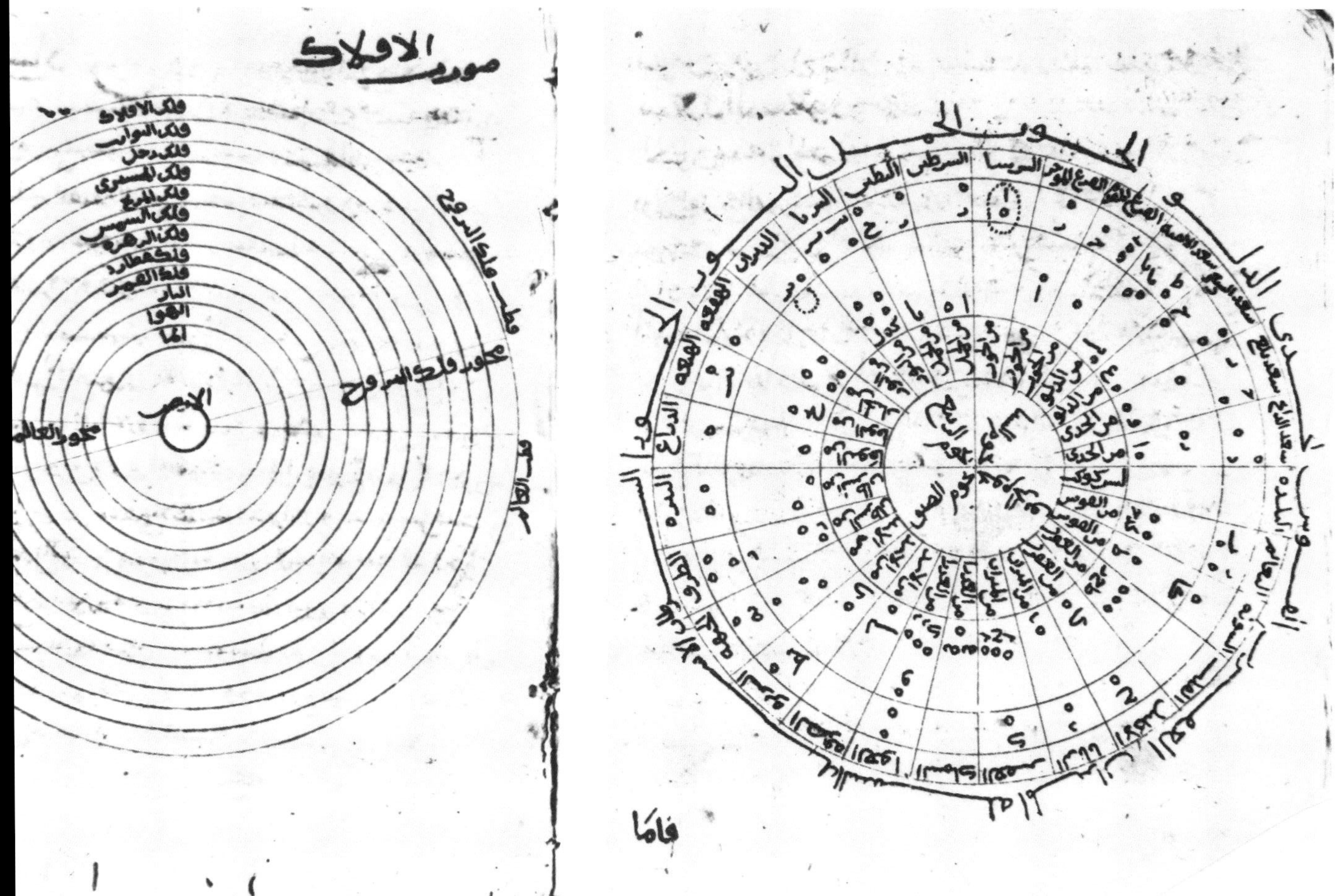

[redacted]he planetary spheres and the lunar mansions in al-Kharaqī’s treatise on theoretical astronomy

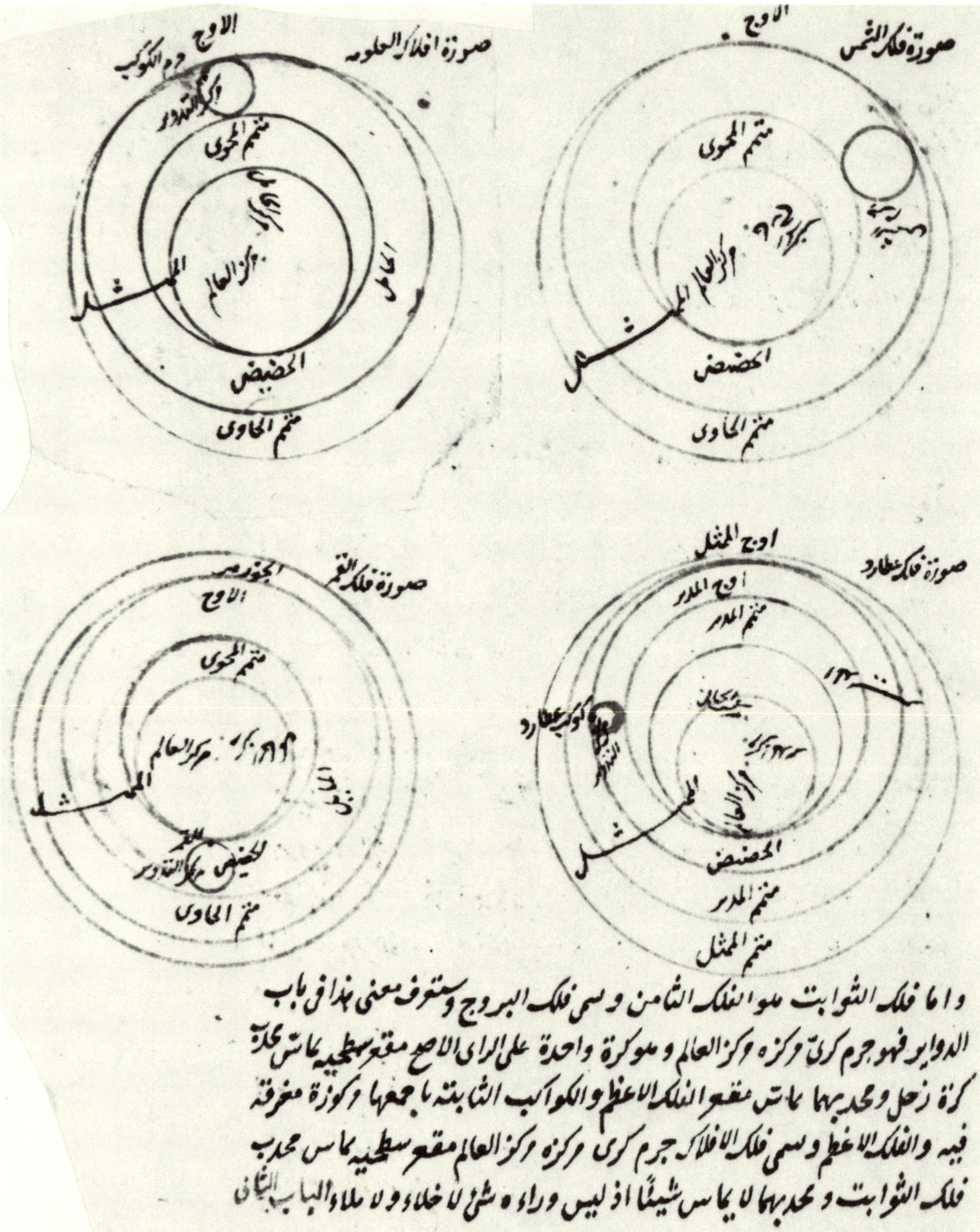

VI Diagrams of Ptolemaic planetary models in al-Jaghmīnī's treatise on theoretical astronomy

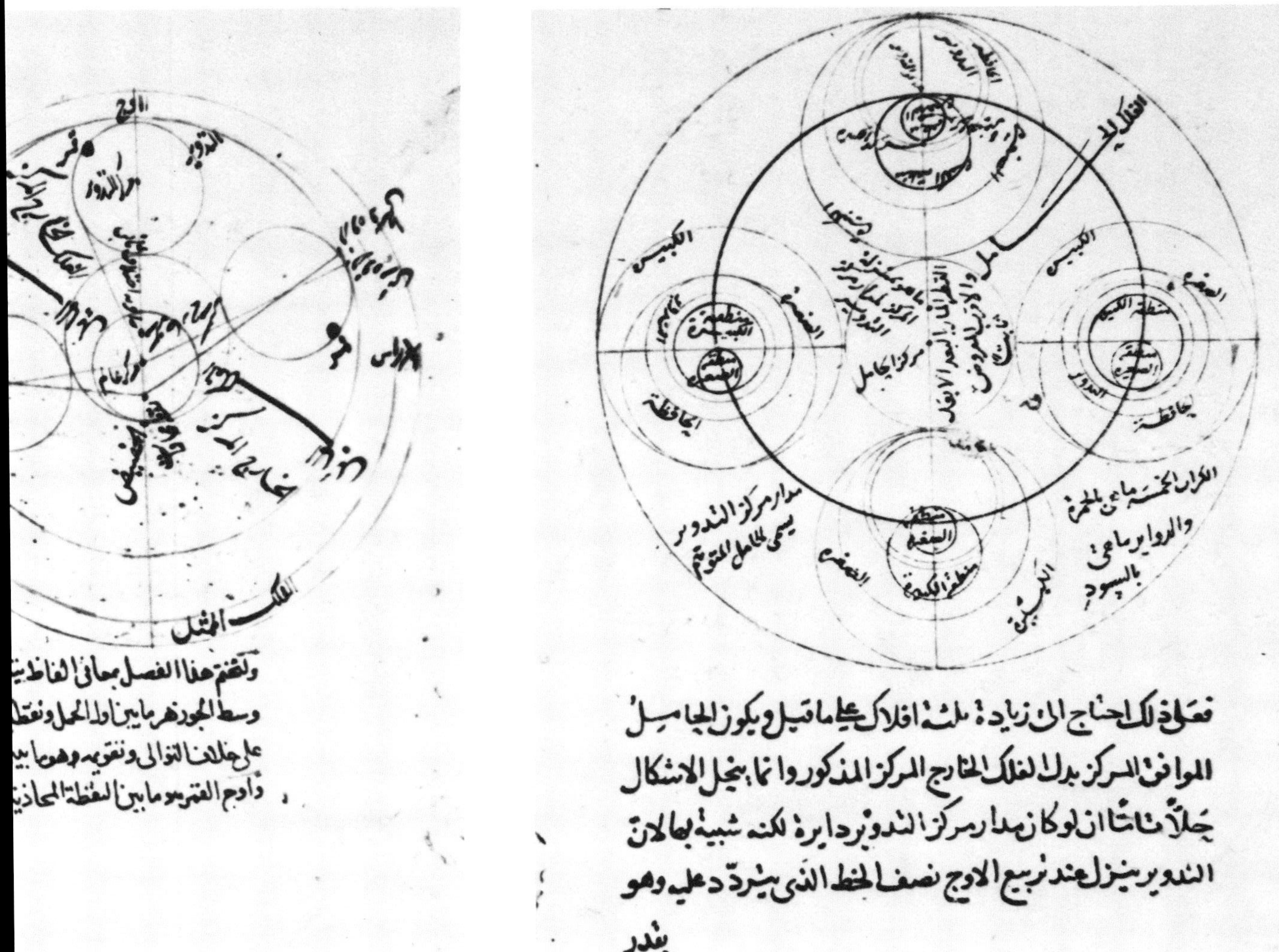

[redacted] of modified Ptolemaic lunar and planetary models in al-Ṭūsī's treatise on theoretical astronomy and al-Farisi's [redacted]ereon

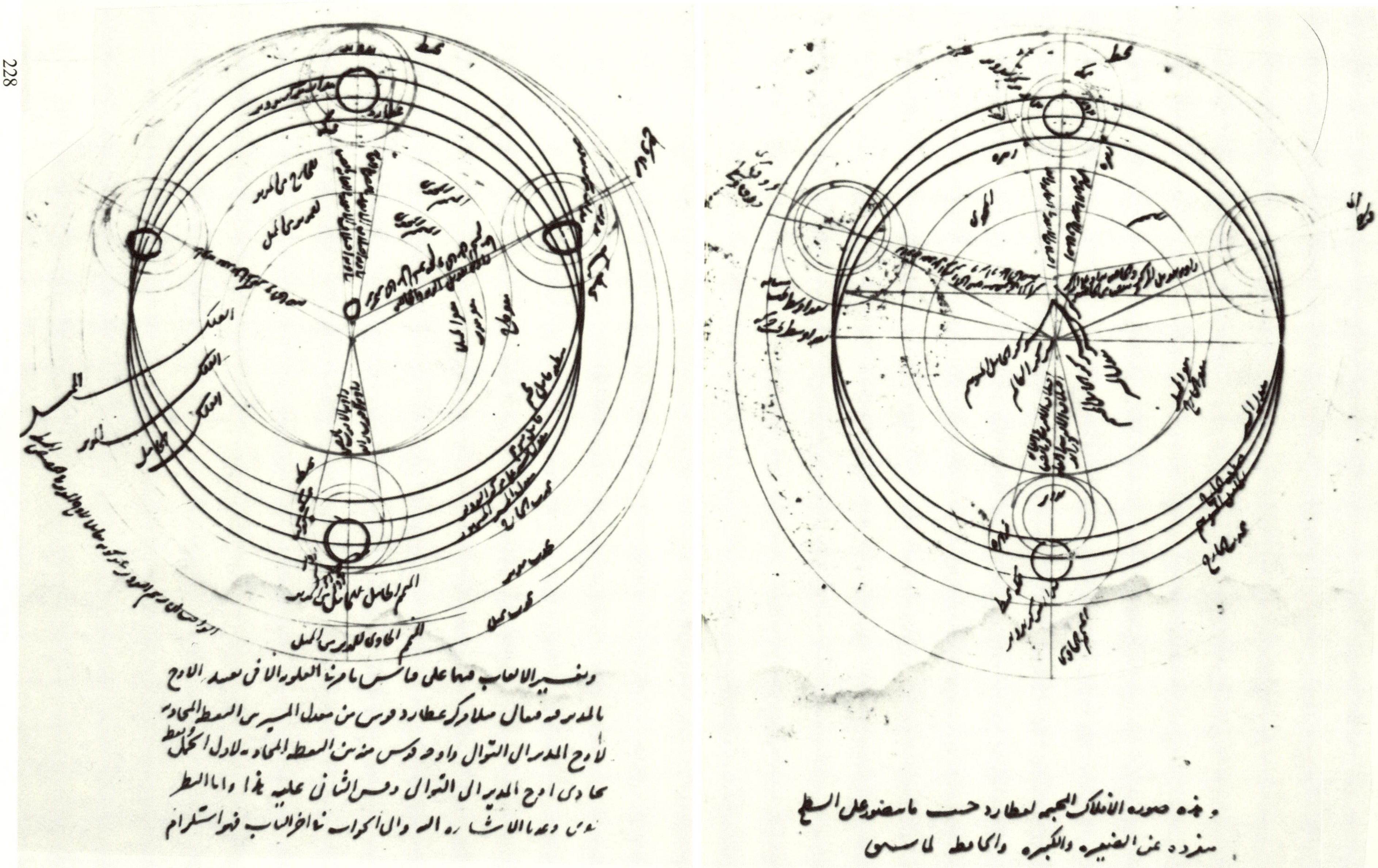

VIII Diagrams of the modified Ptolemaic model for Mercury in al-Shīrazī's treatise on theoretical astronomy

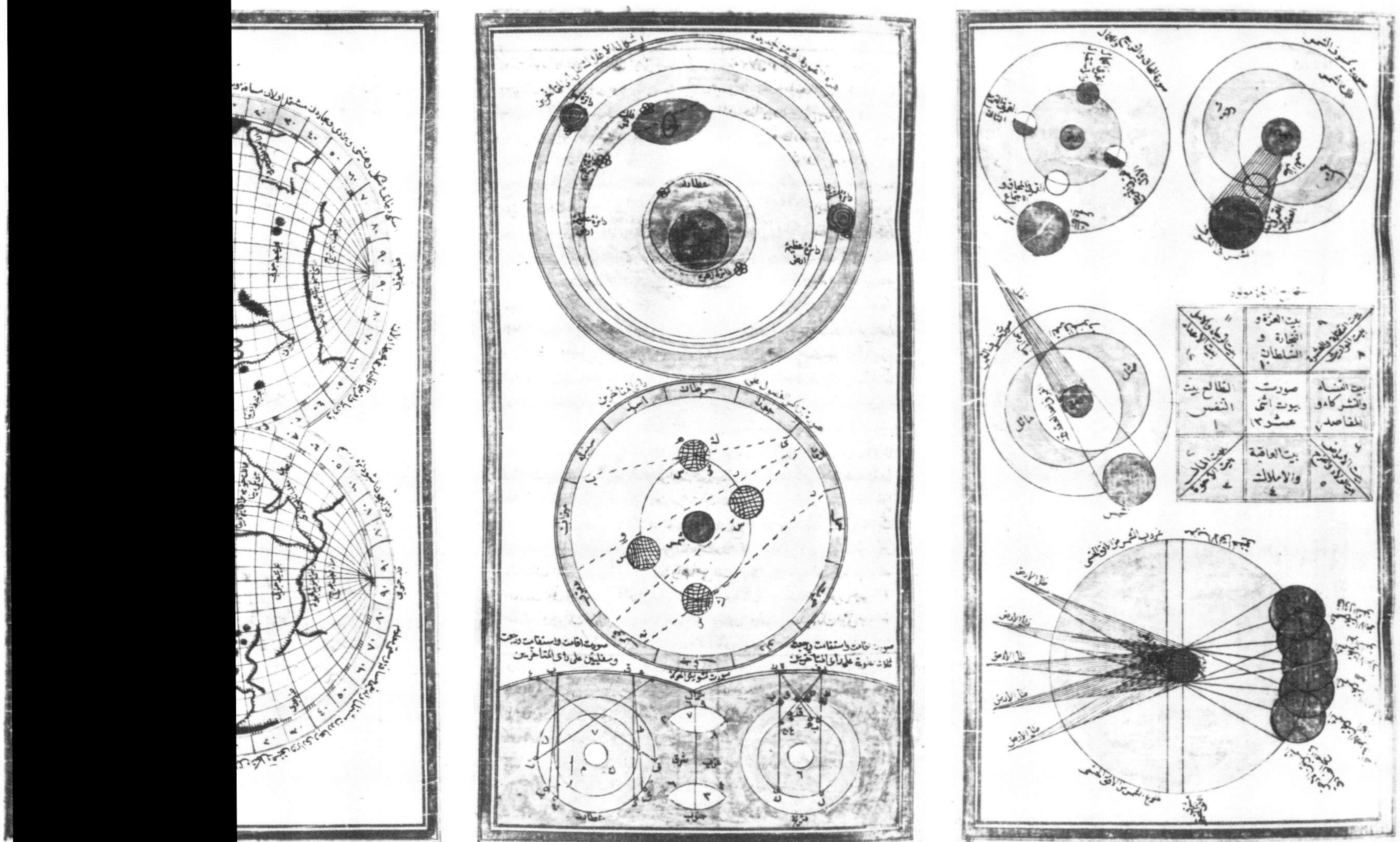

I[redacted]d diagrams of the solar/planetary system in the encyclopaedia of Ibrāhīm Ḥaqqī

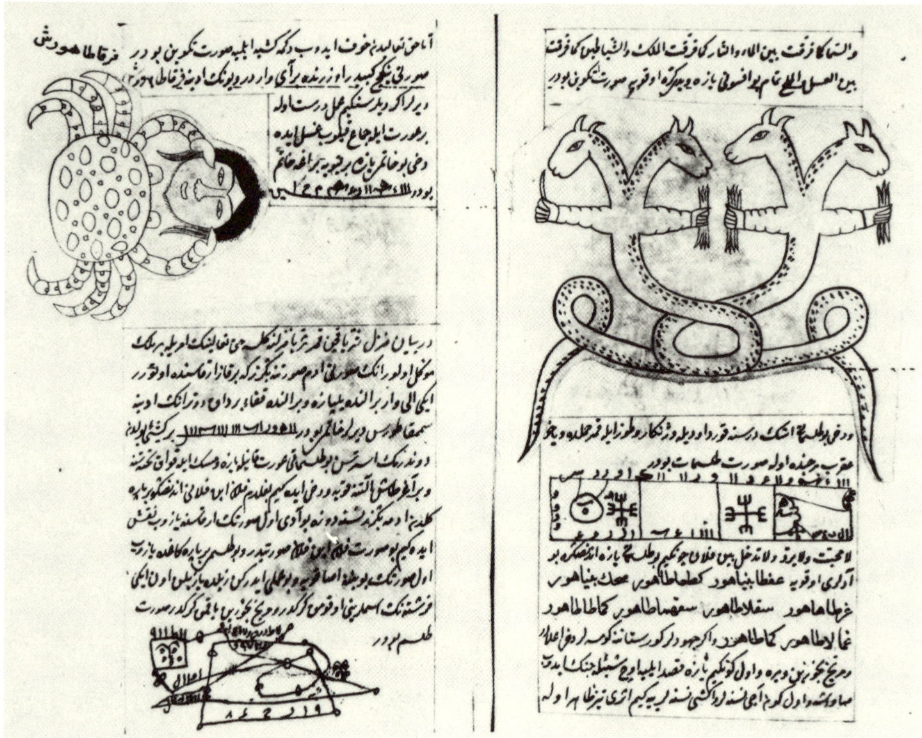

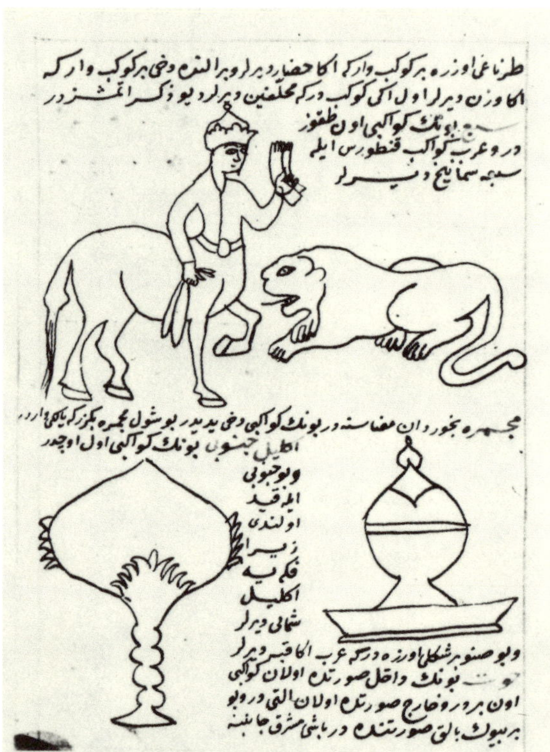

X Illustrations of the constellation figures in a late Turkish work on the fixed stars

[redacted] commentary on the astronomical handbook of al-Khwārizmī

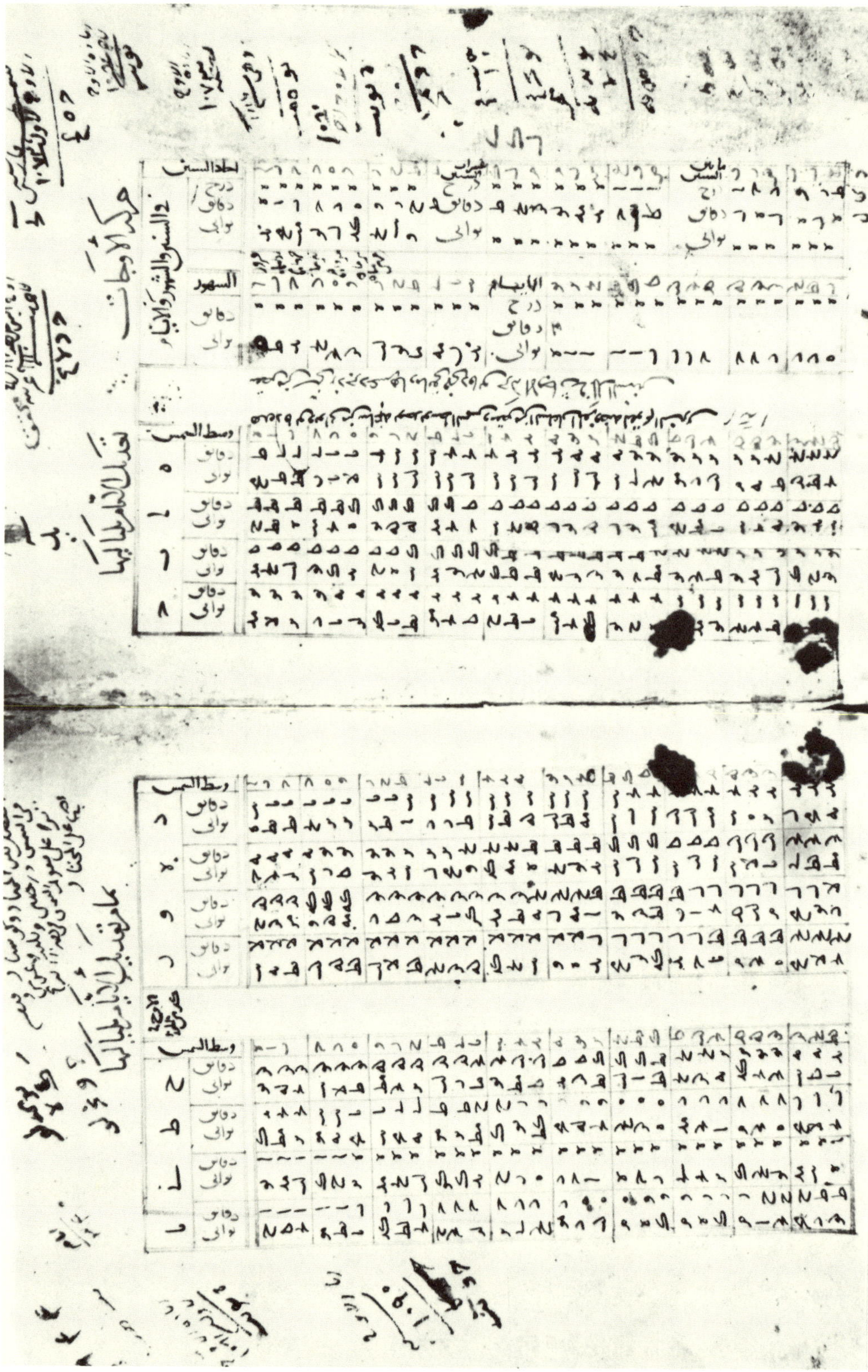

XII Sample tables from the astronomical handbook of Kūshyār ibn Labbān

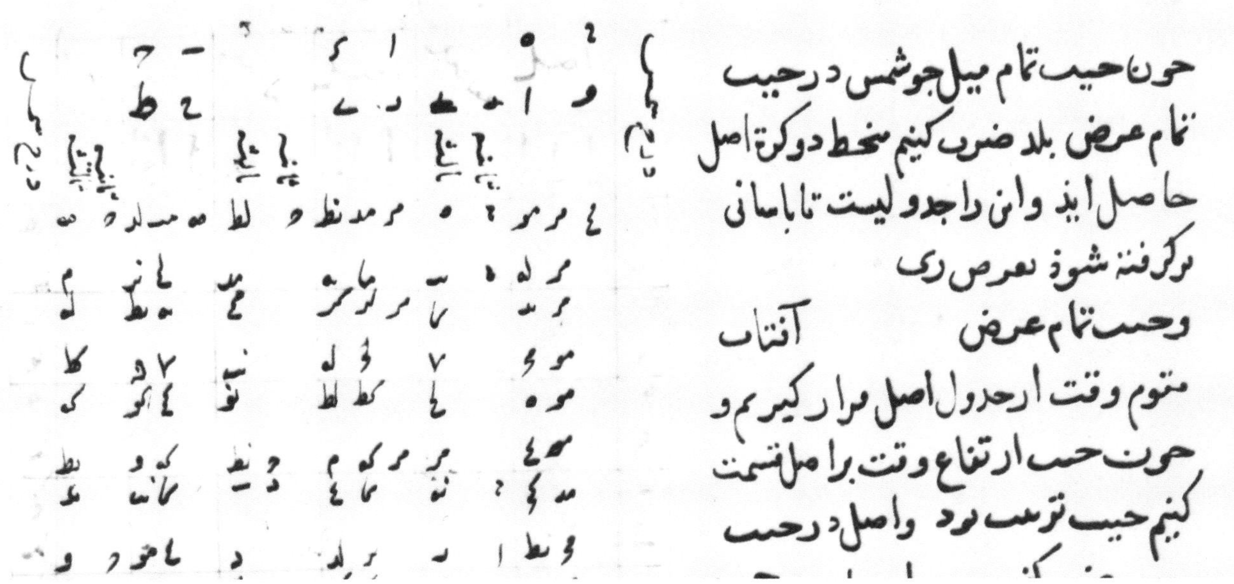

XIIIa A table for timekeeping due to al-Nasawī in the compendium of his student, Shāhmardān Rāzī

[redacted]

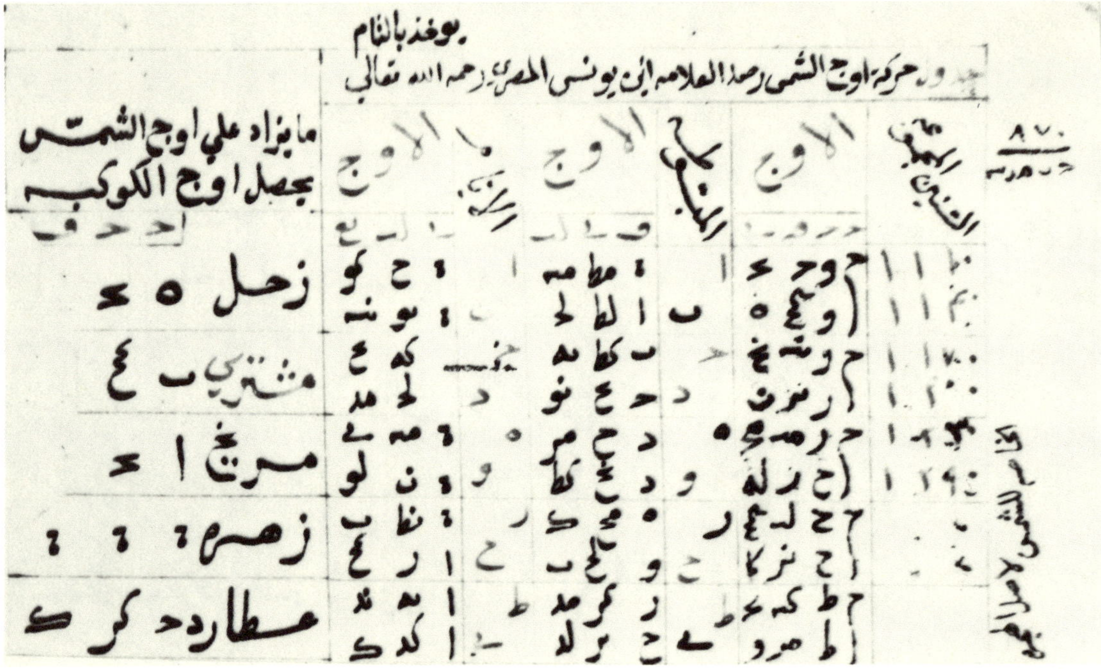

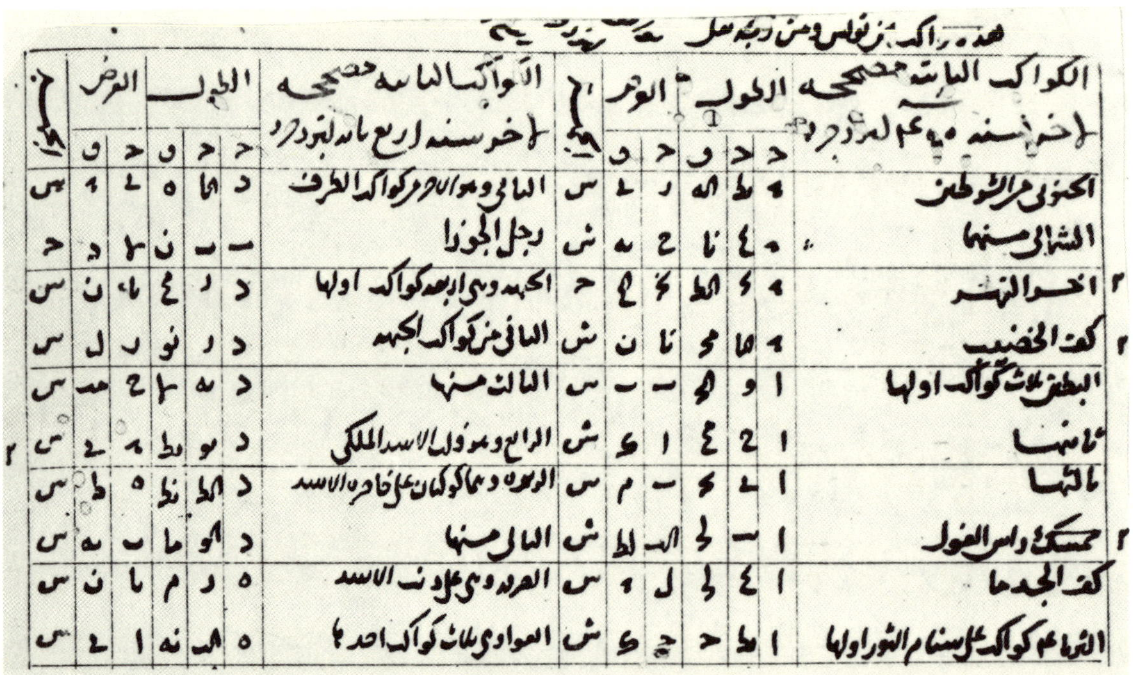

XIV Planetary tables based on the parameters of Ibn Yūnus and his star catalog

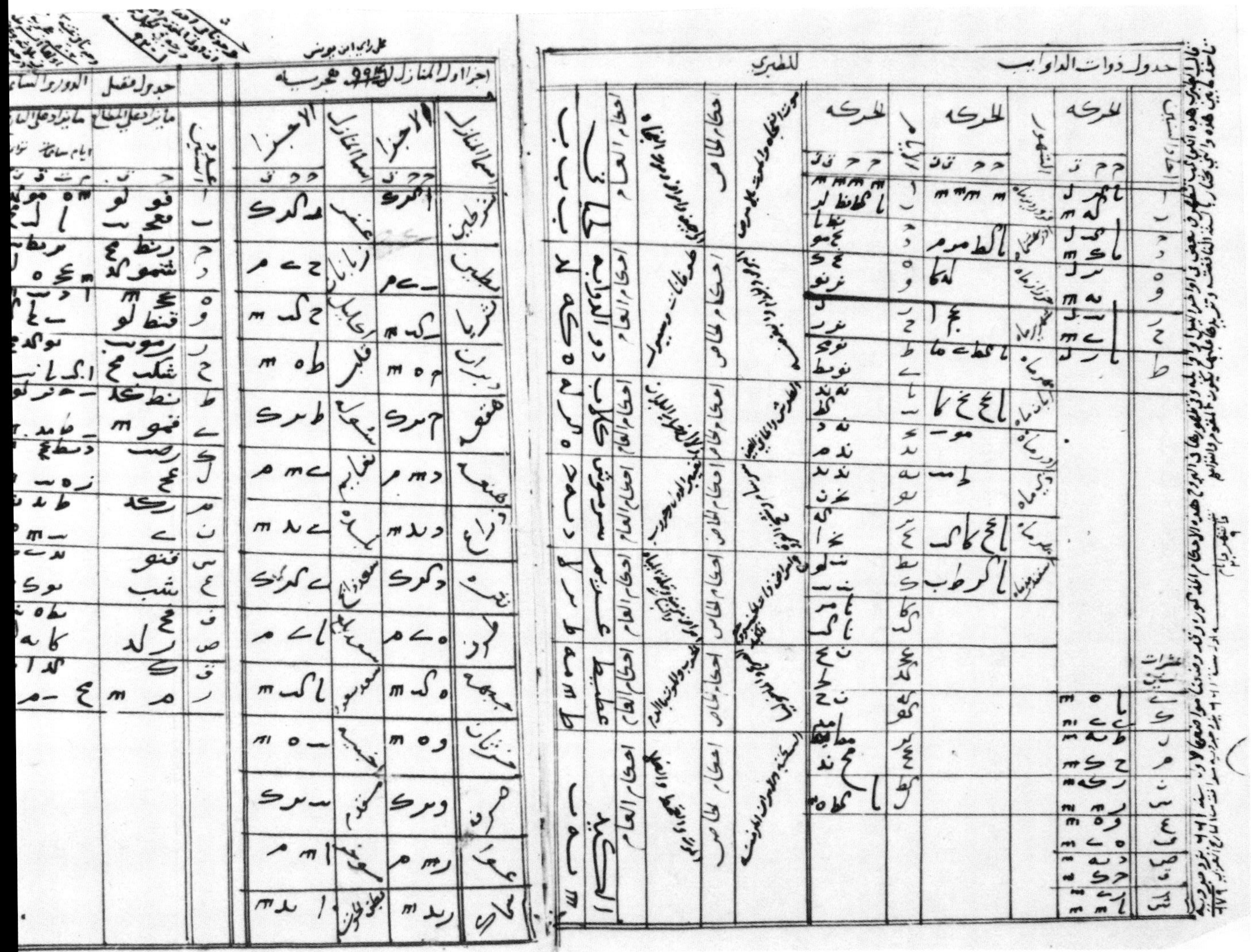

[redacted]tributed to al-Ṭabarī and Ibn Yūnus in a late compendium

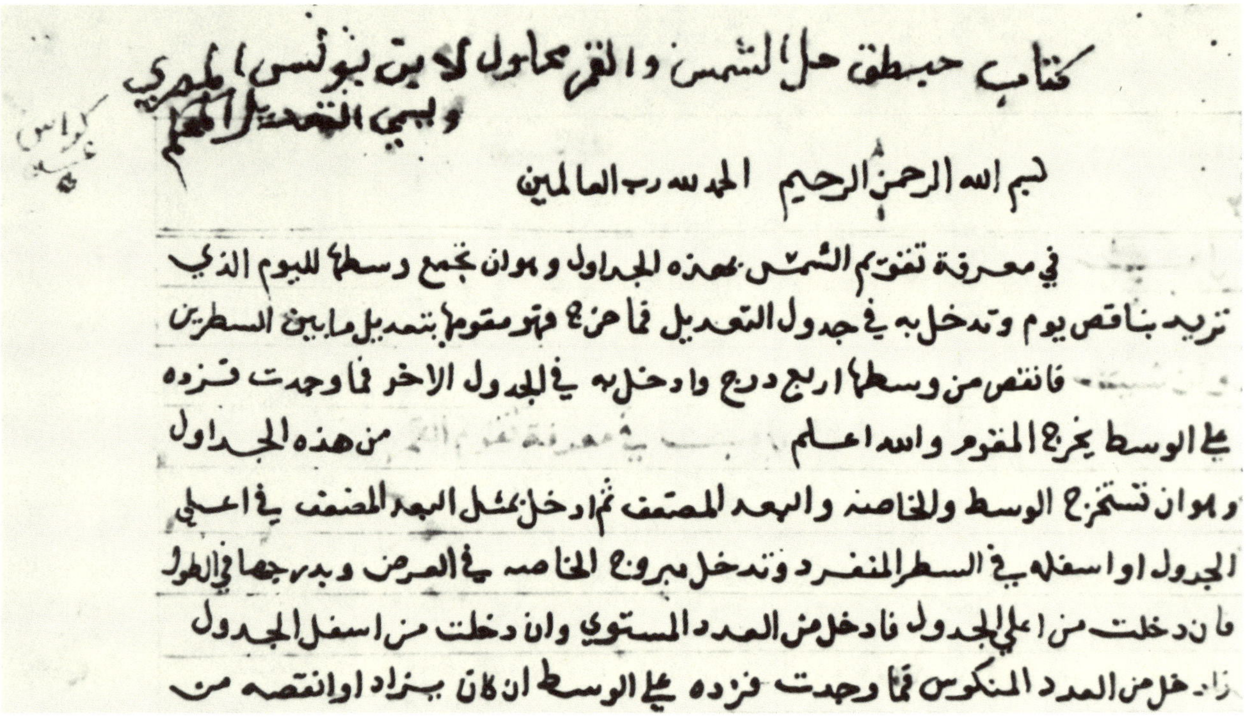

كتاب حبسطق حل الشمس والقمر بجداول لابن يونس المصري

بسم الله الرحمن الرحيم الحمد لله رب العالمين

في معرفة تقويم الشمس بهذه الجداول وهو ان تجمع وسطها لليوم الذي
تريده بناقص يوم وتدخل به في جدول التعديل فما خرج فهو مقوم بتعديل ما بين السطرين
فانقص من وسطها اربع درج وادخل به في الجدول الاخر فما وجدت فزده
على الوسط يخرج المقوم والله اعلم من هذه الجداول
وهو ان تستخرج الوسط والخاصة والبعد المضعف ثم ادخل بمثل البعد المضعف في اعلى
الجدول او اسفله في السطر المنفرد وتدخل ببروج الخاصة في العرض وبدرجها في الطول
فان دخلت من اعلى الجدول فادخل من العدد المستوي وان دخلت من اسفل الجدول
فادخل من العدد المنكوس فما وجدت فزده على الوسط ان كان يزاد او انقصه من

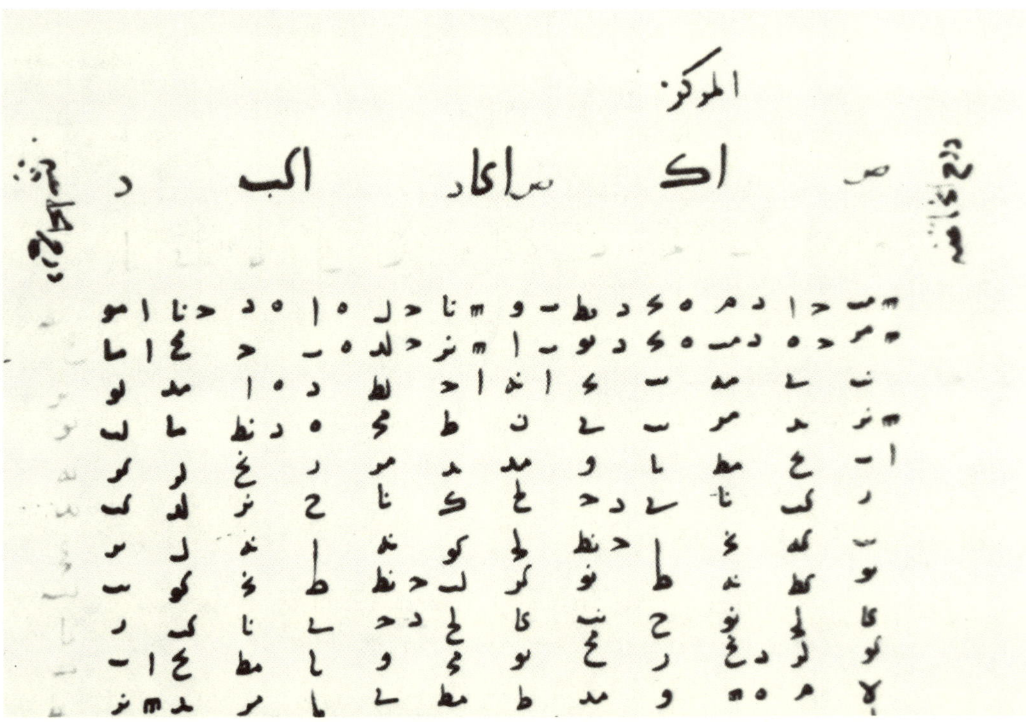

XVI Some double argument tables for the lunar equation attributed to Ibn Yūnus

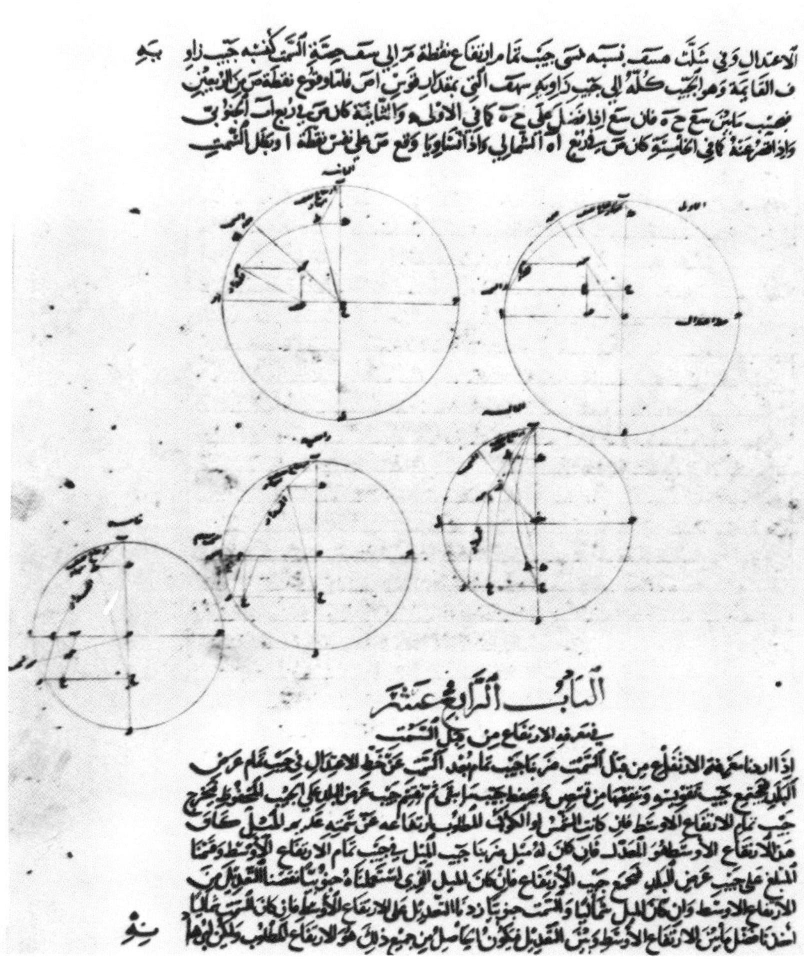

[redacted] for problems of spherical astronomy in the astronomical handbook of al-Bīrūnī

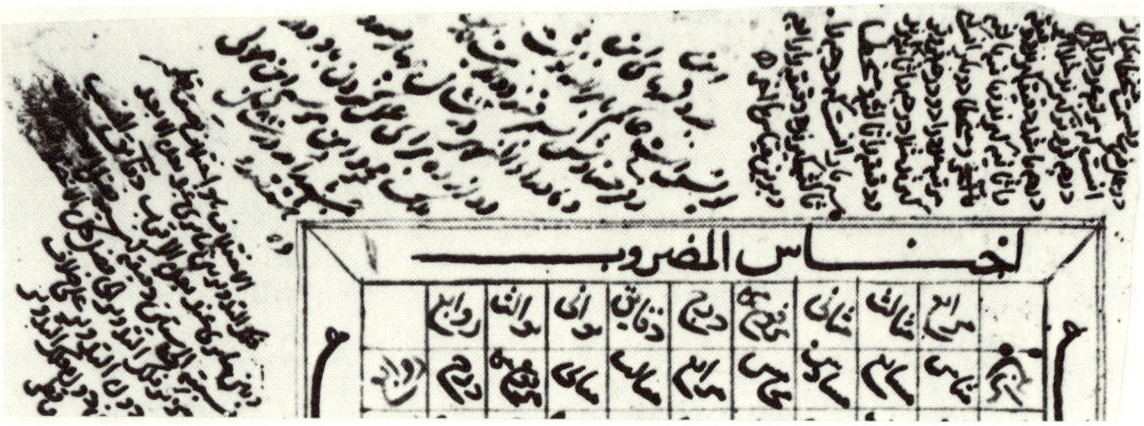

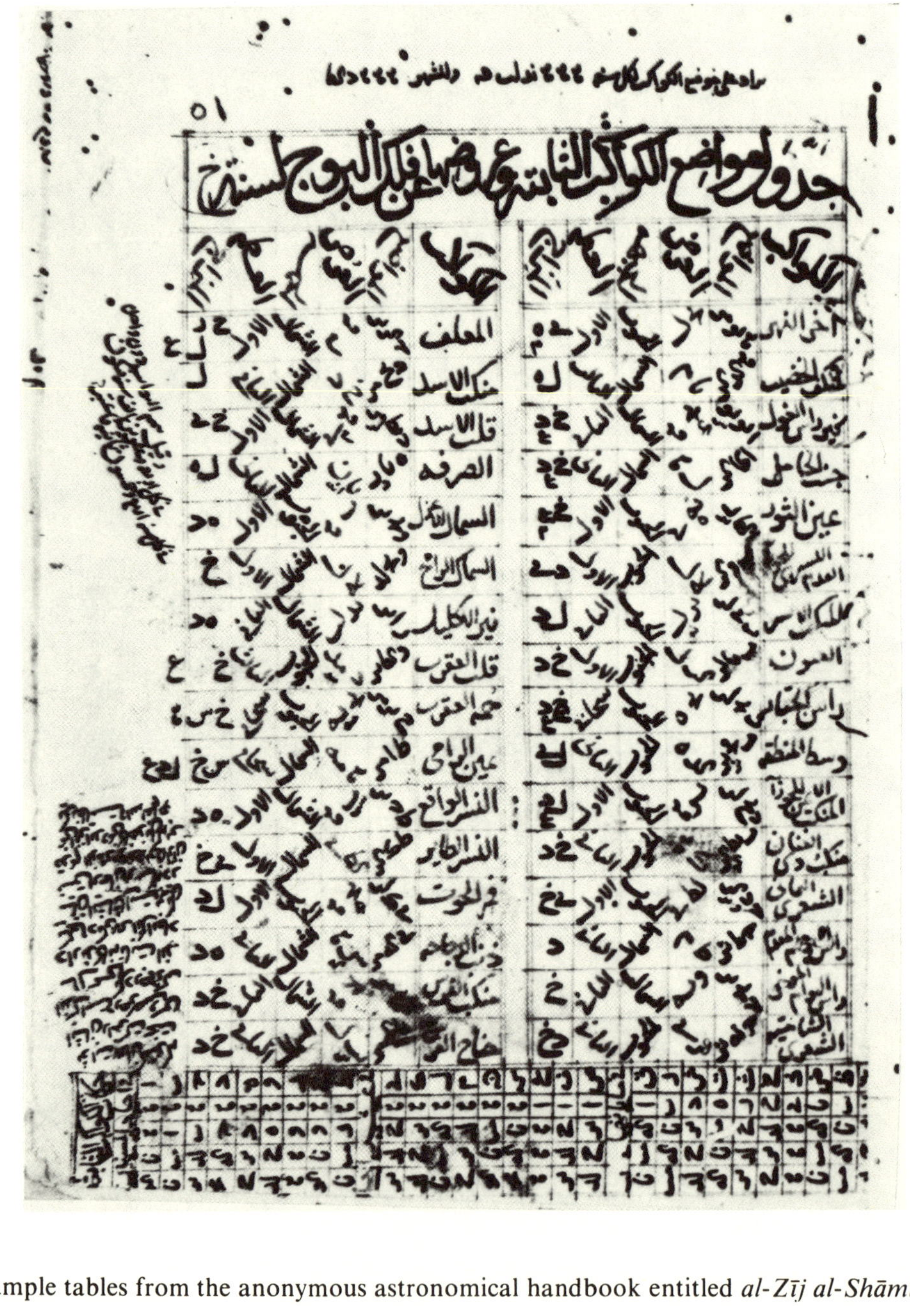

XVIII Sample tables from the anonymous astronomical handbook entitled *al-Zīj al-Shāmil*

XIXa A table for timekeeping for Maragha from the astronomical handbook of Naṣīr al-Dīn al-Ṭūsī

XIXb A late Egyptian star catalog apparently based on an earlier one by al-Saḥīṭī

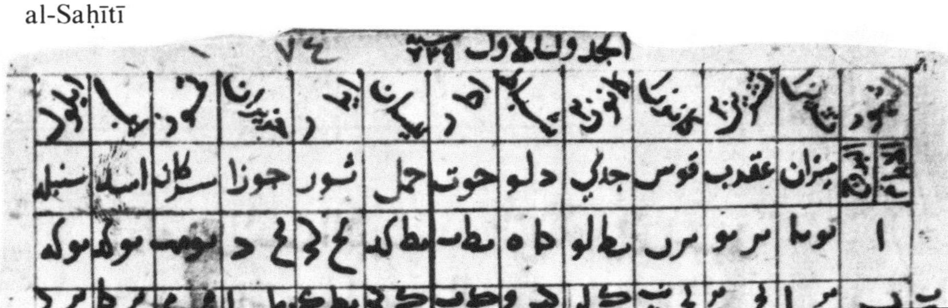

[redacted]

a.

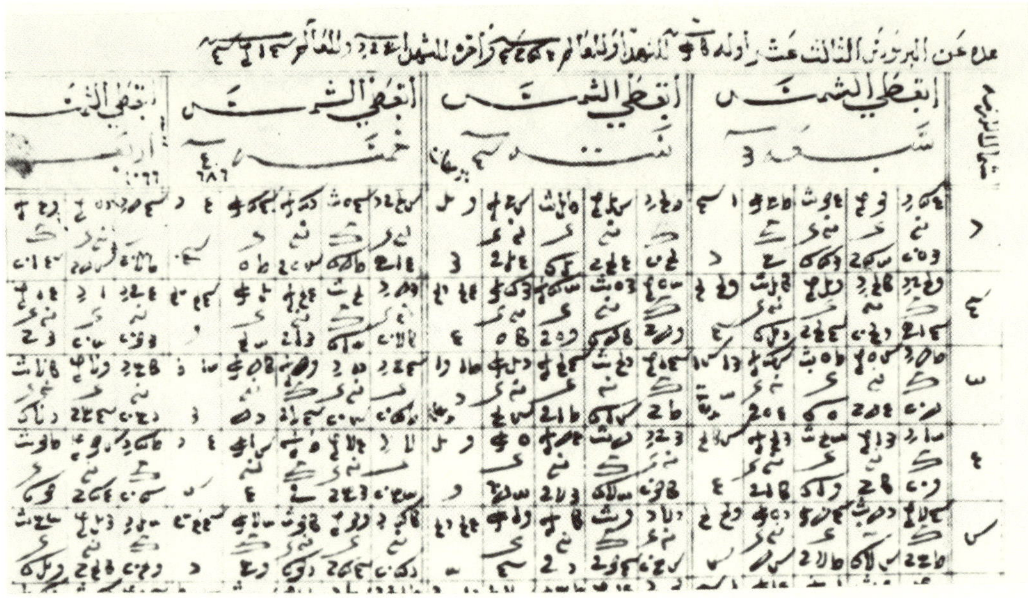

b.

c.

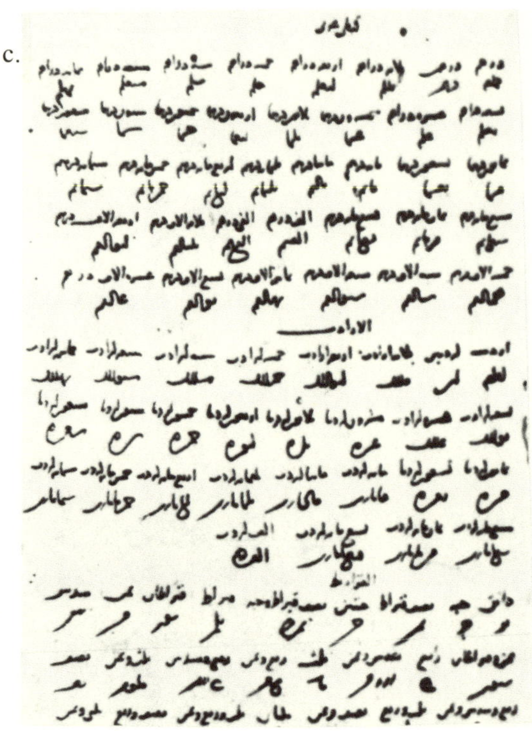

XXa–b The astronomical tables of the Coptic scholar al-As^cad ibn al-^cAssāl

XXc An extract from a list of Coptic weights and measures

[redacted]cts from an anonymous Yemeni ephemeris and almanac

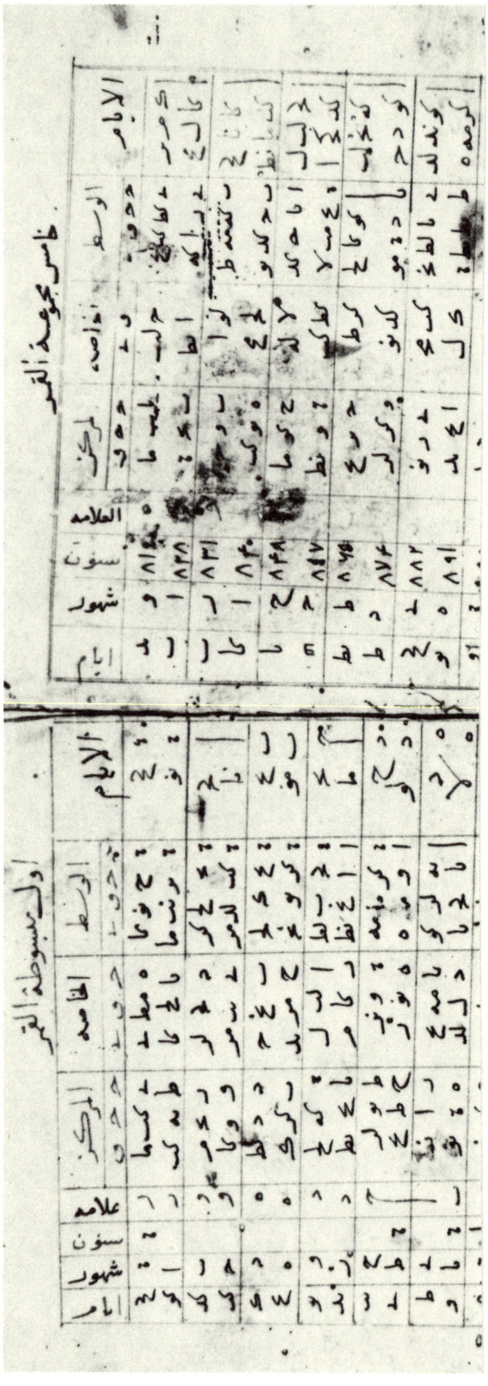

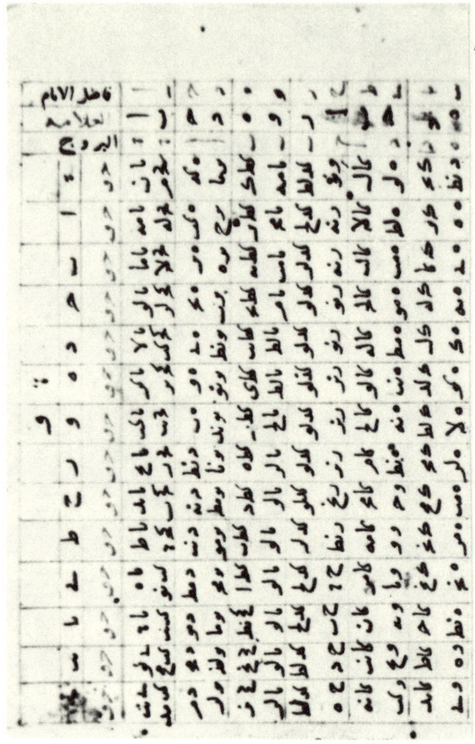

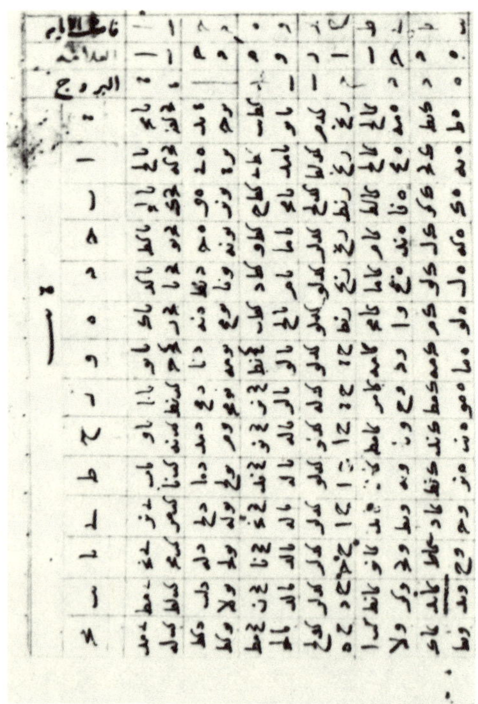

XXII The auxiliary tables for generating ephemerides by Ibn al-Majdī

رسالة أولى على زيج زكوت

بسم الله الرحمن الرحيم وصلى الله على سيدنا ومولانا محمد وآله وصحبه وسلم

الحمد لله مجري الأفلاك والصلاة والسلام على سيدنا ومولانا محمد الذي عم برسالته الجن والانس والأملاك. تسوية البيوت الاثني عشر 1472

الأصل لهذا الزيج ألف وأربعمائة واثنان وسبعون للمسيح عيسى عليه السلام فإذا أردت معرفة الطالع وتسوية البيوت فانظر في أول الجداول تجد اثني عشر جدولا على عدد الشهور وابتداؤها من مارس وآخرها فبراير فإذا أردت الطالع فادخل باليوم من الشهر الذي أنت فيه وخذ ما بحياله من الدرج إن كان الوقت نصف النهار وأما إن كان بعد نصف النهار فانظر كم ساعة مضت بعده وزدها على الساعات الواجبة وادخل الجميع تجد الطالع أمامها وإن كنت قبل نصف النهار فأسقط الساعات منها وادخل في الأيام فإن كانت الساعات التي عندك أكثر من ساعات الجدول فزد لها أربعا وعشرين واطرح من المجموع الساعات التي عندك وما بقي ادخل به في الساعات كما تقدم واعلم أنه إذا مضى مائة وستة وثلاثون سنة بعد تاريخ الزيج اسقط من أيام الشهر الذي أنت فيه يوما واحدا وادخل بما قبله وتعرف باقي البيوت مما بإزاء الطالع في تمام السنة الأولى والسنة الباقية عن يمينها بالمقابلة للانتظار

تعديل موضع الشمس الحقيقي 1472

اسقط الأصل المذكور من تاريخ المسيح واقسم الباقي على أربعة فما خرج فأدوار وما بقي فادخل به في جدوله بالشهر في عرضه وباليوم في طوله وخذ ما يجتمعان عليه واحفظه ثم اضرب عدد الأدوار في دقيقة وستة وأربعين ثانية وزد الخارج على المحفوظ يكن موضع الشمس الطبيعي أو ادخل بعدد الأدوار في الجدول الموضوع لها وهو من واحد إلى أربع وثلاثين وزد الخارج على المحفوظ الأول يكن موضع الشمس وإذا كانت معك ساعات بعد زوال اليوم تمام الأيام فخذ موجب اليوم كما تقدم وموجب [illegible]

[illegible]

وهي في جدول [illegible]

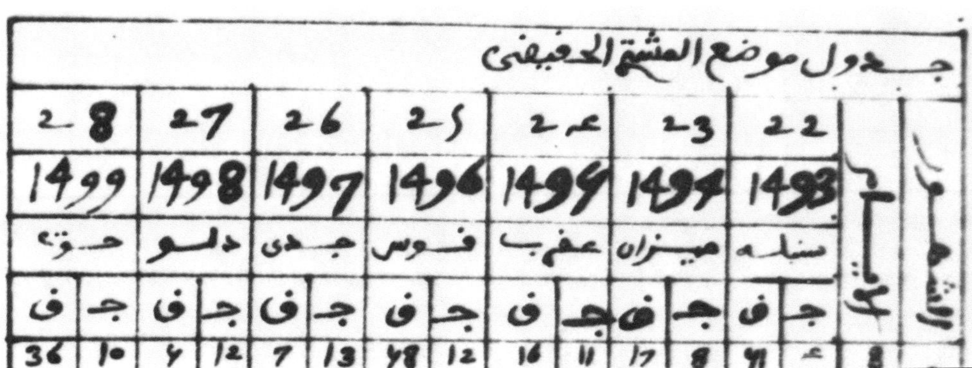

جدول موضع الشمس الحقيقي															
28		27		26		25		24		23		22		تاريخ	الشهور
1499		1498		1497		1496		1495		1494		1493			
حوت		دلو		جدي		قوس		عقرب		ميزان		سنبلة			
ف	ج	ف	ج	ف	ج	ف	ج	ف	ج	ف	ج	ف	ج		

[redacted]

a.

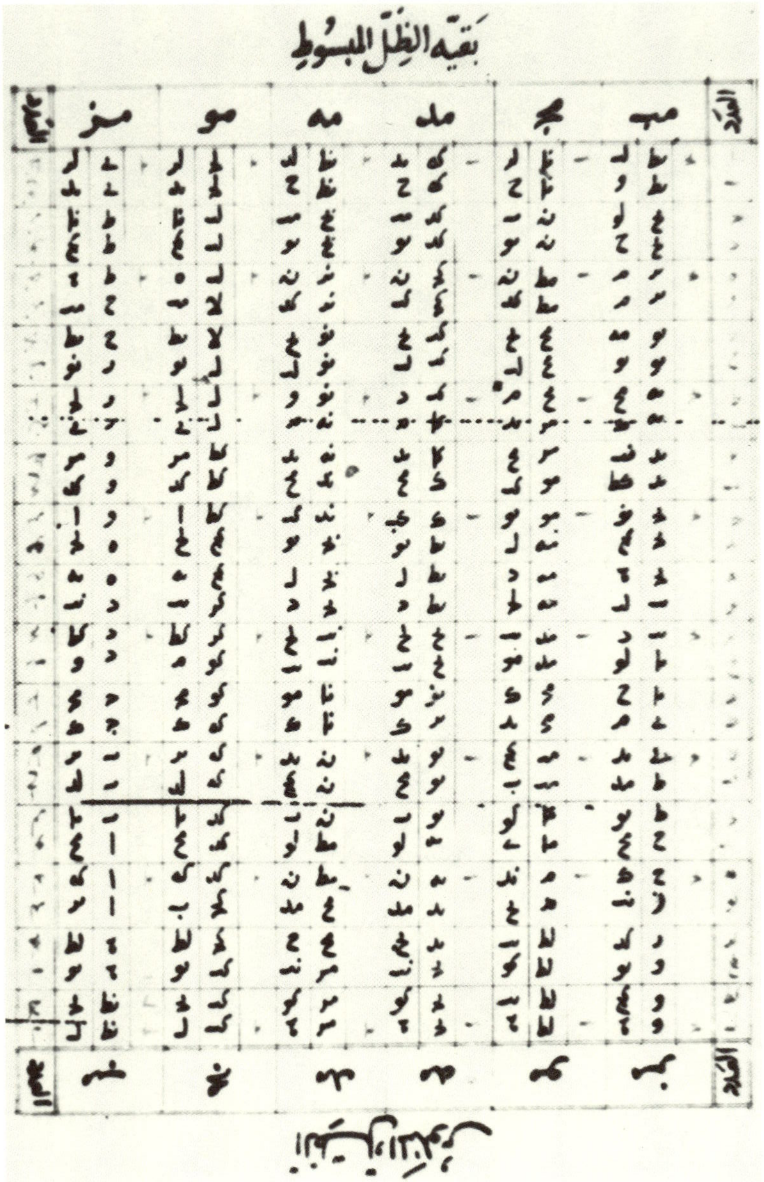

b.

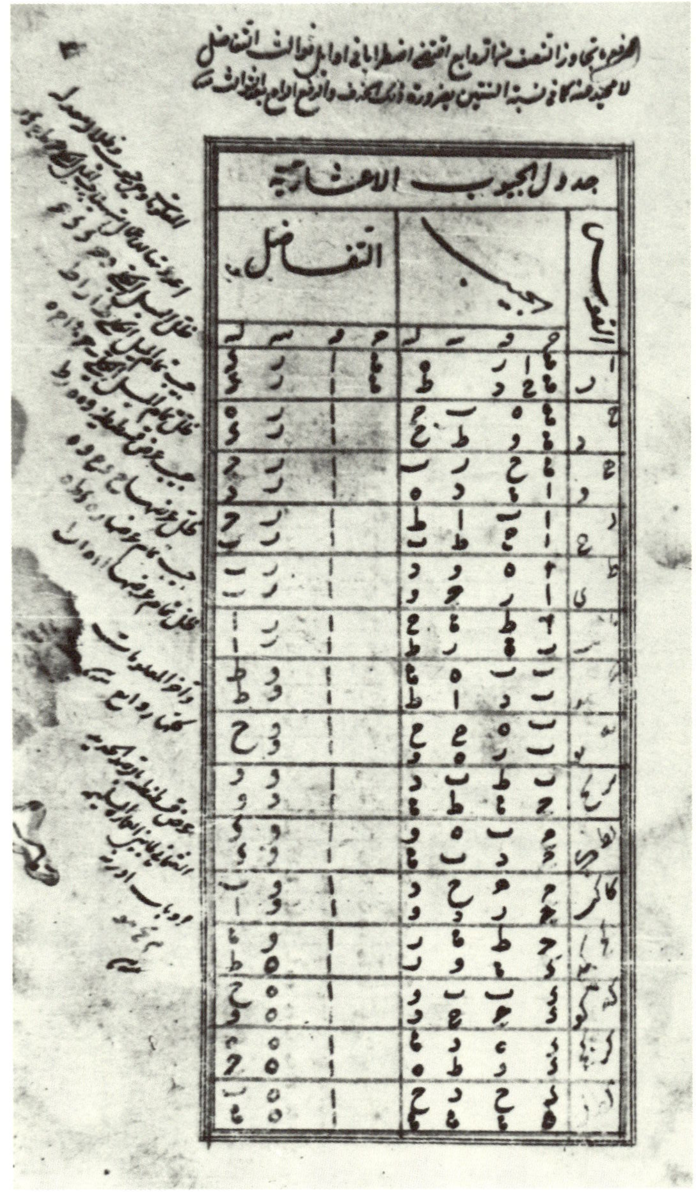

XXIV Extracts from the trigonometric tables of Sāṭilmish and Taqi l-Dīn

a.

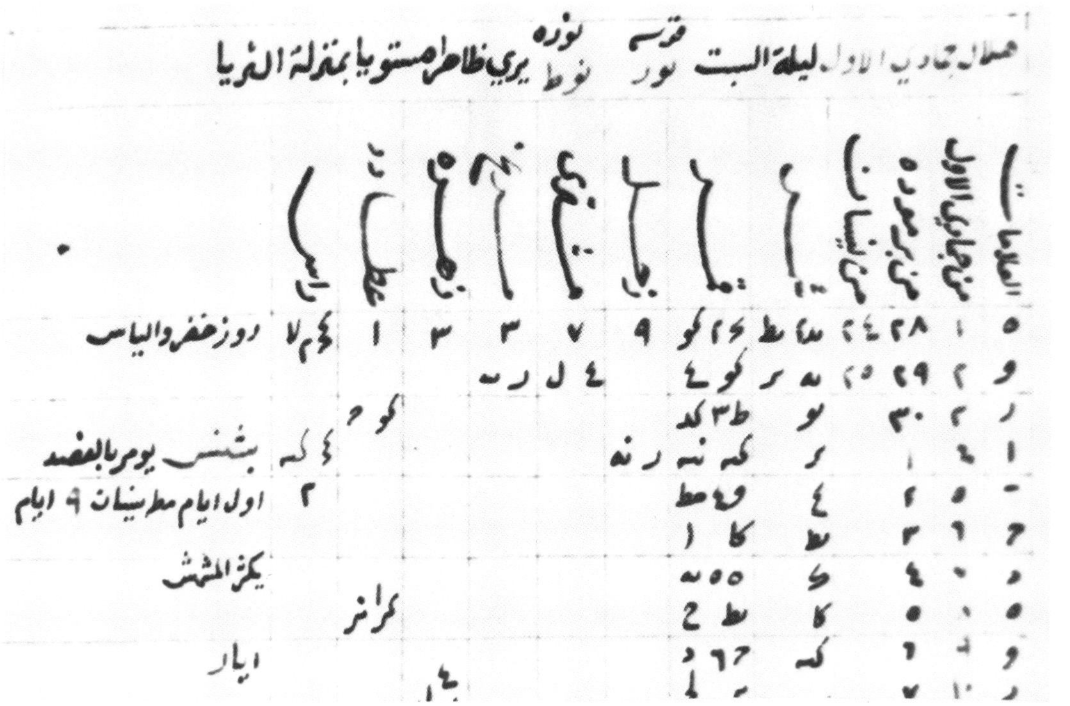
هلال جمادي الاول ليلة السبت نور يري ظاهرا مستويا بمنزلة الثريا
روز خضر والياس
اول ايام مطرينيات ٩ ايام
يكثر المشمش
ايار

b.

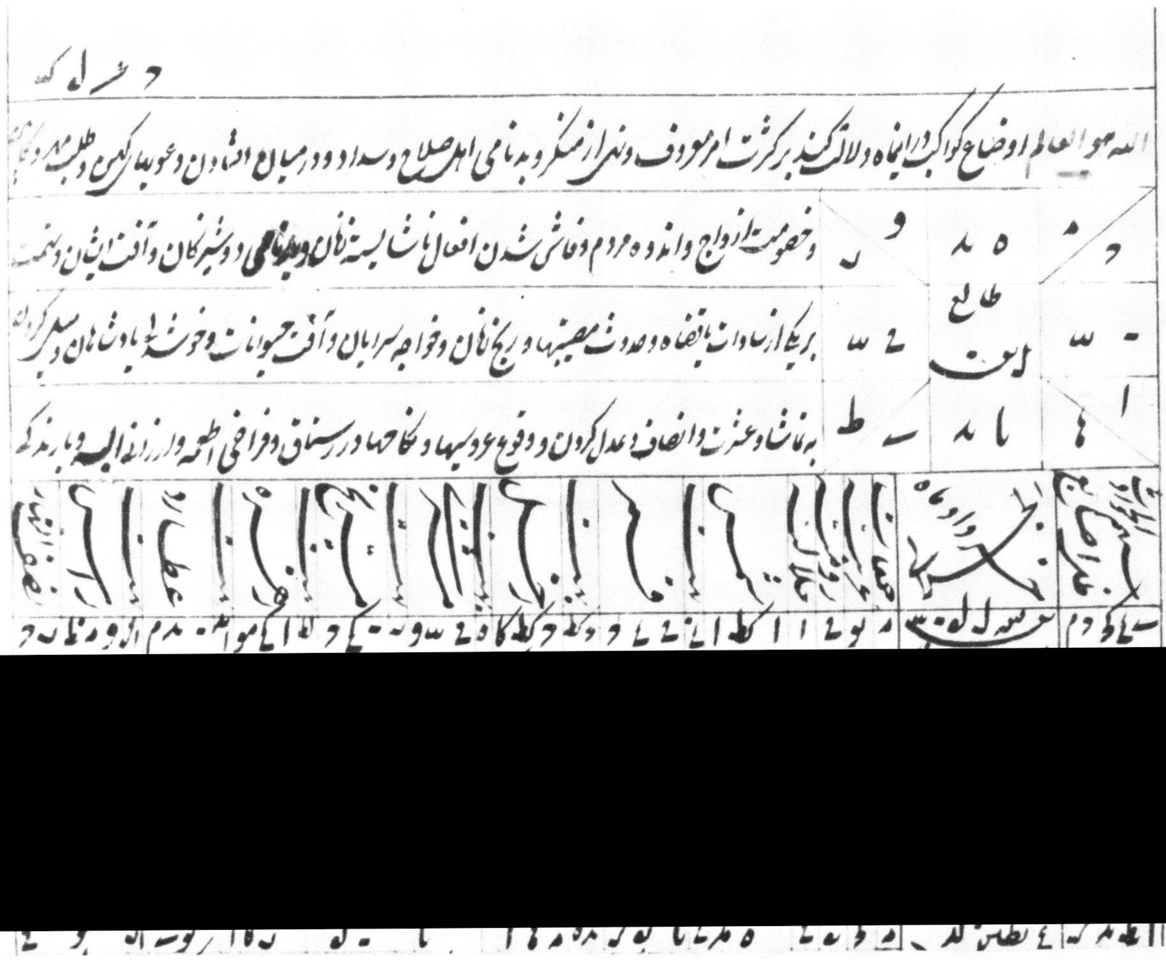

[redacted]

a.

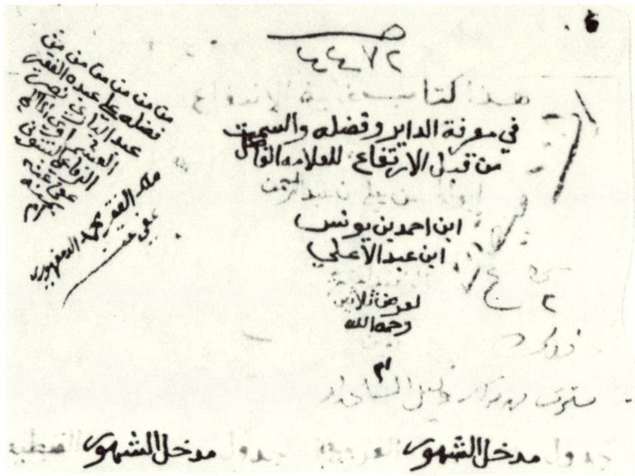

b.

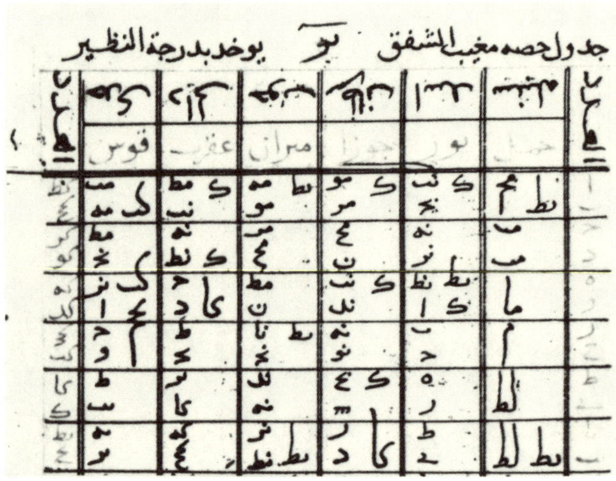

c.

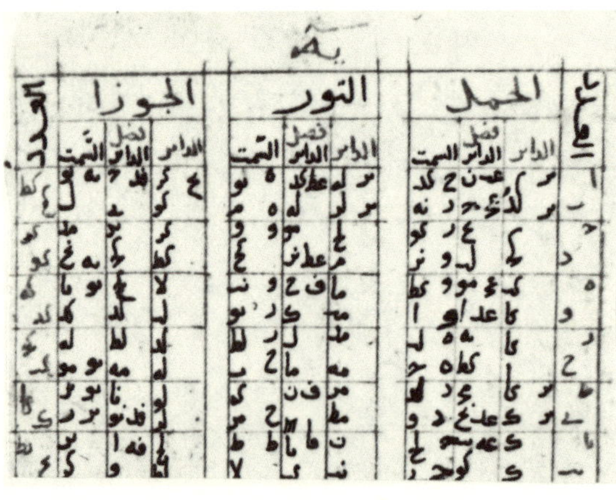

XXVI Extracts from the main corpus of tables for time-keeping used in medieval Cairo

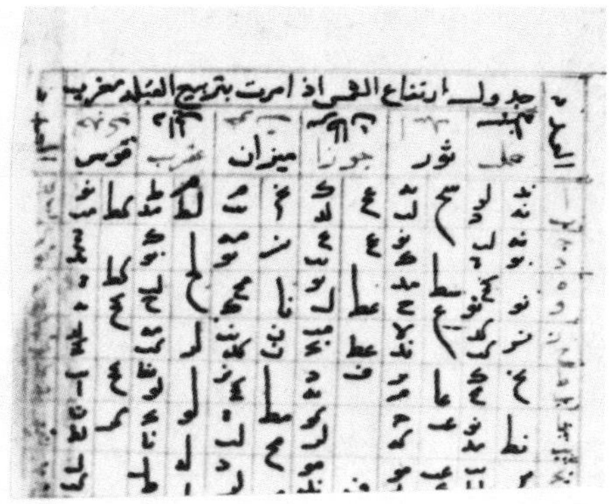
جدول ارتفاع الشمس

جدول الماضي من الزوال

دقايق اختلاف ما بين الافقين لعرض شمال

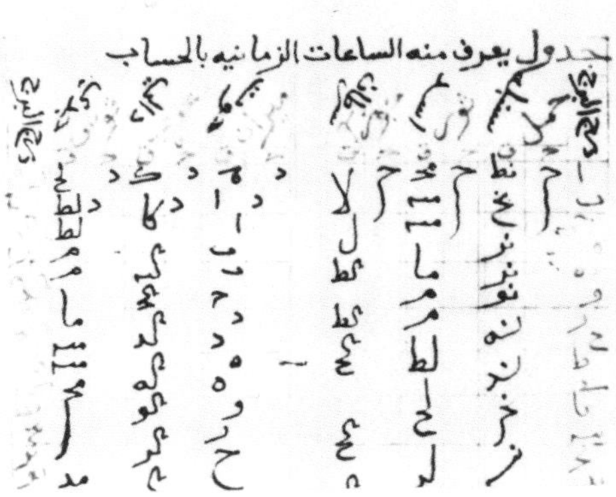
جدول يعرف منه الساعات الزمانيه بالحساب

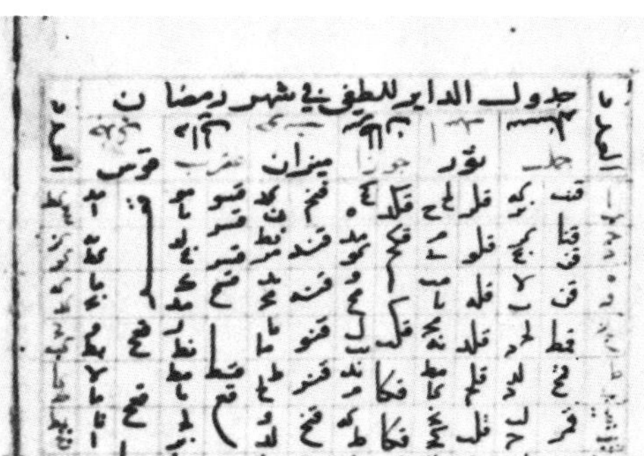
جدول الداير

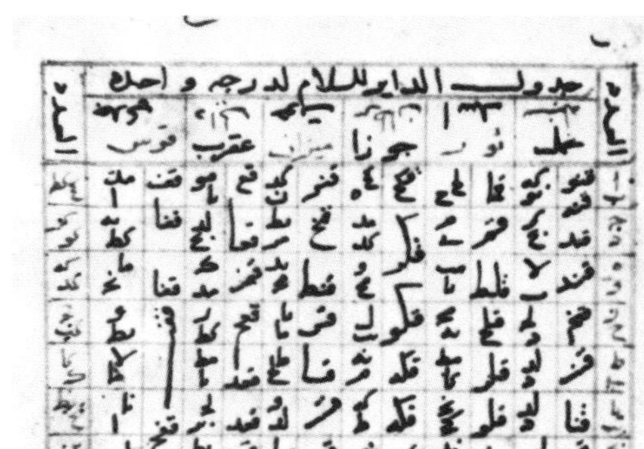
جدول الداير

[redacted]

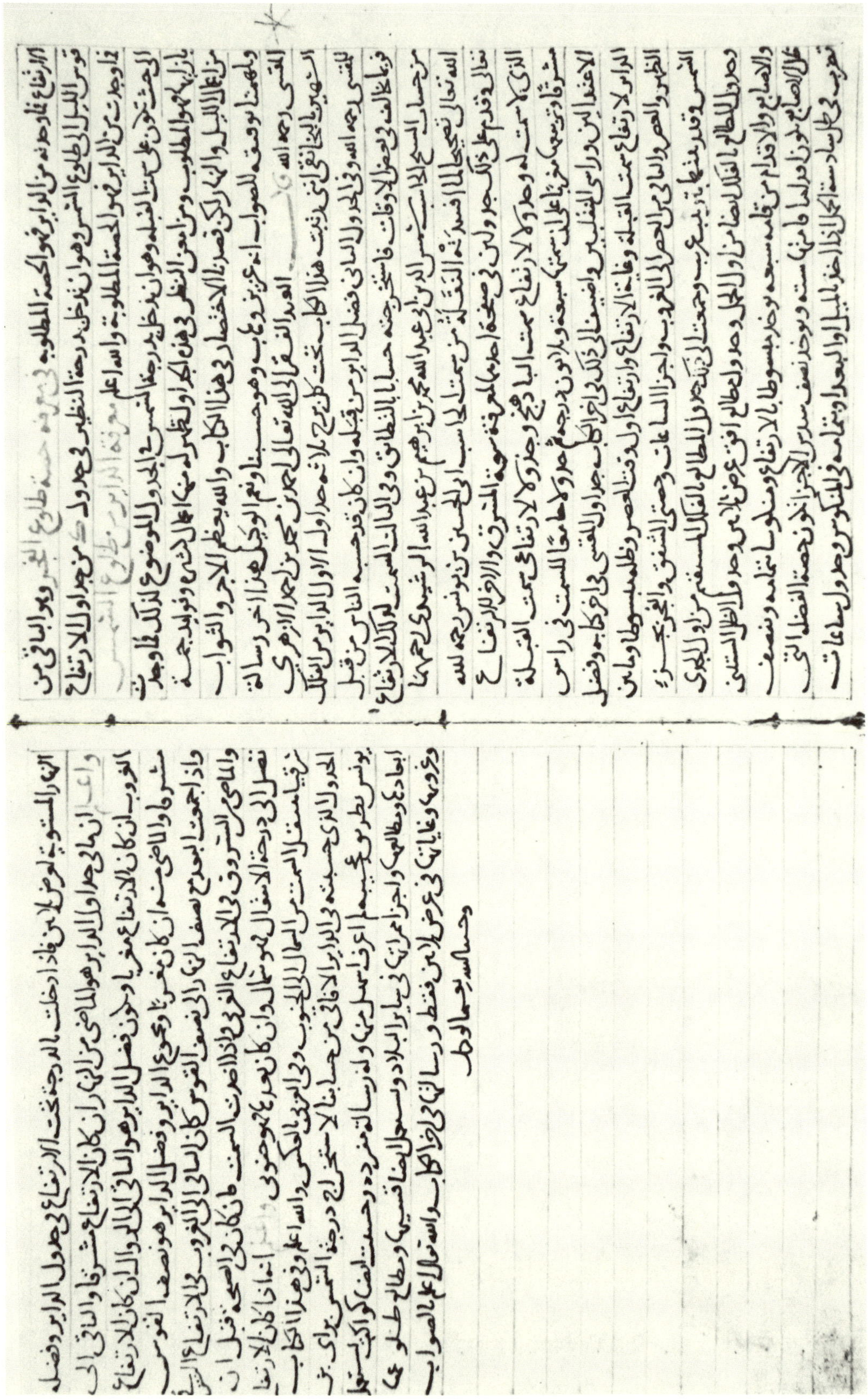

XXVIII al-Bakhāniqī's account of the way in which he put together his edition of the Cairo corpus

[redacted]

[redacted]s universal table for timekeeping by the sun and stars

XXX al-Marrākushī's list of formulae for spherical astronomy

XXXIa al-Khalīlī's tables for timekeeping by the sun for Damascus

[redacted]

XXXII al-Khalīlī's universal tables for solving all of the standard problems of spherical astronomy for any latitude

مسألة في العمل بدقائق اختلاف الافاق المدينية وهي التي حسبها الشيخ علاء الدين
ابن الشاطر المنجم الرصدي وطريق ذلك ان تعرف موضع الشمس عند الزوال وتزيد عليه
سيرها الى وقت الغروب تبقى موضعها عند الغروب فاعرف به نصف قوس نهارها وبتعديل ما بين
المطالع وتزيد دقائق اختلاف الافاق ونحصها المكتوبة تحت برج الشمس يحصل نصف قوس النهار
المحقق الذي اذا كان معك عمل صحيح فقد مراعيه يكون غيبان الشمس ولو زدت هذه الدقائق
على مطالع غروب الشمس المستخرج بمقومها عند الغروب حصل لك مطالع غروبها المحقق
ولو زدت عليها حصة الشفق حصل لك مطالع الشفق المحقق التي اذا كان معك عمل صحيح
من الغروب المشاهد بعمل توسط كوكب مطالع ذلك ولو انقصت دقائق الاختلاف
من الدائر للفجر حصل لك الدائر للفجر الذي اذا كان معك عمل احكام صحيح وعلى اخر منه كوكب
طالعة ولو زدت نصف دقائق الاختلاف على مطالع الشروق حصل لك مطالع الشروق
المحقق الذي اذا كان معك عمل صحيح يتعين مراعيه يكون قرص الشمس على وجه الارض
ولو زدت على نصف قوس النهار المستخرج بموضع الشمس عند الزوال حصل لك
نصف قوس النهار الحقيقي الذي اذا كان معك صحيح فتعين مراعيه يكون استواء الشمس
على خط المسامتة ومتى امعنت النظر في ذلك ظهر لك الحصص الذي يزاد على مقوم
الشمس عند الزوال يحصل موضعها عند الغروب

الحمل الثور الجوزا السرطان الاسد السنبلة الميزان
ه لو نو لو ه لو ه

العقرب القوس الجدي الدلو الحوت الحصص المزاد
لد ج لد

على موضع الشمس عند الغروب يبقى موضعها عند العشا الحمل ثور جوزا
د د د

السرطان الاسد السنبلة الميزان العقرب القوس الجدي الدلو الحوت
ه د د د ذ د ح ه د د

ولو زدت على حصة كل برج دقيقة واحدة حصل لك الحصة الذي
اذا زدتها على موضع الشمس عند الغروب حصل لك موضعها عند الشروق

الحصص التي تزاد على موضع الشمس عند الزوال يبقى موضعها
عند الفجر حمل ثور جوزا سرطان اسد سنبلة ميزان
مر لج لو لو لو لح بز

عقرب قوس جدي دلو حوت
مد مب مد مه مد مب

هذه دقائق اختلاف الافاق المدينية
حمل ثور جوزا سرطان اسد سنبلة ميزان عقرب قوس جدي دلو حوت
[redacted]

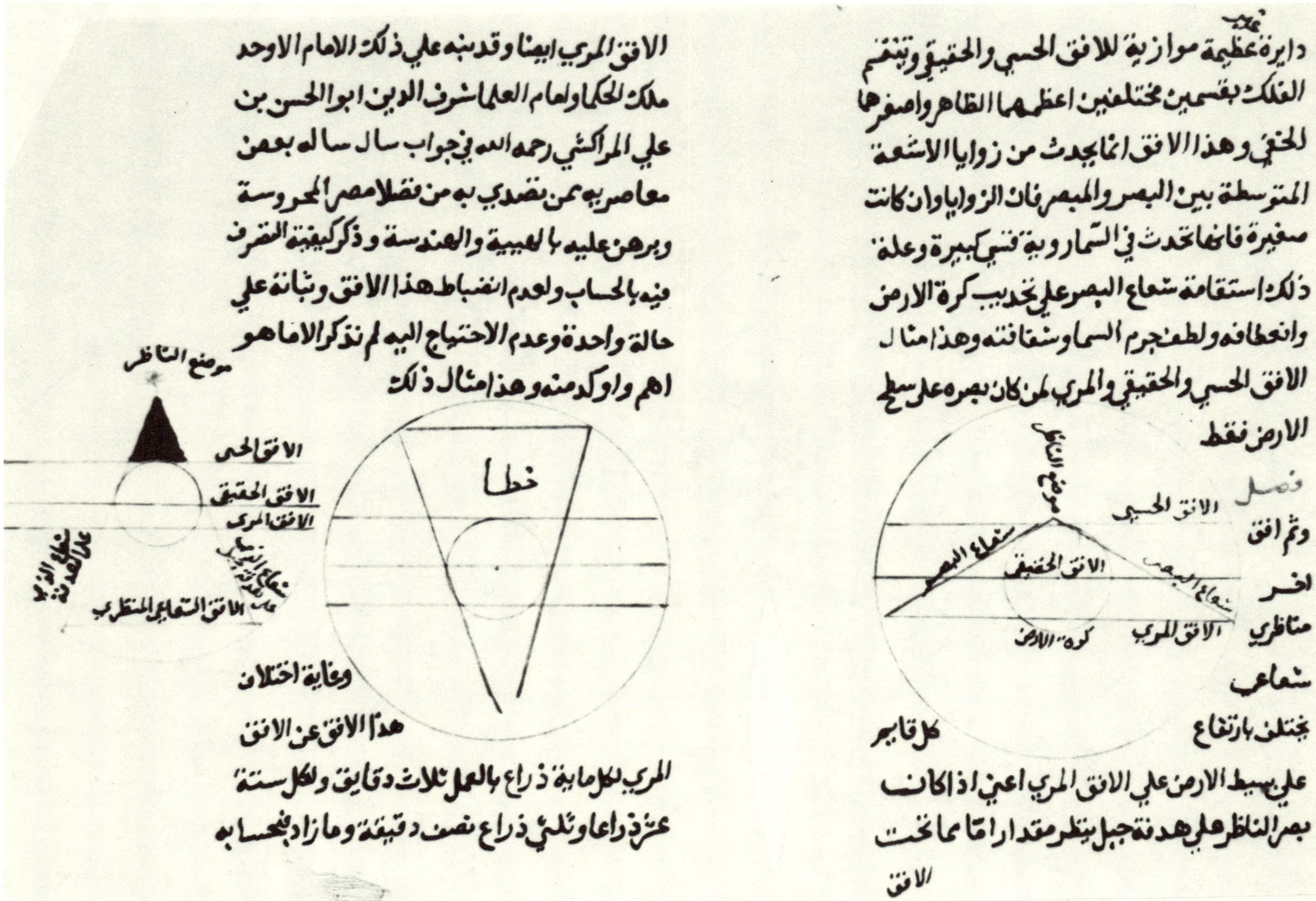

XXXIV Diagrams in a treatise on the effect of refraction at the horizon by al-Minūfī

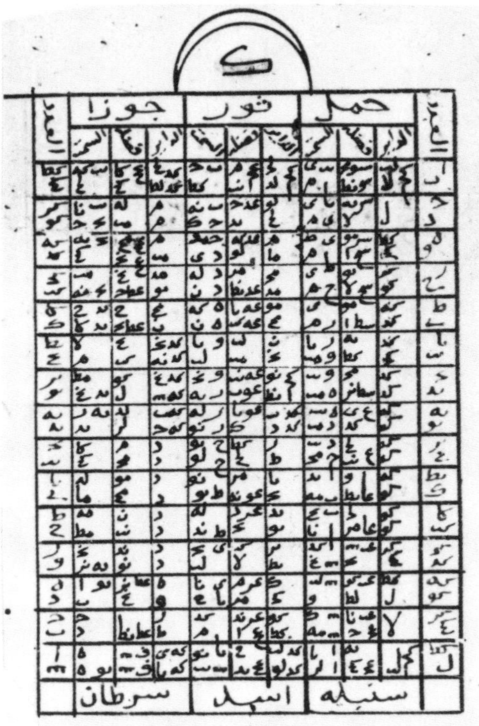
ك
حمل ثور جوزا
سنبله اسد سرطان

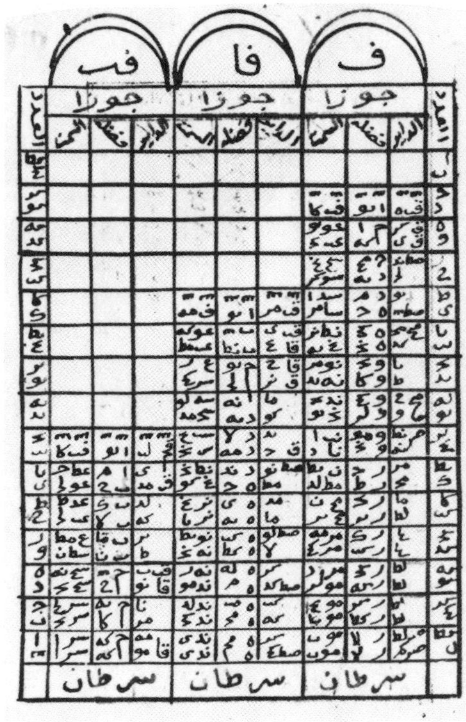
ف فا فب
جوزا جوزا جوزا
سرطان سرطان سرطان

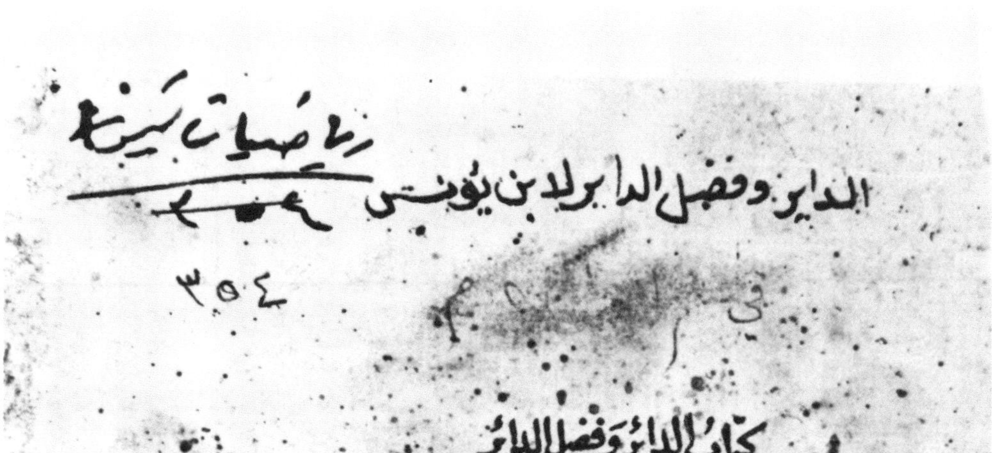
رياضيات تيمور
المدار وفضل الدائر لابن يونس
٣٥٤
كتاب الدائر وفضل الدائر

[redacted]

a.

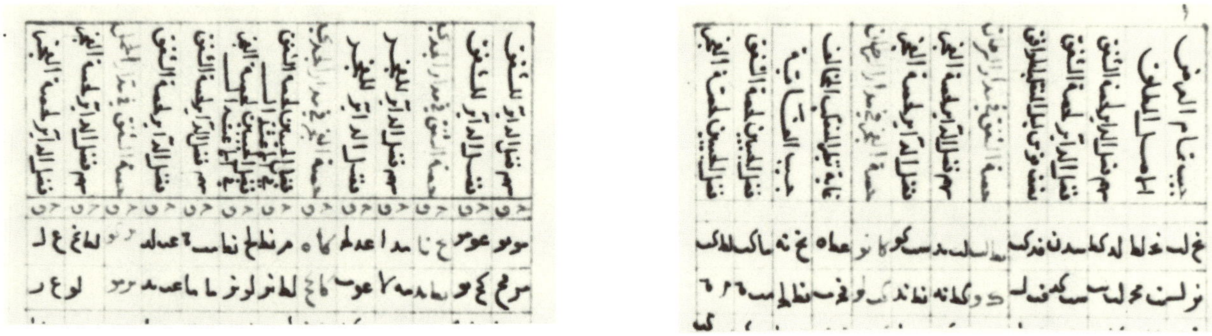

b.

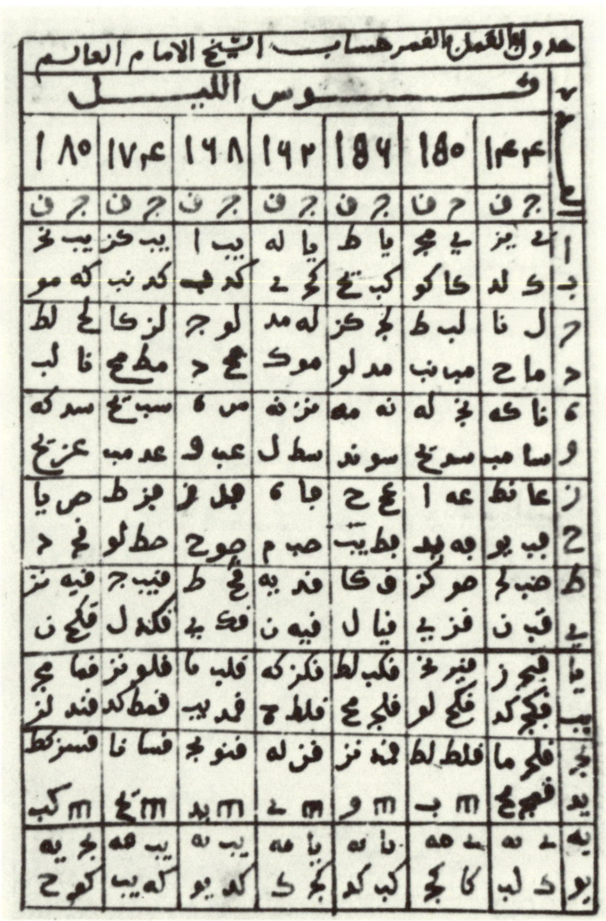

c.

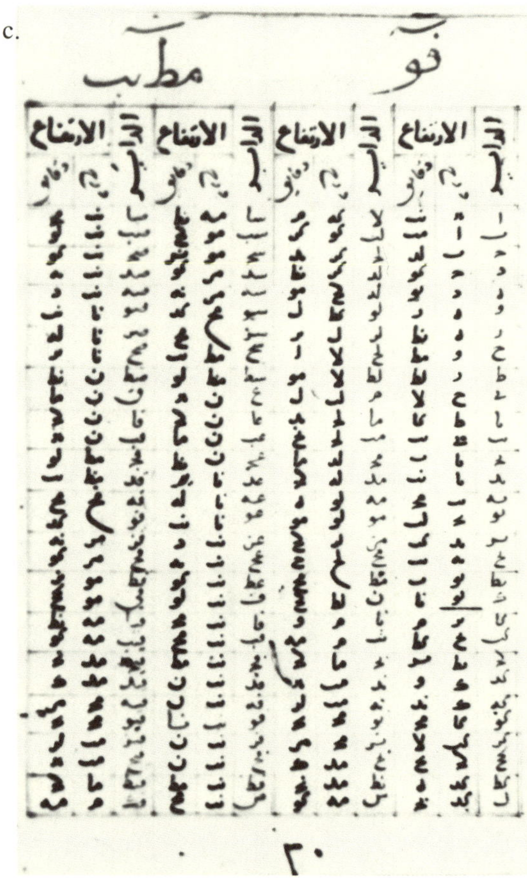

XXXVI Miscellaneous Egyptian tables for timekeeping

الحمد لله جدول بقية مطالب المنازل لحركة [illegible] لمدينة مراكش [illegible] بلد عرضه لا

[illegible]	قوس النهار				قوس الليل				ما بين الزوال والغروب			ما بين الفجر والشروق			ما بين نصف الليل وطلوع الفجر			ما بين نصف الليل والشروق		
1	9	13	11	30	عا	1	[illegible]	30	عم	عا	36	6	6	36	6	9	عم	1	٥	عم
2	9	13	عم3	17	عا	1	[illegible]	33	عم	عا	49	6	6	48	6	9	12	7	٥	12
3	9	13	61	عم	عا	1	ے	66	عم	عا	69	6	6	69	6	9	1	7	٥	1
4	9	13	61	61	عا	1	عم	9	عم	عا	11	6	6	11	6	9	[illegible]	7	٥	وعم
5	9	عا	عا	38	عا	٥	17	[illegible]	عم	عا	[illegible]	6	6	[illegible]	6	9	31	7	٥	37
6	9	عا	6	24	عا	٥	[illegible]	36	عم	عا	[illegible]	6	6	36	6	9	26	7	٥	[illegible]
7	9	عا	[illegible]	12	عا	٥	31	[illegible]	عم	عا	[illegible]	6	6	[illegible]	6	9	[illegible]	7	٥	16
8	9	عا	40	49	عا	٥	9	1	عم	عا	66	6	6	46	6	9	عم	7	٥	عم
9	9	عا	3	[illegible]	عا	٥	[illegible]	عا	عم	عا	8	6	6	8	6	9	62	7	٥	62

الحمد لله جدول مطالب الحوت لحركة [illegible] لمدينة [illegible] مراكش [illegible] عرضه لا

[illegible]	الميل		الغاية		ارتفاع العصر		جيبه		بعد الأفق			الدائر المطلق			الفضلة			ما بين الزوال والعصر		
1	23	30	36	0	[illegible]	30	22	62	12	[illegible]	30	56	3	40	31	[illegible]	0	2	11	0
2	23	7	36	23	[illegible]	60	23	61	12	46	26	56	7	47	31	64	10	2	11	10
3	23	[illegible]	36	66	22	40	23	0	12	38	20	56	14	8	31	23	[illegible]	2	11	20
4	23	21	36	9	22	0	23	10	12	26	14	56	16	19	31	62	30	2	11	30
5	23	68	36	32	22	61	23	21	12	14	10	56	20	26	30	41	40	2	11	40
6	23	37	37	66	22	62	23	32	12	2	6	57	26	37	30	40	40	2	11	40
7	22	12	36	18	22	33	23	23	11	40	0	57	28	[illegible]	30	26	0	2	11	0
8	22	9	36	41	22	23	23	16	11	37	44	57	32	43	30	49	10	2	11	10

[redacted]

XXXVIII Diagrams and calendrical tables in the almanac of Shaykh Vefā

[redacted]

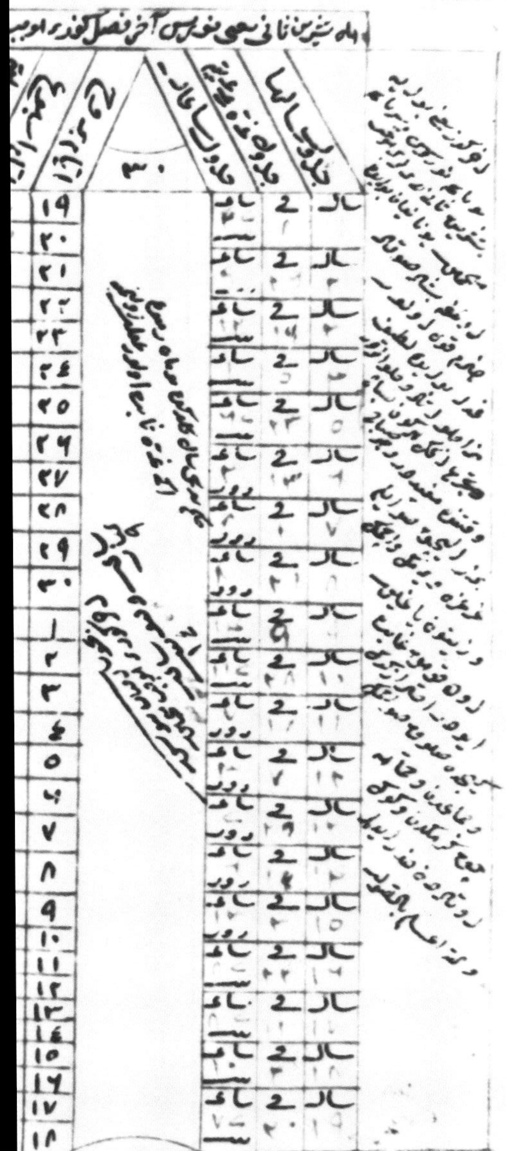

b.

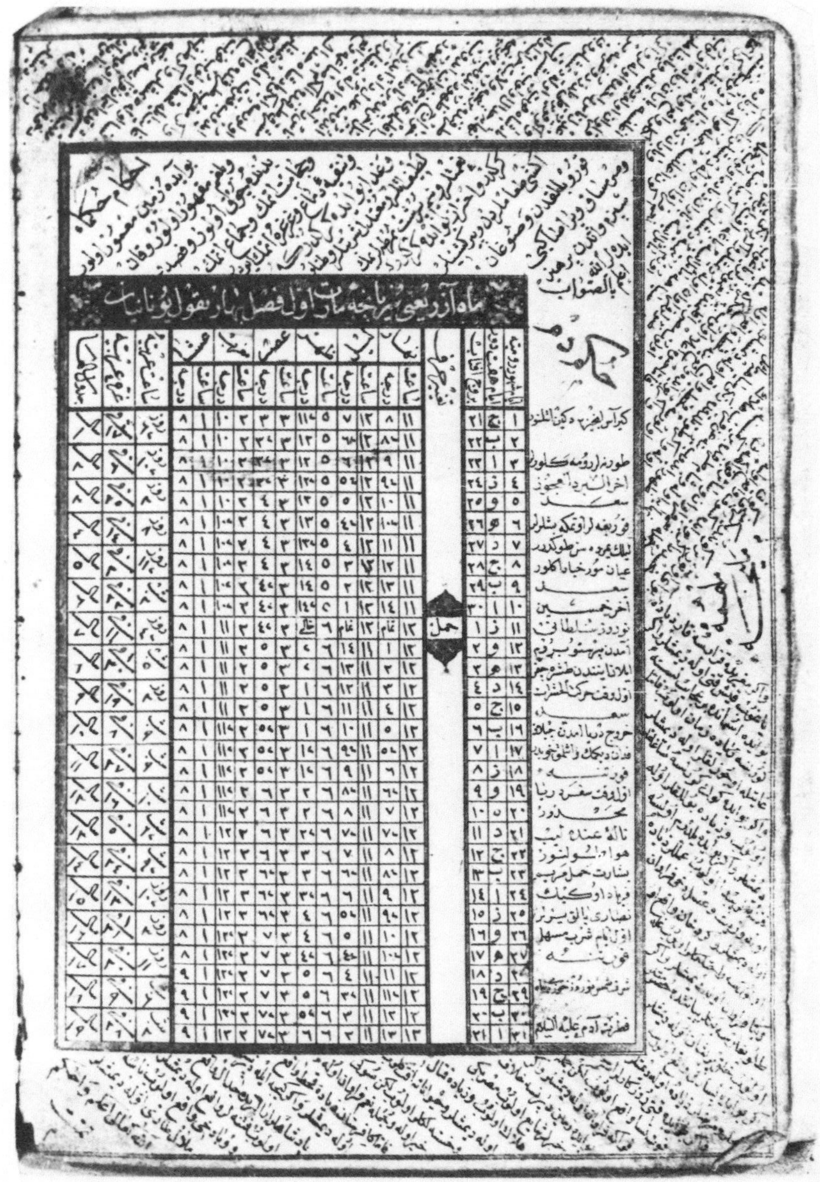

[redacted]s for Istanbul and calendrical tables in the almanac of Shaykh Vefa

XLa Prayer tables for Istanbul in the almanac of Darendeli

XLb A simple rule for telling time of day by means of shadows cast on a hand

XLIa Anonymous Tunisian auxiliary tables for timekeeping

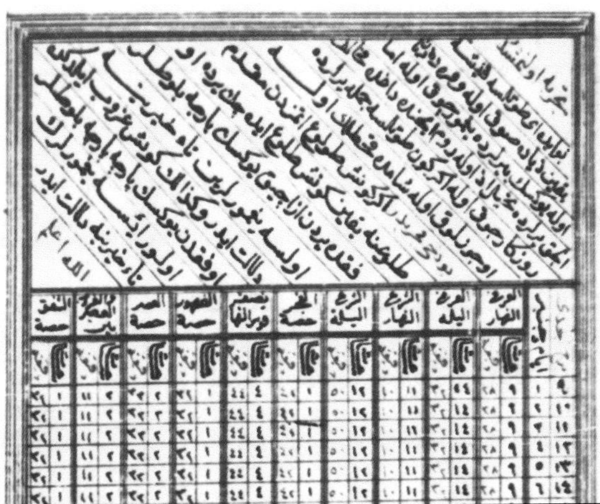

[redacted]

a.

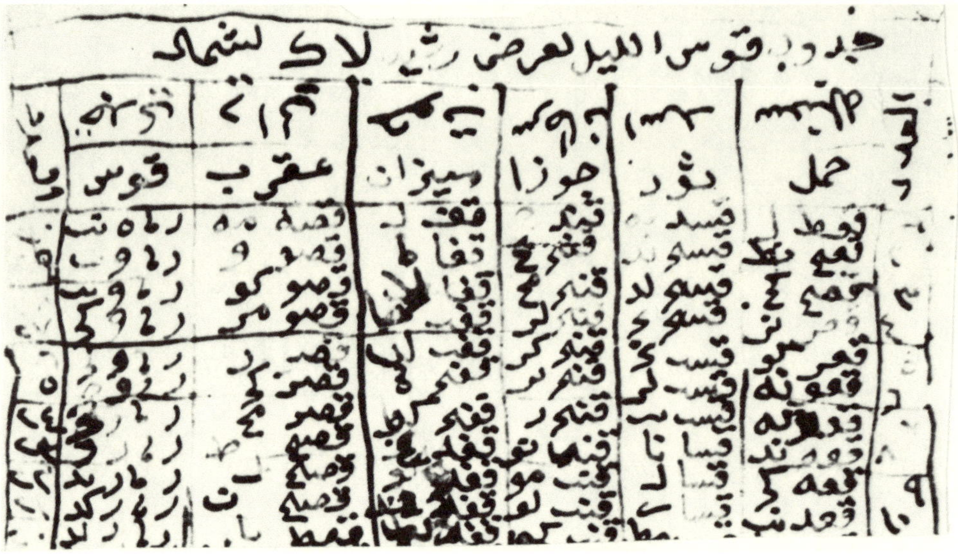

b.

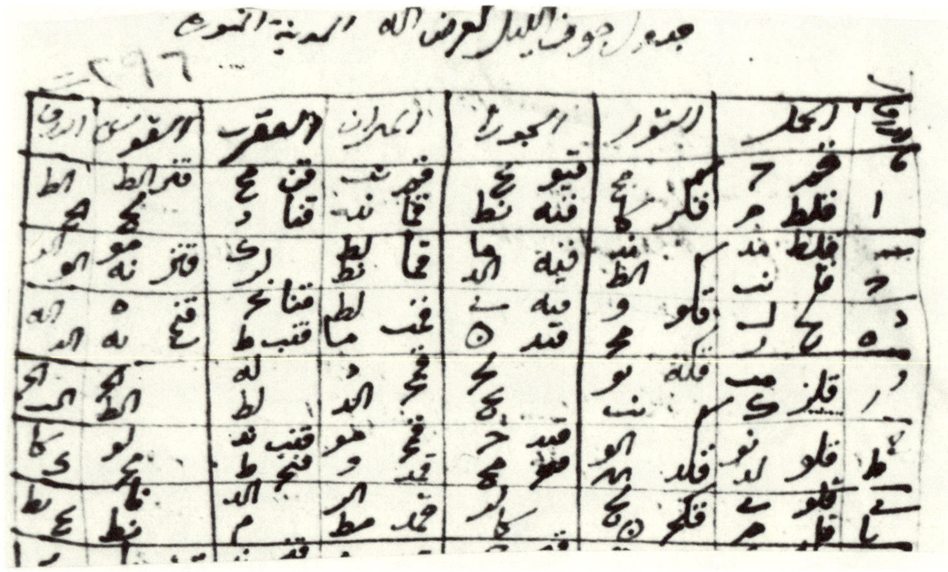

c.

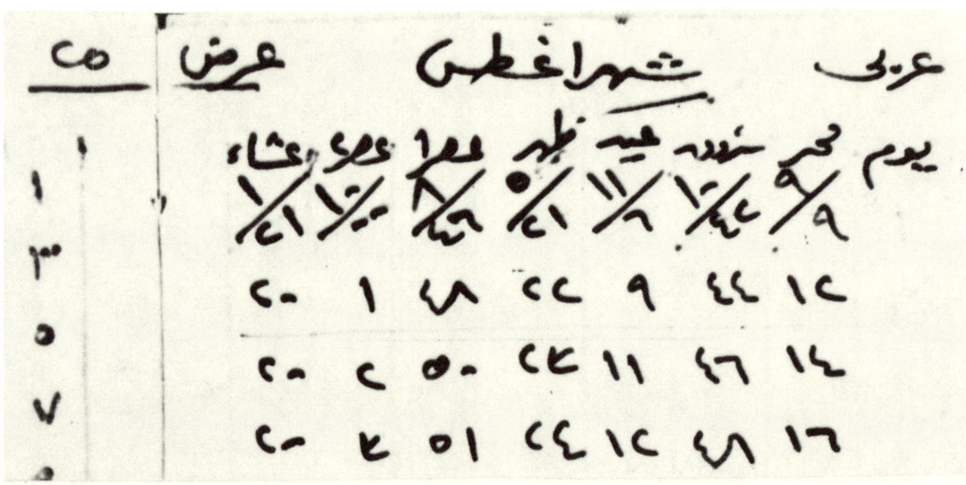

XLII Miscellaneous late tables for timekeeping for Rosetta, Medina and various latitudes in Egypt and the Hejaz.

[redacted]

s notes of historical interest

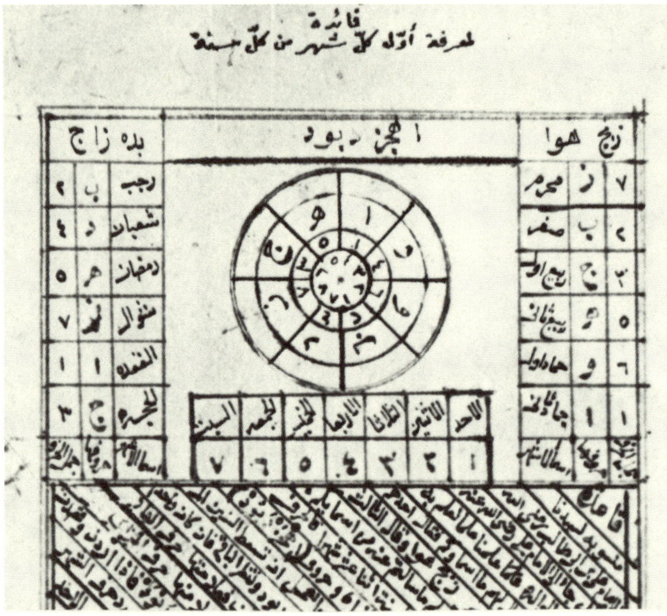

XLIVa Simple rules for regulating the lunar calendar attributed to ᶜAlī ibn Abī Ṭālib

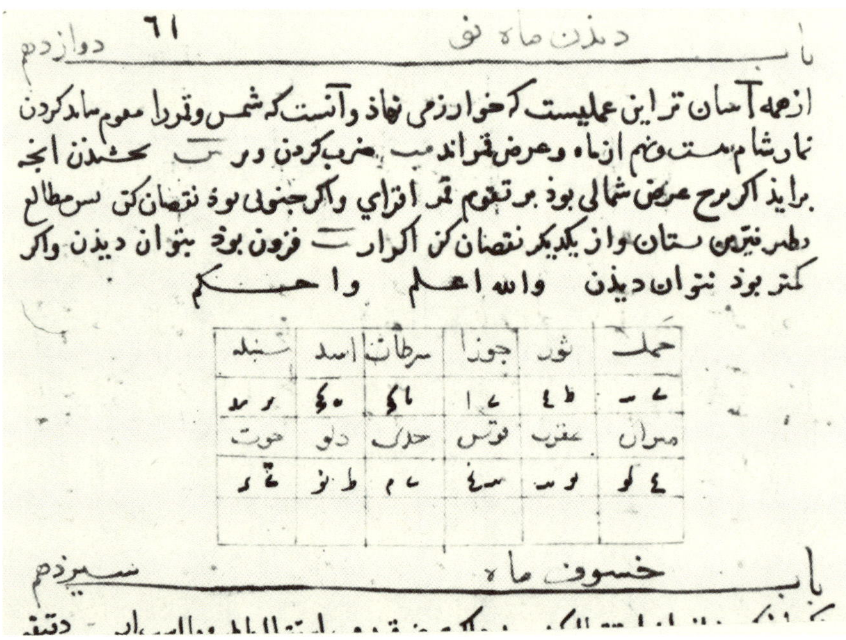

XLIVb al-Khwārizmī's table for predicting lunar crescent visibility

a.

[redacted]

b.

c.

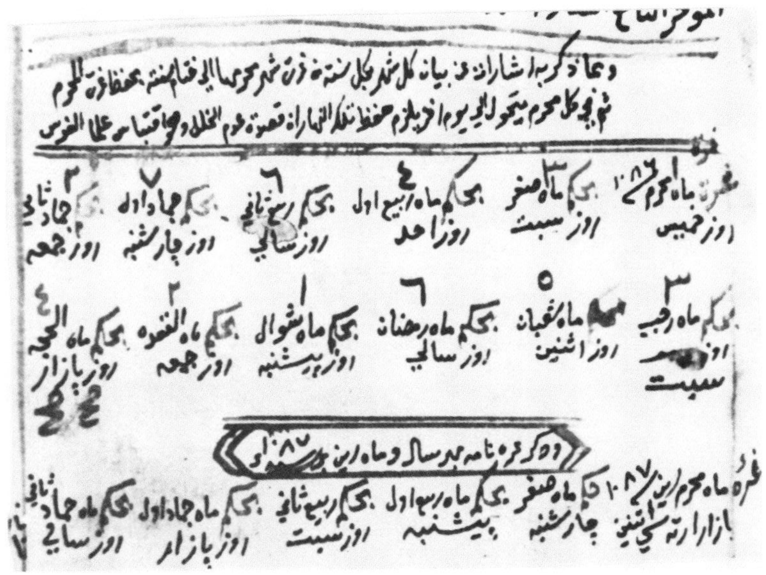

[redacted] tables associated with lunar crescent visibility

a.

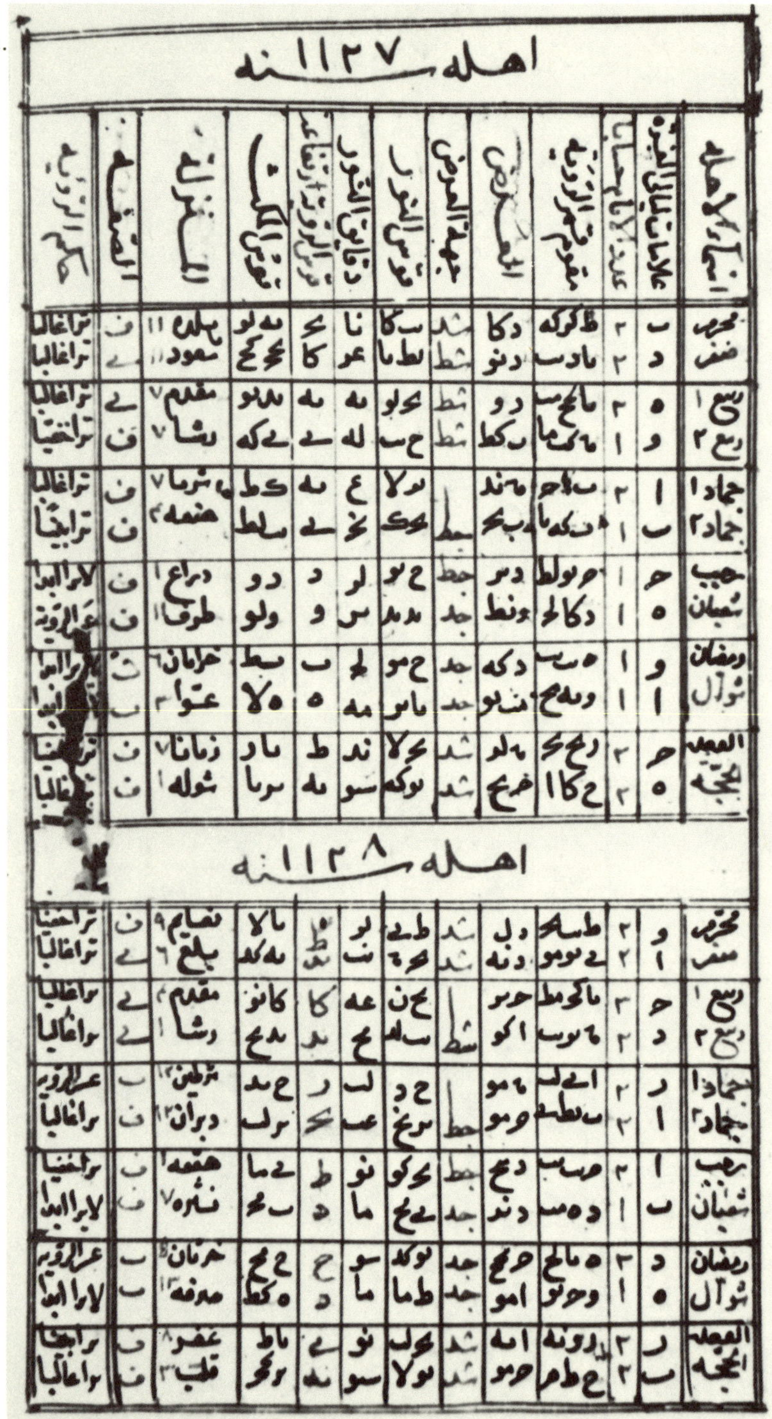

b.

XLVI Calculations of the possibility of lunar crescent visibility for each month of the year

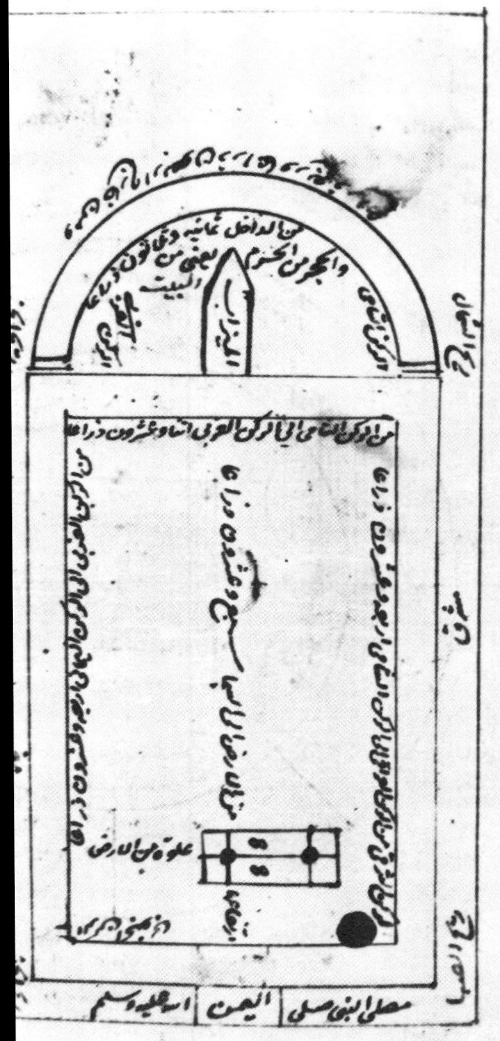

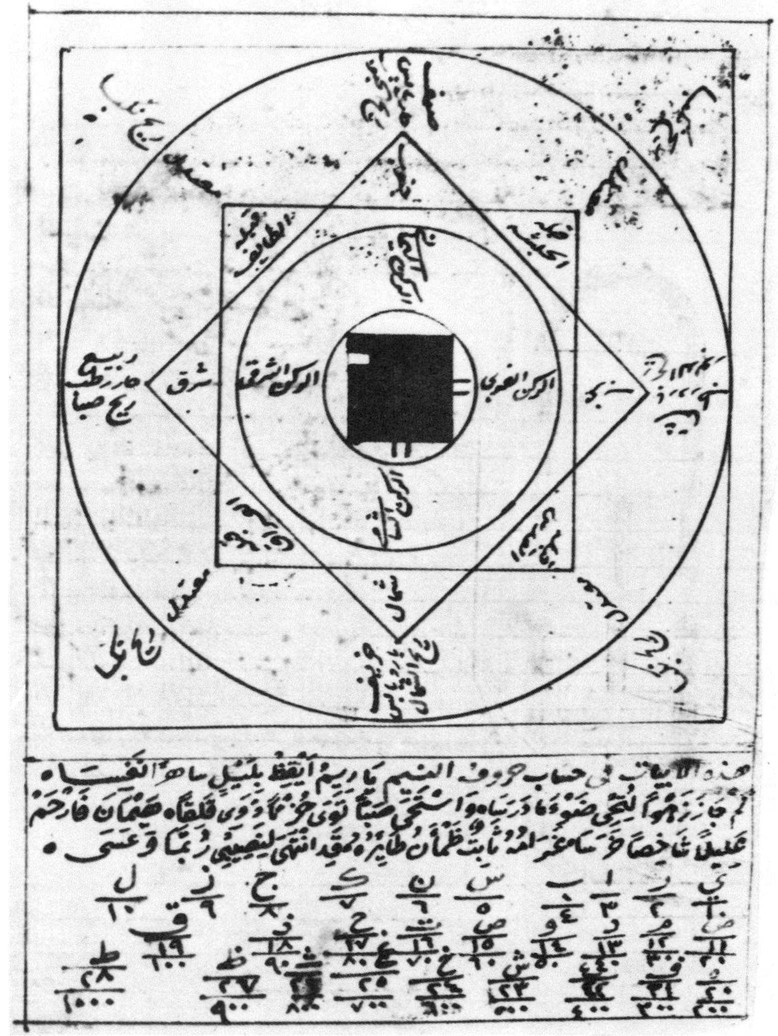

[redacted] the rectangular base of the Kaʿba and its associations with the winds

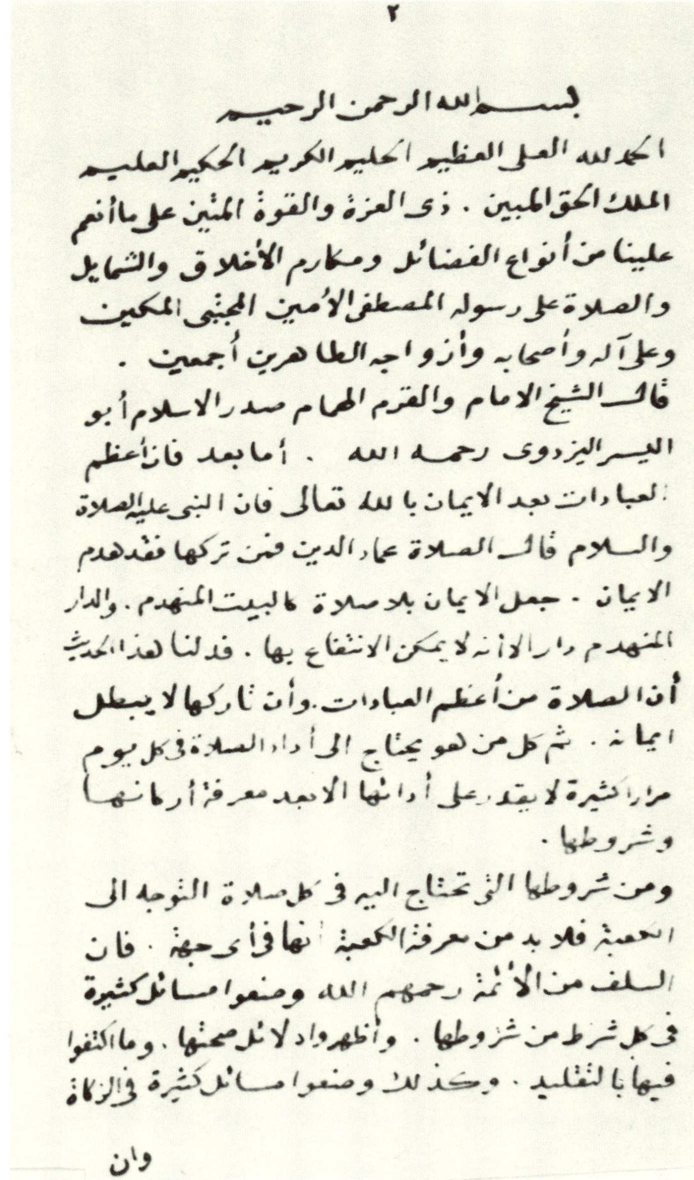

٢

بسم الله الرحمن الرحيم
الحمد لله العلي العظيم الحليم الكريم الحكيم العليم
الملك الحق المبين . ذي العزة والقوة المتين على ما أنعم
علينا من أنواع الفضائل ومكارم الأخلاق والشمايل
والصلاة على رسوله المصطفى الأمين المجتبى المكين
وعلى آله وأصحابه وأزواجه الطاهرين أجمعين .
قال الشيخ الامام والقرم الهمام صدر الاسلام أبو
اليسر البزدوي رحمه الله . أما بعد فان أعظم
العبادات بعد الايمان بالله تعالى فان النبي عليه الصلاة
والسلام قال الصلاة عماد الدين فمن تركها فقد هدم
الايمان . جعل الايمان بلا صلاة كالبيت المنهدم . والدار
المنهدم دار الا انه لا يمكن الانتفاع بها . فدلنا هذا الحديث
أن الصلاة من أعظم العبادات . وأن تاركها لا يبطل
ايمانه . ثم كل من هو يحتاج الى أداء الصلاة في كل يوم
مرارا كثيرة لا يقدر على أدائها الا بعد معرفة أركانها
وشروطها .
ومن شروطها التي تحتاج اليه في كل صلاة التوجه الى
الكعبة فلا بد من معرفة الكعبة انها في أي جهة . فان
السلف من الأئمة رحمهم الله وضعوا مسائل كثيرة
في كل شرط من شروطها . وأظهروا دلائل صحتها . وما اكتفوا
فيها بالتقليد . وكذلك وضعوا مسائل كثيرة في الزكاة
وان

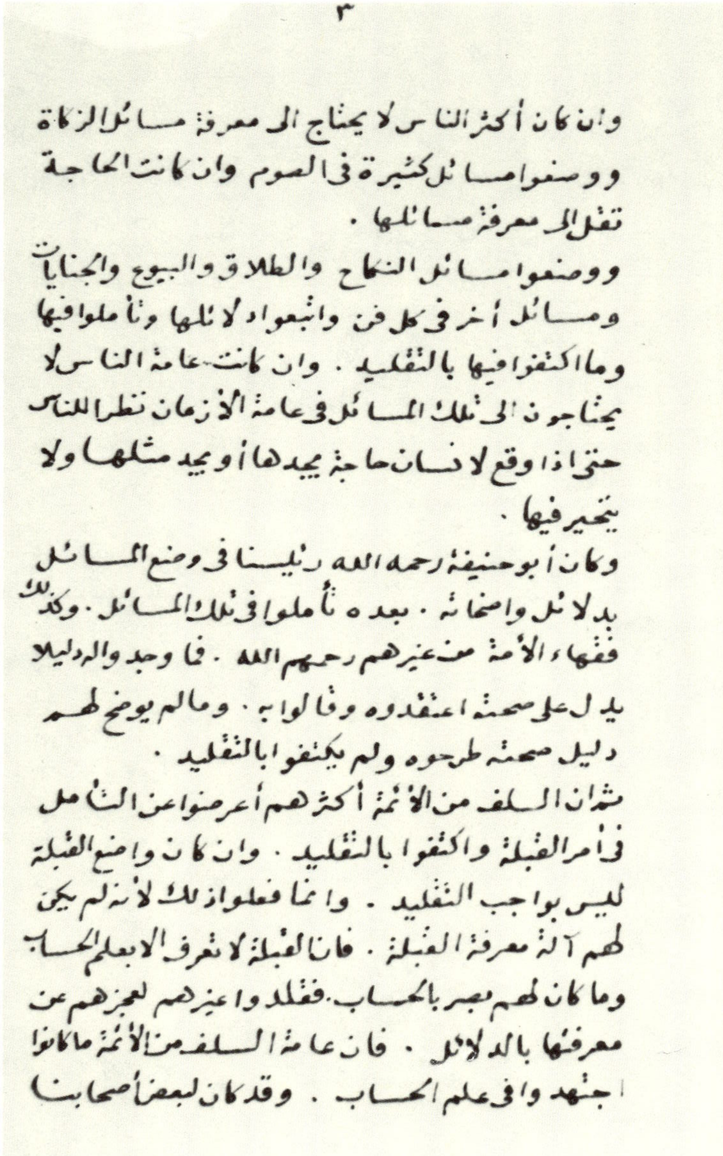

٣

وان كان أكثر الناس لا يحتاج الى معرفة مسائل الزكاة
ووضعوا مسائل كثيرة في الصوم وان كانت الحاجة
تقل الى معرفة مسائلها .
ووضعوا مسائل النكاح والطلاق والبيوع والجنايات
ومسائل أخر في كل فن واتبعوا دلائلها وتأملوا فيها
وما اكتفوا فيها بالتقليد . وان كانت عامة الناس لا
يحتاجون الى تلك المسائل في عامة الأزمان نظرا للناس
حتى اذا وقع لانسان حاجة يجدها أو يجد مثلها ولا
يتحير فيها .
وكان أبو حنيفة رحمه الله رئيسنا في وضع المسائل
بدلائل واضحاته . بعده تأملوا في تلك المسائل . وكذلك
فقهاء الأمة من غيرهم رحمهم الله . فما وجدوا الدليلا
يدل على صحته اعتقدوه وقالوا به . وما لم يوضح لهم
دليل صحته طرحوه ولم يكتفوا بالتقليد .
ثم ان السلف من الأئمة أكثرهم أعرضوا عن التأمل
في أمر القبلة واكتفوا بالتقليد . وان كان واضع القبلة
ليس بواجب التقليد . وانما فعلوا ذلك لأنه لم يكن
لهم آلة معرفة القبلة . فان القبلة لا تعرف الا بعلم الحساب
وما كان لهم بصر بالحساب . فقلدوا غيرهم لعجزهم عن
معرفتها بالدلائل . فان عامة السلف من الأئمة ما كانوا
اجتهدوا في علم الحساب . وقد كان لبعض أصحابنا

XLVIII A treatise by al-Bazdawī on the qibla in early Islamic Central Asia

XLIX A diagram of a scheme of sacred geography, in which various regions of the world are centered on the Ka^{c}ba

[redacted]

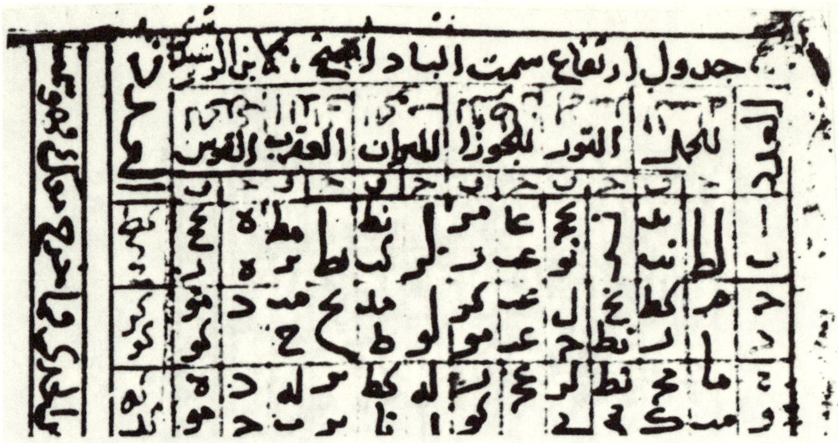

La A table for orienting ventilators found in the main Cairo corpus of tables for timekeeping

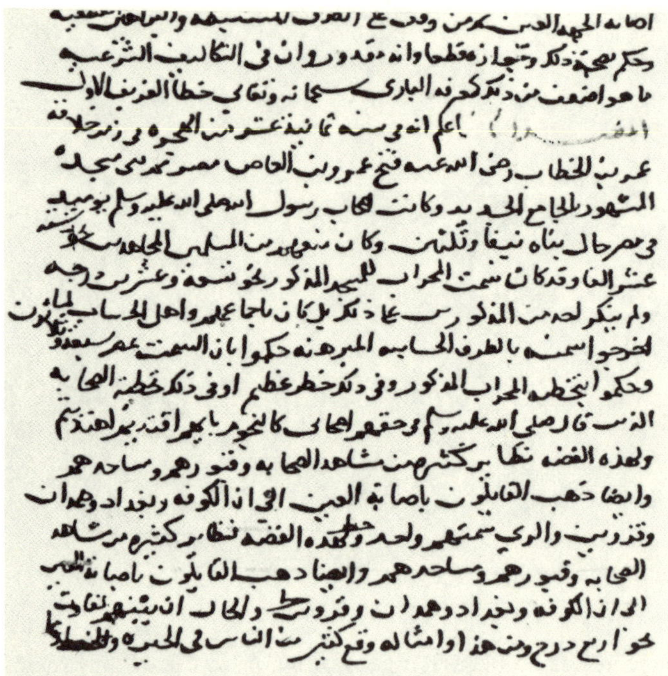

Lb Remarks on the qiblas used in Fustat and Cairo by Ghars al-Dīn al-Ḥalabī

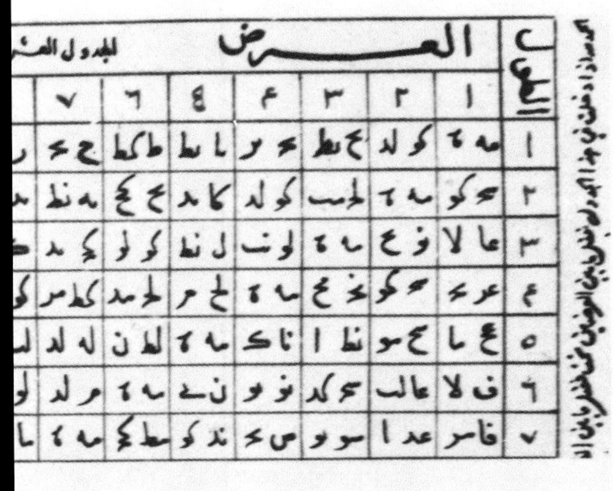

c.

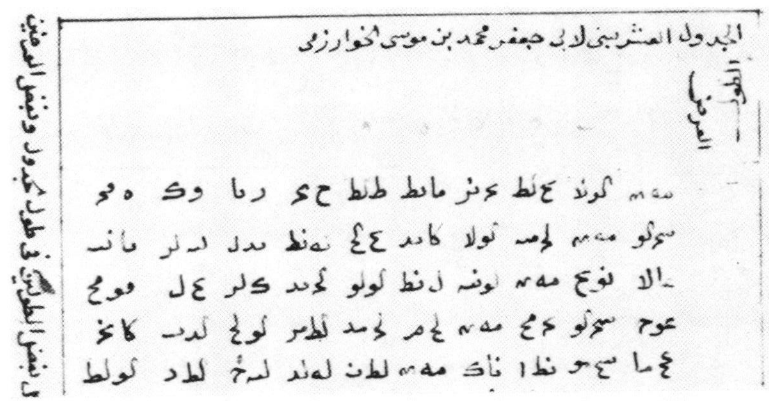

[redacted]ous tables displaying the qibla as a function of terrestrial latitude and longitude

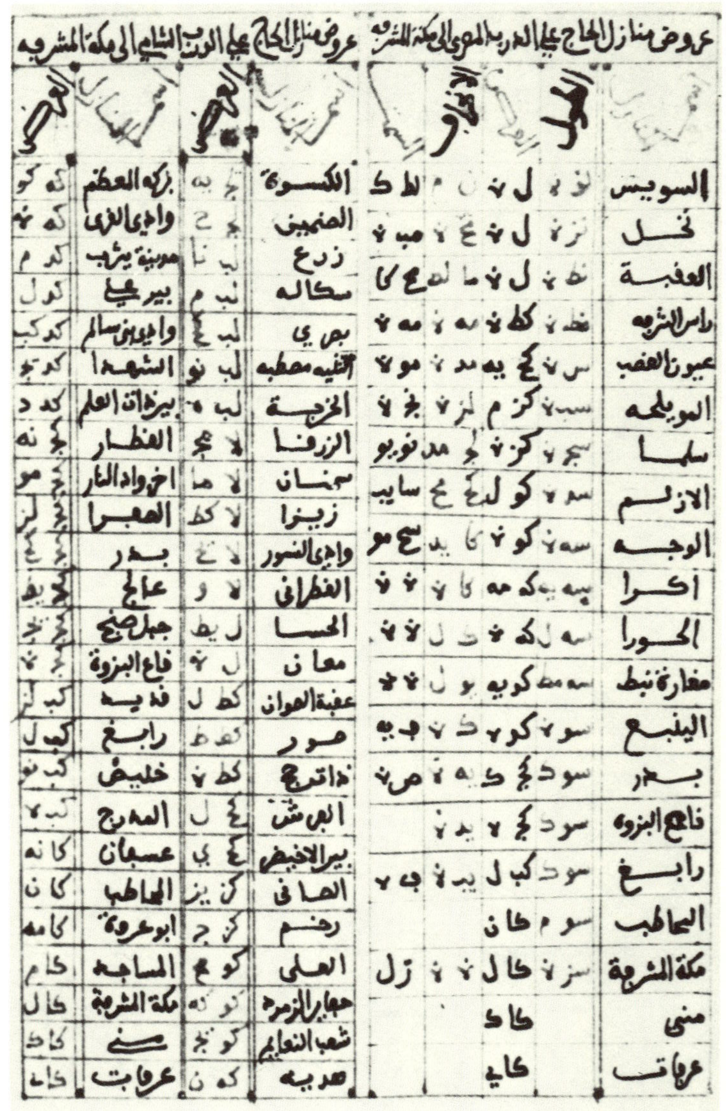

LII Lists of localities on the pilgrim routes from Cairo and Damascus, and of various localities with their qiblas

...عه المرئي من السموت وهو ما بينه وبين دائرة اول السموت على تسعين يحصل فضل الداير
...بر الخيط من القوس هو السمت وهو شمالي في هذه الحالة دايما. والله اعلم

اسطرلاب ومعناه [illegible]
كلمة يونانية [illegible] ابن [illegible]
ميزان الشمس [illegible] النجوم [illegible] الفلك [illegible]
وقال [illegible] الكرة على [illegible] الليل والنهار [illegible] معرفة
اصطرلاب [illegible] فقال له لاب [illegible] اسطرلاب و[illegible]
الاصبهاني [illegible] هذا
ان العلماء [illegible] واتخذه عليه
على [illegible] الكرة [illegible] لاب [illegible] منه
بها الطالع [illegible] الفلك فضل [illegible] والعمادة [illegible]
[illegible] فقال [illegible] والاصل [illegible]
الى ابيه [illegible] والاول اصح
الاسم [illegible]
لمكان النطق [illegible]

روایت چنین [illegible] که بادکاری [illegible] اسطرلاب [illegible] ارسطاطالیس [illegible]
اسکندر [illegible] حکما [illegible] پس [illegible] روشن و [illegible]
در جهان [illegible] حکما [illegible] اسطرلاب [illegible]
ساخته که [illegible] لاب نام حکیم دیگر [illegible] که [illegible]
[illegible] و بعضی گفته‌اند لاب نام حکیمی [illegible] که لاب نام [illegible]
[illegible] ساخته و قول دیگر آنست که [illegible] است [illegible]
اسکندر را اسطرلاب را ساخته [illegible] ارسطو [illegible] که اسطرلاب [illegible]
[illegible] است [illegible] در علم حکمت [illegible]
چهارم [illegible] لاب نام [illegible] است [illegible]
[illegible] اسطرلاب [illegible] اصطرلاب و [illegible]

[redacted]

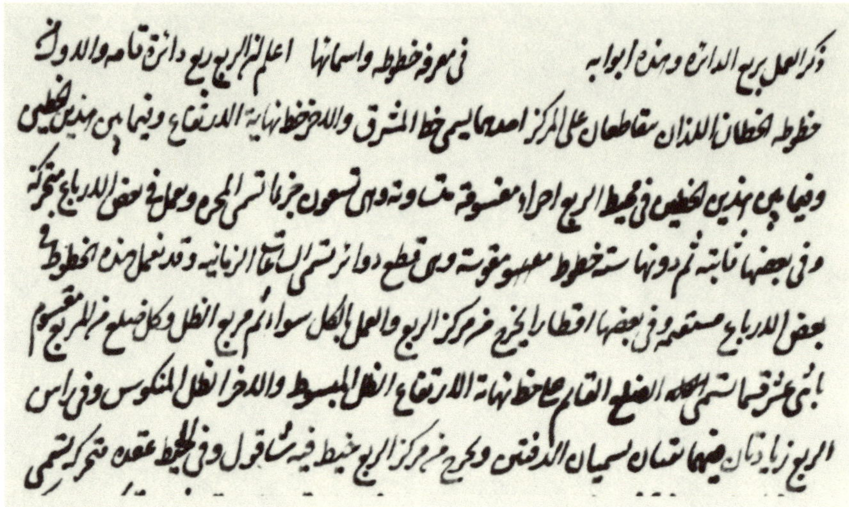
ذكر العمل بربع الدائرة وهذه ابوابه في معرفة خطوطه واسمائها اعلم ان الربع ربع دائرة تامة والدوائر
خطوطه الخطان اللذان يتقاطعان على المركز احدهما يسمى خط المشرق والاخر خط نهاية الارتفاع وفيما بين هذين الخطين
وفيما بين هذين الخطين في محيط الربع اجزاء معشوقة منه ونه وهي تسعون جزءا تسمى المجرى ويعمل في بعض الارباع متحركة
وفي بعضها ثابتة ثم دونها ستة خطوط معقوفة وهي قطع دوائر تسمى الساعات الزمانية وقد نعمل هذه الخطوط في
بعض الارباع مستقيمة وفي بعضها اقطارا تخرج من مركز الربع والعمل بالكل سواء ثم مربع الظل وكل ضلع من المربع مقسوم
اثنى عشر قسما تسمى الضلع القائم على خط نهاية الارتفاع الظل المبسوط والاخر الظل المنكوس وفي راس
الربع زيادتان فيهما ثقبتان يسميان الهدفتين ويخرج من مركز الربع خيط فيه شاقول وفي الخيط عقدة متحركة تسمى

LIVa An early treatise on the horary quadrant with movable or fixed cursor

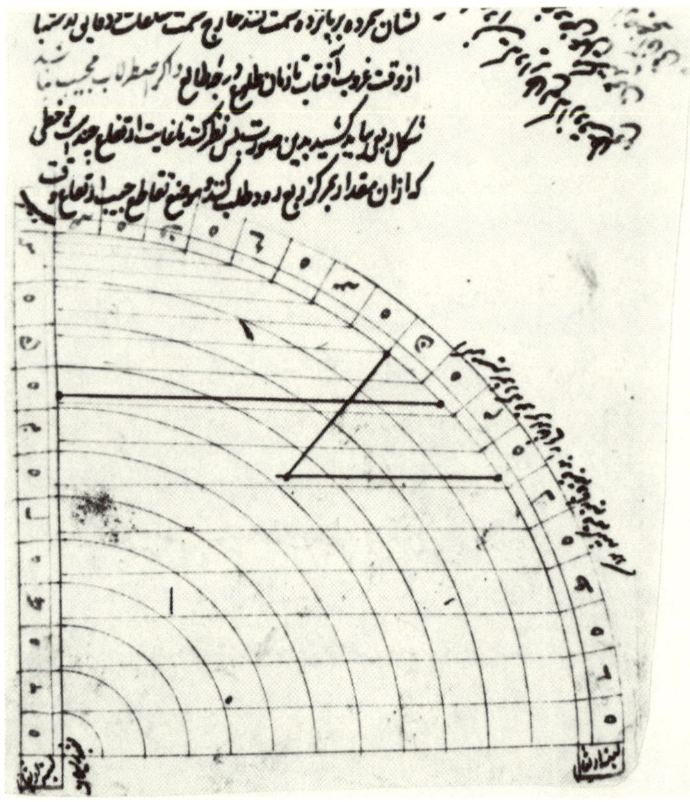

LIVb An illustration of the sine quadrant (invented by al-Khwārizmī)

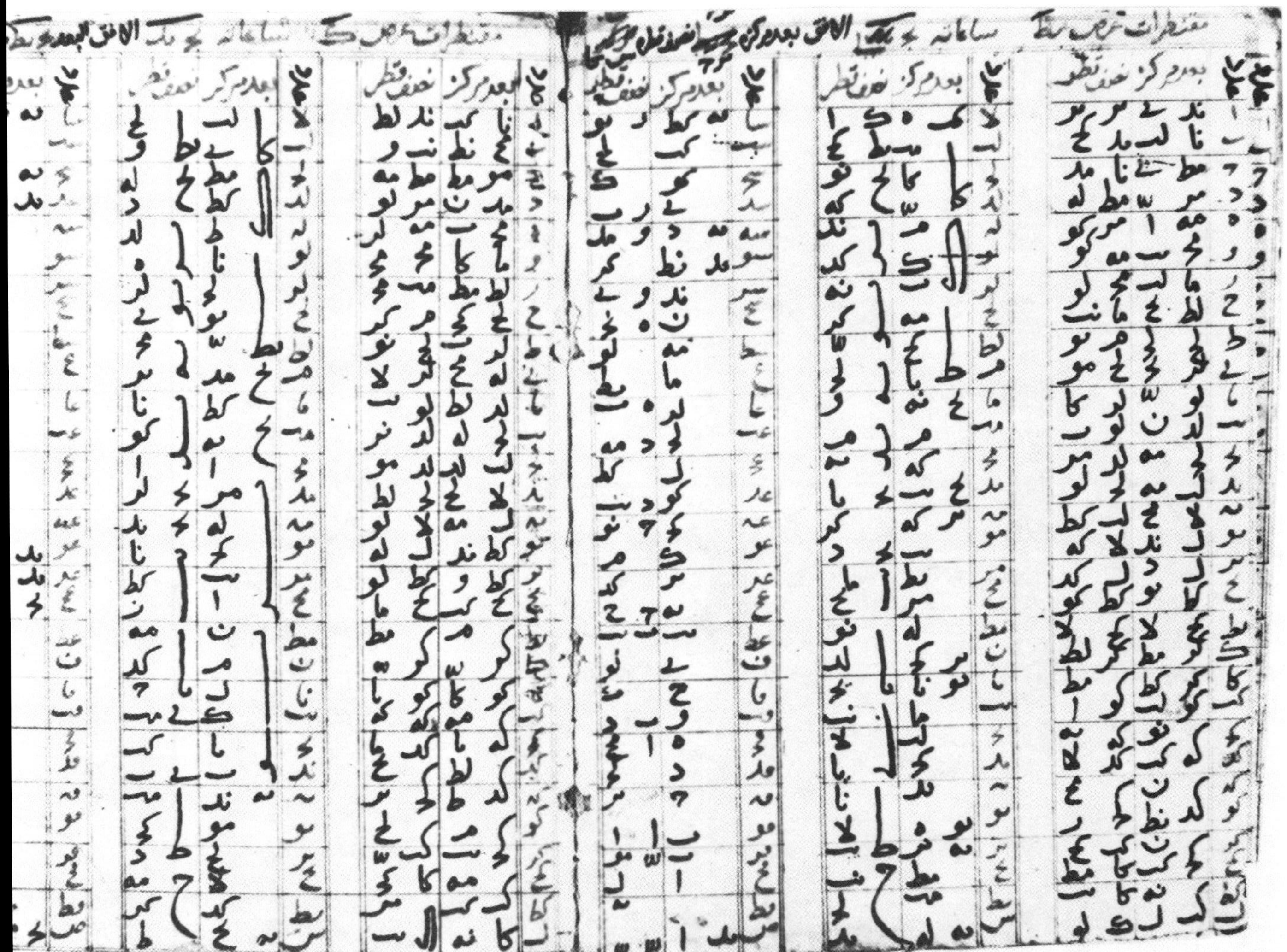

[redacted]nī's tables of coordinates for constructing the curves **on astrol**abe plates

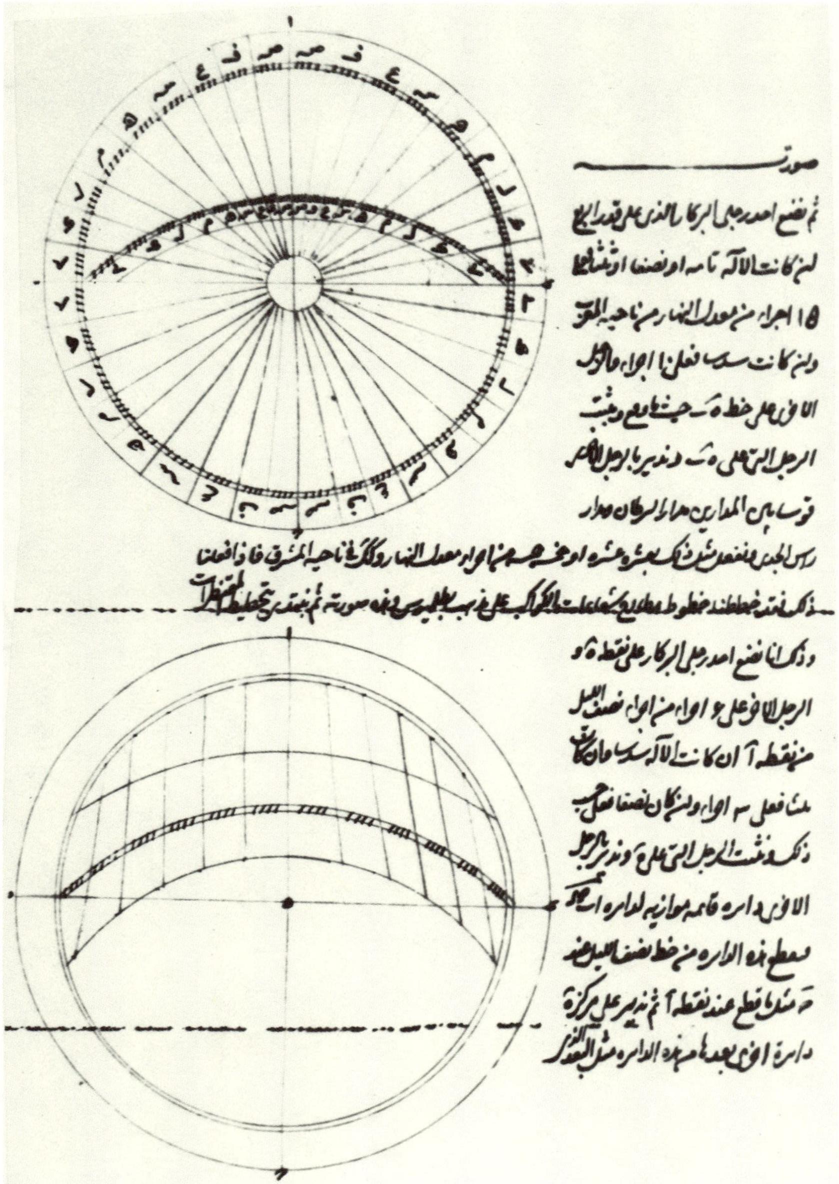

LVIa al-Khujandī's universal instrument

LVIb al-Zarqāllu's universal plate

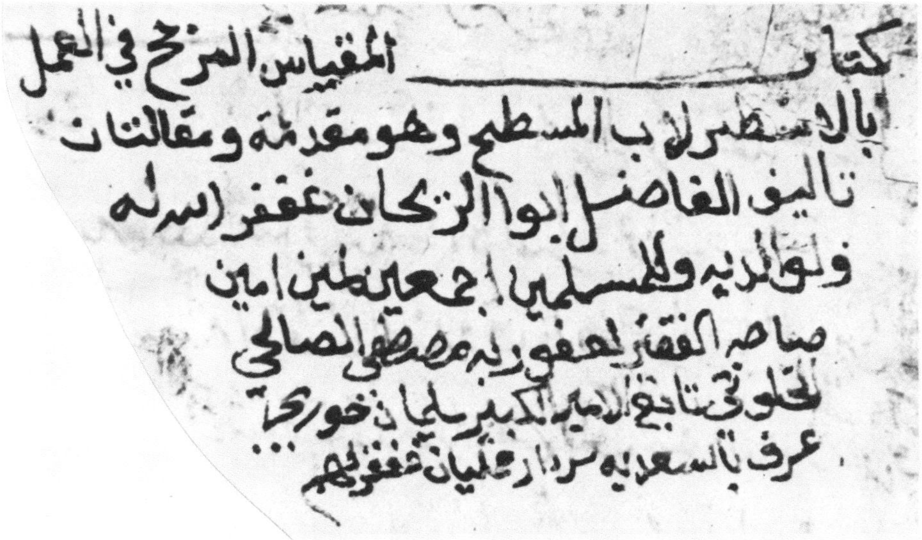
كتاب المقياس المرجح في العمل
بالاسطرلاب المسطح وهو مقدمة ومقالتان
تاليف الفاضل ابو الريحان غفر الله له
ولوالديه وللمسلمين اجمعين امين
صاحبه الفقير لعفو ربه مصطفى الصالحي
الخلوتي تابع الامير الكبير سليمان خورجيا
عرف بالسعديه سردار عثمانيان غفر لهم

بسم الله الرحمن الرحيم وبه استعين
المقياس المرجح في العمل بالاسطرلاب المسطح وهو مقدمة ومقالتان
وكل اسم اجتمع السين فيه اصلا وفيه طاء كالصراط والاصطرلاب او خاء
كصخر او عين كسعنه او قاف كصندوق فانه يجوز فيه السين والصاد و
والاصطرلاب اسم عجمي واشتقاق معناه من العربية بعيد وذكر ابو الحسن
ثابت بن قرة في العمل بالاسطرلاب له ان ابرخس وهو قبل بطلميوس وضع
الاسطرلاب وسطحه على مثال ما وضعه لاب بعد ان كان كريا وان الذي
دعاه الى ذلك انه راى الكرة كثير اعناؤها قليلا نفعها فاراد ان يضع آلة
قريبة يسيرة جامعة لكثير من الاعمال يوضح بها ما غمض في الالة المقببة الكلام
[redacted]

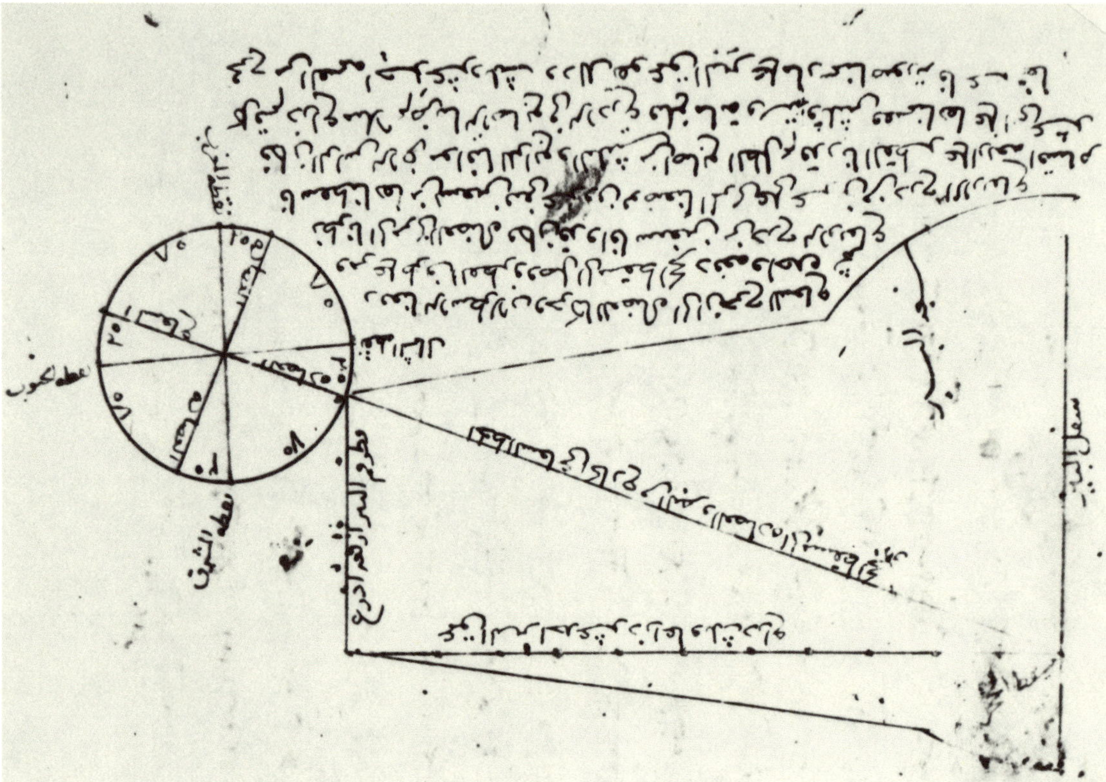

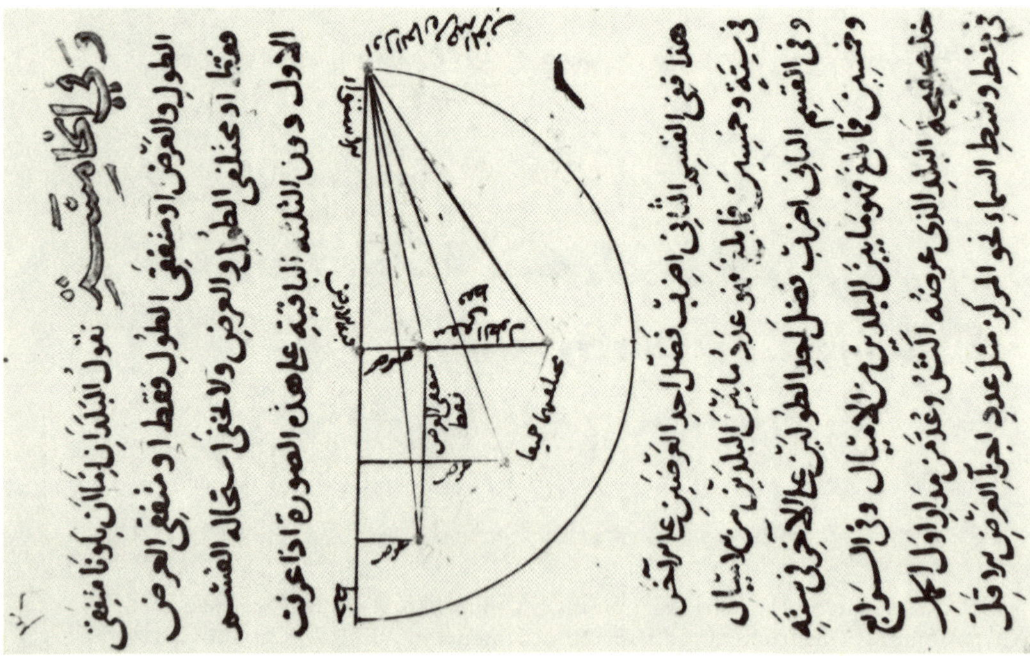

LVIII Illustrations of the use of the astrolabe in surveying

[redacted]

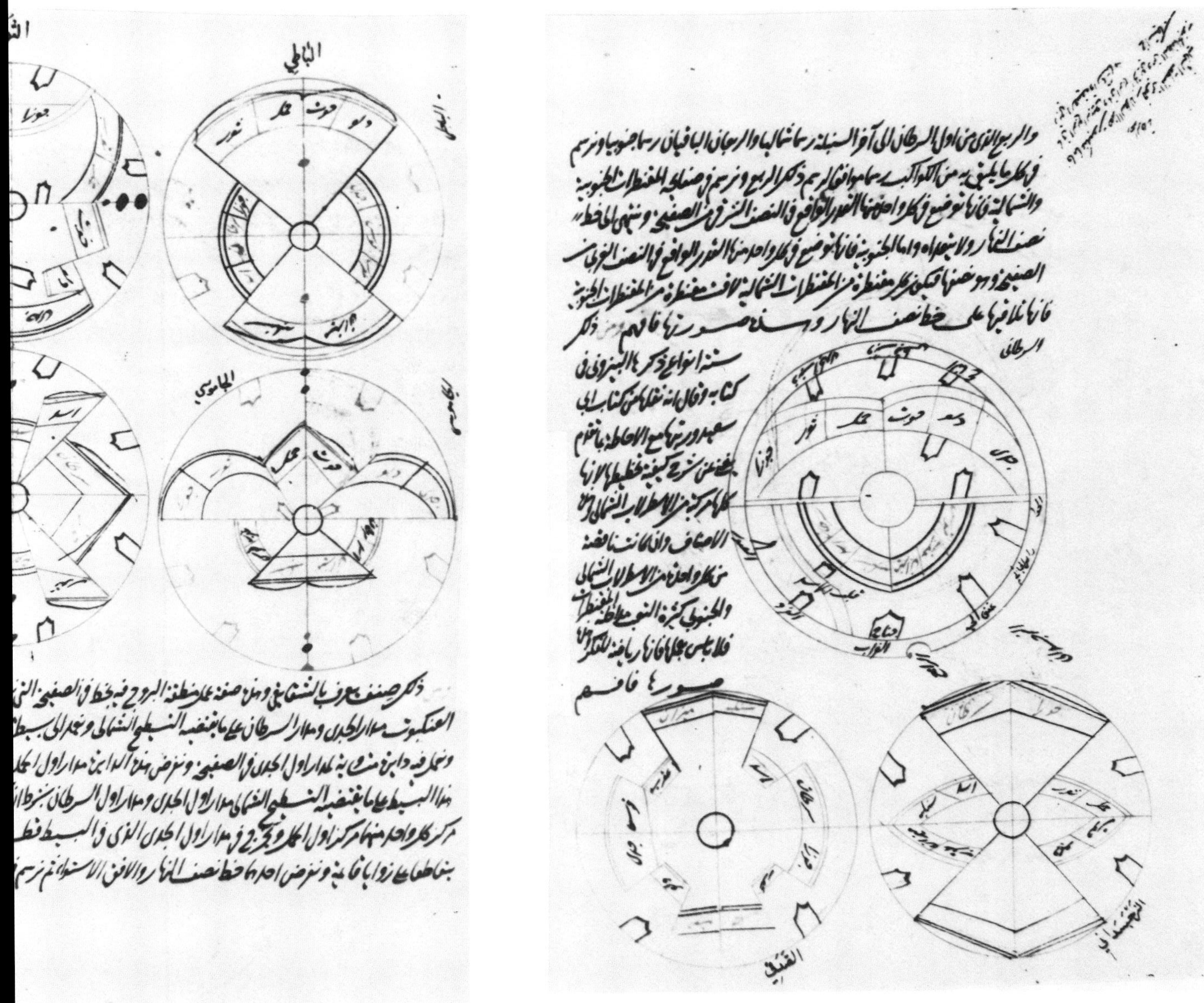

[redacted]rolabe retes in the compendium on spherical astronomy and instruments by al-Marrākushī

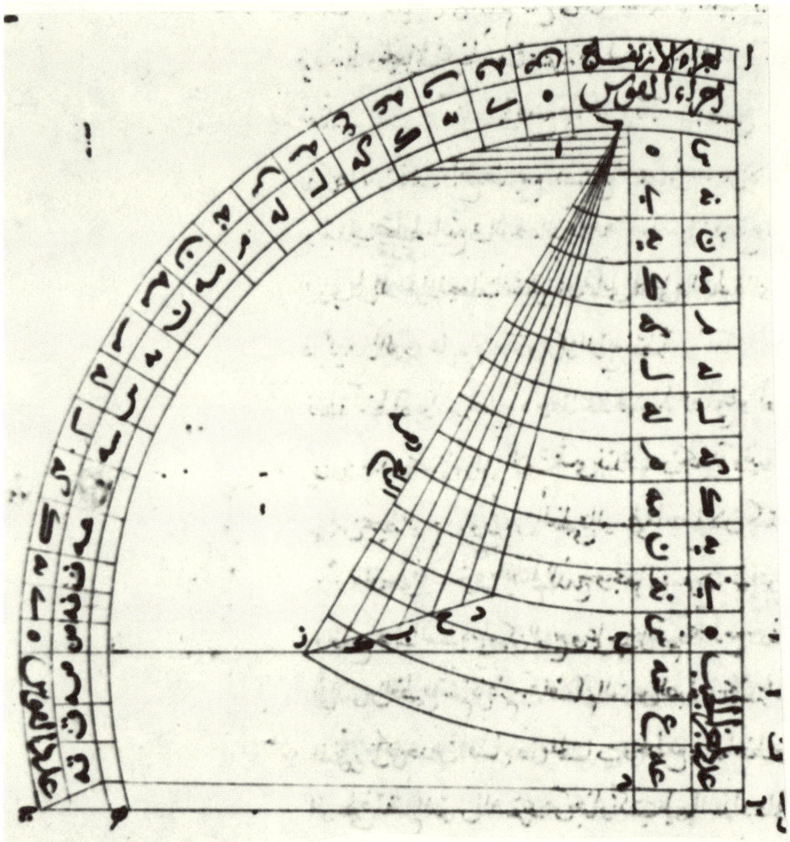

LX A quadrant and geared mechanism for reproducing the relative motions of the sun and moon, from al-Marrākushī's compendium

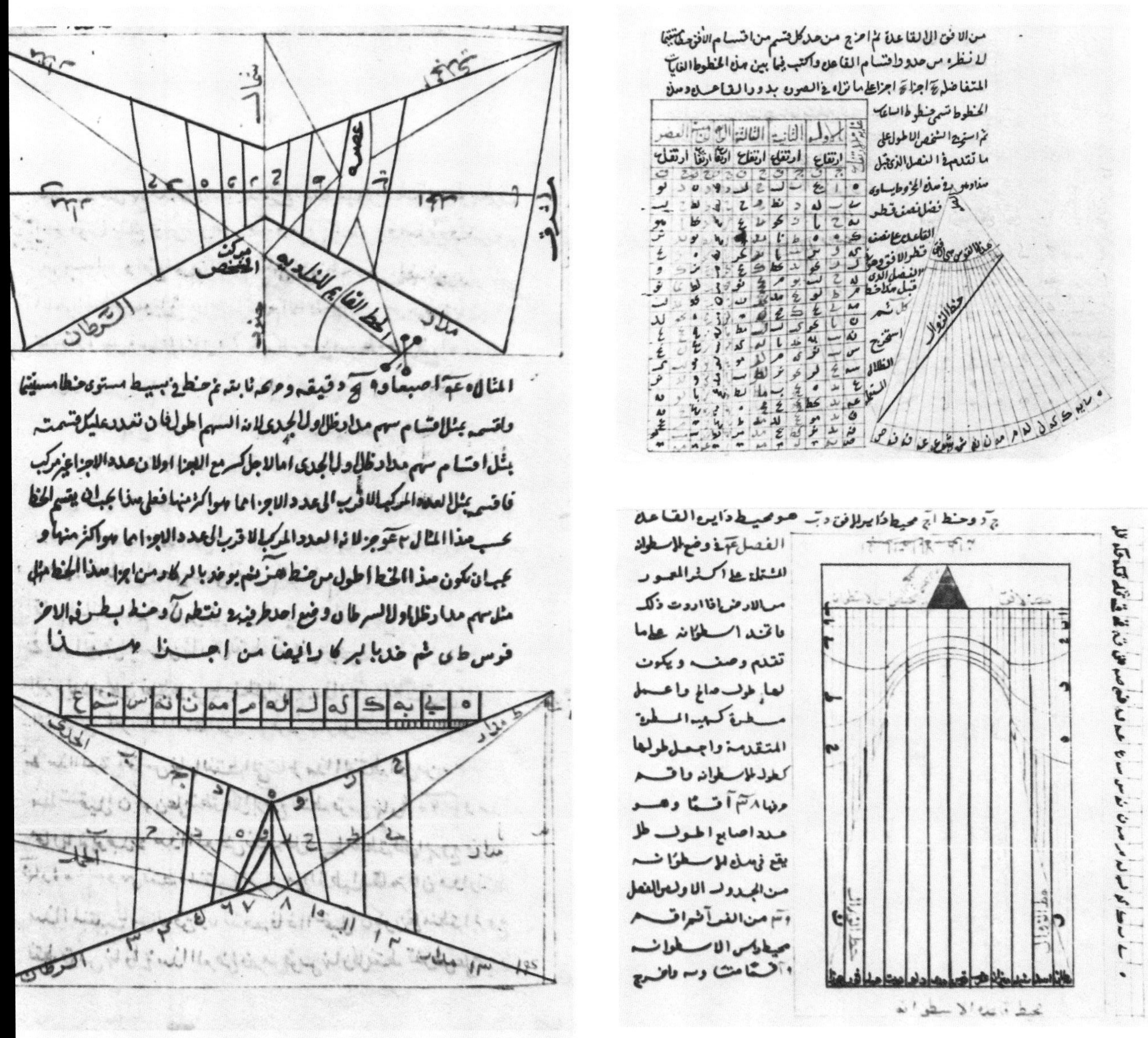

[redacted]ical and cylindrical sundials illustrated in al-Marrākushī's compendium

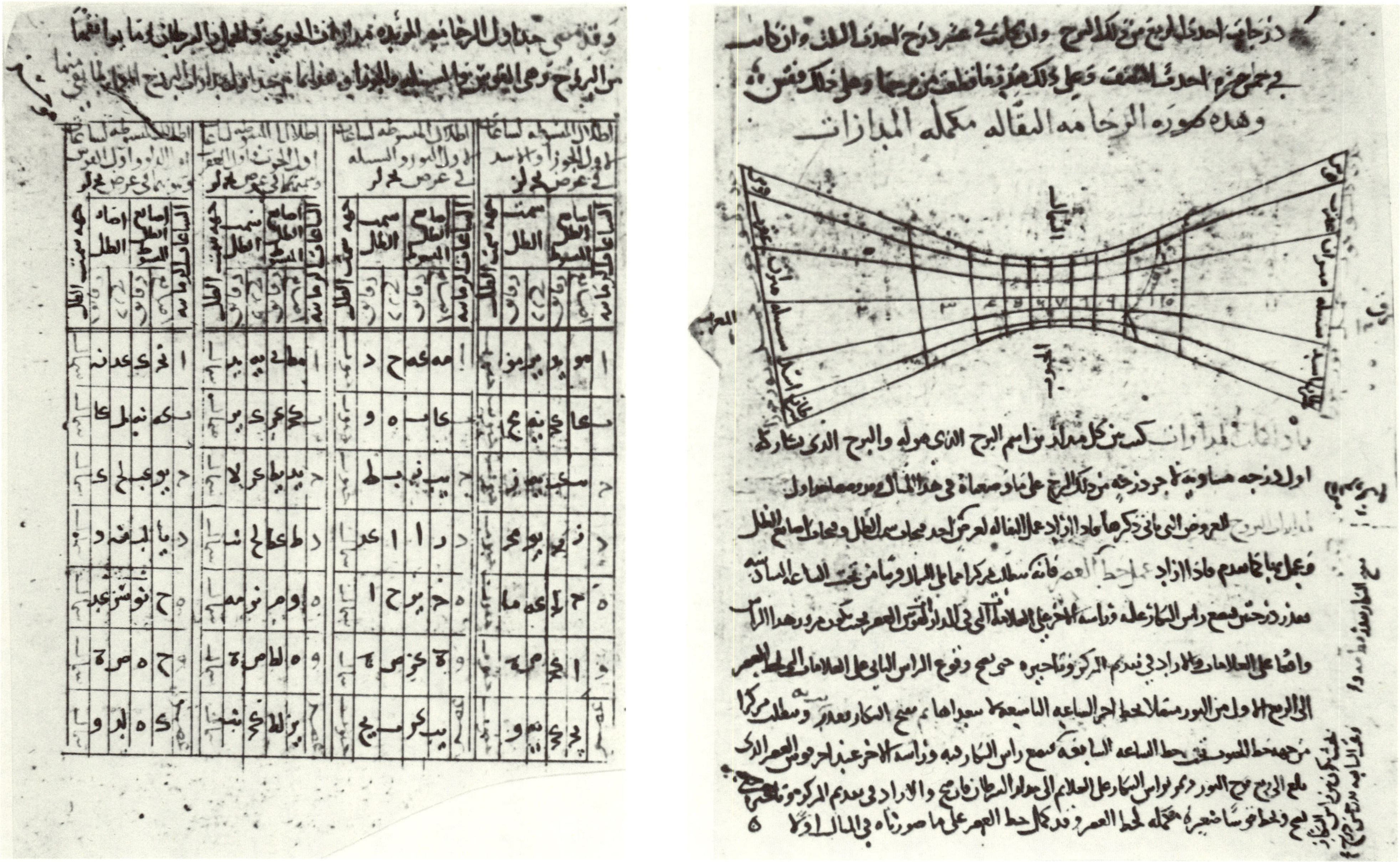

LXII Tables for constructing horizontal sundials in the compendium on instruments by the Yemeni Sultan al-Ashraf

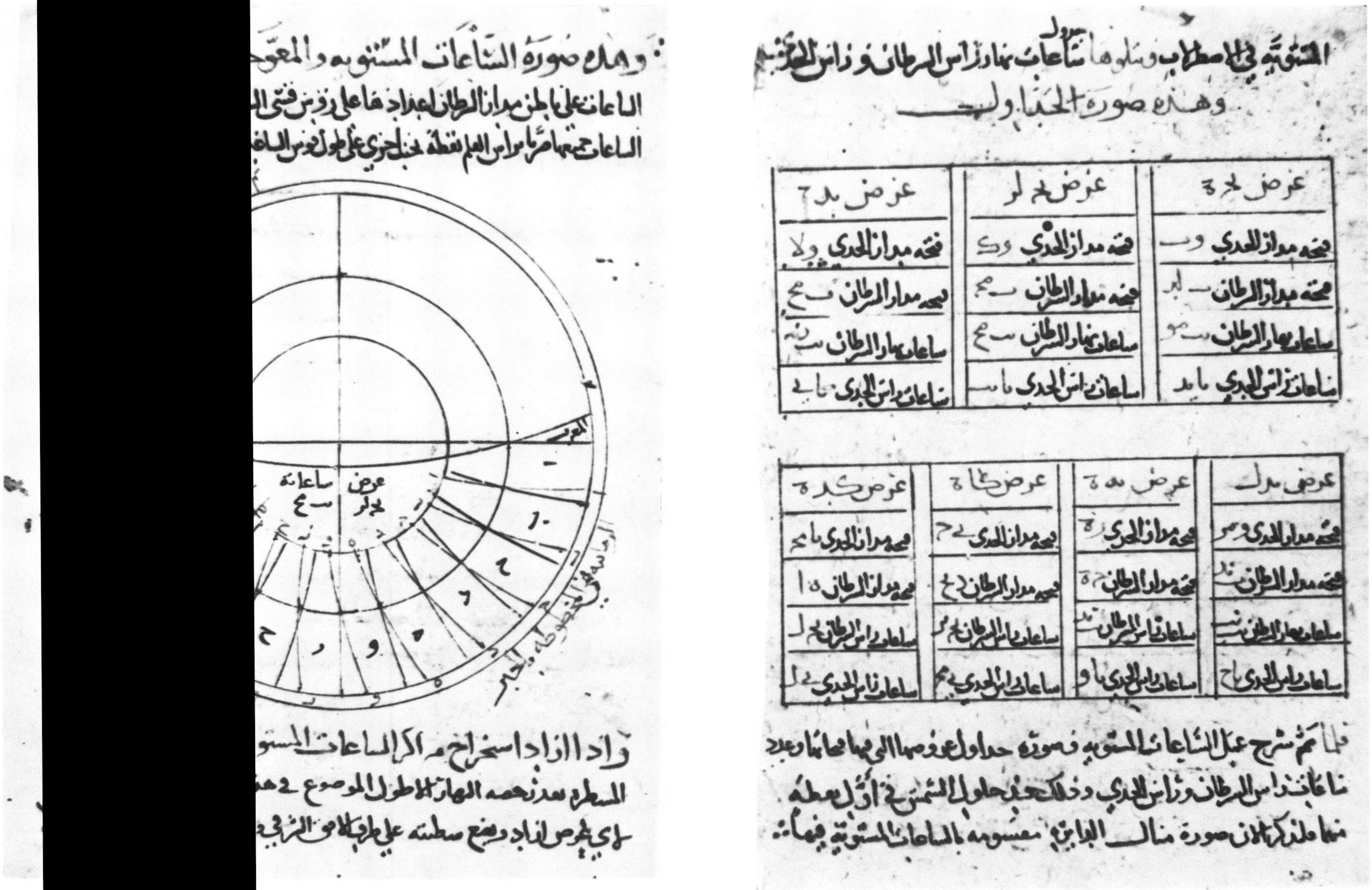

D[redacted]cting astrolabe plates in al-Ashraf's compendium

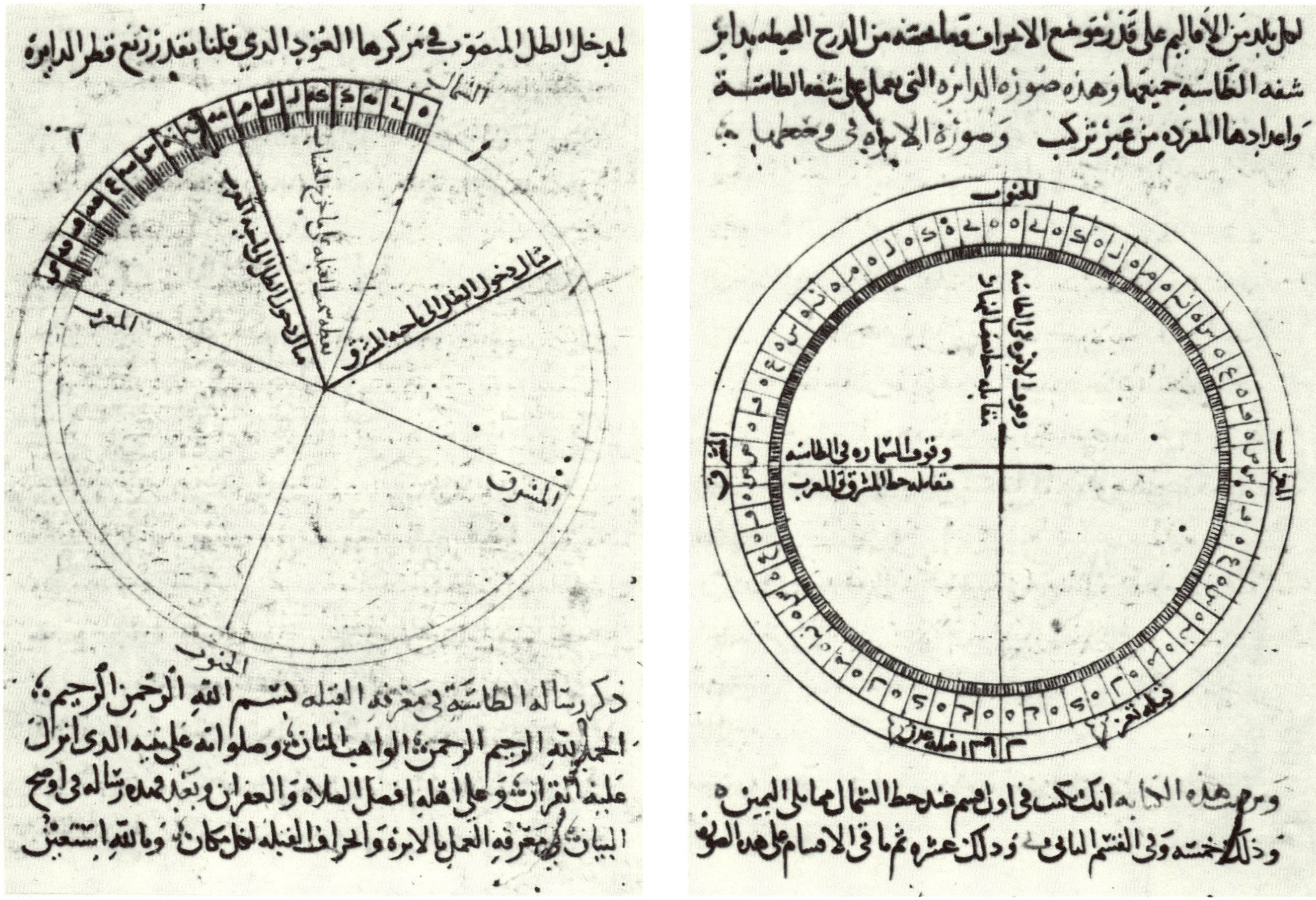

LXIV Illustrations of a magnetic compass in al-Ashraf's compendium

بسم الله الرحمن الرحيم رب يسر برحمتك

قال العبد الفقير الى الله تعالى احمد بن ابي بكر بن علي بن السراج بعد حمد الله والثنا
عليه والصلوه والسلام على نبيه محمد واله وصحبه وسلم اعلم اني لما وقفت
على كيفية الدواير العظام المشهوره التي ذكرها علما الهيه وعلى كيفية تسطيحها
كما ذكره علما الهندسه سنح لي استنباط آله من الشكازيه المشهوره وذلك
عند وقوفي على كيفية استخراج فضل الداير منها ما لا استقرا والتتبع مرات
عديده فانا شرع في الكلام على تسميه الاتها ورسومها فمن ذلك جرم شبيهه بجرم
الاسطرلاب وفيها شبكه من غير صفايح وفي ارض الجرم رسوم شكازيه من غير بروج
وفي احدى نصفي الشبكه نصف رسوم شكازيه وفي النصف الاخر نصف شبكه اسطرلاب
آسي وظهر الجرم على عاده الاسطرلابات من العضاده وقوس الارتفاع والظل والجيوب
وسميتها الاله السراجيه في استخراج الاعمال الافاقيه واعلم ان محيط الجرم
يسمى داير نصف النهار والخط المار بالعلاقه يسمى مدار الاستوا والدواير التي عن
جنبيه تسمى المدارات اليوميه فالتي عن يمين الناظر شماليه والتي عن يساره جنوبيه
والنقطتان اللتان في وسط اصغر المدارات يسميان القطبين والدواير المارة
بهما تسمى الممرات والافاق المايله والخط المستقيم الواصل بينهما يسمى افق الاستوا
والخطان الخارجان من المركز الى مثل الميل الاعظم من داير نصف النهار عن جنبي مدار
الاستوا يسميان خطي البروج على كل واحد البروج التي في جهته والعضاده التي من
الشبكه تسمى الافق المايل والمدارات الموازيه له تسمى مقنطرات الارتفاع والنقطه
التي في وسط اصغرها تسمى سمت الراس والخط الواصل بينه وبين مركز الشبكه يسمى
اول السموت والدواير الا[illegible] من سمت الراس الى الافق تسمى السموت ومركز الشبكه

[redacted]

LXV Ibn al-Sarrāj's treatise on the simple universal astrolabe invented by him

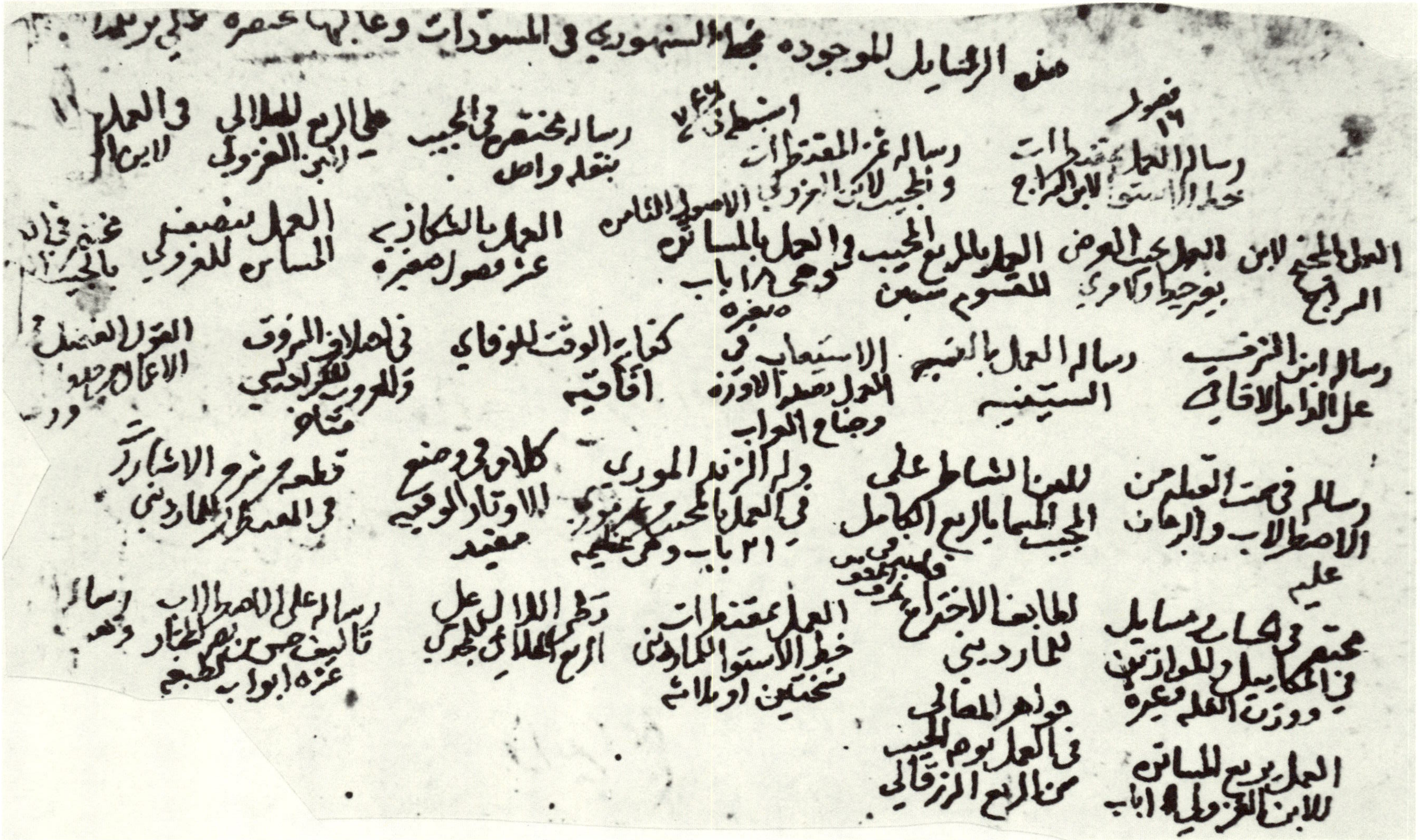

LXVI A list of treatises on the quadrant, several of which are no longer extant

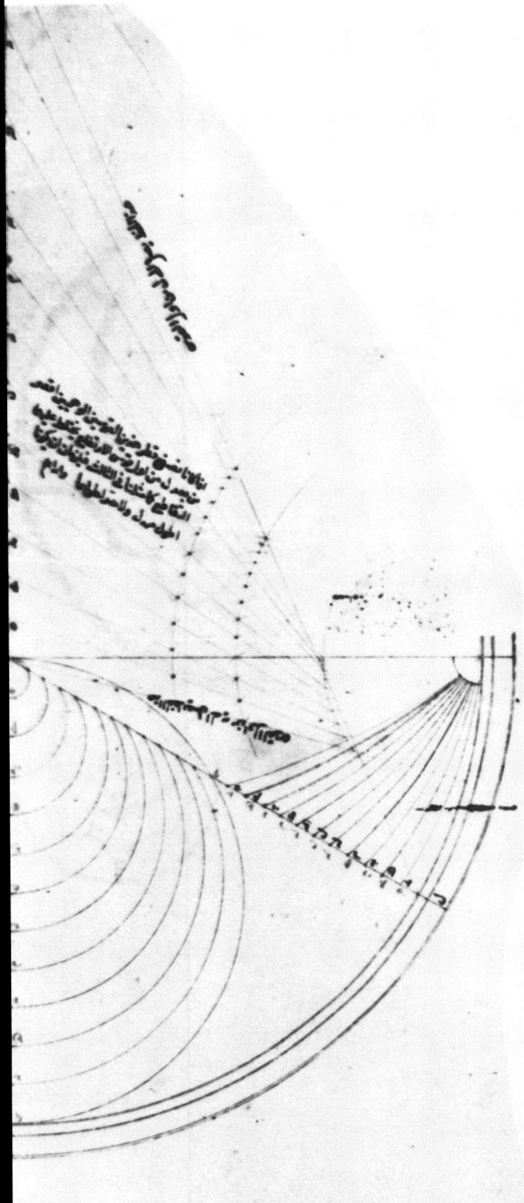

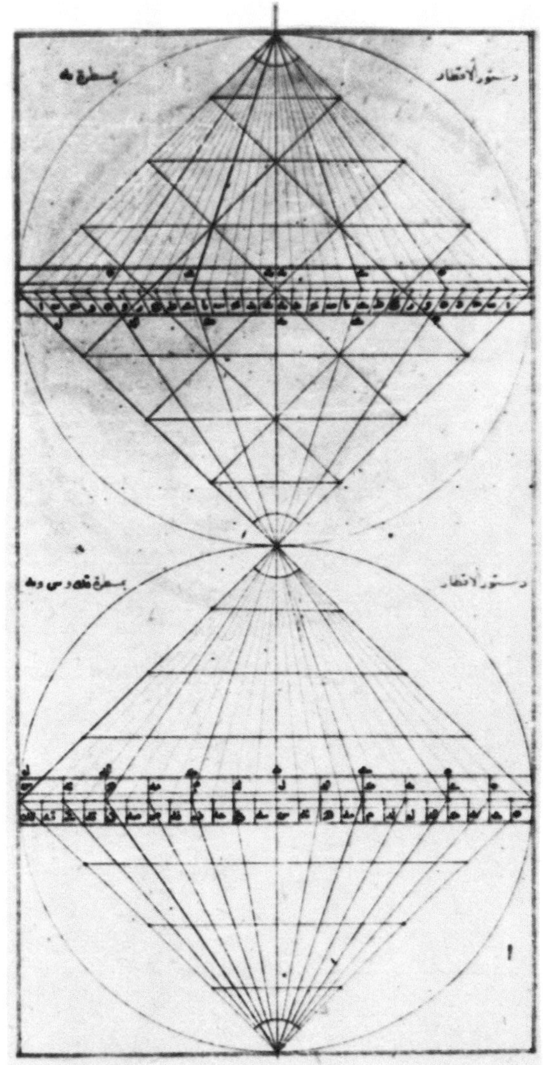

[redacted]LXVII Illustrations of the constructions for marking unusual astronomical instruments

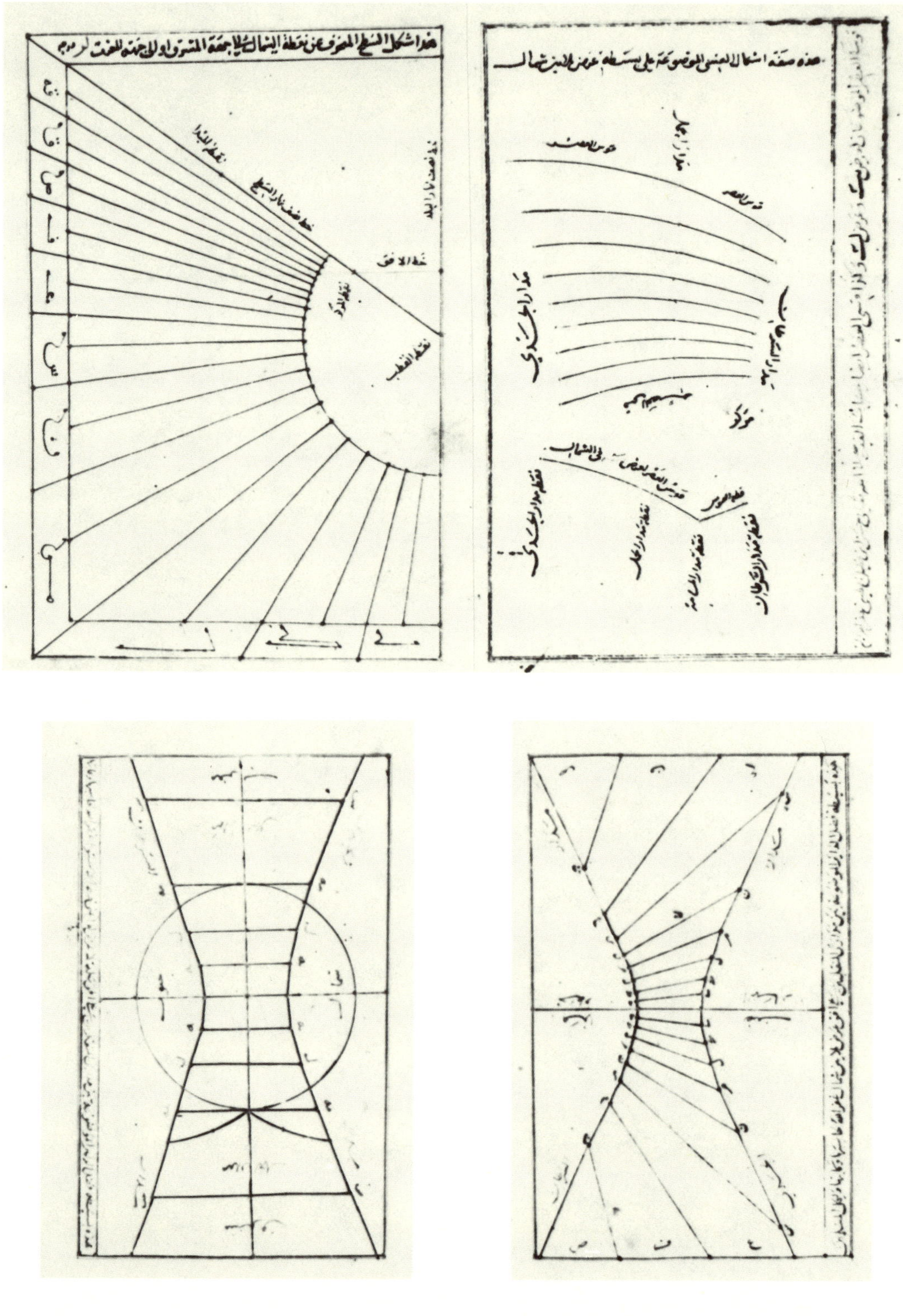

LXVIII Illustrations of vertical, horizontal and polar sundials in the treatise of al-Karādīsī

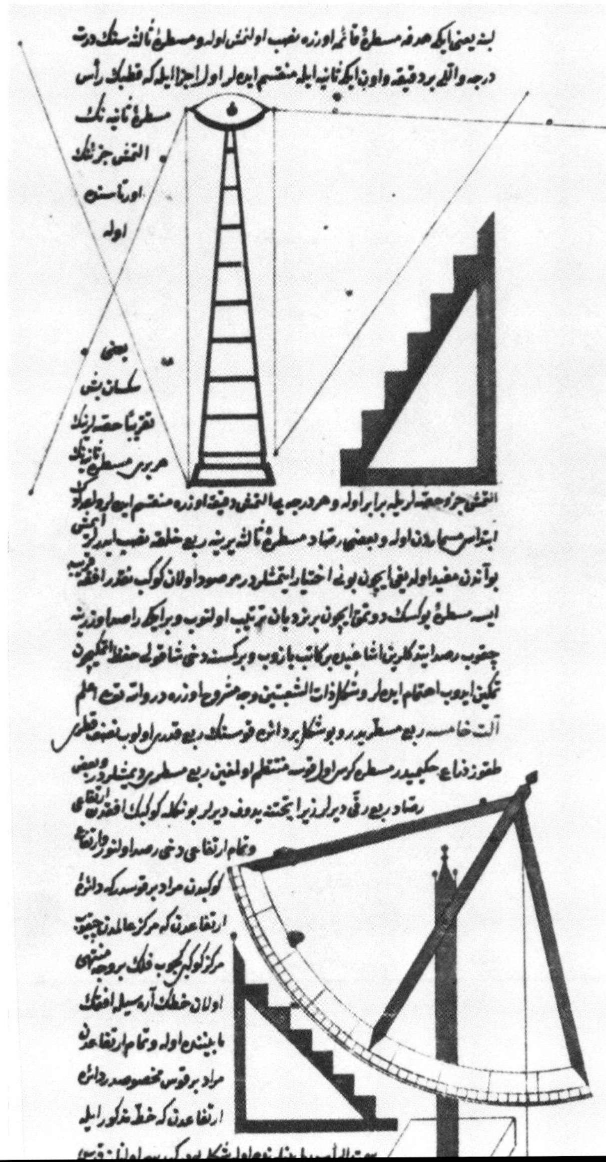

[redacted]

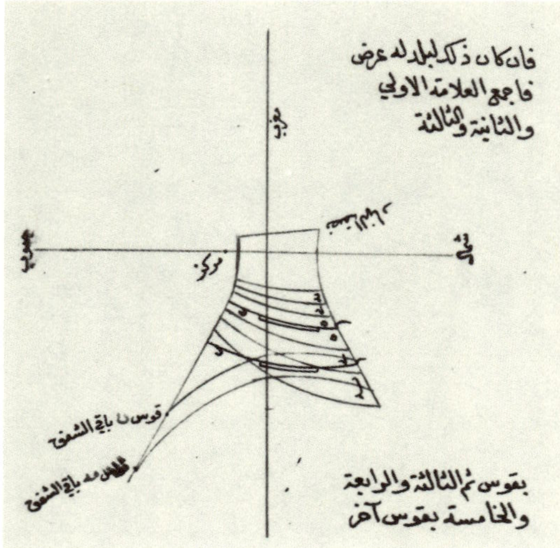

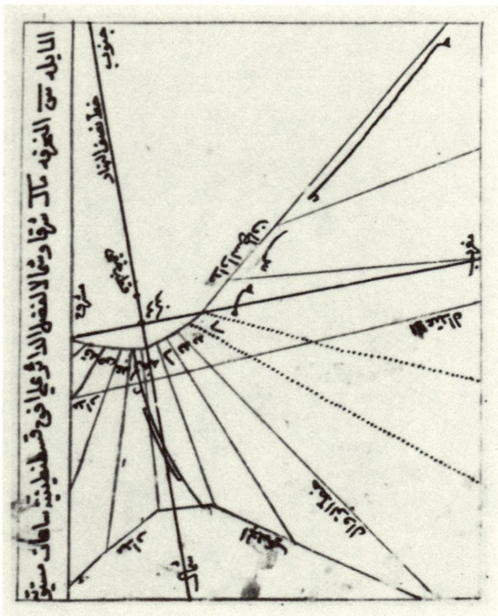

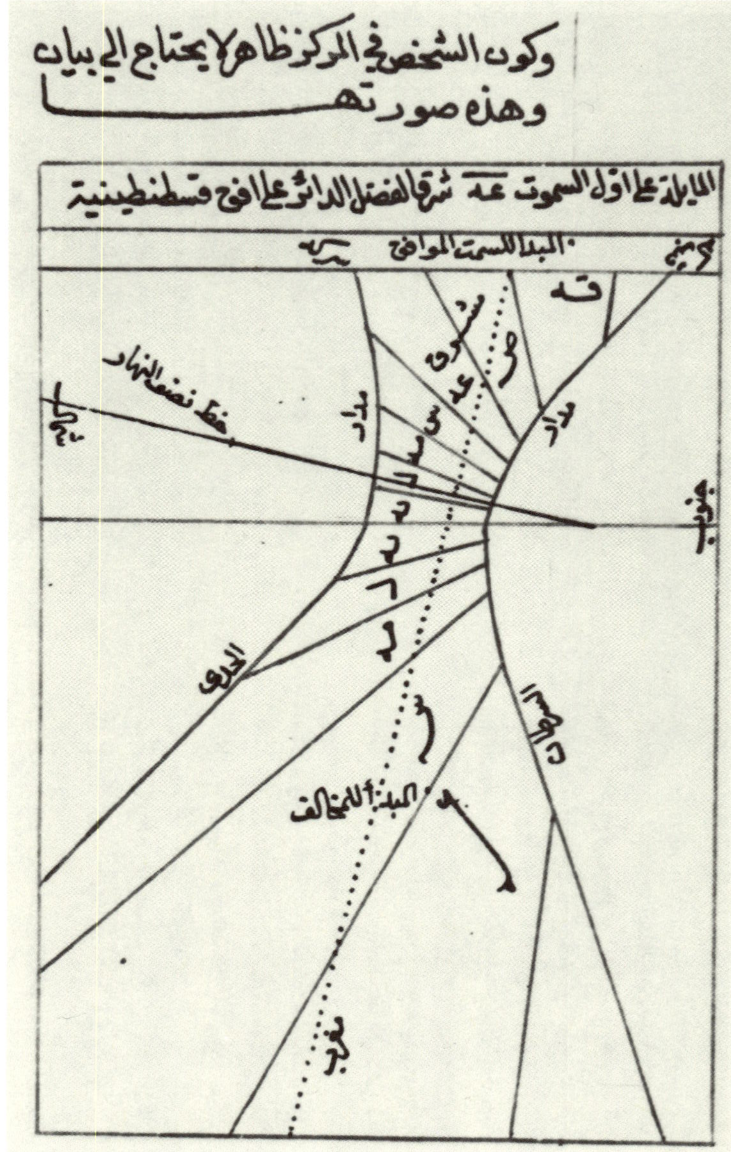

LXX Illustrations of different kinds of sundials in the treatise of Taqī l-Dīn

بسم الله الرحمن الرحيم . اللهم صل على سيدنا محمد النبي وسلم
من كلام دانيال وهرمس الحكيم وذي القرنين وبلعم بن باعور
في الحوادث السماويه كالرعد والبرق والرياح والكواكب والهالات
والكسوفات والمواليد والاشكال وغيره ذلك مرتبا على شهور الروم
والله الموفق للصواب وقد رتبته على اثنين وعشرين بابا
الباب الأول في ذكر كسوف الشمس الباب الثاني
في ذكر كسوف القمر الباب الثالث في ذكر هالة الشمس الرابع
في ذكر هالة القمر الباب الخامس في ذكر اغتلال الهلال السادس
في ذكر البروق الباب السابع في ذكر الرعد وهو اربعة انواع
الباب الثامن في ذكر الرياح والزوابع والغيوم
الباب التاسع في ذكر وقوع المطر الباب العاشر
في ذكر دلايل البرد الباب الحادي عشر في ذكر قوس قزح الباب الثاني عشر
في ذكر الزلازل الباب الثالث عشر في انقضاض الكواكب الباب الرابع عشر
في ذكر الذوابة والذنب الباب الخامس عشر في الحمرة التي تظهر شبه الرمح
الباب السادس عشر في الاشكال التي تظهر بالجو

الباب السابع عشر في ذكر تحويل [redacted]
الباب الثامن عشر في ذكر مولد الس[redacted]
الباب التاسع عشر في ذكر مواقع الك[redacted]
الباب العشرون في ذكر المريخ [redacted]
الباب الحادي والعشرون في ذكر ك[redacted]
الباب الثاني والعشرون في ذكر دخو[redacted]
ايام الاسبوع وها نحن نشرع في الكلام [redacted]
الباب الاول في ذكر كسو[redacted]
اذا انكسفت الشمس في مبدا طلوعها في [redacted]
وقوع السبي في بعض المدن مما يلي الشرق و[redacted]
وكثرة السقط وهلاك قوم لهم صيت [redacted]
الغلا واتصال الاحداث وان كانت عند الغروب [redacted]
الا السبي يقع في البلاد المحاذيه المغرب ويوجب [redacted]
بغض السلاطين وزوال ملك وتشتيت خوا[redacted]
عنه دانيال وهرمس ان انكسفت في [redacted]

[redacted]e on astrology attributed to Hermes

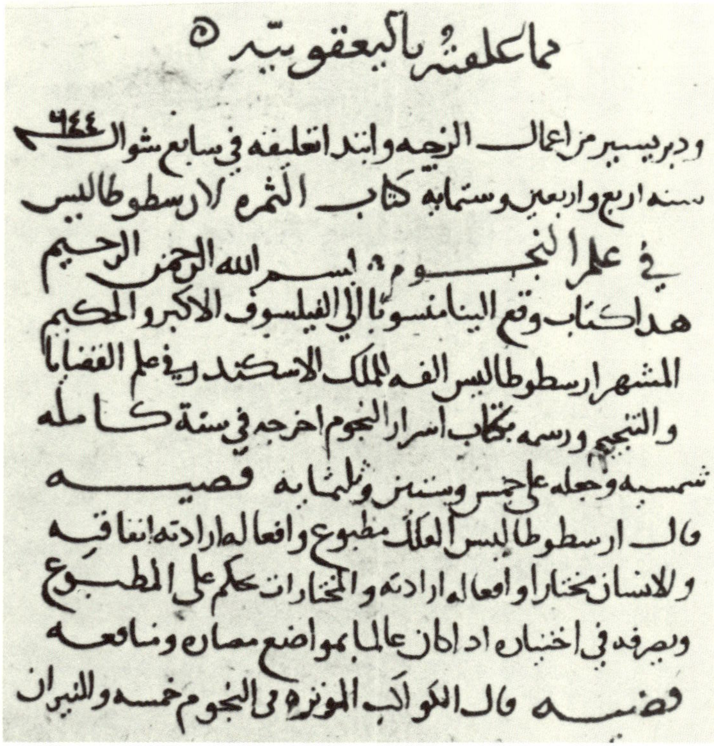

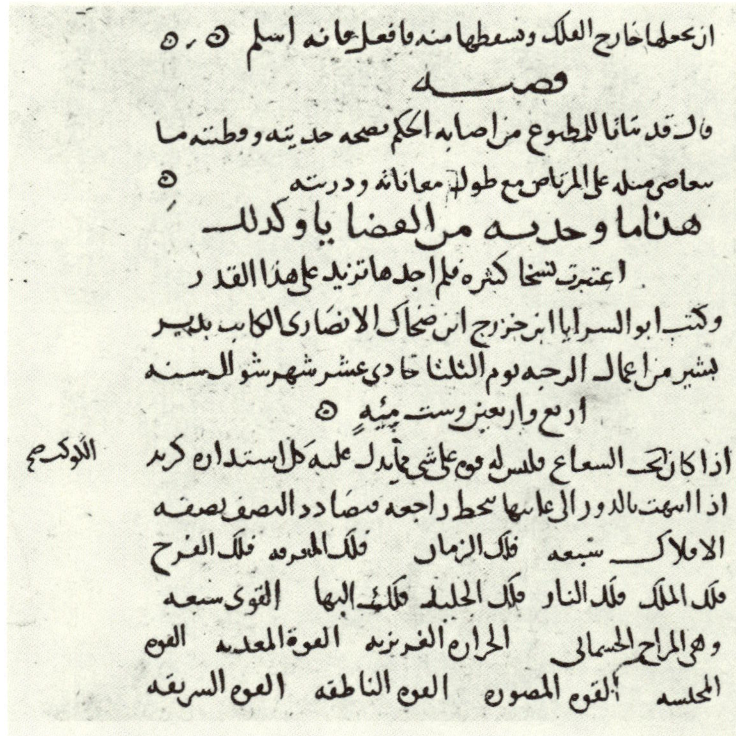

LXXII A treatise on astrology attributed to Aristotle

b.

[redacted]ns of the seating arrangements in royal courts, associated with Ptolemy and Jnh "the Indian."

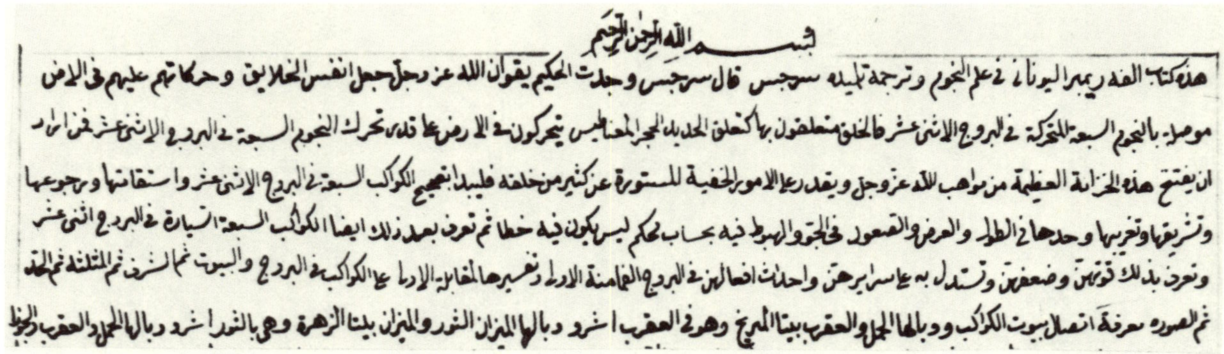

LXXIVa A treatise on astrology attributed to Rymbr "the Greek"

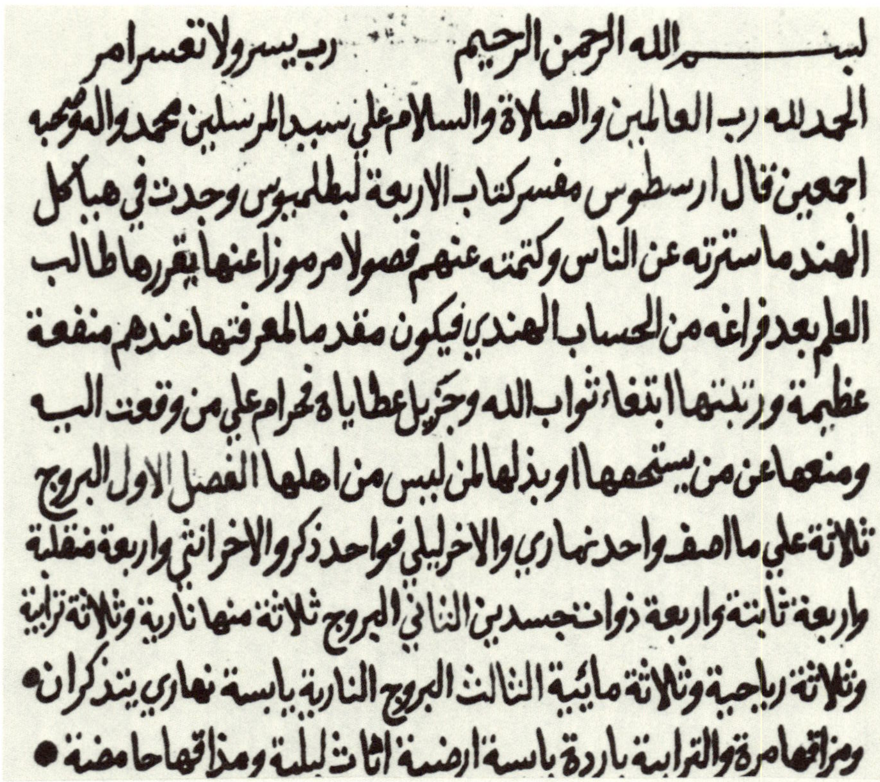

LXXIVb–c A treatise on astrology variously attributed to Aristus and Eutocius

[redacted]scope of the Prophet Muḥammad

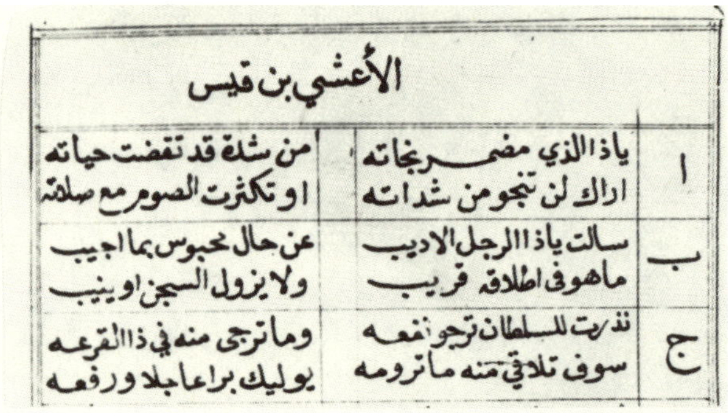

	الأعشي بن قيس	
ا	ياذ االذي مضمر نجاته اراك لن تنجو من شداته	من شلة قد تقضت حياته او تكثرت الصوم مع صلاته
ب	سالت ياذ االرجل الاديب ماهو في اطلاقه قريب	عن حال محبوس بما اجيب ولا يزول السجن او ينيب
ج	نذرت للسلطان ترجو نفعه سوف تلاقي منه ماتر ومه	وما ترجى منه في ذا القرعه يوليك براعاجلا ورفعه

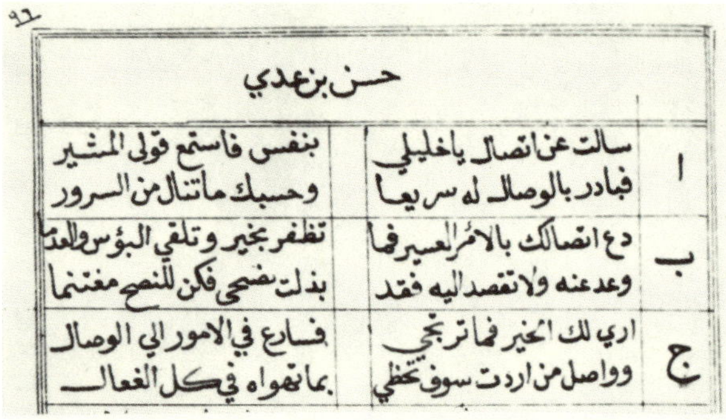

	حسن بن عدي	
ا	سالت عن اتصال ياخليلي فبادر بالوصال له سريعا	بنفس فاستمع قولي المشير وحسبك ماتنال من السرور
ب	دع اتصالك بالامر العسير فما وعدعنه ولا تقصد اليه فقد	تظفر بخير و تلقي البؤس والعدما بذلت نصحي فكن للنصح مغتنما
ج	اري لك الخير فما ترتجي وواصل من اردت سوف تحظي	فسارع في الامور الي الوصال بما تهواه في كل الفعال

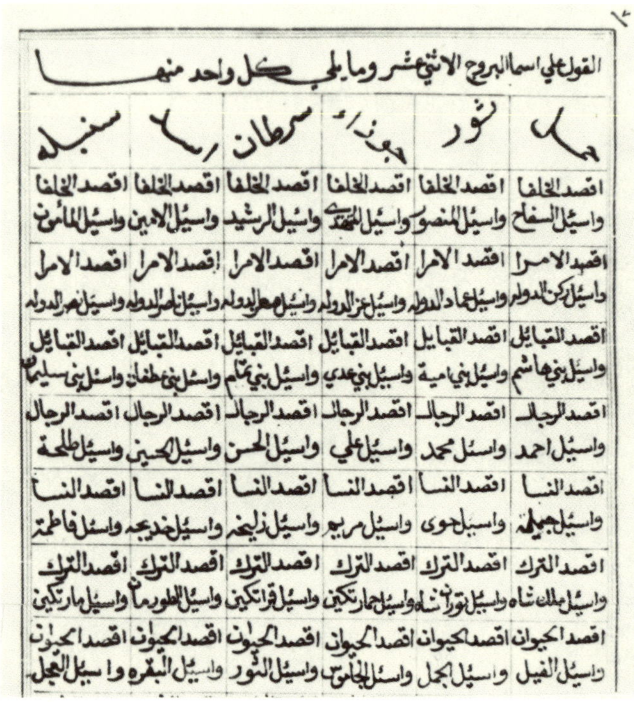

القول علي اسما البروج الاثني عشر وما يلي كل واحد منها

حمل	ثور	جوزا	سرطان	اسد	سنبله
اقصد الخلفا واسيل السفاح	اقصد الخلفا واسيل المنصور	اقصد الخلفا واسيل المهدي	اقصد الخلفا واسيل الرشيد	اقصد الخلفا واسيل الامين	اقصد الخلفا واسيل المامون
اقصد الامرا واسيل ركن الدوله	اقصد الامرا واسيل عماد الدوله	اقصد الامرا واسيل عز الدوله	اقصد الامرا واسيل معز الدوله	اقصد الامرا واسيل ناصر الدوله	اقصد الامرا واسيل نصر الدوله
اقصد القبايل واسيل بني هاشم	اقصد القبايل واسيل بني امية	اقصد القبايل واسيل بني عدي	اقصد القبايل واسيل بني تمام	اقصد القبايل واسيل بني غطفان	اقصد القبايل واسيل بني سليمان
اقصد الرجال واسيل احمد	اقصد الرجال واسيل محمد	اقصد الرجال واسيل علي	اقصد الرجال واسيل الحسن	اقصد الرجال واسيل الحسين	اقصد الرجال واسيل طلحة
اقصد النسا واسيل جميلة	اقصد النسا واسيل حوى	اقصد النسا واسيل مريم	اقصد النسا واسيل زليخا	اقصد النسا واسيل خديجه	اقصد النسا واسيل فاطمة
اقصد الترك واسيل ملك شاه	اقصد الترك واسيل توران شاه	اقصد الترك واسيل احمار تكين	اقصد الترك واسيل قراتكين	اقصد الترك واسيل الطوره	اقصد الترك واسيل مارتكين
اقصد الحيوان واسيل الفيل	اقصد الحيوان واسيل الجمل	اقصد الحيوان واسيل الجاموس	اقصد الحيوان واسيل الثور	اقصد الحيوان واسيل البقره	اقصد الحيوان واسيل العجل

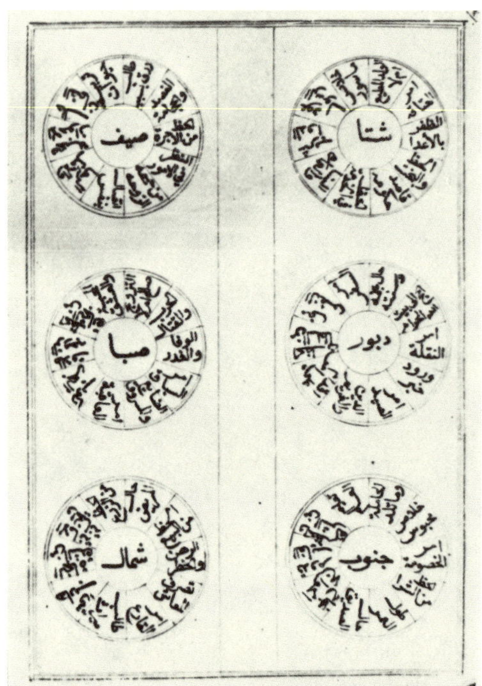

LXXVI Extracts from a book for making prognostications attributed to the Abbasid Caliph al-Maʾmūn

[redacted]

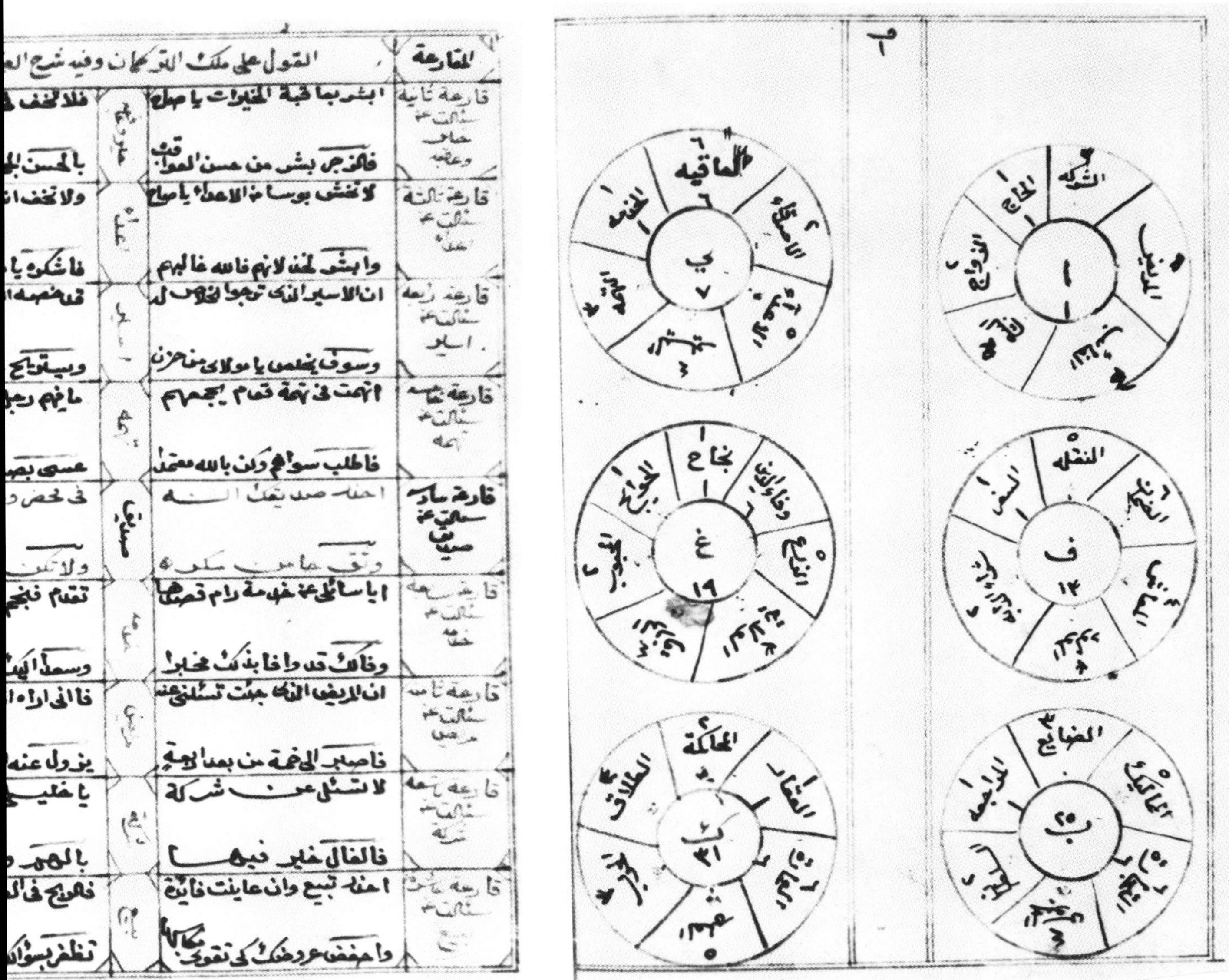

cts from a book for making prognostications attributed to Abū Maʿshar

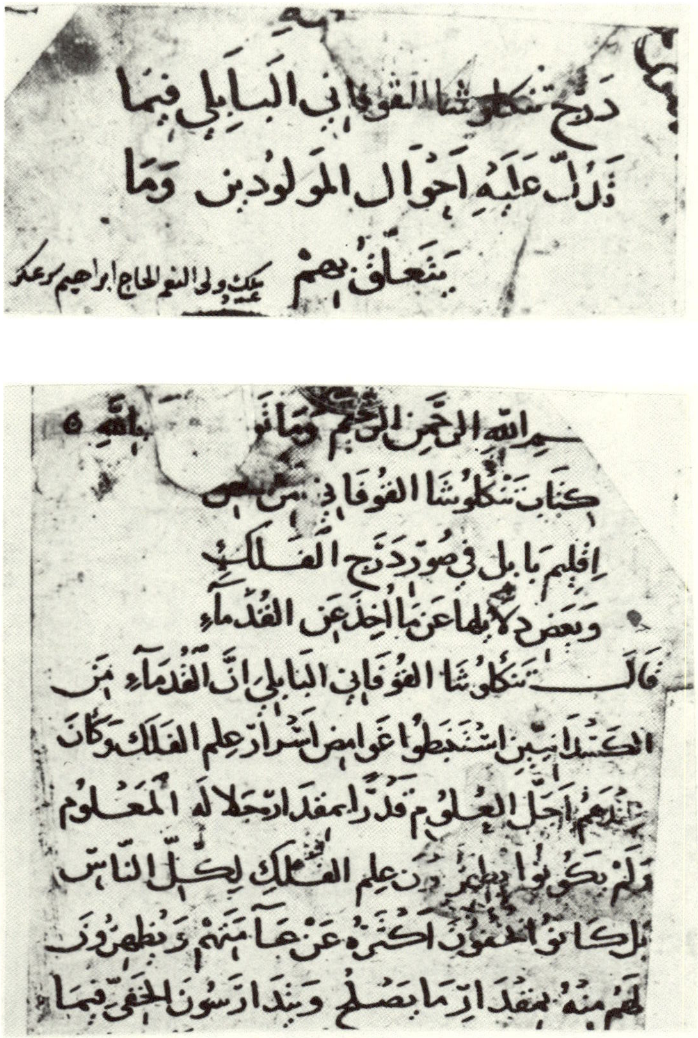

LXXVIIIa An astrological treatise attributed to Teucros

LXXVIIIb An astrological treatise by al-Kindī

[illegible] / ه	حمل / سنبلة / عقرب	ثور / قوس	جوزاء / جدي	سرطان / دلو	اسد / ميزان / حوت
ا ب	مد و	كو كح	ح ى	ك ك	ب د
ج د	يح ك	ل ب	ب مد	كد كو	و ح
ه و	كب كد	د و	يو يح	يح ل	ى ب
ز ح	ح ى	ك كب	ب د	مد و	كو كح
ط ى	ب مد	كد كو	و ح	يح ك	ل ب
يا يب	يو يح	يح ل	ى ب	كب كد	د و
يج يد	ب د	مد و	كو كح	ح ى	ك كب
يه يو	و ح	يح ك	ل ب	ب مد	كد كو
يز يح	ى ب	كب كد	د و	يو يح	يح ل
يط ك	كو كح	ح ى	ك كب	ب د	مد و
كا كب	ل ب	ب مد	كد كو	و ح	يح ك
كج كد	د و	يو يح	يح ل	ى ب	كب كد
كه كو	ك كب	ب د	مد و	كو كح	ح ى
كز كح	كد كو	و ح	يح ك	ل ب	ب مد
كط ل	يح ل	ى ب	كب كد	د و	يو يح

الحمد لله هذا جدول الحياة وملك عمر المولود
كما وجدته في كتاب الكامل في النجوم بخزانة عبد الله
بن طاهر بن الحسين من الملكون مكتوبا بالذهب
وصورة العمل بهذا الجدول ان تأخذ اجزاء ساعات
درج الطالع واضربه في ب واحفظه ثم انظر
الجدول الحياة وخذ ما بازائه من درجة
الطالع من العدد فما وجدته انسبه من س
وخذ مثل تلك النسبة مما خرج من الضرب
المحفوظ فهو حياة المولود من السنين والشهور
مثال ذلك ولد مولود وله من الطالع خمس
درج من الجوزاء فاخذنا ما بازائه من الجدول
فوجدناه يو وهي من س سدس وعشر
واجزاء ساعات درج الطالع يو مد مضروبا في
ب الخارج ل ب مح سدس ذلك وعشره
نوبه وهي ستة وخمسين سنة
وتسعة اشهر بالتقريب والله اعلم بالغيب
وجد بخط ابن ابي الفتح خذ قوس نهار الطالع
واحفظه وادخل بالدرجة الطالعة في الجدول
وخذ ما وجدته اضربه في المحفوظ واقسم الحاصل
على س تخرج القسمة سنين والله اعلم

هكذا وجدته وكما وجدته نقلته

[illegible]

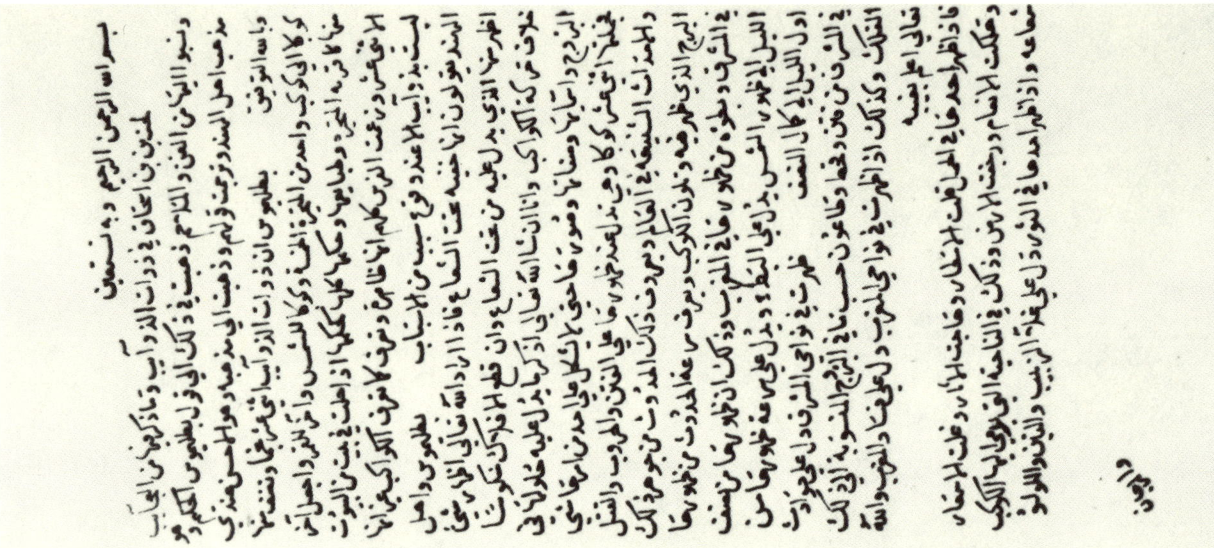

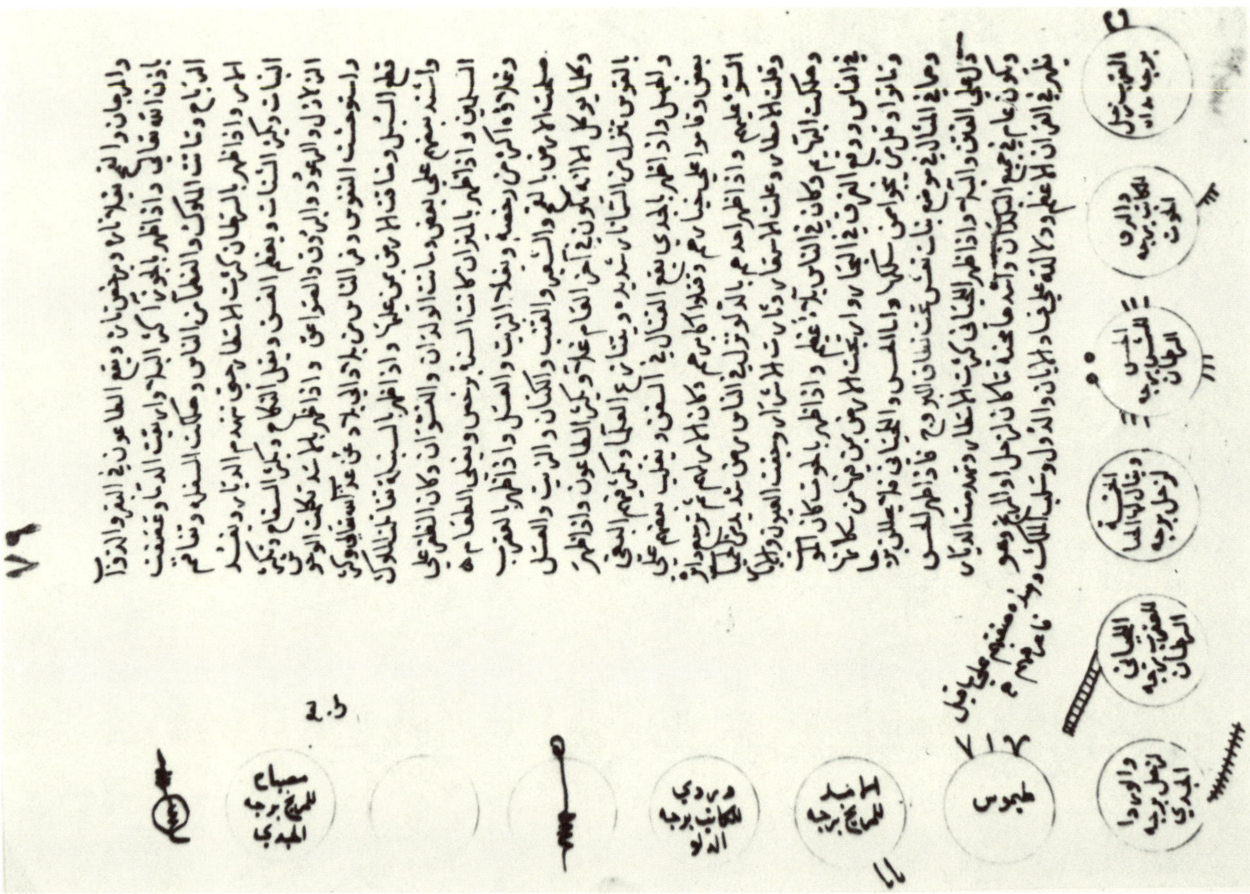

LXXX Ḥunayn ibn Isḥāq's treatise on comets

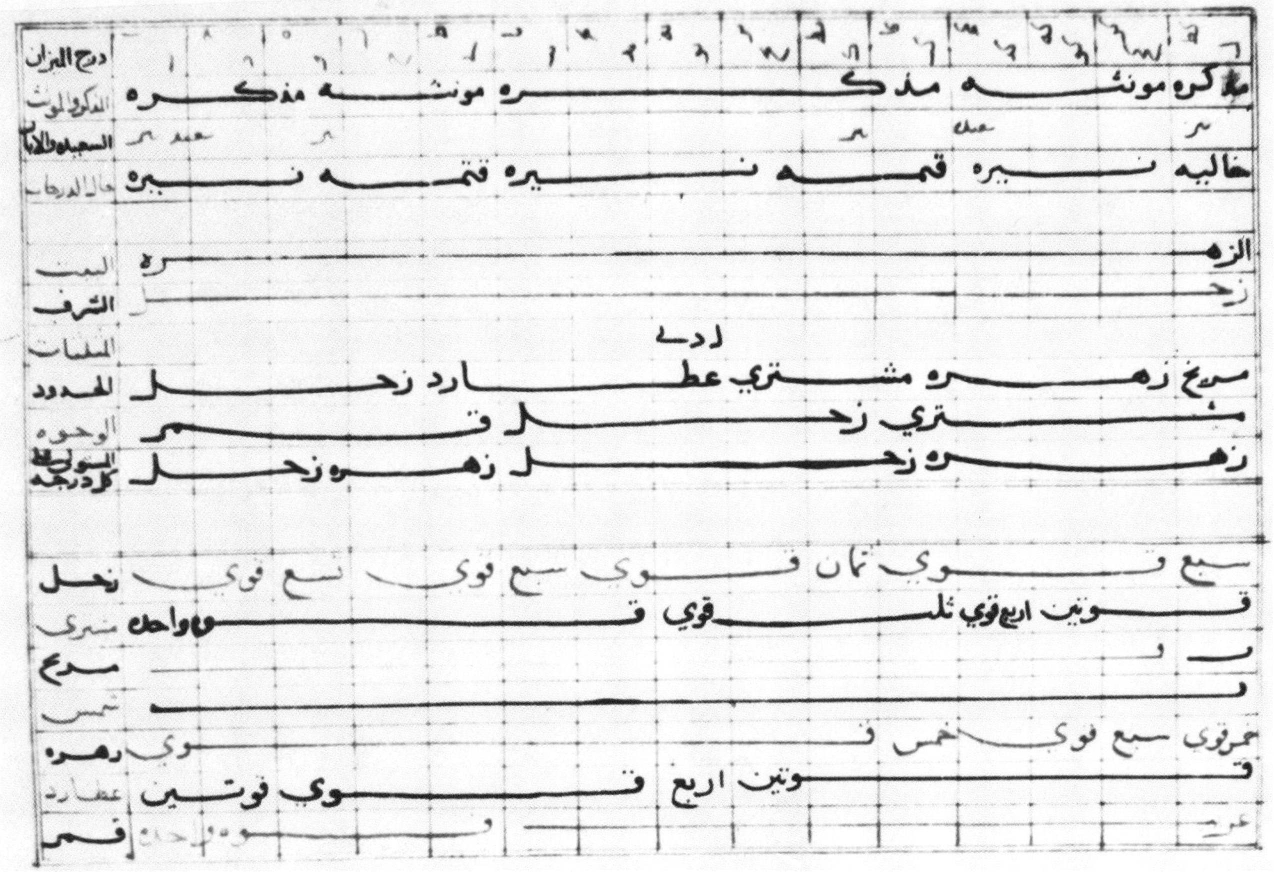

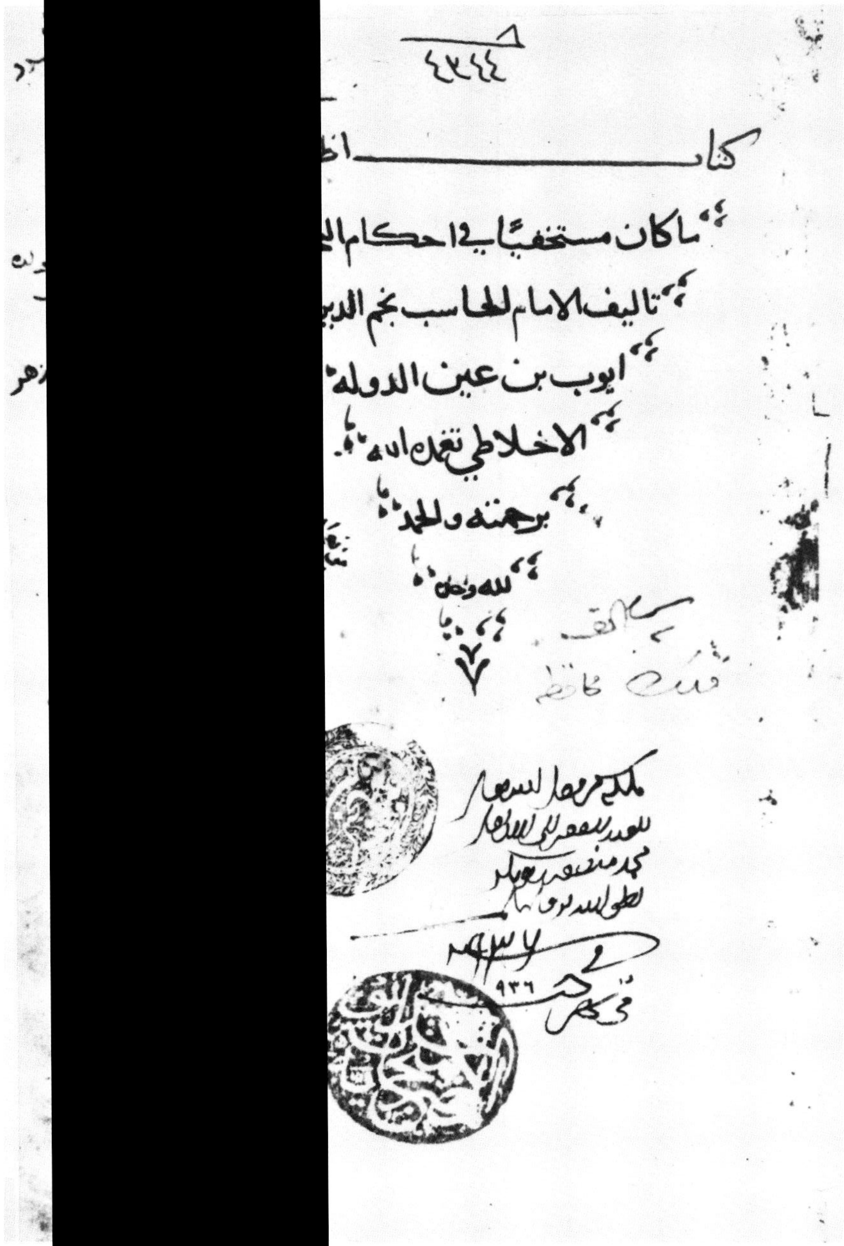

[redacted]tise on astrology

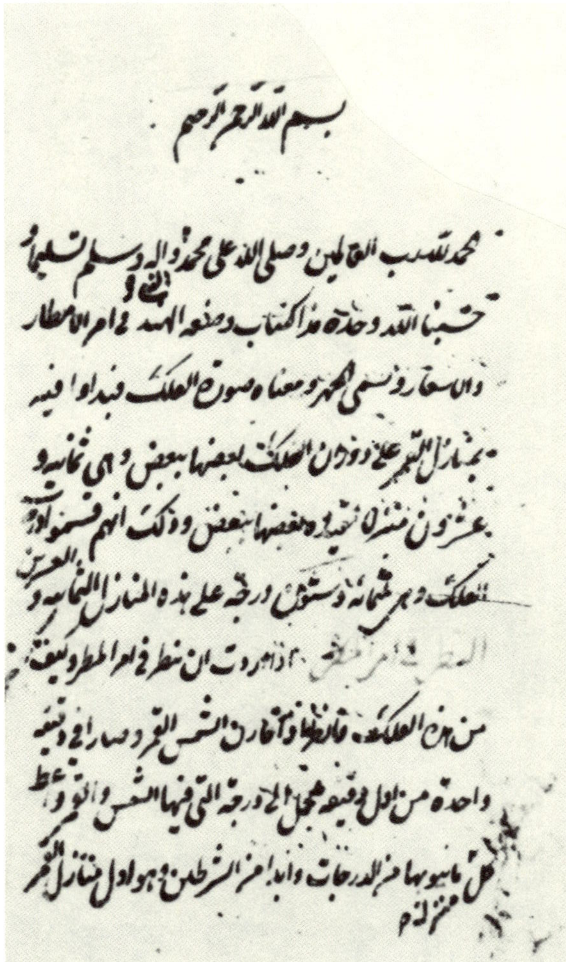

LXXXIIa A treatise on astrology apparently based upon a Sanskrit original

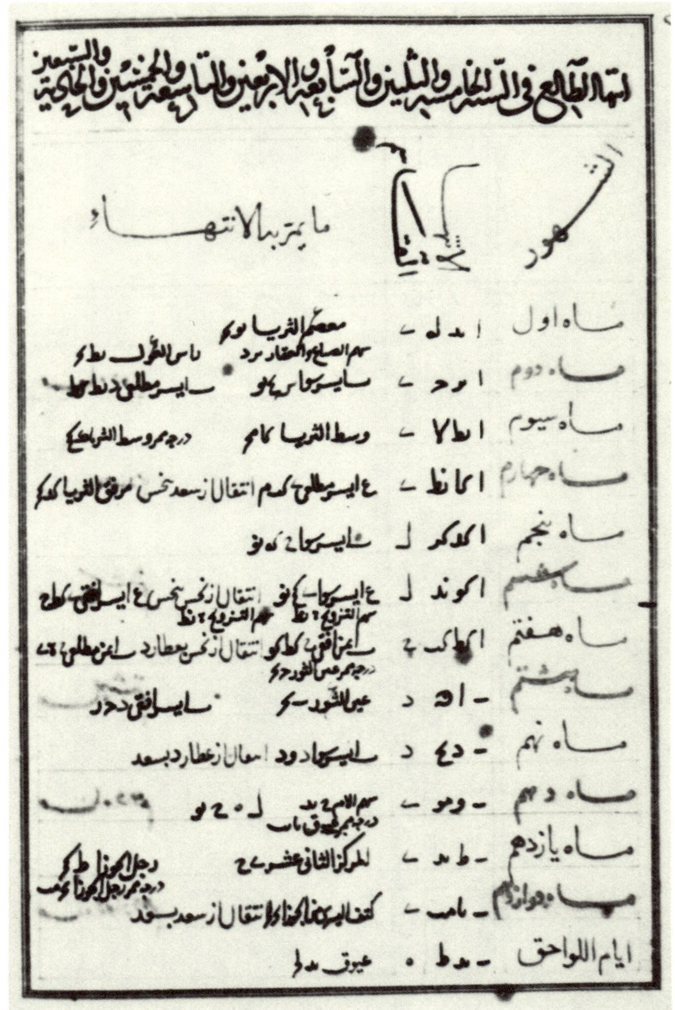

LXXXIIb A table of prognostications in a Safavid nativity book

[redacted]

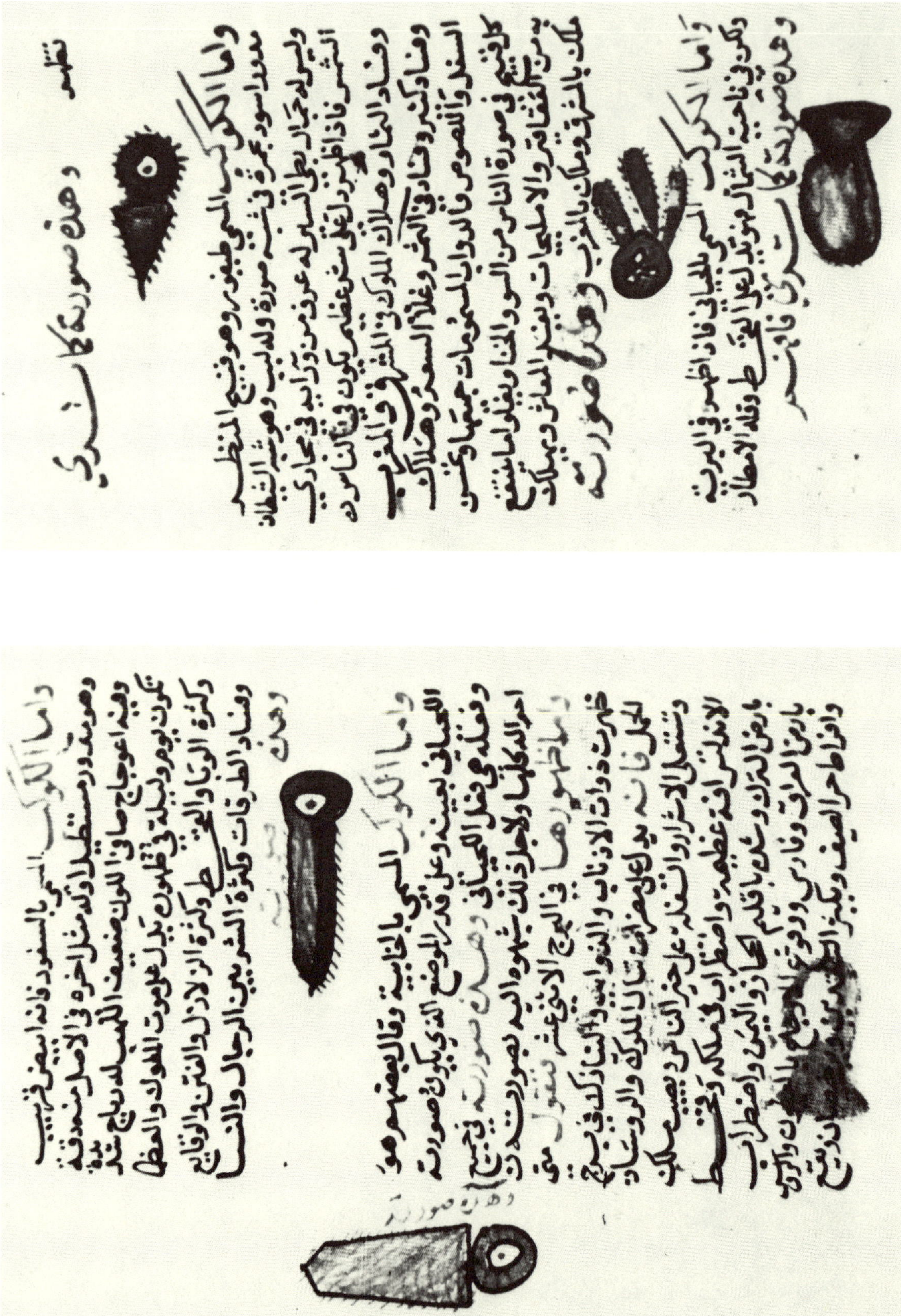

LXXXIV Illustrations of pseudo-comets in a late Egyptian treatise

[redacted]

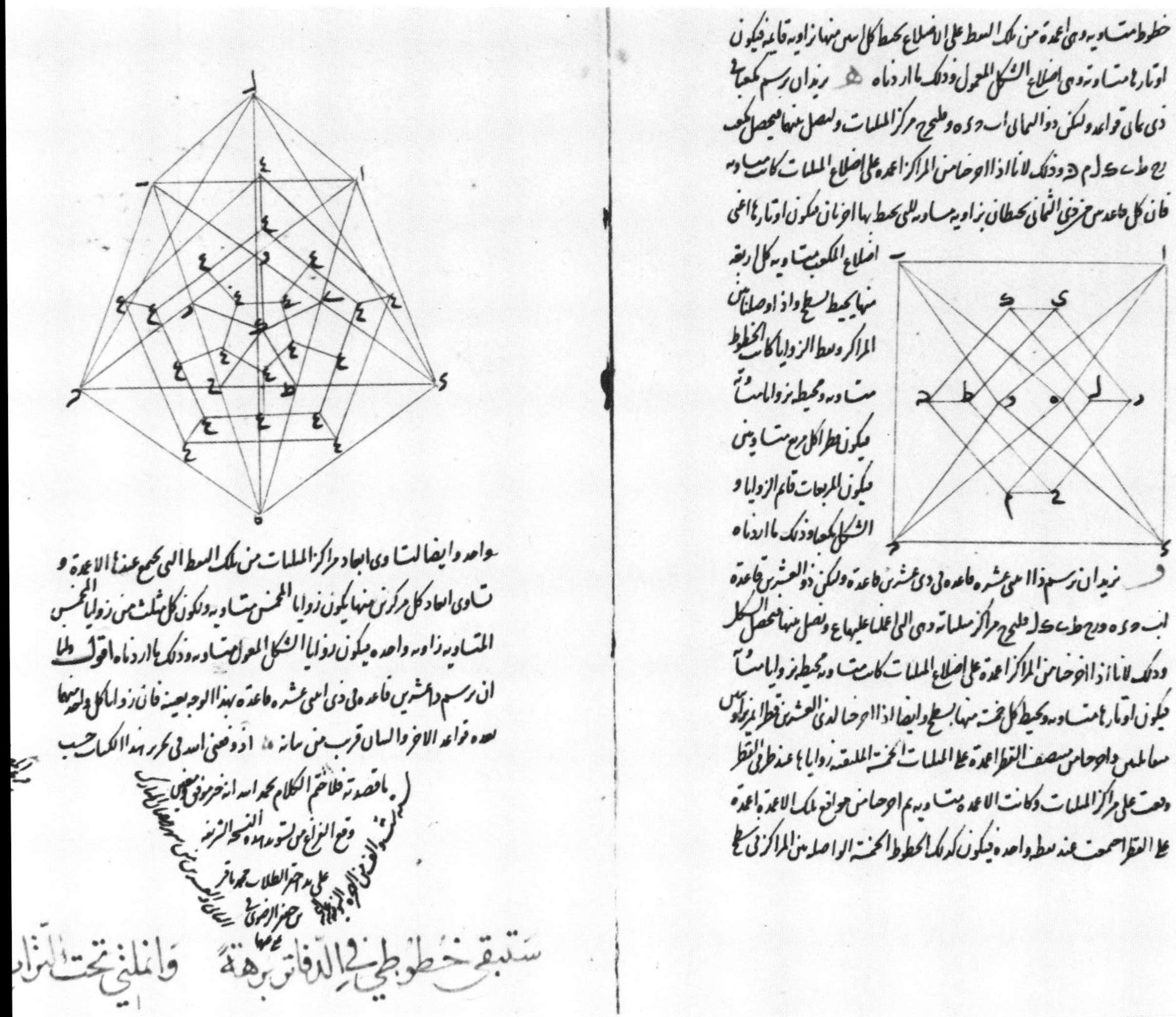

The colophon of al-Ṭūsī's recension of Euclid's *Elements*

منقول من كتاب شرح مختصر الشيخ

شرح الفقيه الأوحد الفاضل أبي عبد الله

ذكر بعد أن فرغ من شرح باب التكملة

وآخر مسائله قال الشيخ رحمه الله

أن الثلث نصيب وخمس خماسيات فهو اذا ثلاثة عشر

فاجعل الثلث ثلاثة عشر فيكون المال تسعة وثلاثين والنصيب

ثمانية والوصية تسعة والله أعلم وهذه صورته

LXXXVI al-Khuzāʿī's commentary on the section on inheritance in al-Khwārizmī's *Algebra*

[redacted]

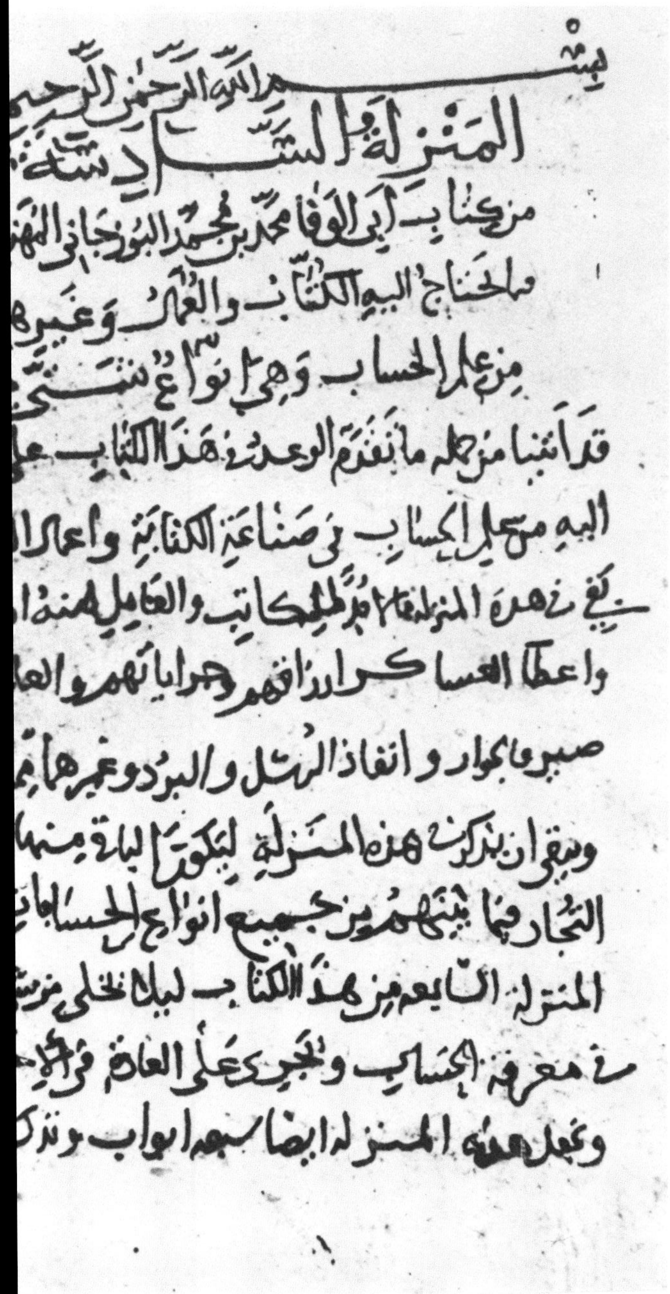

[redacted]st folio of al-Būzajānī's treatise on arithmetic

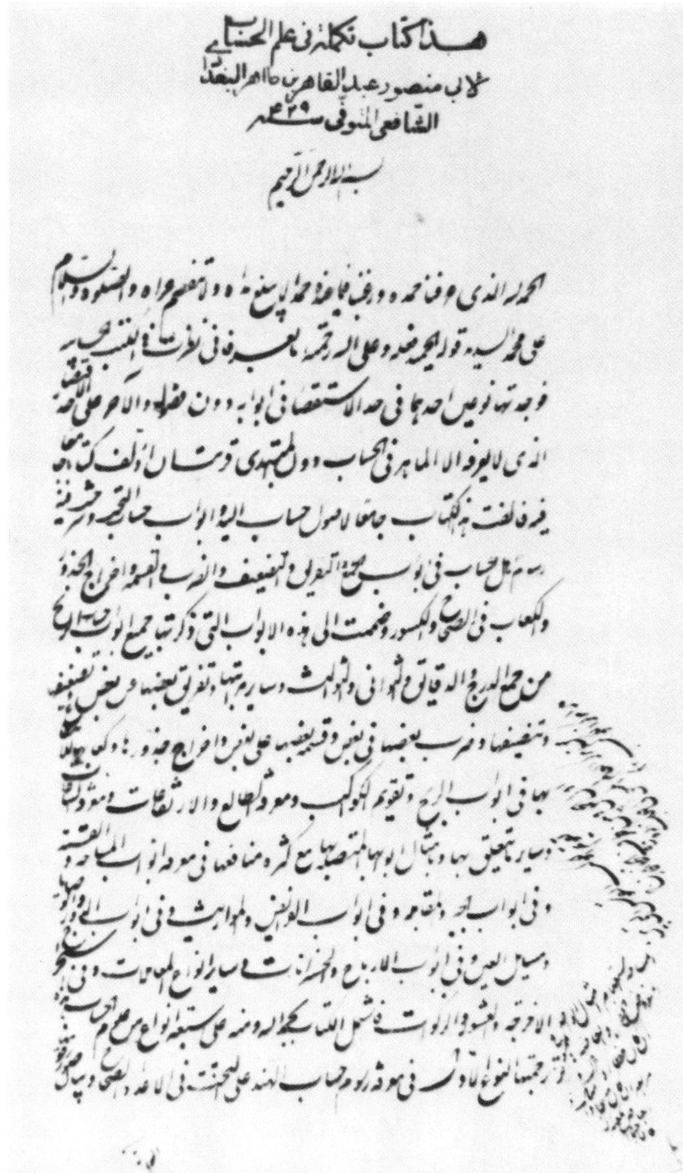

LXXXVIIb The first folio of al-Baghdādī's treatise on arithmetic

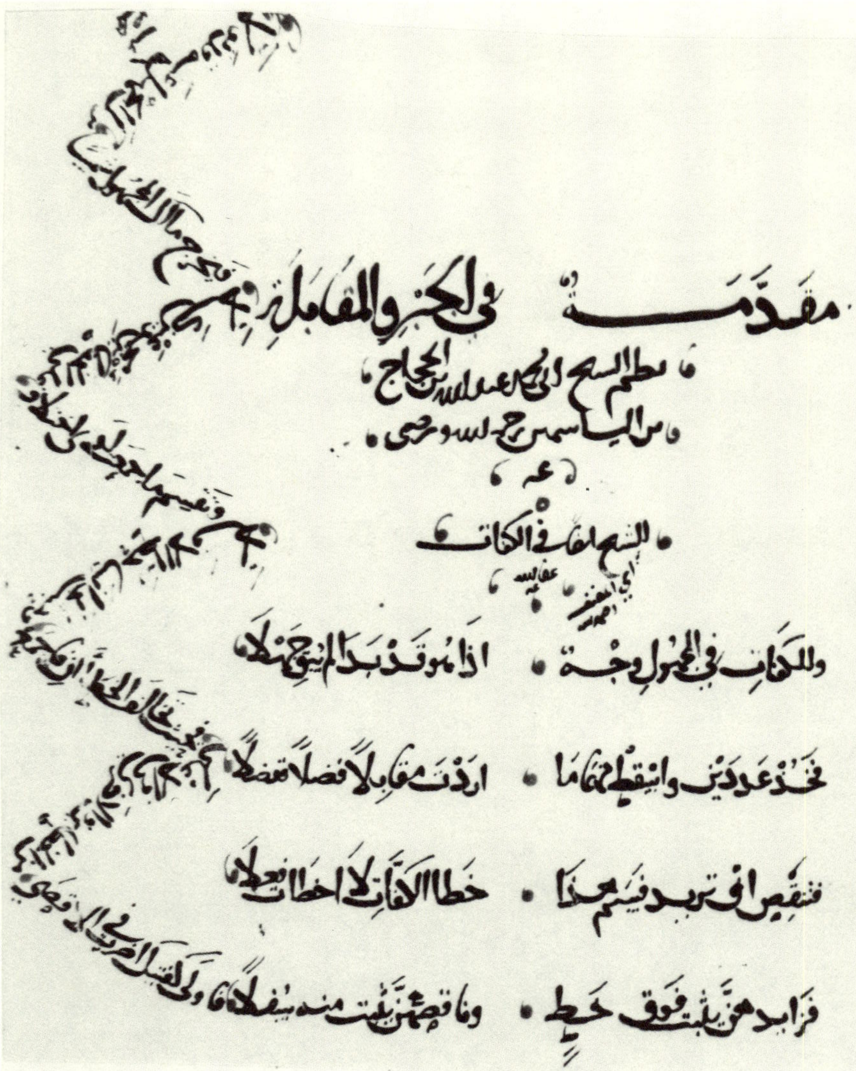

LXXXVIIIa The beginning of Ibn al-Yasamīn's poem on algebra

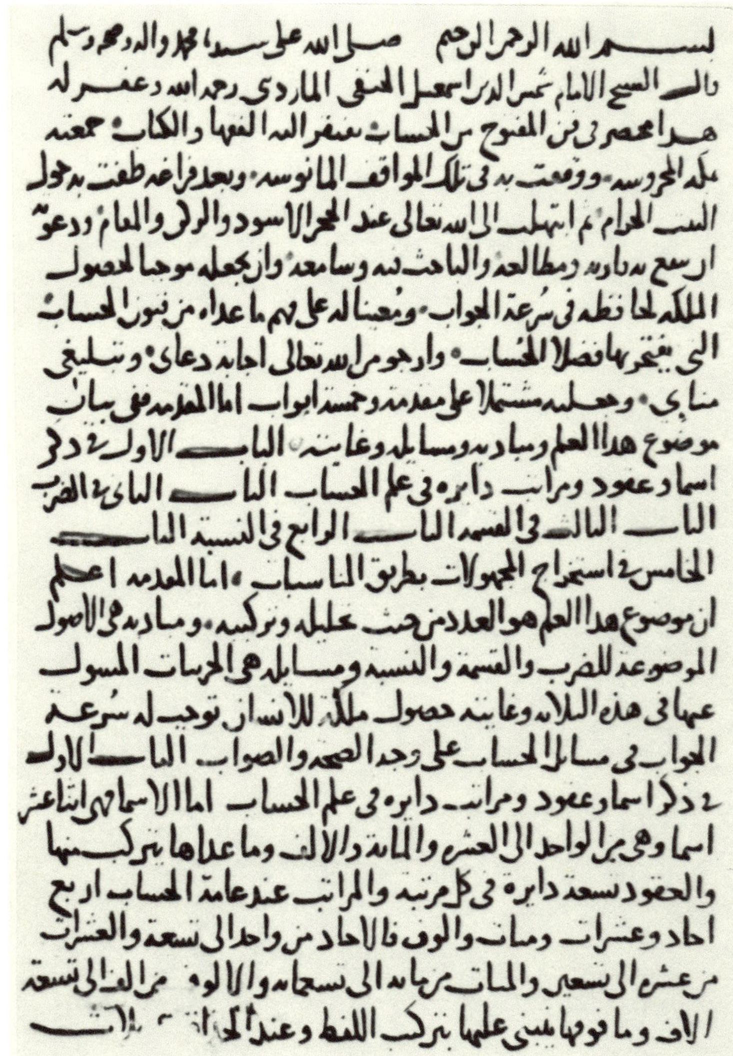

LXXXVIIIb The beginning of a treatise on arithmetic by Ibn Fallūs

[redacted]

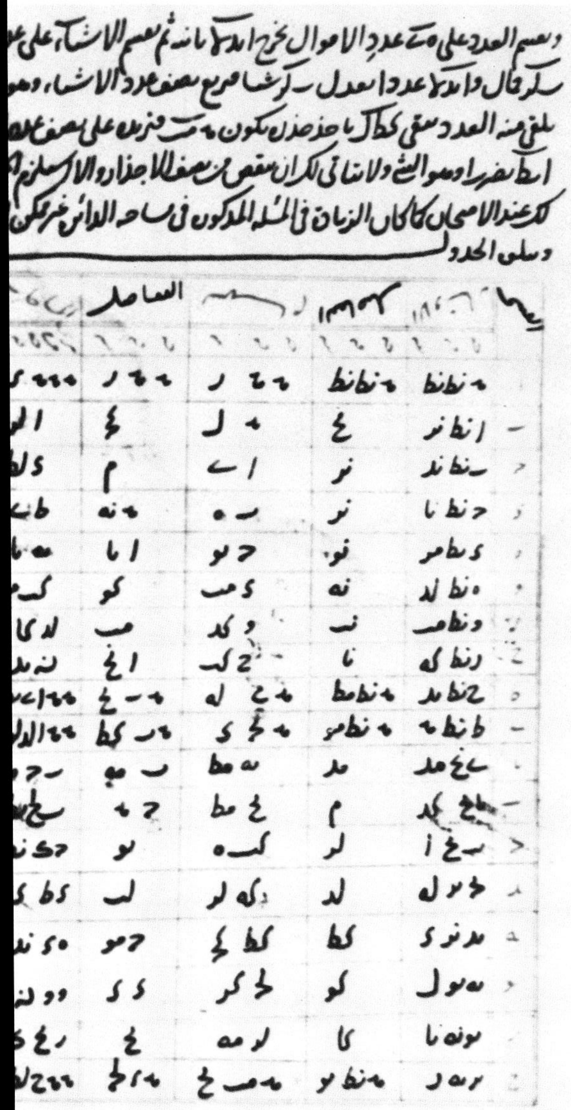

Tables in al-Fārisī's treatises on mathematics

XC The table of contents of al-Ghaznawī's treatise on mathematics

[redacted]

[redacted]ultiplying sexagesimal numbers

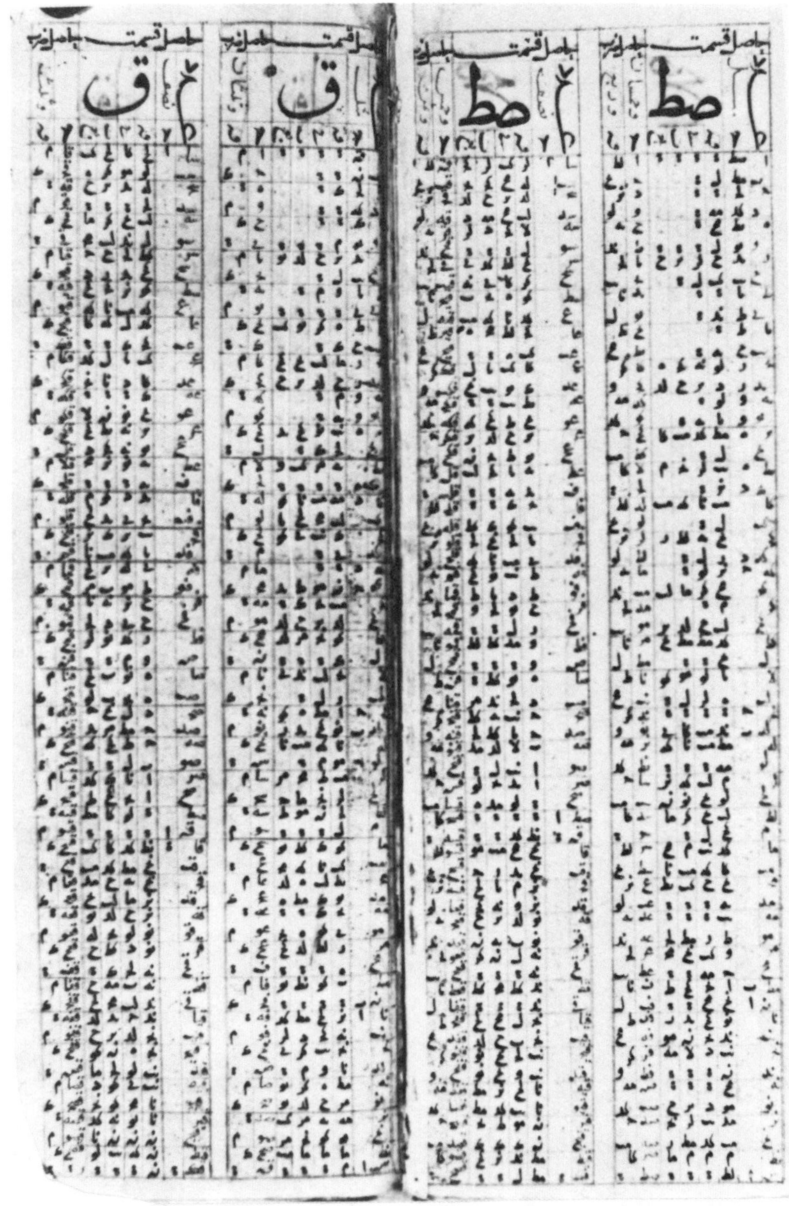

XCIb A table of sexagesimal quotients

XCIIa An extract from one of Ibn al-Hā'im's treatises on inheritance

XCIIb A table from the treatise on arithmetic by al-Kāshī

[redacted]

[redacted]athematical notation in the treatise on algebra
[redacted]ṣādī

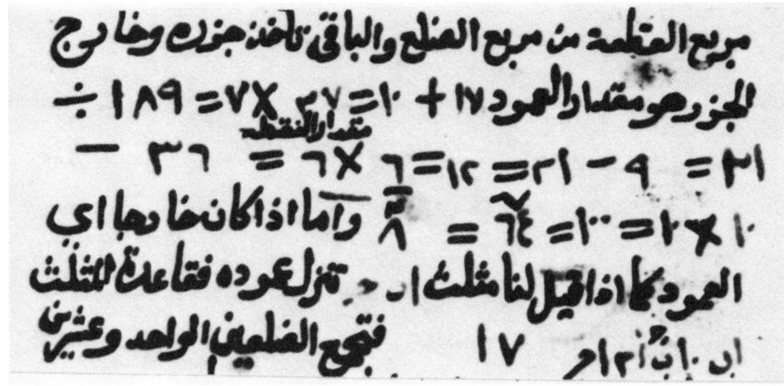

XCIIIb Unusual mathematical notation in a late Egyptian treatise

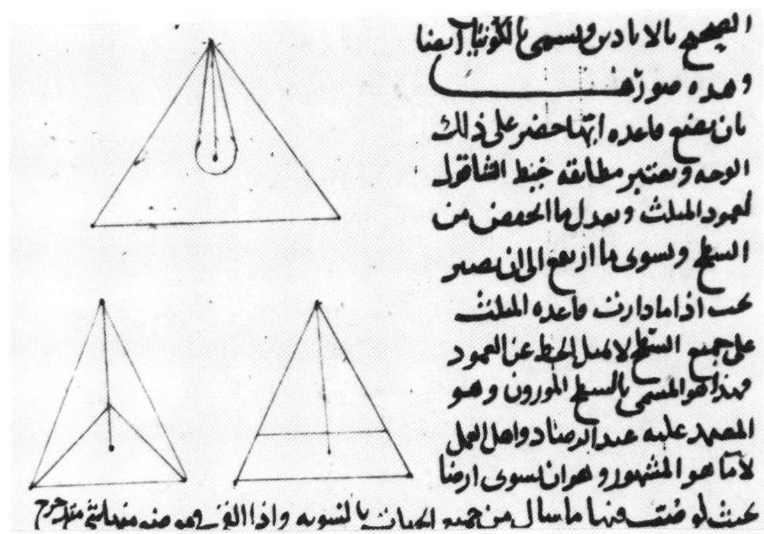

XCIIIc Illustrations of plumb-lines in a treatise by Quṭb al-Dīn al-Shīrazī

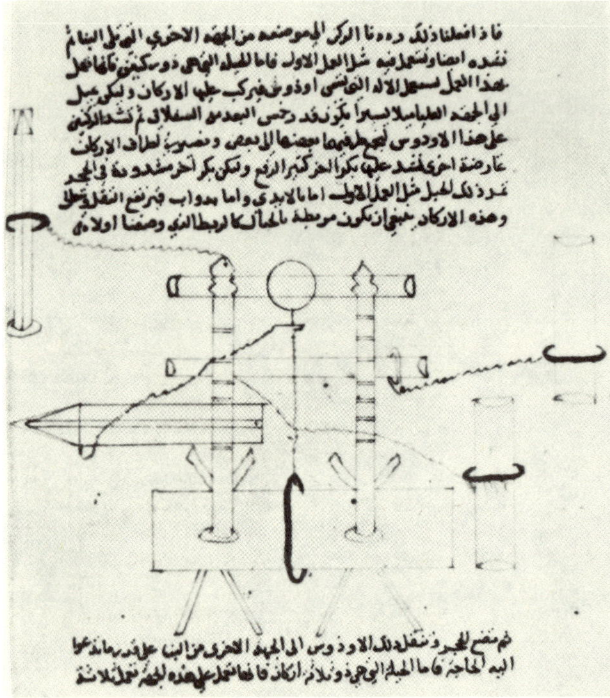

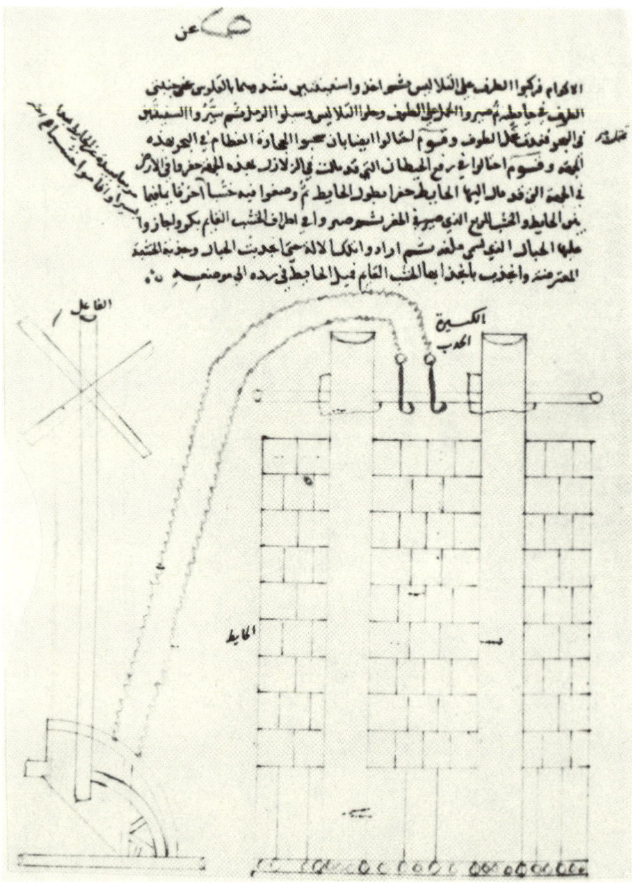

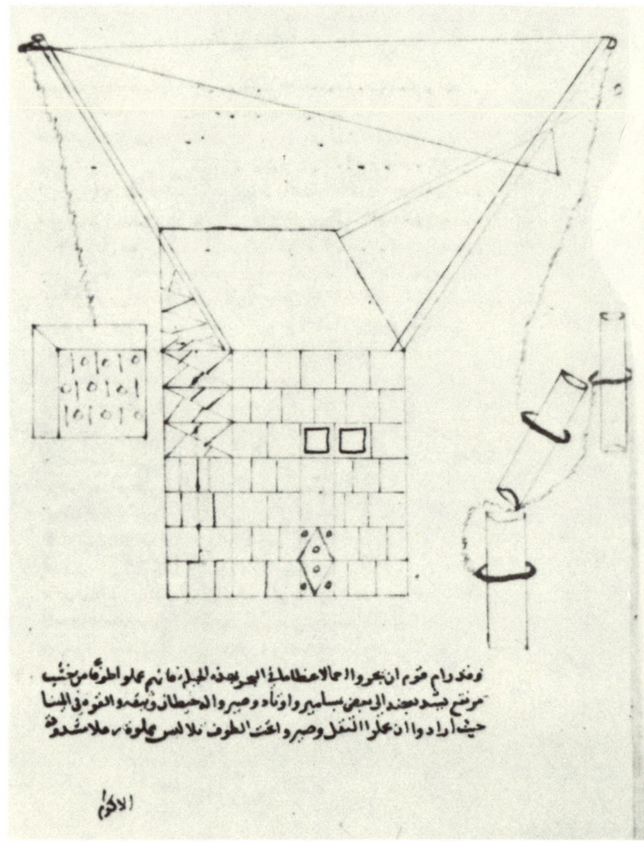

XCIV Mechanical devices illustrated in the Arabic translation of the treatise by Heron

[redacted]

[redacted]stration of a mechanical clock in the [redacted]l-Dīn

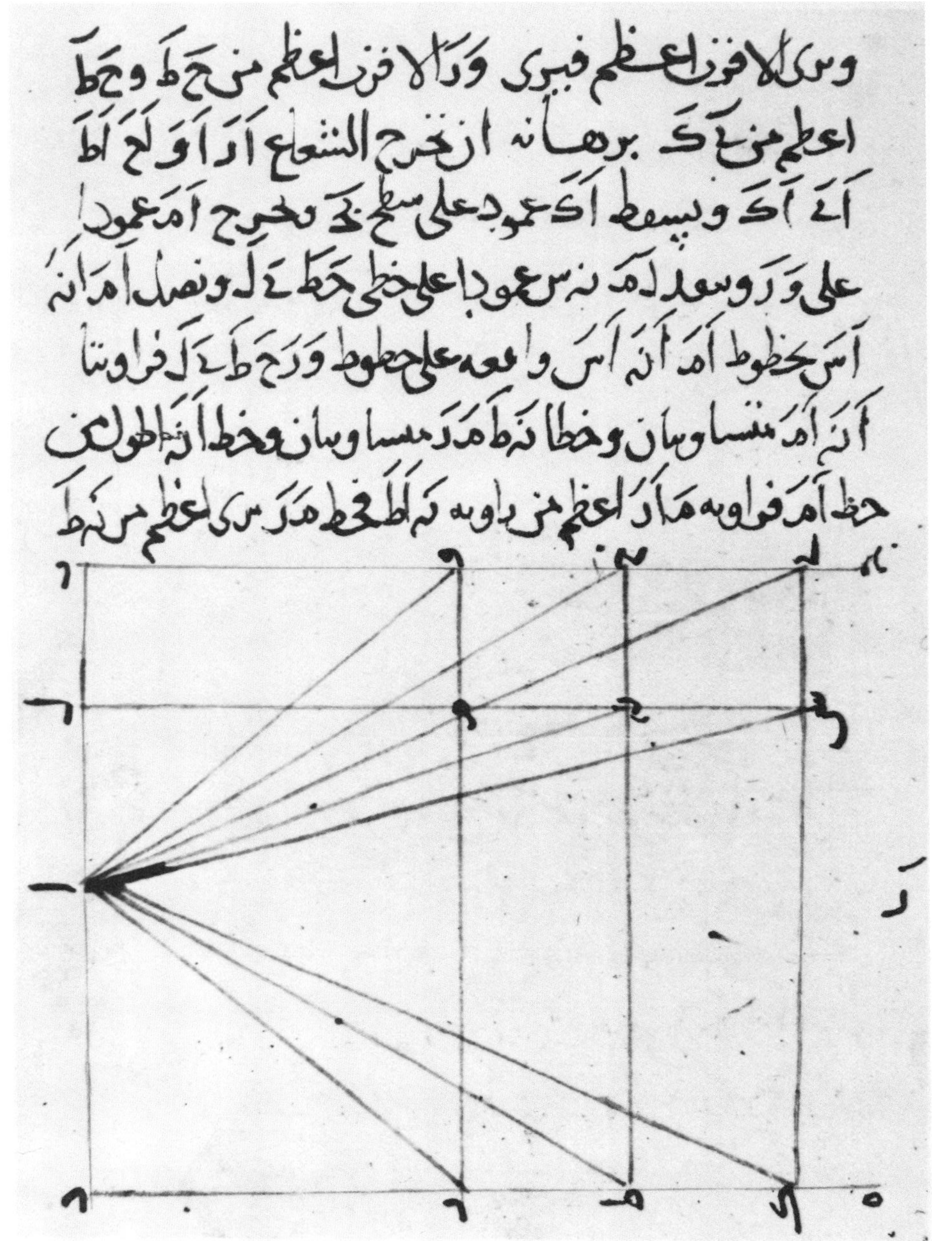

XCVb An illustration in the Arabic translation of Euclid's *Optics*

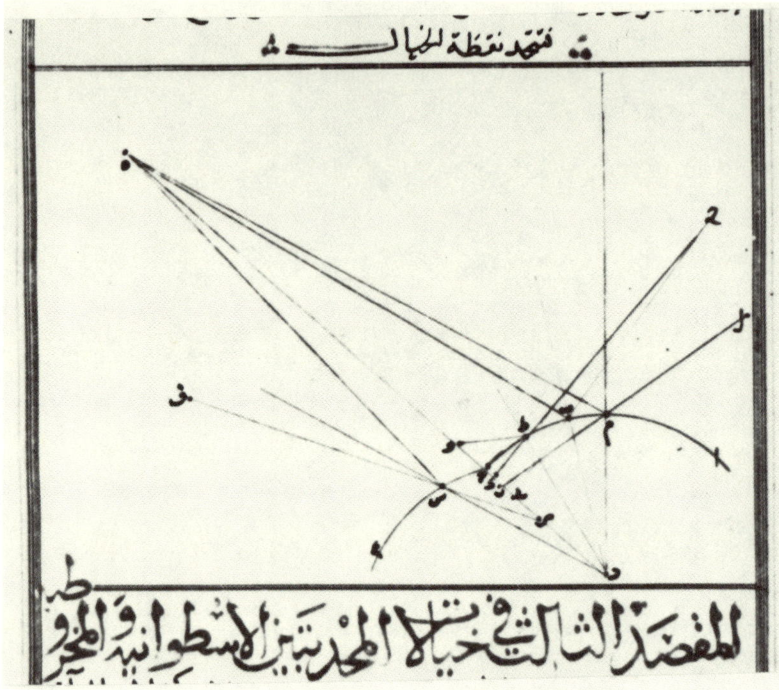

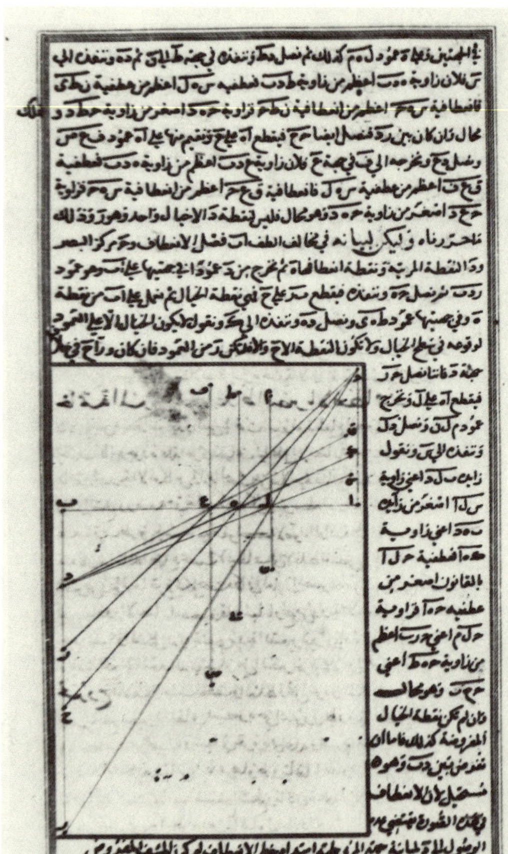

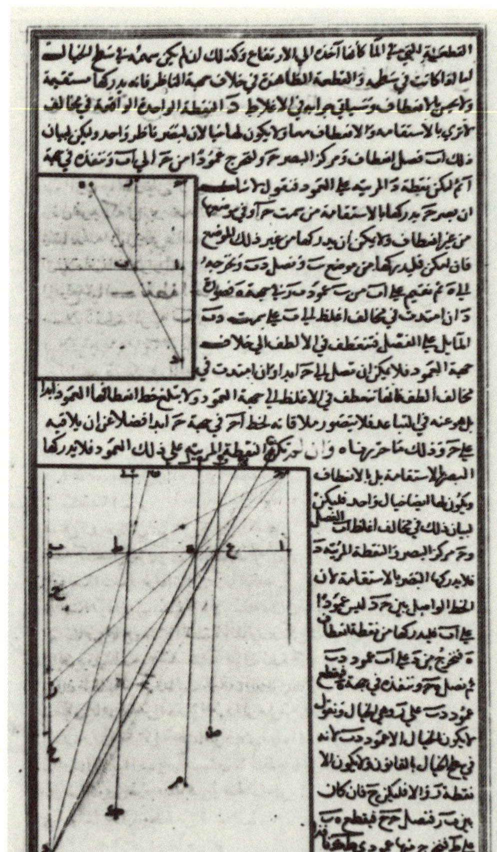

XCVI Illustrations from Taqi l-Dīn's treatise on optics

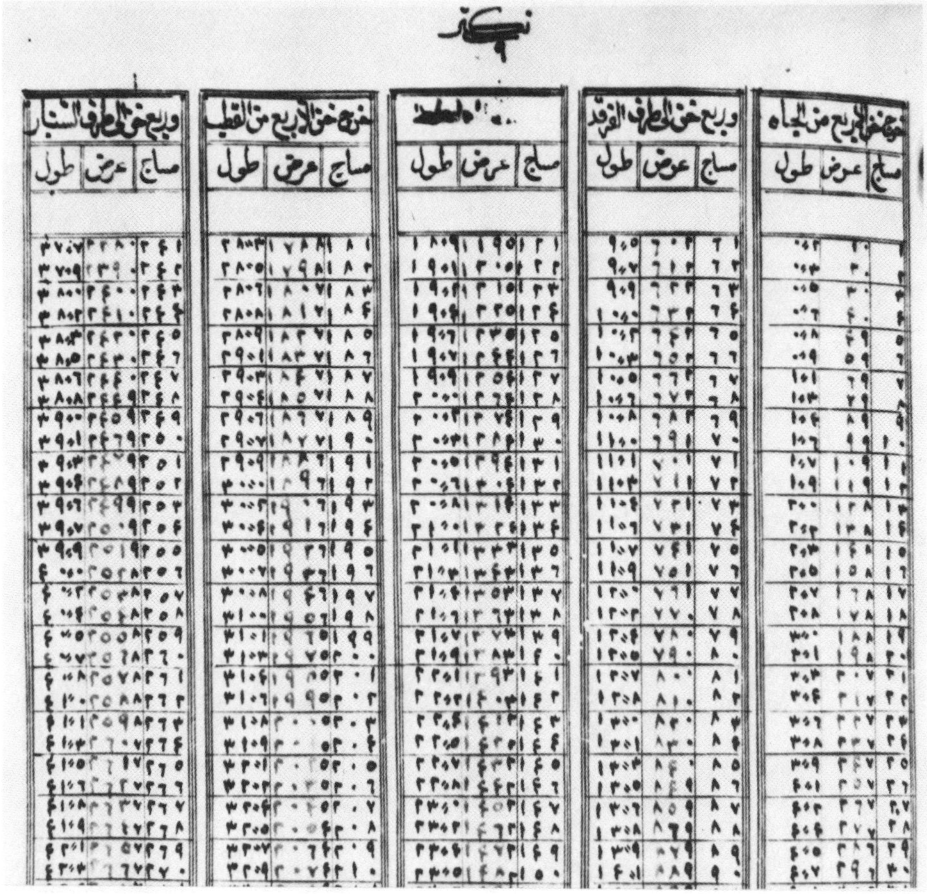

XCVII Extracts from an anonymous Ottoman treatise on navigation

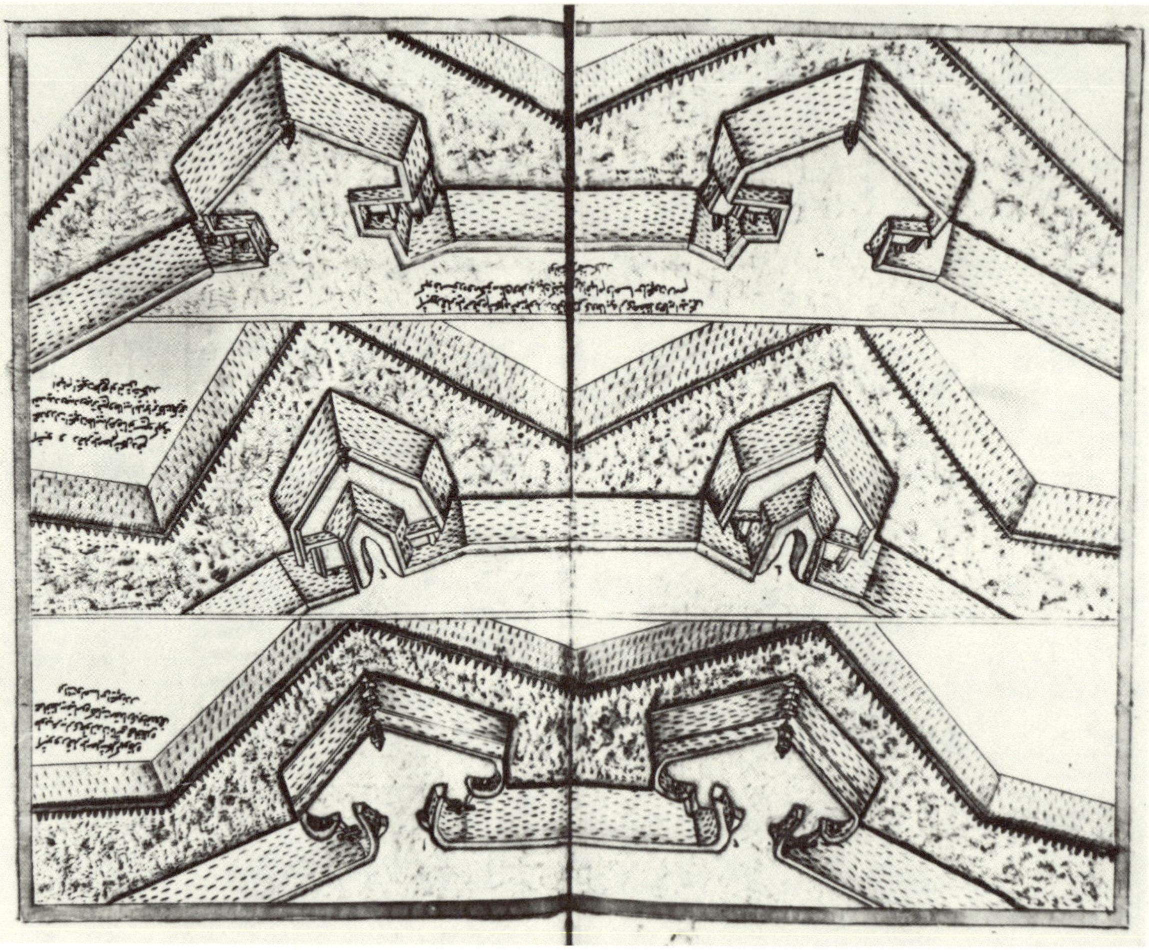

XCVIII Illustrations from an Ottoman treatise on the "arts" of warfare

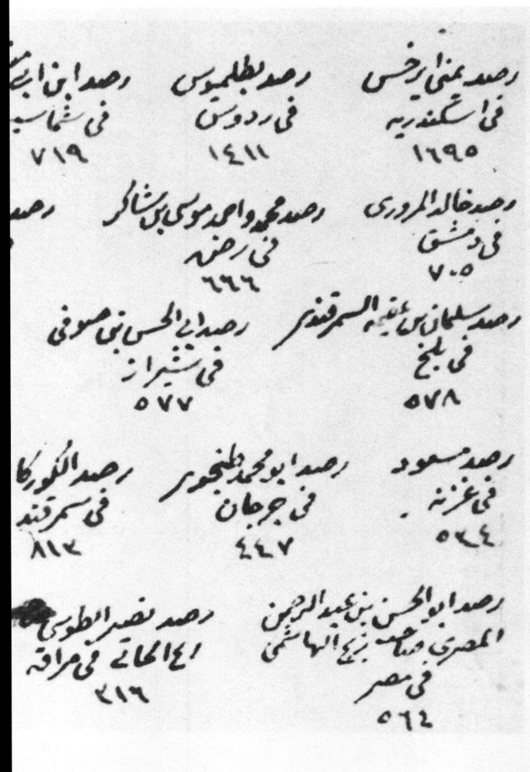

[redacted]onomers and observational programs

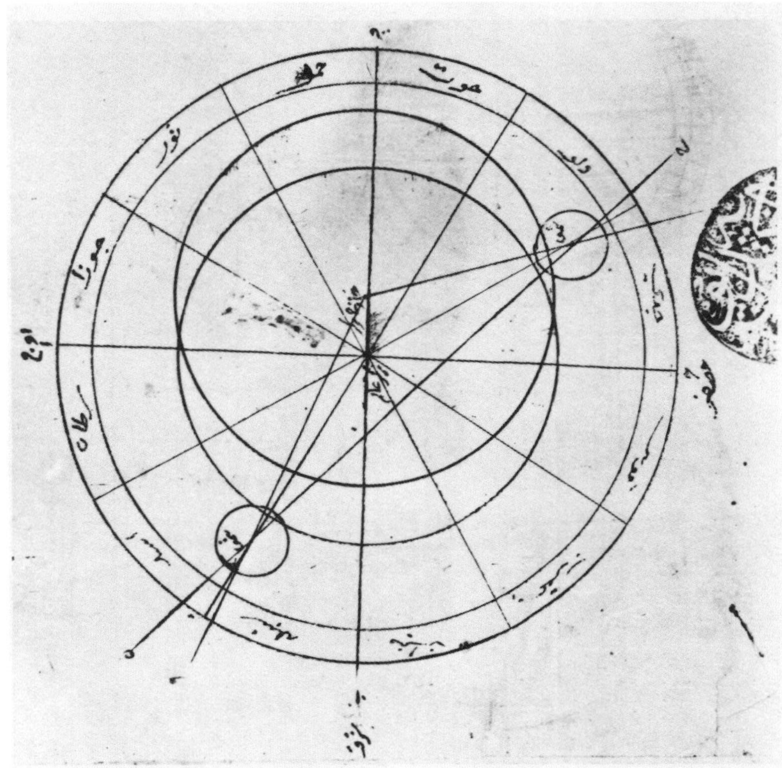

XCIXb A curious model for the sun

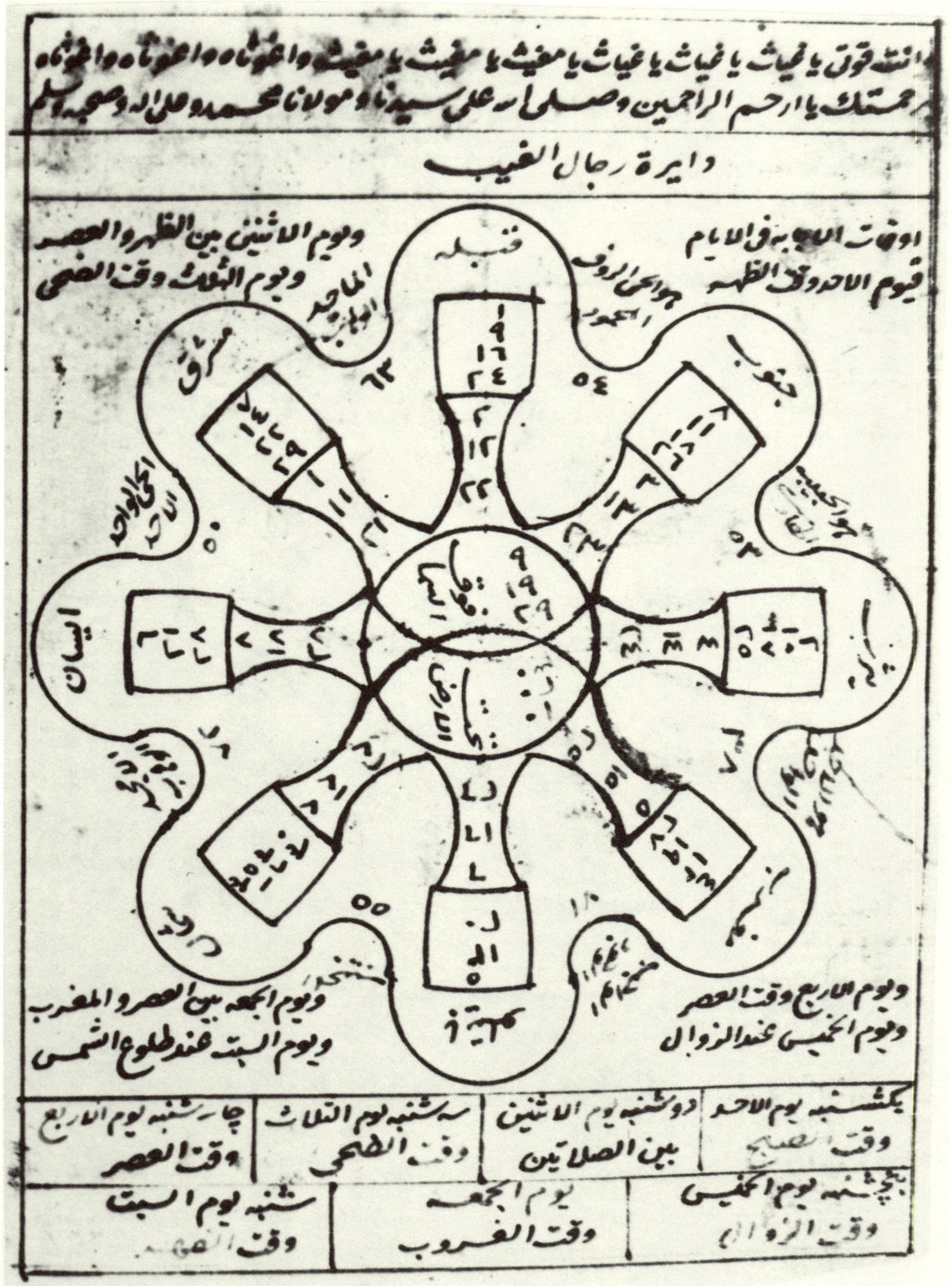

Ca A circle for locating the *rijāl al-ghayb*, intermediaries between man and God

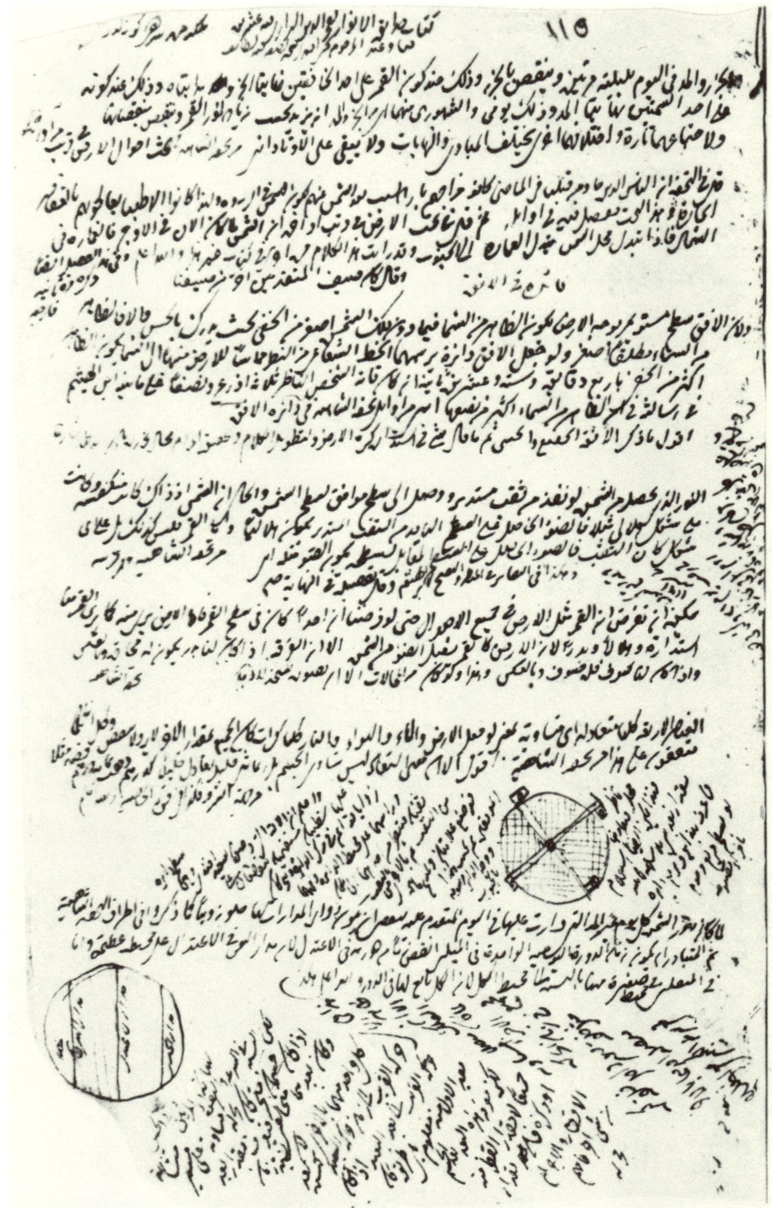

Cb A student's notes on astronomy

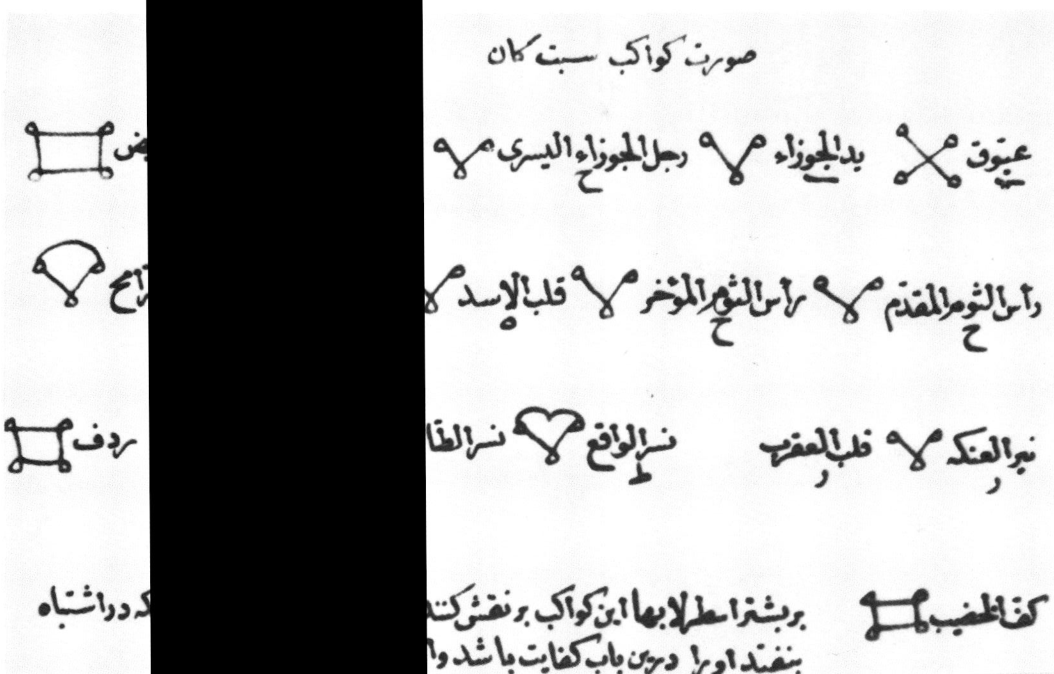

CIa Symbo[redacted]

CIc An account of a curious celestial phenomenon and a table for constructing sundials for the latitude of Marrakesh

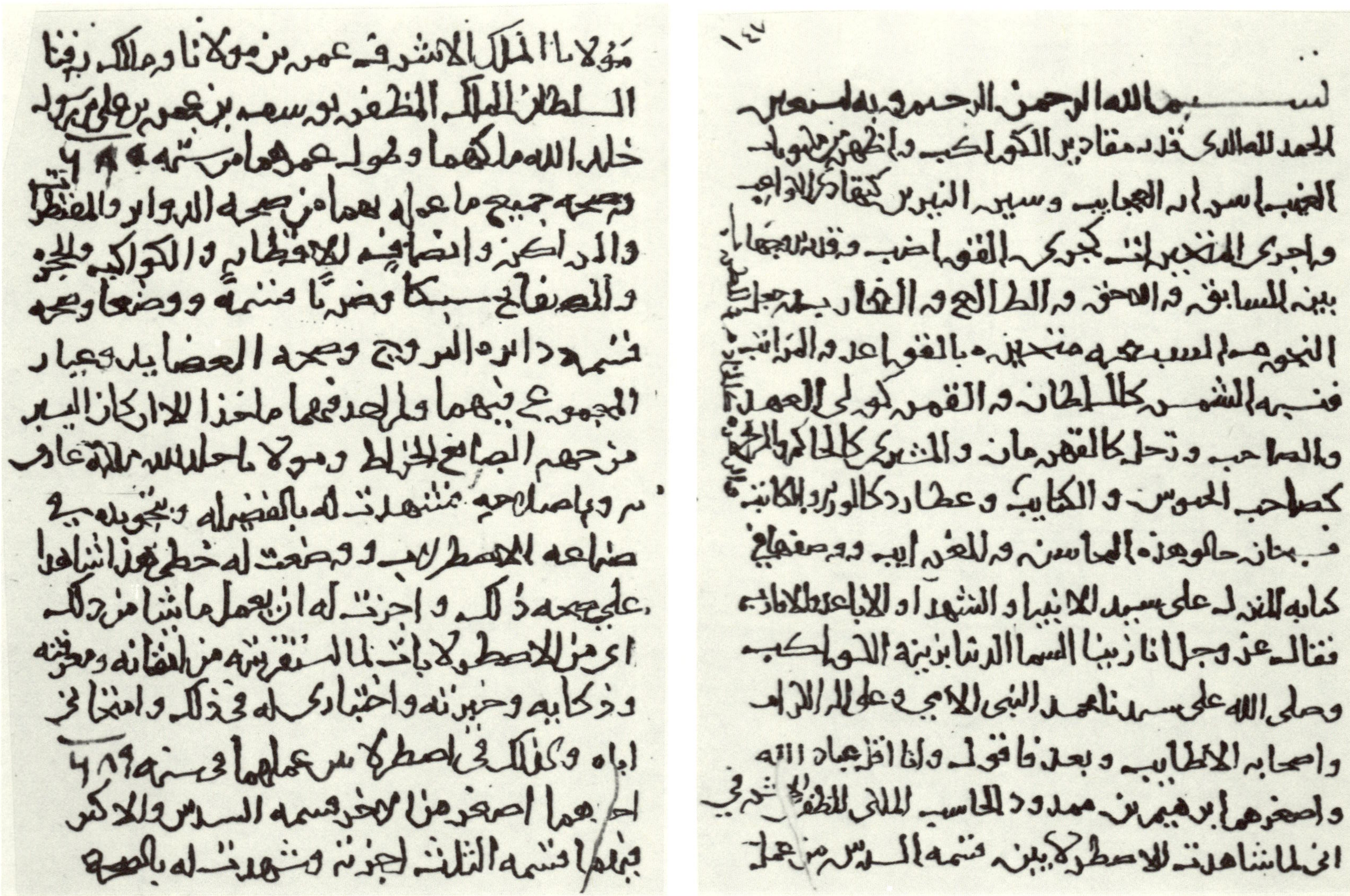

CII Notes by the teacher of the Yemeni Sultan al-Ashraf on some astrolabes constructed by the Sultan

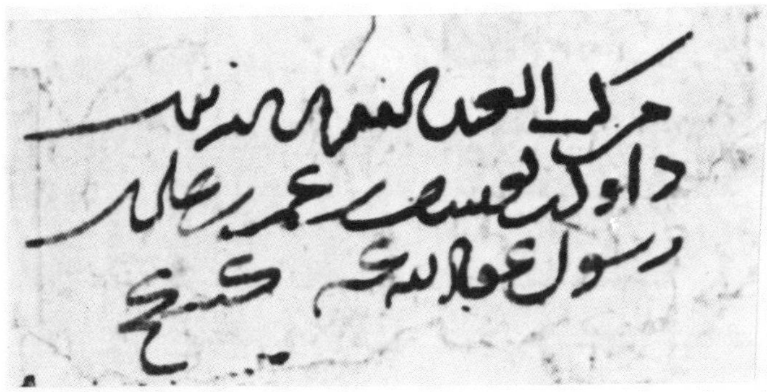

CIIIa A mark of ownership by the Yemeni Sultan al-Muʾayyad

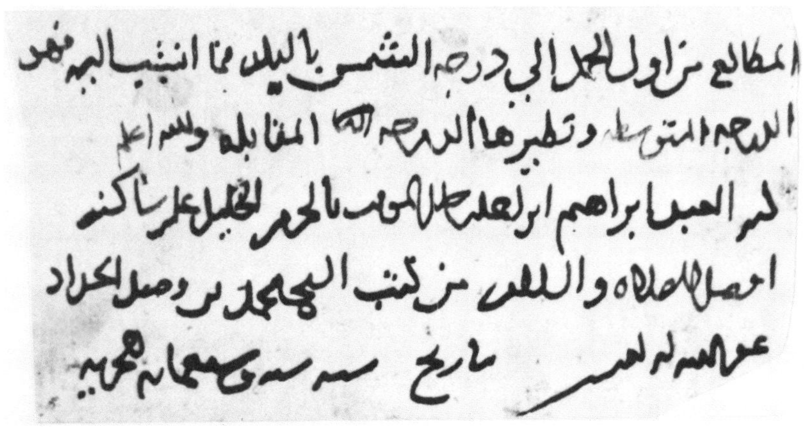

CIIIb A signature of a *muwaqqit* in Hebron in the early fourteenth century

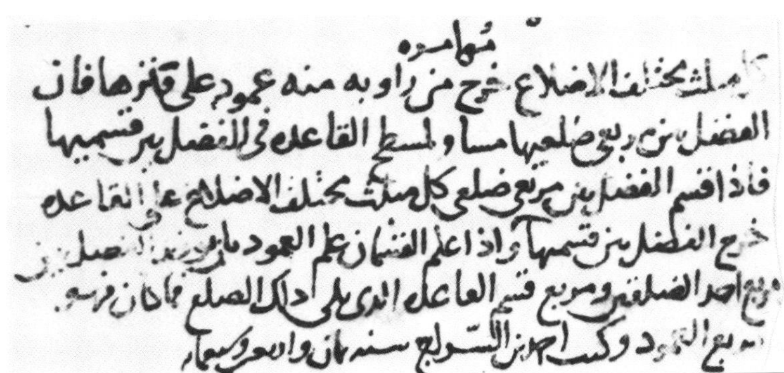

[redacted]

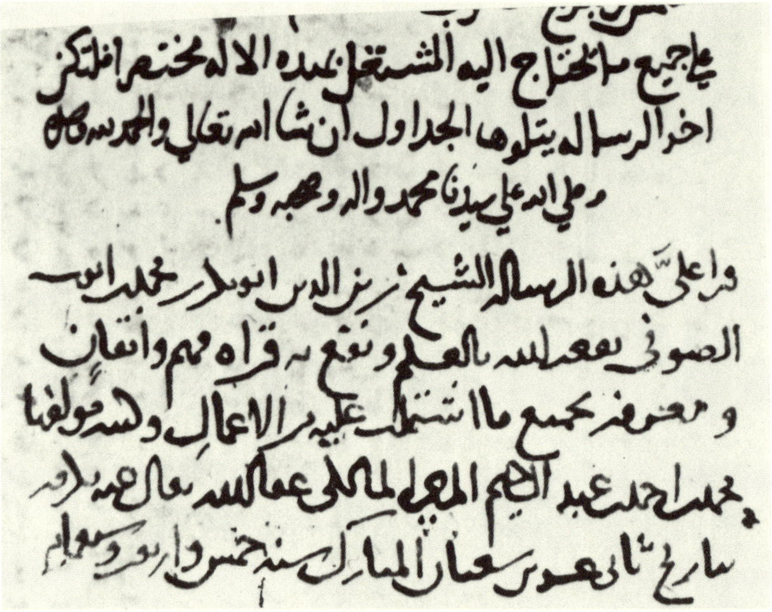

على جميع ما يحتاج اليه المشتغل بهذه الالة مختصرا فلنكتف
اخر الرسالة ويتلوها الجداول ان شاء الله تعالى والحمد لله وحده
وصلى الله على سيدنا محمد واله وصحبه وسلم
قرا علي هذه الرسالة الشيخ زين الدين ابو بكر محمد بن …
الصوفي نفعه الله بالعلم ونفع به قراءة فهم واتقان
ومعرفة بجميع ما اشتملت عليه من الاعمال وكتبه مولفها
محمد بن احمد عبد الرحيم المزي … عفا الله تعالى عنه
بتاريخ ثاني عشر شعبان المبارك سنة خمس واربعين وسبعمائة

CIVa The handwriting of al-Mizzī

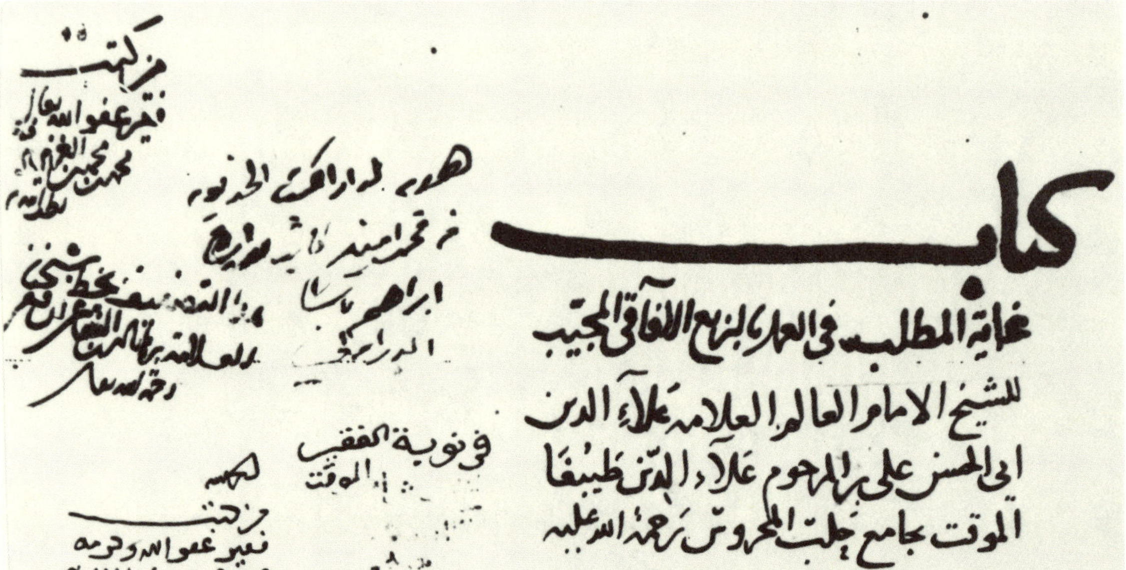

كتاب
غاية المطلب في العمل بالربع الآفاقي المجيب
للشيخ الامام العالم العلامة علاء الدين
ابي الحسن علي بن المرحوم علاء الدين طيبغا
الموقت بجامع حلب المحروس رحمة الله عليه

CIVb The title folio of ᶜAlāᵓ al-Dīn ibn Ṭībughā's treatise on the sine quadrant

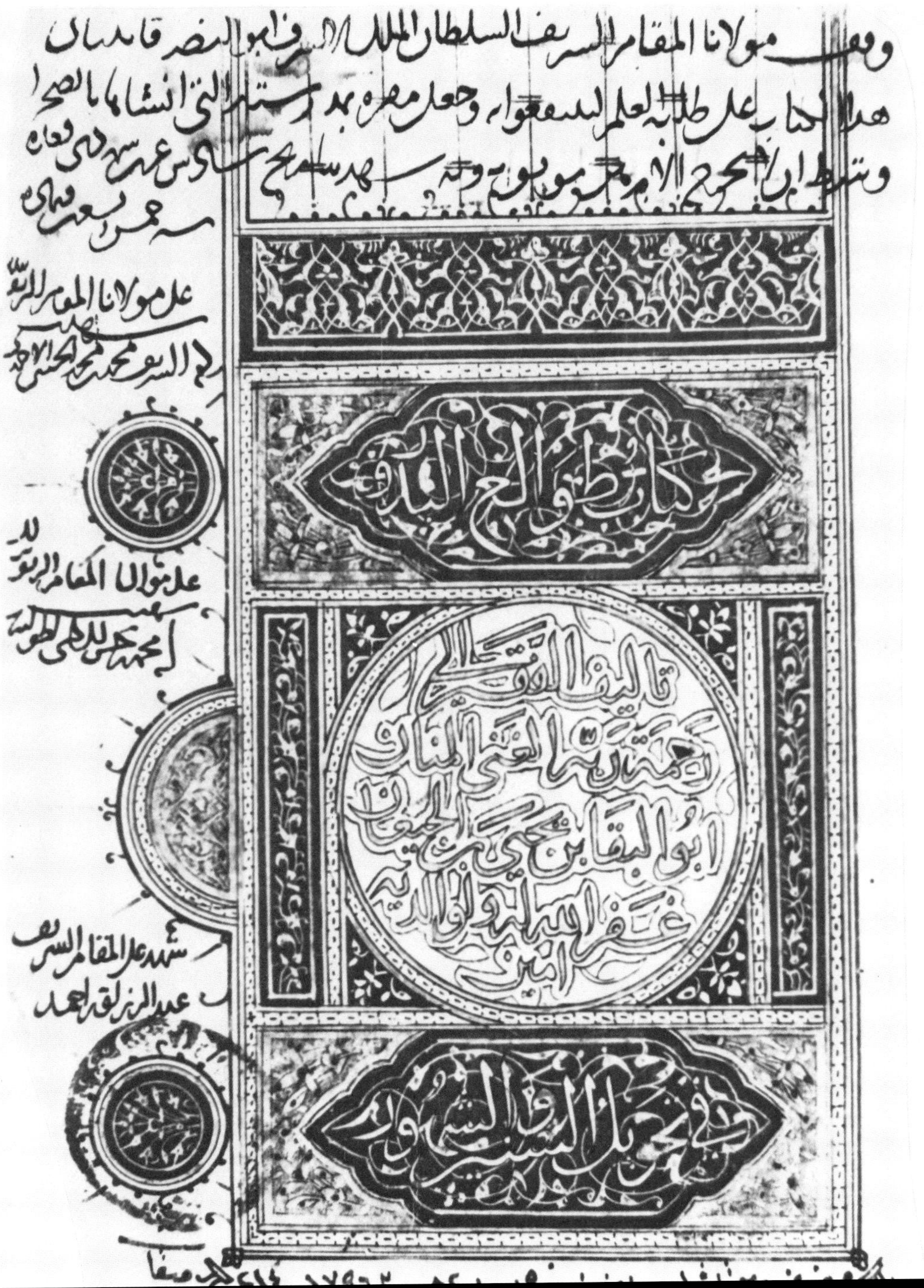
وقف مولانا المقام الشريف السلطان الملك الاشرف ابو النصر قايتباي
تأليف الفقير الى رحمة ربه الغني المنان
ابو البقا بن يحيى بن الجيعان
غفر الله له ولوالديه
امين

[redacted]

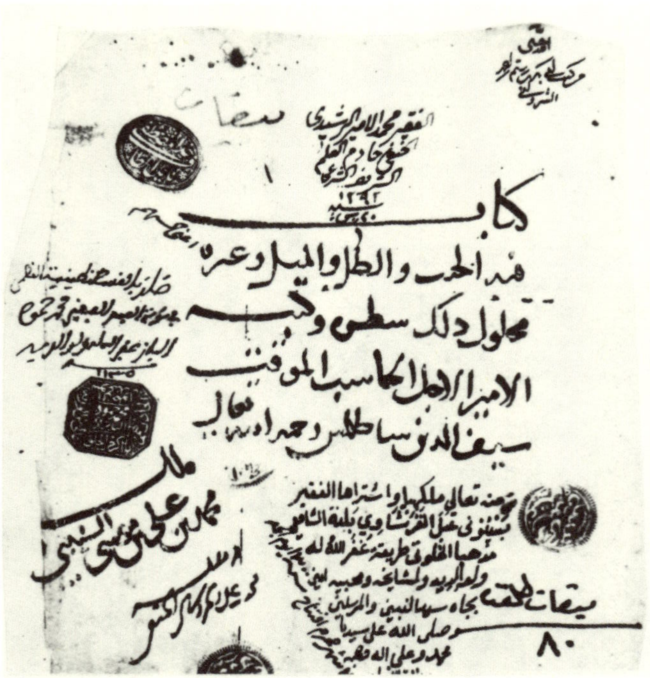

CVIa The title folio of the trigonometric tables of Sāṭilmish

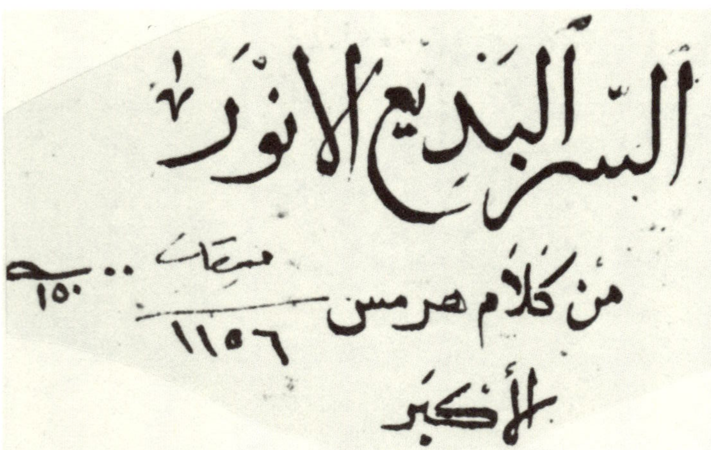

CVIb The title of a treatise attributed to Hermes

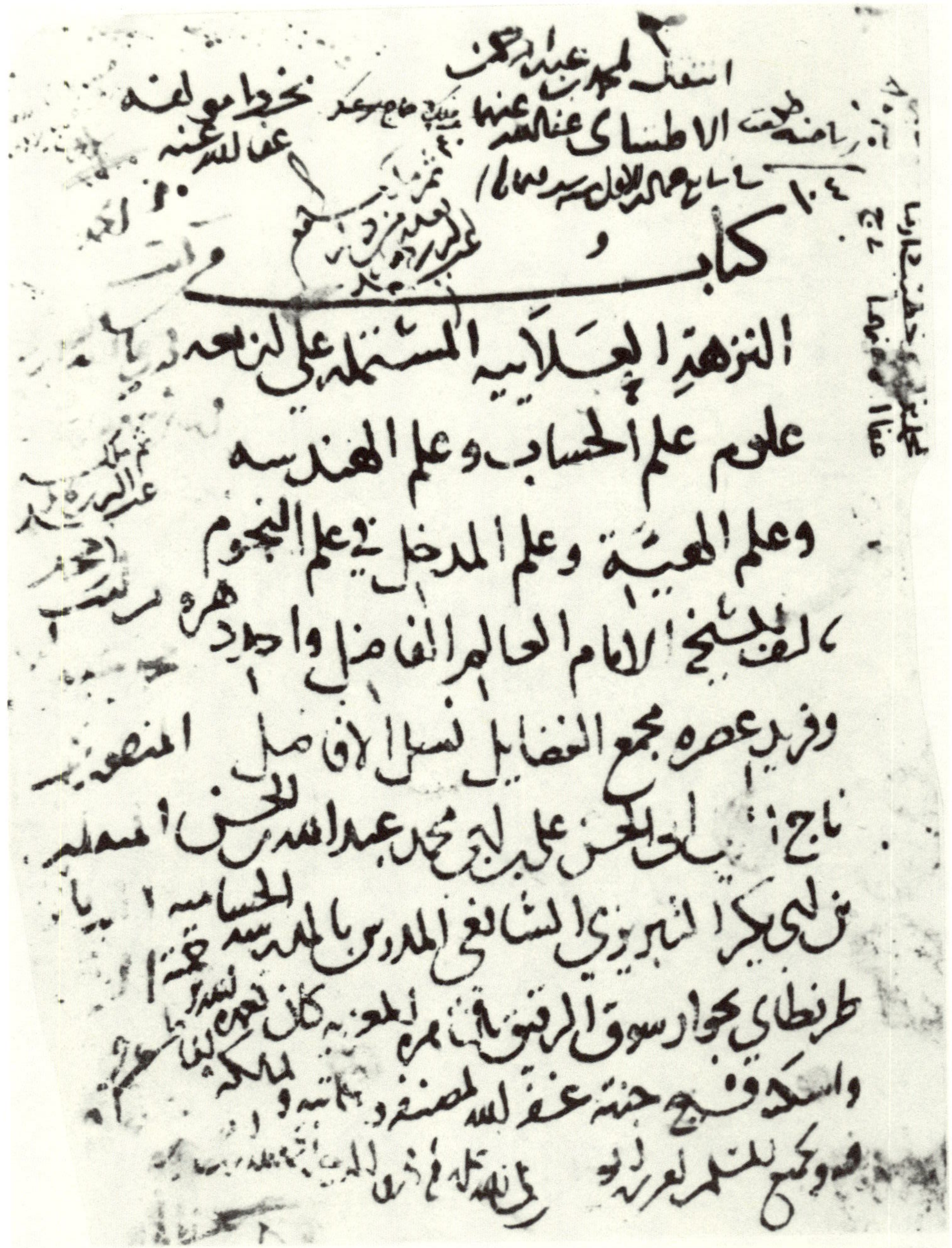

CVIc The title folio of a treatise on astronomy, astrology and mathematics by al-Tabrīzī

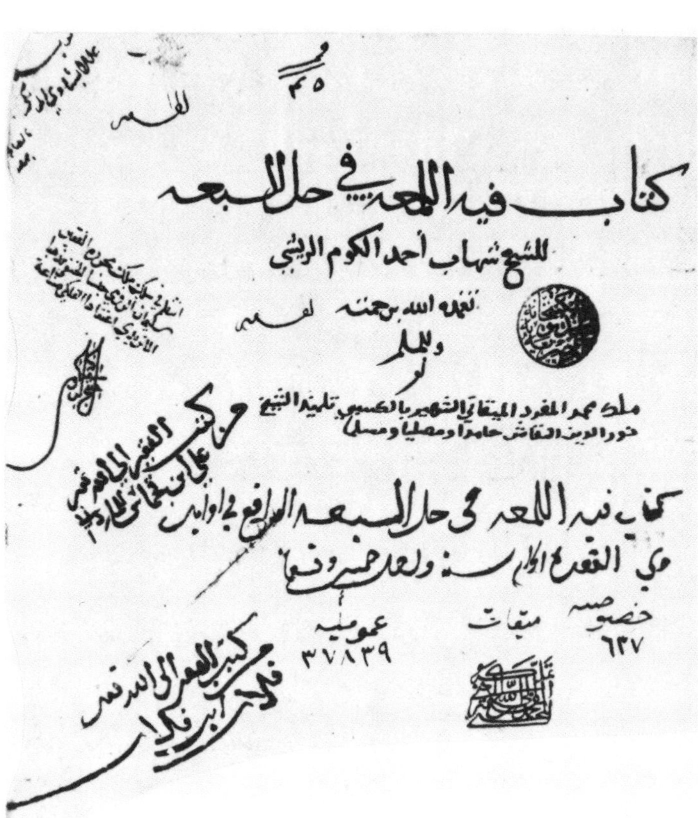

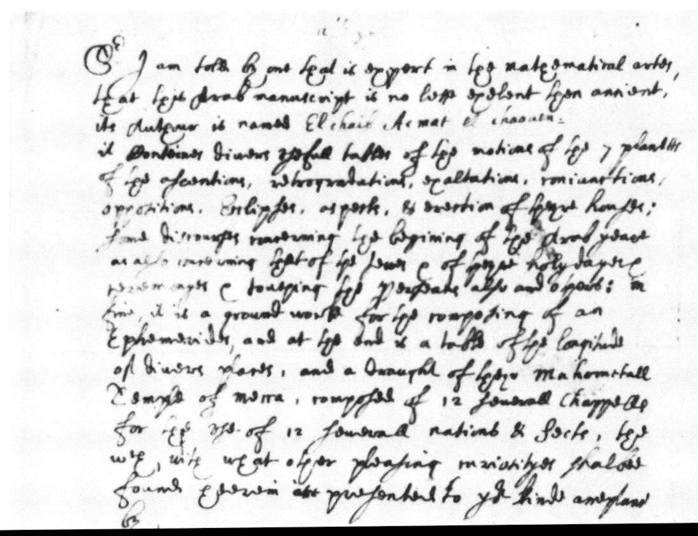
Sr I am told by one that is expert in the mathematicall artes, that this Arab manuscript is no lesse excelent than ancient, its Authour is named Elcheich Acmat el choumi.
it containes divers usefull tables of the motions of the 7 planetts of the ascentions, retrogradations, exaltations, coniunctions, oppositions, Eclipses, aspects, & direction of their houses; some discourses concerning the begining of the Arab years as also concerning that of the Jewes & of their holy dayes & newmoones & touching the paschale ... : in fine, it is a ground work for the composing of an Ephemerides, and at the end is a table of the longitude of diverse places, and a draught of their Mahometall Temple of mecca, composed of 12 severall chappells for the use of 12 severall nations & sects, the which, with what other pleasing varieties shalbe found therein are presented to yr kinde acceptance

[redacted]

CVIIIa The handwriting of ᶜIzz al-Dīn al-Wafāʾī

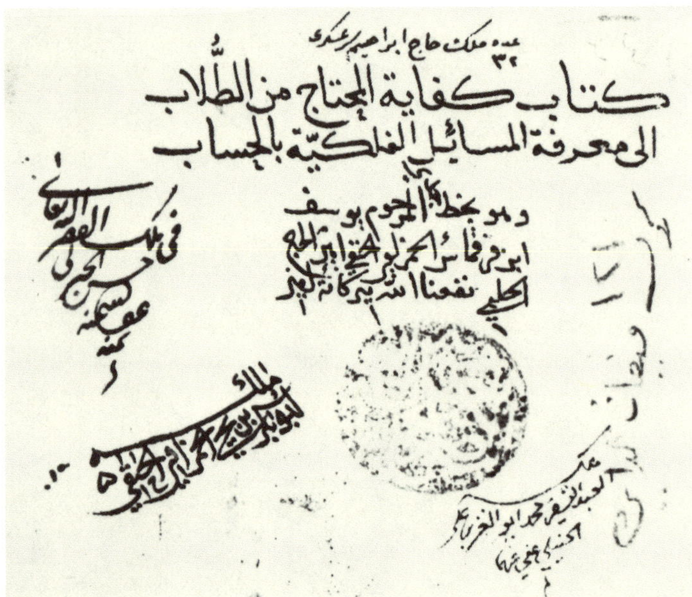
كتاب كفاية المحتاج من الطلاب
الى معرفة المسائل الفلكية بالحساب

CVIIIb The title folio of a manuscript in the hand of Yūsuf al-Ḥamzāwī

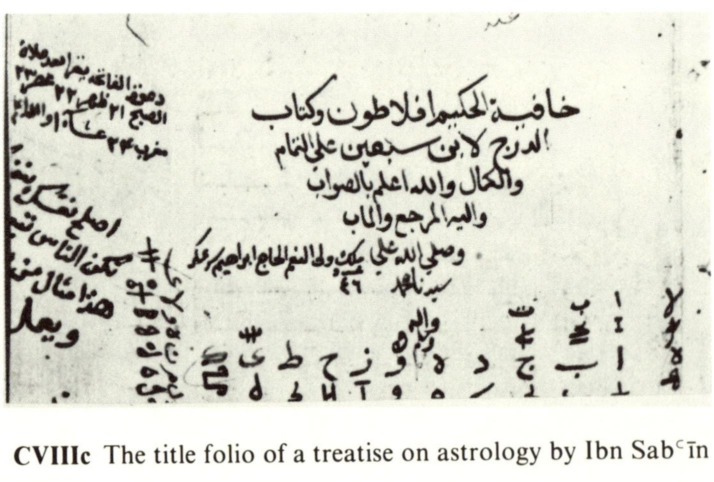
خافية الحكيم افلاطون وكتاب
الدرج لابن سبعين على التمام
والكمال والله اعلم بالصواب
واليه المرجع والمآب

CVIIIc The title folio of a treatise on astrology by Ibn Sabᶜīn

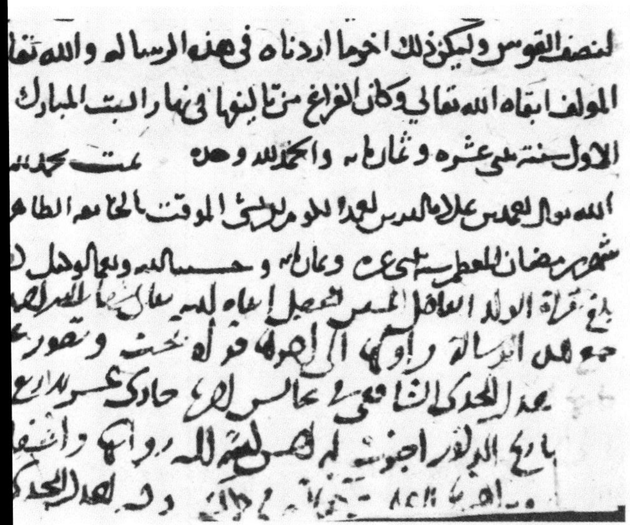

[redacted]g of Ibn al-Majdī

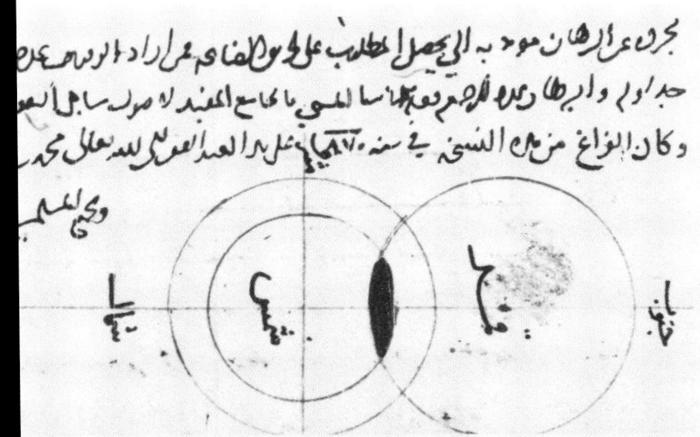

[redacted]g of Ibn Abi l-Fatḥ al-Ṣūfī

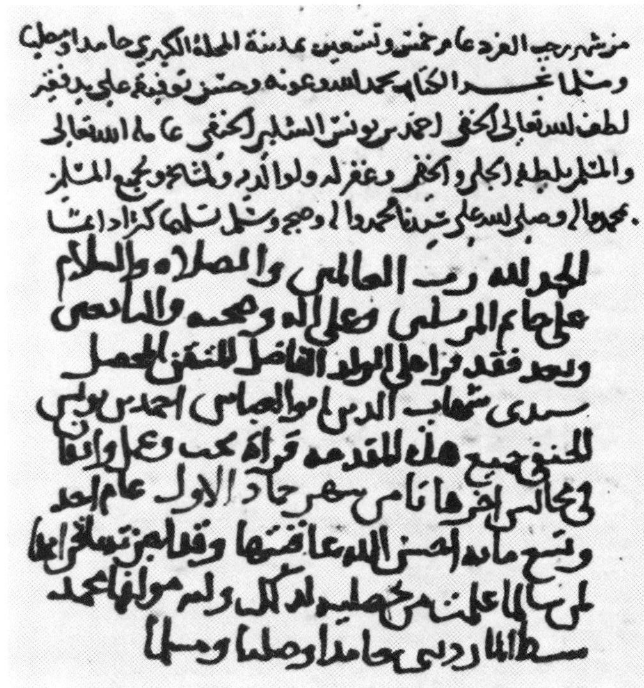

CIXc The handwriting of Sibṭ al-Māridīnī

a.

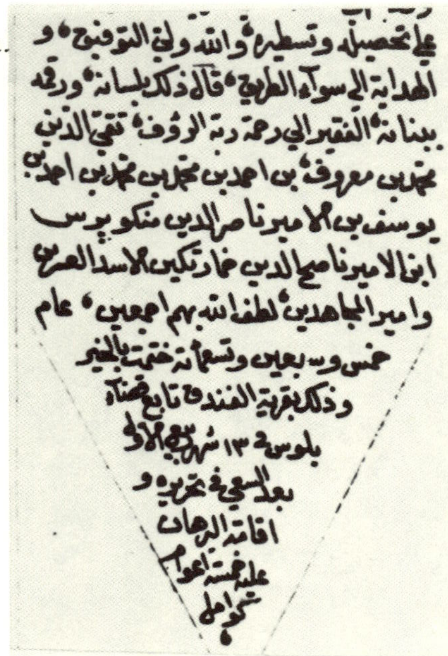

c.

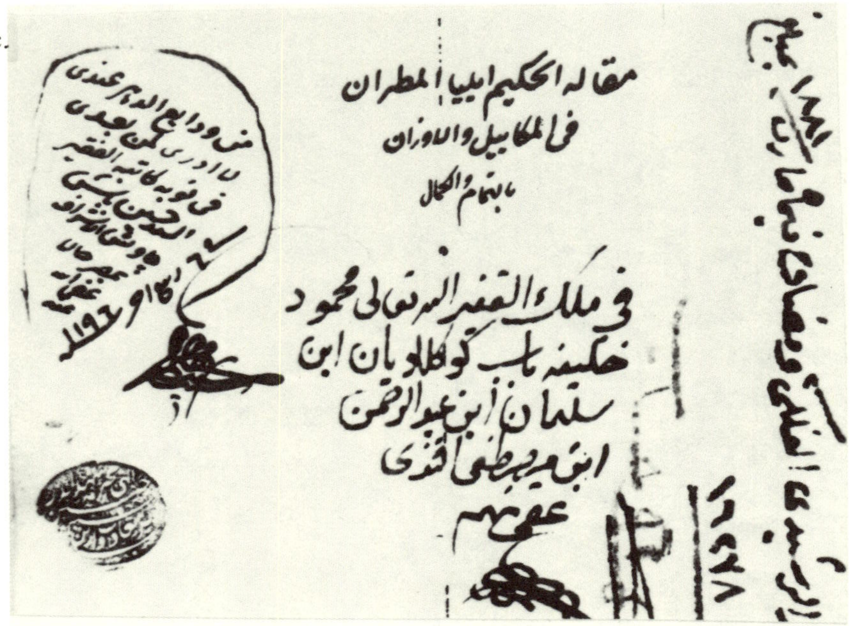

b.

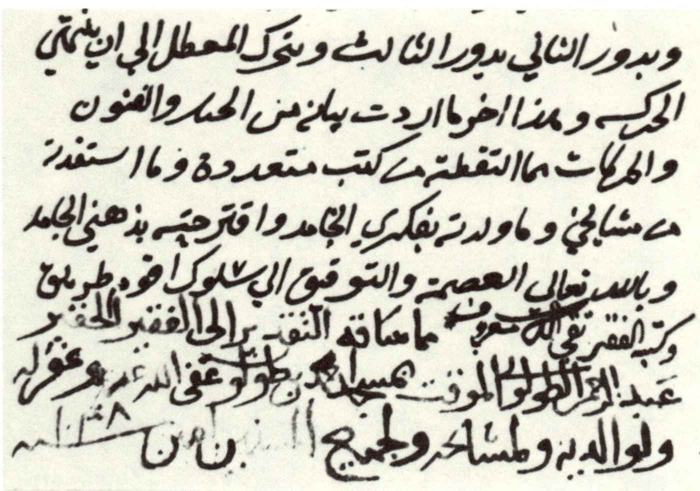

d.

CX Miscellaneous signatures

a.

b.

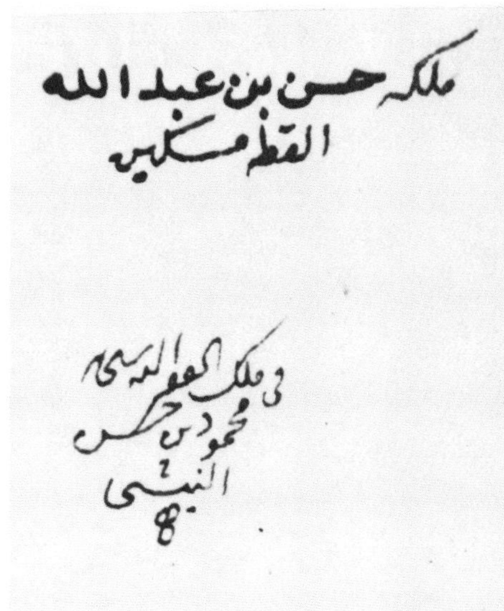

c.

[redacted]

[redacted]

[redacted]